MECHANICS OF TECTONIC FAULTING

Models and Basic Concepts

Developments in Structural Geology

H.J. Zwart (Series Editor)

Developments in Structural Geology, 1 (Series Editor: H.J. Zwart)

MECHANICS OF TECTONIC FAULTING

Models and Basic Concepts

G. Mandl

Formerly: *Koninklijke/Shell Exploratie en Produktie Laboratorium, Rijswijk, Z.H., The Netherlands*

At present: *Technical University, Graz, Austria*

ELSEVIER

Amsterdam — Oxford — New York — Tokyo 1988

ELSEVIER SCIENCE PUBLISHERS B.V.
Sara Burgerhartstraat 25
P.O. Box 211, 1000 AE Amsterdam, The Netherlands

Distributors for the United States and Canada:

ELSEVIER SCIENCE PUBLISHING COMPANY INC.
52, Vanderbilt Avenue
New York, NY 10017, U.S.A.

ISBN 0-444-42946-8 (Vol.1)
ISBN 0-444-42947-6 (Series)

Printed in The Netherlands

Dedication: *To B.*
and KSEPL's
Structural Geology Group

*"Science is made out of facts as a house is made out of stones,
but an accumulation of facts is no more a science than
a heap of stones a house."*

(Henri Poincaré)

PREFACE

Tectonic faults and folds – the predominant deformation features in the Earth's crust – are the results of mechanical processes, the understanding of which has become a major concern of modern Structural Geology. This book deals with the mechanics of tectonic faulting and its central theme is the formation of faults in the tectonic stress field and the interaction between faults and stresses.

This branch of geomechanics requires a multidisciplinary approach that combines continuum mechanics, rock mechanics and structural geology, and applies theoretical, numerical and experimental methods and involves the study of field cases. The book has evolved from such multidisciplinary research and is written for structural geologists, engineering geologists, rock engineers and geophysicists whose work demands a similar approach. In addition, the book is intended to encourage mechanical engineers and even mathematicians, to enter the fascinating and in parts still untilled fields of tectono-mechanical processes in the Earth's crust.

Obviously, in addressing potential readers of such different scientific backgrounds and preferences, I have been confronted with serious problems in terms of the presentation and organisation of the material. I have tried to achieve a reasonable compromise by dividing the book into two parts: Part I presents the tectono-mechanical models of tectonic faulting without dealing with the underlying mechanical principles and theories in more than a very cursory way, while all the basic mechanical concepts and theories required are presented in Part II. This should enable a reader who is already familiar with basic continuum mechanical and rock mechanical concepts to enter the discourse on tectonic faulting without much delay, while readers less conversant with the underlying mechanical concepts may at least have a first reconnaissance of the material presented in Part I. Many cross references in the text should make it easy for the reader to consult the relevant section in Part II as the need arises. Non-geological readers are expected to have already acquired some familiarity with the basic concepts and nomenclature of Structural Geology, such as presented in the booklet by R.G. Park (1983).

In the introductory chapter of Part I the basic phenomena of tectonic faulting in the brittle crust are discussed, and these motivate the choice of the mechanical theories to be applied. Although no mechanical theory exists which may cope with *all* aspects of tectonic faulting, the theory found most adequate and practicable is the mechanical theory of plasticity. It forms the backbone of this book. I hasten to add that the deformation behaviour dealt with in the theory of plasticity is quite different from the 'ductile', i.e. viscous, material behaviour that is meant by the term 'plastic' in common geological jargon. A full and self-contained introduction into the theory of rock plasticity is presented in Chapter II.7, which is the central chapter of Part II. It is not only intended as an introduction for readers with little previous exposure to the mechanical theory of plasticity, but also for readers who are already familiar with the theory in the form that is well-established for metals and presented in current textbooks. The reasons for this are the modifications and generalisations necessary when dealing with brittle rocks, and the difference between tectonic and engineering problems, which requires a shift of emphasis to aspects that are paid less attention in the engineering literature.

The second chapter of Part I deals with mechanical aspects of tectonic faulting that are not only typical for a specific fault type, but may apply to various, if not all, types of tectonic faults. Typical examples of such more general aspects are the curvature of faults, the role of antithetic faults and the formation of precursory and secondary faults. The following three chapters of Part I elaborate on these more general aspects within the context of specific tectono-mechanical models for extensional faulting, wrench faulting and thrusting. Besides critical reviews of already published mechanical models, Part I of the book contains a fair number of new or hitherto unpublished results obtained by myself and my associates, mainly from model experiments and numerical simulations of faulting processes. These results also demonstrate the great potential that lies in a combination of model experiments and numerical analyses, two approaches that complement each other.

I have tried to restrict the mathematical formalism to a minimum, even though this implies some loss in completeness and elegance which the mathematically oriented reader may regret. Wherever feasible, I have replaced analytical derivations by the extremely useful graphical method of Mohr's stress circle. The reader will therefore need a certain familiarity with this elegant graphical method, which, for his convenience, is presented in a separate chapter of Part II (Ch. II.4).

Basic mechanical concepts and theories are presented in Part II in more or less self-contained articles, with emphasis placed on conceptual aspects rather than mathematical formalism. Geological readers who wish a systematic introduction into the mechanical backgrounds of Part I are advised to study these articles in the sequence in which they are presented, starting with the crucial concept of stress, which is our instrument to trace the influence of boundary loads and displacements through the interior of a rock mass.

Much attention is given in Part II to the mechanical interaction of pore fluids and the solid skeleton of porous rocks. The space devoted to the discussion of topics such as the principle of effective stress, buoyancy and seepage flow, reflects the vexations of a physicist confronted with seeming ambiguities in the formulation of basic mechanical equations for fluid saturated rocks. To clarify these matters and to save others similar experiences as far as possible, the discussions in Part II occasionally go farther than strictly necessary for the purpose of some readers.

The presentations of Part II are illustrated by applications to geological topics. These examples, which are listed at the end of the table of contents, not only help to enable a full grasp of the mechanical concepts and laws discussed, but also provide results needed and referred to in Part I.

I hope that the presentation in the book strikes a reasonable balance between results of direct practical use and materials that are still in early stages of development, but offer promising fields of future research with numerical and experimental methods. For the benefit of researchers some topics are discussed further in small print and in additional notes at the end of chapters; these may be bypassed by other readers without the thread of the argument becoming lost.

The book ends with a bibliographical list of all papers referred to and with a subject index.

The book obviously reflects the petroleum-geological research environment in which it has grown as a result of many years of team-work. My thanks therefore go first of all to my former associates in the Koninklijke/Shell Exploratie en Produktie Laboratorium at Rijswijk (The Netherlands): W. Crans, J. Haremboure, W.T. Horsfield, L.N.J. de Jong, F.K. Lehner, G.H. Mäkel, A. Maltha, M.A. Naylor, D.A. Nieuwland, W.F.H. Pilaar, G.K. Shippam, J.N. Thomas, J.V. Walters and H.W.M. Witlox.

I would also like to give special thanks to my friends and discussion partners through many years, R.H. Harkness, N.J. Price and S.B. Spijer.

I am grateful for the initiation of an interdisciplinary team for Structural Geology Research in 1966 by the former director of the Laboratory, J. van Heiningen, and for the continual support and encouragement which the project received from the research directors C. van der Poel, H.R. Grunau, A. Boomer and, in particular, from the late Koos van Wijnen.

I finally thank the present management of the Laboratory, and the editing, typing and drafting departments for their support in the preparation of this book.

CONTENTS

Part II: Basic concepts

GEOLOGICAL TOPICS DISCUSSED IN EXAMPLES OF PART II

Part I

TECTONO-MECHANICAL MODELS

I.1. INTRODUCTION

Basal thrust plane

Fig. I.1-1 GLARUS OVERTHRUST, SWITZERLAND (View from Flims) Malm limestone overthrusted from left to right by Verrucano sandstone

1. GEOMECHANICS IN STRUCTURAL GEOLOGY

In the structural interpretation of geological data, geologists have to rely on models and concepts developed in Structural Geology. These structural models indicate how a rock mass may deformationally comply, mostly by faulting and folding, with an imposed shortening or extension, or with some more complex changes in the external configuration. Traditionally, these models and concepts are of a predominantly geometric/kinematic nature as demanded by the enormous variety of fault and fold structures that have to be described and classified, and by the need to unravel their kinematic genesis. Unravelling the kinematic genesis implies that, in general, a large variety of alternative deformation processes may be postulated that are compatible with the geological data and which, from a purely geometrical point of view, would account for a certain change in the external configuration of a rock mass.

In selecting one of the feasible models for the structural interpretation of the data, the geologist is in practice largely guided by analogies with better documented and analysed cases. These cases, naturally, have some features in common but the comparison can provide reliable guidance in the structural interpretation only when the common features are expressions of an essential correspondence between the tectonic processes involved rather than indications of merely superficial similarities.

In general, it will be difficult to assess by geometrical or kinematical considerations alone the significance of the features on which to base structural analogies, since tectonic processes are not of a geometrical nature but of a mechanical one, and are produced by the interplay of forces and deformations. Accordingly, the features that constitute reliable structural analogies will usually be of a mechanical nature.

This was recognised by a few pioneers more than half a century ago and it is now widely accepted that mechanical and rock-mechanical principles and theories should be combined with the traditional methods of structural geology to obtain a sound basis for the comparison of structures and structural styles and to improve our understanding of tectonic processes.

The 'geomechanical' approach aims at a direct theoretical analysis and computational and experimental modelling of tectonic processes. In comparison with field studies, which it is to complement, this 'laboratory' approach has the advantage of precisely specified controlling parameters, which, moreover, can be varied at will to assess their relevance for the process being considered. Obviously, a geomechanical model will apply to an *idealised* process which in nature would take place when the postulated boundary conditions and material properties were strictly met. The assumptions made in a tectono-mechanical model will always imply considerable simplification of the geological reality, not only because the model has to remain practicable, but also because only through focussing upon a few, yet important, parameters may one reasonably hope to acquire a fundamental insight into the mechanism of a tectonic process.

The last point is sometimes misunderstood and tectono-mechanical models have been discredited as being much too simple to cope with the complexities of the 'real world'. It is certainly true that a single tectono-mechanical model will hardly ever account for *all* features of a specific structural development, simply because the process is approached from a particular point of view that emphasises certain aspects and deliberately neglects others – the general way natural sciences and engineering approach the real world. For instance, in many tectonic processes the rock has deformed by both faulting and folding. Field data may often suggest that folding merely accompanied faulting as some adjustment deformation of moving fault blocks. In this case the mechanical model may concentrate on the development of the faults by utilising a mechanical theory that does not make allowances for the creep-like deformational behaviour inherent in many processes of folding. In spite of this, the model may account for the fault features very well.

It is evident that an adequate analysis of complex structures may require the application of several models that complement each other by dealing with different aspects of the same tectonic process. Comparison with the field data may then show whether the mechanical models incorporating idealising assumptions still account for the essential features of the real processes; if so, they may serve as reference processes in the interpretation of the – usually quite fragmentary – field data. Such tectono-mechanical process models have the advantage of being free of structural 'noise' caused by accidental and unpredictable local irregularities in material, stress field or boundary conditions; instead, they focus upon the variation of average properties and major trends of the tectonic processes.

Another great advantage of models based on mechanical theory is that they are not restricted by scale. When a geometrical dimension, say the thickness of a layer, is an important factor, it is taken into account as a variable parameter.

Understanding the mechanics of a structural process implies an insight into the development of the associated stress field. This can be of great practical importance. The paleostresses associated with ancient tectonic processes – on a regional or more local scale – for instance, play a key role in the migration of mineral solutions and hydrocarbons during the formation of fractured hydrocarbon reservoirs or systems of mineral veins. Even when the stresses themselves have relaxed long after the termination of a tectonic process, the principal directions of the paleo-stress system may still be preserved. When associated with still active tectonic processes, the stresses inferred from mechanical analyses may provide the most important parameters in the field of rock engineering where the present-day in-situ stresses are major controlling factors, such as in the construction of underground cavities, the subsurface storage of fluids, in certain drilling operations and artificial hydraulic fracturing, or in assessing possible incitement of slippage.

Geomechanical models of fold and fault structures rely to a large extent on rock-mechanical concepts and theories. Unfortunately, in spite of the impressive amount of research into the microscopic and macroscopic deformation processes in rocks, rock mechanics cannot yet provide mechanical theories that are sufficiently comprehensive and practicable to deal with all the different processes of faulting or folding. Therefore, in attempting to elucidate the mechanical genesis of tectonic structures and to establish the associated controlling parameters, one has to resort to a synthesis of various mechanical models of restricted validity, model experiments, computer simulations and field observations.

This is the approach followed in Part I of this book. The underlying mechanics will be dealt with only in a very cursory way, since all the basic mechanical concepts and principles required are presented and illustrated by geological applications in Part II, to which the reader will often be referred.

2. TECTONIC FAULTING – BASIC PHENOMENA

Tectonic faulting is the most prominent mode of crustal deformation and its mechanical aspects are the objective of this book. We shall start from a few basic phenomena of faulting that can be observed in the field and in the laboratory; these will be directives for the mechanical approach to be followed.

Tectonic faults are zones of *concentrated shearing* along which the adjacent rocks have been offset. The narrowness of the shear zones (fault zones proper) is indeed the most striking feature of tectonic faults. As an illustration, Fig. I.1-1 shows a basal thrust plane, characteristic of many large overthrusts. The shear zones may be separated from practically undeformed rock by sharp interfaces or the offset rock may show shear deformation for a distance of tens or hundreds of metres from the fault zone proper. However, in the latter case a closer inspection will reveal that the major part of the tectonic shear has been accommodated in narrow shear bands, as shown, for instance, in Fig. I.1-2 or Figs. II.7-20,21. A closer look reveals that not even in these narrow bands was the shear distributed continuously but that it was concentrated along slip surfaces, sometimes only being visible in samples studied in the laboratory. In this context, it is of little relevance whether these slip surfaces have the character of 'brittle' fractures accompanied by sudden rupturing of the material and complete loss of cohesive strength or whether they result from crystal-plastic gliding without interruption of existing bonds. Between these two extremes, cataclastic shearing takes an intermediate position, combining brittle fracturing on a grain scale with smooth plastic shearing on a somewhat larger scale. With respect to the width of individual cataclastic shear bands, it is worth mentioning an empirical rule of rock and soil engineering which states that shear displacement in sandy and clayey materials takes place in bands not wider than 10 – 20 mean grain diameters.

While in an ideally brittle material faulting would occur as fracturing preceded only by some small and purely elastic deformation, in reality inelastic processes contribute to the pre-

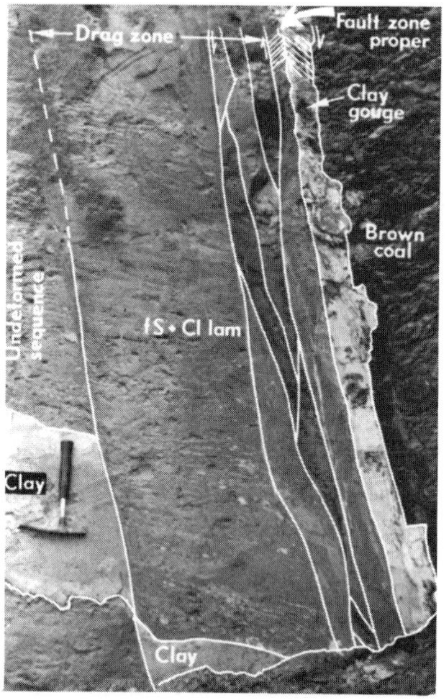

Fig. I.1-2
TYPICAL SHEAR ZONE OF MAJOR NORMAL FAULT IN MODERATELY CONSOLIDATED SEDIMENTS (Open-cast lignite mine of 'Rheinische Braunkohle', Frechen, W. Germany; Courtesy F. Lehner and W.F.H. Pilaar.)
Note the extremely narrow zone where most of the 70 m shear displacement has taken place (fault zone proper).

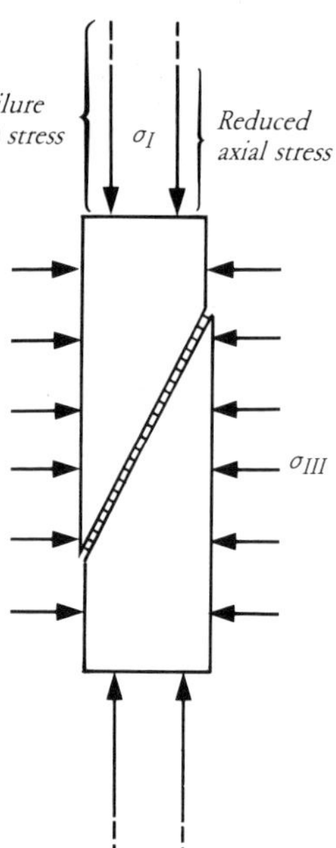

A Biaxial loading

B Sample with shear band

Fig. I.1-3 PLAIN STRAIN, BIAXIAL LOADING OF DENSE SAND PRODUCED ALMOST PLANE SHEAR BAND (after I. Vardoulakis et al., 1978; by permission of John Wiley & Sons, Ltd) faulting deformation of rocks (see Section II.7.7). Therefore, the width of the sheared zone surrounding a fault zone proper will increase with the relative amount of the inelastic pre-faulting shear strains that is characteristic of the individual rocks subject to the same conditions. For example, a fault plane in a sandstone will be embedded in a narrower zone of continuous shearing than, say, a fault plane in a limestone or soft shale under the same loading conditions and temperature.

In the 'brittle' upper crust the inelastic pre-faulting deformation processes (for example, the growth of microcracks, or the breakage and sliding of grains) are normally only slightly affected by the rate of loading and straining, but they become highly rate-sensitive at the elevated pressures and temperatures of the lower crust and thus exhibit viscous fluid behaviour. (See also Section II.7.9, "The ductile regime".) It is generally concluded that this basic difference in the nature of inelastic pre-faulting deformation causes tectonic shear zones to broaden when crossing the transition from the brittle upper crust to the ductile lower crust. Yet one has to expect, even in such wide ductile shear zones, the shearing to concentrate in narrow bands or fault zones proper. This may be concluded from many observations on uplifted deep crustal material (see, for example, Ramsay, 1980) and from the striking evidence provided by deep seismic reflection profiling of major faults extending into the lower crust and even across the Moho discontinuity rather than terminating in diffuse deformation (Smithson et al., 1979; Blundell et al., 1985).

Naturally, the intensive shearing along tectonic faults will produce changes in the texture of the fault zone material, the character of which will depend on whether the shearing is of a cataclastic nature, in which intergranular movements, abrasion and crushing of rock particles predominate, or of a crystal-plastic nature with predominantly intragranular movements, or whether shear deformation is accommodated by recrystallisation processes. All these processes have in common that they *completely consume the mechanical work* that the tectonic forces perform on the fault zone proper. Of course, this does not exclude that elastic strain energy is stored *outside* the fault zone proper, as is evidenced by the energy released in earthquakes.

Shear bands or shear fractures have been produced in numerous rock tests covering the whole range from very brittle to highly plastic behaviour. Fig. I.1-3 shows a cataclastic shear band in a 'vacuum-packed' sample of medium-grained dense sand which was loaded as indicated in Fig. I.1-3A. The test was carried out under plane strain conditions, i.e. the material was confined between two fixed plates parallel to the directions of the loads applied. Plane strain conditions approximate various tectonic situations quite closely (see Chapter II.5, Figs. II.5-10,12,13). As the axial load was gradually increased above the lateral load, the sample first deformed in a uniform way – with the vertical cross-sections remaining rectangular. The mode of deformation then suddenly changed and a thin shear band formed bounded by two 'rigidly' behaving sand blocks. As in numerous other experiments on rocks reported in the literature, the shear band formation took place when the *differential stress*, i.e. the difference between the greatest normal stress σ_I and the smallest one σ_{III}, had reached its *maximum* – the 'ultimate shear strength'. (Note that, following rock mechanical usage, compressive stresses are counted *positive* in this book.)

The maximum differential stress, i.e. twice the maximum shear stress (Section II.1.3), of a geological material depends on various conditions. To some degree it depends on the type of test used. With reference to the testing procedures shown schematically in Fig. I.1-4, one therefore speaks of uni-axial, bi-axial (plane strain) and tri-axial shear strength. More importantly, the shear strength is affected by the temperature, the rate of straining, the control of the pressure of the pore fluid, but in particular by the *confining pressure*, i.e. the mean normal stress.

When discussing the influence of the confining pressure on the shear strength of a porous fluid-filled rock, one has to consider that numerous experiments have established that the pressure of the pore fluid has no effect on the shear strength of the rock. It is therefore the mean of the *effective* normal stresses, which are defined as the total normal stresses transmitted across bulk cross-sections minus the fluid pressure inside the rock (see Sections II.3.2 and II.7.2, "The effective stress principle"), which affects the shear strength of the rock and, when increased within certain limits, strengthens the rock by raising its ultimate shear strength. This is clearly shown by the stress/strain curves of Fig. I.1-5 which were obtained for a limestone in the confined compression test of Fig. I.1-4B under various effective confining pressures.

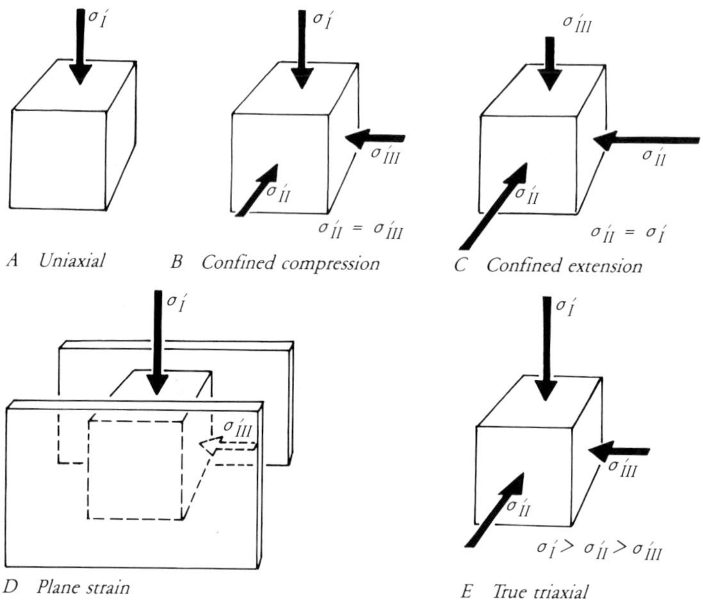

Fig. I.1-4 UNIFORM STRESS STATES IN TESTING ROCK SAMPLES

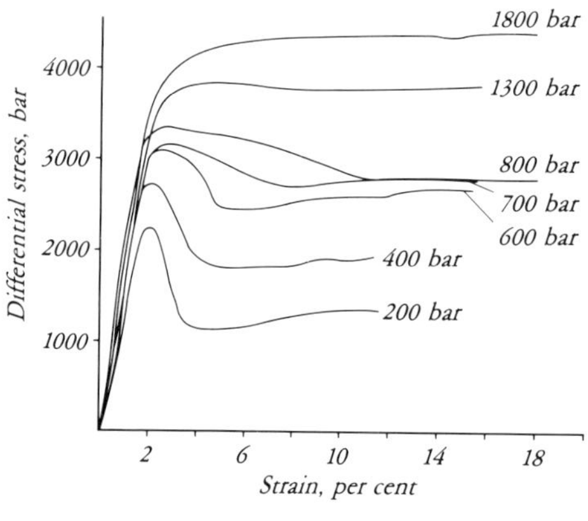

Fig. I.1-5
STRESS/STRAIN CURVES FOR AN OOLITIC LIMESTONE (After Donath, 1967). Note the increase of maximum differential stress and associated contractive strain with increasing effective confining pressure.

The shear band in the experiment of Fig. I.1-3 makes an acute angle with the direction of the maximum compressive stress – the principal stress σ_I. There is indeed vast evidence, from both laboratory experiments and field observations, that tectonic shear zones (faults) which develop when the differential stress has reached the limit value are not randomly oriented but that their orientation is controlled by the directions of the principal stresses. A striking feature of the angular relationship between fault planes and principal stress directions is the occurrence of *conjugate sets* of parallel faults, which often cooperate in accommodating imposed deformations. An example is shown in Fig. I.1-6 (see also Figs. II.7-16,20), in which the σ_I-direction, which bisects the angle between the conjugate shears, is perpendicular to the bedding of the rock. Because of this symmetry, a lower shear strength of bedding planes cannot give rise to the preferential development of one of the two conjugate sets. Although of small scale, the fault structures in Fig. I.1-6 are typical for the overwhelming abundance of faults that operate at any scale, ranging from microscopical size to fault systems of continental dimensions, in accommodating deformations.

In inelastic shearing the rock structure changes and the rock loses some of its strength. This *shear softening* of geological materials is well-known to soil and rock engineers. Returning to the stress/strain curves of Fig. I.1-5, we notice that the differential stress that the limestone can support under moderate effective confining pressures drops sharply to a lowest value ('residual shear strength') when straining has gone beyond the state of maximum differential stress. At high effective confining pressures the post-peak decrease in shear strength is less and occurs gradually. At the highest effective confining pressure the stress/strain curve does not show any decrease in strength and strain-softening seems to be suppressed. However, we expect that even at these high confining pressures the curve will eventually bend downwards if the testing device were to permit higher and more realistic, slower straining of the specimens (see also Section II.7.7, "Shear softening"). It should be mentioned at this point that, although the stress/strain curves of Fig. I.1-5 were obtained for a limestone, the curves published for various sandstones, shales, sand (Bishop, 1972), etc. show similar trends.

Naturally, the question arises as to the nature of the softening processes that operate on a 'micro'-scale in a rock subject to differential stress under confining pressure. In view of the complex microstructure of rocks, the mechanical differences between the rock constituents, and the variety of physical and chemical environments, our knowledge in this field is still very limited. Nevertheless, we can roughly subdivide the softening processes into two classes: *brittle* and *ductile* softening mechanisms. In the brittle mechanisms, by definition, the reduction in shear strength depends on the accumulated inelastic strain, but not, or only to a negligible extent, on the rate of the imposed straining. Examples of such processes are the loosening of grains in uncemented or slightly cemented sediments with a concomitant volume increase known as 'dilatancy', the abrasion and breakage of grains, their shifting and reorientation, changes in grain size, and the growth of microcracks in strong rocks. Likewise, the structure of a jointed rock mass will change in a brittle way as joints are opened up and grow, and block corners are crushed.

In the ductile processes, which operate at high temperatures and pressures, by definition the strength reduction is strongly rate-dependent. This group includes all microstructural changes that take place by means of mass diffusion processes, such as pressure solution and dynamic recrystallisation.

The evidence from field and laboratory work warrants the general conclusion that active shear zones on all scales are *zones of weakness and localised strain softening*.

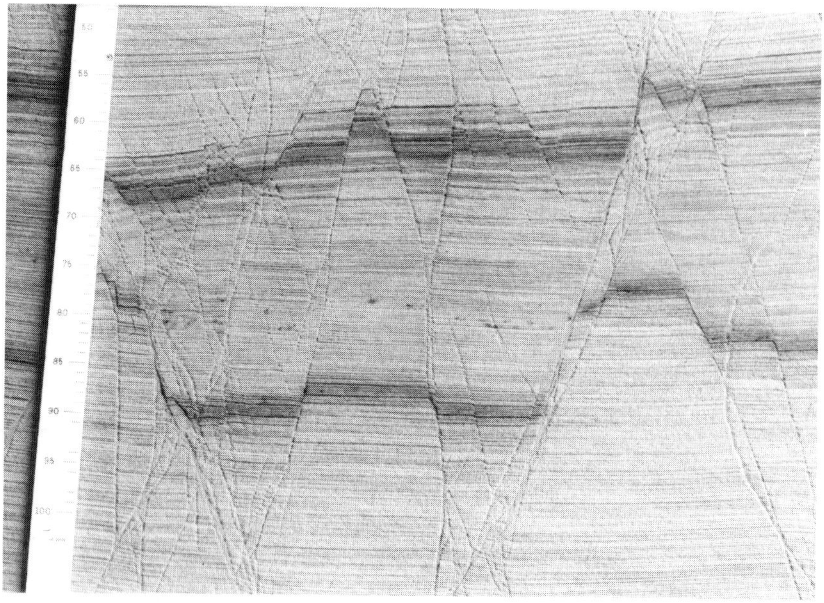

Fig. I.1-6 CONJUGATE NORMAL FAULTS IN PLEISTOCENE GLACIAL SEDIMENTS (Münster Basin, W. Germany; Courtesy D. Meier, Clausthal)

3. THE PLASTICITY APPROACH

Summarising the preceding section, we note that a full mechanical analysis of the genesis, shape, pattern and operation of tectonic faults should take account of the following general characteristics. Tectonic faults develop as very narrow zones of concentrated shearing inside rock masses, which themselves have undergone some elastic and permanent inelastic deformation. In the fault zone proper, all the mechanical work performed on the fault is consumed. The faults form when the differential stress attains a critical value which, within certain limits, increases with the effective confining pressure. The faults make an acute angle with the direction of the greatest compressive (principal) stress and frequently operate in conjugate sets. While active, faults are zones of reduced shear strength.

Furthermore, when the inelastic deformation processes that precede and accompany the formation of a fault are being considered, it is important to distinguish between deformations that are the result of brittle micro-processes and deformations caused by micro-processes of a general viscous character. Although both types of deformation consume all mechanical energy, there is an important difference between them: an individual grain breaks or slides along frictional contacts with its neighbours, or a brittle microcrack grows in jerky movements, when the local stresses reach a critical level, but the rate at which such a micro-process takes place is practically independent of the magnitude of the local stresses. In contrast, viscous micro-processes take place at a rate that increases strongly with the magnitude of the local shear stresses. Correspondingly, when viewed on a macro-scale, which is large in comparison with, say, grain size, the associated deformation processes will differ too. In the first one, the strain rate will not depend on the magnitude of the stresses, and the shear strength of the material and its reduction during faulting will not depend on the rate of straining. This property is referred to as *rate independence*. In contrast, the viscous deformation mechanisms constitute strong rate sensitivity; stresses and strain rates are strongly related and the shear strength of the material will vary – and, in general, increase – with the rate at which the strain is applied.

Energy dissipation, shear strength and rate independence characterise a class of idealised materials, referred to as *plastic* in continuum mechanics. In accordance with this, we use the term in a more restricted and precise sense than is common in the geological literature, where it is applied synonymously with 'ductile' to any permanent continuous deformation. The attribute *ductile* will be assigned to strain-rate-dependent material behaviour associated with viscous deformation processes. (The difference between 'plastic' and 'viscous' or 'ductile' is discussed more extensively in Sections II.7.1 and II.7.4 ('Rate-independence of Plastic Deformation') in the chapter on 'Plasticity of Rocks'.)

Mechanical theories of plasticity were originally developed for materials, in particular metals, whose shear strength could be considered as independent of the confining pressure. Application of the theories to geological materials with pressure-sensitive shear strength therefore required some generalisations. This generalised plasticity concept has been named *'frictional plasticity'* since it was first applied to soil the shear strength of which stems mainly from interparticle friction. The special attractiveness of frictional plastic models for the analysis of tectonic faulting lies not only in the fact that they account for shear strength and permanent deformations, but rather in their extraordinary capability to predict fault-type features, either as (slip) surfaces across which the tangential displacement component can discontinuously change, or alternatively, in a more advanced model, as narrow shear zones of weakened material. Moreover, these 'faults' do not arise out of a purely elastic state but, more realistically, are either embedded in or bordering on a region in the plastic state, or emerge from a plastically deforming mass.

A full account of frictional plasticity theory, as relevant to tectono-mechanics, is given in Chapter II.7, to which the reader may turn when in need of more details on the underlying mechanical theory. Here, it may suffice to illustrate the three basic models of frictional plasticity, all of which will be applied in this book, by the simple analogue models of Fig. I.1-7. The essential feature in all three analogue models is the solid block in frictional contact with a solid base onto which it is pressed by a vertical force. The vertical pressure determines the frictional resistance and, consequently, the pulling force required to start frictional sliding and to represent the critical differential stress that induces plastic deformation. The model of Fig. I.1-7A represents a rigid/frictional plastic material, a model that does not account for any elastic deformation prior to the onset of plastic deformation. More realistically, in model B allowance is made for an elastic component by inserting a linearly elastic spring. Obviously, before the block starts moving, the spring will have to extend. When the pulling force is reduced below the critical value and movement of the block stops, the spring will contract somewhat and release elastic strain energy. Reality is even more closely approached by model C which accounts for plastic *strain hardening and softening*. The model allows plastic deformation (movement of the frictional block) to start before the differential stress (pulling force) has reached its maximum. The block, initially resting partly on a bed of frictionless rollers, will experience increasing frictional resistance when being moved from the roller bed until it is in full contact with the frictional base. This part of the movement corresponds with a deformation phase of plastic strain hardening, so called because an

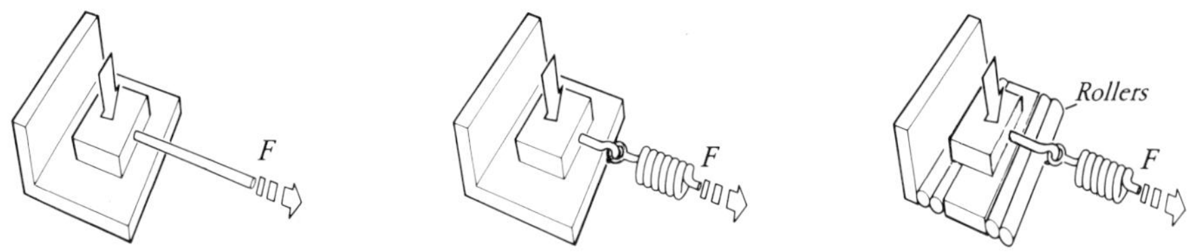

A Rigid/frictional plastic (Coulomb) *B Elastic/frictional plastic* *C Elastic/frictional plastic with strain hardening and softening*

Fig. I.1-7 PLASTICITY MODELS

increase in differential stress is required to continue straining. A second bed of frictionless rollers in the model allows shear resistance and pulling force to decrease again as the block is moved further. This phase then represents the process of strain softening that starts once the differential stress has reached its peak.

Figures I.1.7 not only represent analogues of plasticity models of increasing complexity, they also indicate the limitations of the plasticity approach. The absence of strain-rate-depending mechanisms – the frictional resistance of the block is meant to be independent of sliding rate – also implies that time-dependent phenomena, such as rock creep at constant load or relaxation of stresses at constant state of strain, are not taken into account. Therefore, ductile (i.e. viscous) deformations which might precede or succeed the formation of faults are outside the range of this theory (see also Section II.7.9).

It is usually justified to consider tectonic processes as *quasi-static*, i.e. as successions of static equilibrium states, and to neglect inertial forces. This assumption is also made throughout this book. In particular, when modelling the growth of a tectonic fault zone we consider only stable growth. The modelling of unstable, i.e. accelerated growth of frictional plastic zones (plastic 'run-away' zones) has apparently not yet been successfully undertaken.

Although we consider the formation of a tectonic fault to be, in general, dominated by inelastic processes, rather than envisaging it as the growth of a brittle shear crack inside an ideally elastic continuum, we do not exclude that purely elastic analyses may in certain cases provide valuable insight into the development of the stress field and its principal directions. In the main, however, an appropriate application of elastic-brittle fracture mechanics to the field of faulting will remain limited to analysing the elastic field changes and the associated energy release that accompany fast local slip movements along existing fault planes or any other planes of weakness. However, such problems of earthquake mechanics are outside our scope.

Some elementary results on faulting can be obtained without making any assumptions on the mechanical character of the solid material. Such results are derived from the balance of forces or from an energy balance, i.e. from general principles which are not restricted by the choice of the material laws. With such a result we shall begin the next chapter.

I.2. GENERAL FAULT FEATURES

In this chapter we address the mechanical aspects of genesis and operation that are shared by all or, at least, by a wide variety of tectonic faults.

1. ENERGY DISSIPATION AND THE SENSE OF MOVEMENT

We have stated in the introduction that the work performed by tectonic forces on the fault zone proper is completely consumed within this zone. This imposes an important restriction on the sense of displacement along a fault. The mechanical work expended on an active fault per cm^2 of its surface is, by definition (see for example, Section II.8 and Fig. II.8-5), the product of a shear stress on the fault plane and the relative displacement of the fault blocks. The shear stress in this product is the component parallel to the line of the relative displacement. Since the mechanical work is supplied to the fault zone proper, the product considered has to be positive; otherwise, mechanical work would be extracted from the fault. Both shear stress and displacement must therefore have the same sign, and hence point in the same direction. This implies further that the direction of the greatest compressive stress (i.e. the greatest principal stress σ_I) has to make an acute angle with the direction of the displacement along the fault. This almost obvious conclusion can be readily verified by considering the shear stress orientations on planar elements whose normals vary in orientation but lie in the same plane as the greatest (σ_I) and smallest (σ_{III}) principal stresses (see, for example, Fig. II.8-7).

To illustrate the tectonic relevance of this elementary result, we consider schematically in Fig. I.2-1A a family of parallel imbricate thrusts which accommodate a horizontal shortening of the crust, obviously produced by the action of a (sub-)horizontal maximum compressive stress. Although all faults contribute to the shortening, the advanced position of the shaded fault block in the figure would be energetically impossible, since it would have required the generation of mechanical work along fault x.

A second example (Fig. I.2-1B) shows a schematic set of parallel normal faults at a graben boundary. In this case the greatest principal stress, caused by the weight of the overburden,

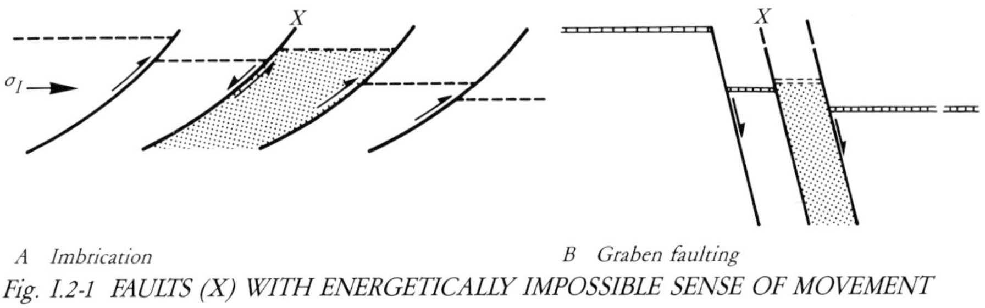

A Imbrication *B Graben faulting*

Fig. I.2-1 FAULTS (X) WITH ENERGETICALLY IMPOSSIBLE SENSE OF MOVEMENT

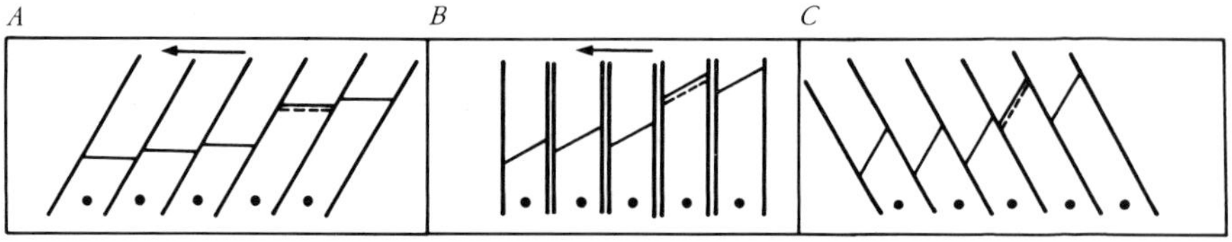

Fig. I.2-2 ROTATION OF PARALLEL FAULTS TO STEEPER POSITION POSSIBLY CAUSING ANOMALY IN FAULT DISPLACEMENT (Full circles indicate fixed axes of rotation)

is (sub-)vertically oriented. This makes it energetically impossible for a fault block to have lagged behind its two neighbours, although this would be admissible from a purely kinematic point of view.

It is not difficult to find structural interpretations in the literature that show elements that seem to contradict the simple rule of energetic admissibility. But before discarding such an interpretation, one has to make sure that the out-of-step offset along a fault is not caused by a strike-slip component or by the rotation of parallel faults. The displacement of dipping bedding planes along the strike of a fault can easily feign an energetically impossible dip-slip displacement. Similarly, the simultaneous steepening of parallel normal faults or the flattening of parallel thrust faults can reverse the displacement along a single fault when its original displacement deviated considerably in magnitude from the displacement on neighbouring faults. This is illustrated schematically in Fig. I.2-2, in which parallel normal faults are simultaneously rotated around fixed points into a steeper position.

2. THE ONSET OF FAULTING

The Critical Stress State. From the observations of small-scale faulting in a great variety of soil and rock tests it is generally concluded that the onset of large-scale faulting also requires that the differential stress $\sigma_I - \sigma_{III}$ attains a certain critical value. Or, phrased differently, since the differential stress is twice the maximum shear stress τ_{Max} (see for example, Section II.1.3) associated with the greatest and smallest principal stresses, a fault can form only after τ_{Max} has reached the highest value the material can withstand. However, in geological materials, this critical value – the 'ultimate shear strength' – is not a material constant, but it is affected by certain factors. We have already mentioned (Section I.1.2) that these factors depend on whether the material deforms in a predominantly brittle or ductile way. The primary factor in the brittle deformation mode is the effective confining pressure σ'

$$\sigma' = \sigma - p = (\sigma_I + \sigma_{II} + \sigma_{III})/3 - p \qquad \text{(I.2-1)}$$

i.e. the total confining pressure σ minus the full [1)] pressure p of the pore fluid. For a given geological material in the brittle deformation range, the ultimate shear strength increases monotonically with the effective confining pressure. However, in the ductile range the shear strength is primarily affected by temperature and rate of straining, and tends to decrease with increasing temperature and/or decreasing strain rate (see also 'Shear softening' in II.7.7, and 'The ductile regime' in Section II.7.9).

We should recall that here the terms *'brittle'* and *'ductile'*, which are often used in a somewhat vague and loose way, are given the following precise meaning: the term 'brittle' characterises not only a material that deforms predominantly elastically up to the point of sudden shear failure (a shear failure that occurs at a rate that is practically independent of the state of stress and the rate at which the material was strained prior to failure), but also comprises any strain-rate-independent inelasticity, i.e. any inelastic deformation process in which the strains are somehow associated with the imposed stresses, but are not affected by the rate of straining. In contrast, 'ductile' refers to a viscous material for which, by definition, the stresses are primarily related to the strain rate, be it in a linear or non-linear way. 'Brittle', in the wider sense we use this concept to indicate a combined elastic and strain-rate independent inelastic deformation behaviour of a material, is synonymous with 'elastic/frictional plastic' – even though the terms may suggest a different meaning. Therefore, the theory of frictional plasticity, summarised in Section II.1.3 and dealt with extensively in Chapter II.7, is an adequate tool for dealing with brittle deformations [2)].

Generally, in the critical state the stresses have to satisfy a limit condition, which in principal stress components σ_I ($\equiv\sigma_{Max}$), σ_{II} ($\equiv\sigma_{Int}$), σ_{III} ($\equiv\sigma_{Min}$) has the form

$$(\sigma_I - \sigma_{III})/2 = f(\sigma') \qquad \text{(I.2-2)}$$

where f is a monotonically increasing function of the effective confining pressure (eq. I.2-1). This limit condition may be approximated by a simpler expression. First, experimental rock and soil mechanical evidence indicates that the intermediate principal effective stress σ'_{II} has much less effect on the shear strength of the geological material than the maximum and minimum effective stresses (see 'Independence from intermediate principal stress' in Section II.7.4). We therefore replace eq. I.2-2 by the condition

$$|\tau_{Max}| \equiv (\sigma_{Max} - \sigma_{Min})/2 = f[(\sigma'_{Max} + \sigma'_{Min})/2] \qquad \text{(I.2-3)}$$

which is *independent of the intermediate principal stress* ($\sigma'_{II} \equiv \sigma_{Int}$).

Secondly, for most rocks this function is approximately linear in the compression range up to say 1 kbar effective confining pressure. For higher confining pressures the increase in τ_{Max} declines. In Mohr's stress plane (Chapter II.4) of normal effective stresses $\sigma'_\perp$ and shear stresses τ the limit stress function therefore plots as a pair of straight lines – accounting for positive and negative shear stresses – which at high effective pressures may curve and become more nearly parallel to the $\sigma'_\perp$-axis. In Fig. I.2-3 such a limit line is shown for positive shear stresses. Any point Q on this line defines a limit state of stress graphically represented by the stress circle (Chapter II.4) with centre on the $\sigma'_\perp$-axis and vertex Q. The Set of all limit stress circles has two envelopes – the *'Coulomb-Mohr limit lines'* – the straight parts of which are described by

$$|\tau| = \tau_o + \sigma_\perp' \tan\varphi \qquad \text{(I.2-4)}$$

The angle φ is usually referred to as 'angle of internal friction' – a term reminiscent of the soil mechanical origin of the limit condition, indicating interparticle friction as the source of the pressure-sensitive shear resistance of soils. Typical values of φ range from, say, 10° for a very soft clay to 35° – 50° for dense sands, and they may attain even higher values for strong rocks. The parameter τ_o is the 'cohesive shear strength' which the material would still exhibit at vanishing effective confining pressure.

The Coulomb-Mohr limit lines distinguish two points of tangency (T, T' in Fig. I.2-3) on a limiting stress circle. Since a stress circle associates tangential and normal stresses with the planar elements upon which these stresses act in the real material (Chapter II.4), the two tangential stress points are associated with two such elements, the normals of which lie in the plane of the greatest and smallest principal stresses, and which therefore intersect along the axis of the intermediate principal stress. As shown in Fig. I.2-4, the two elements that are acted upon by the critical stress components of T and T′ in Fig. I.2-3, include the angle $90° - \varphi$ (i.e. half the angle between the corresponding radii of the Mohr circle in Fig. I.2-3). This angle is bisected by the axis of the greatest (compressive) principal stress.

In the sense of Coulomb-Mohr the limit condition therefore simply states that the onset of faulting requires that on a pair of planar material elements the expression $|\tau - \tau_0|/\sigma'_\perp$ attains the maximum value $\tan\varphi$.

It should be understood that we have so far refrained from any statement concerning the mode of faulting itself. Instead, we have considered Coulomb-Mohr's limit condition as a purely static statement on the state of stress that gives rise to faulting. In this sense, the limit condition has been approved by numerous successful applications in rock and soil mechanics. The condition is of striking simplicity which is mainly a consequence of the absense of the intermediate principal stress – an approximation which appears quite acceptable for most structural geological applications. For a more detailed discussion of the limit condition, the reader is referred to Section II.7.3.

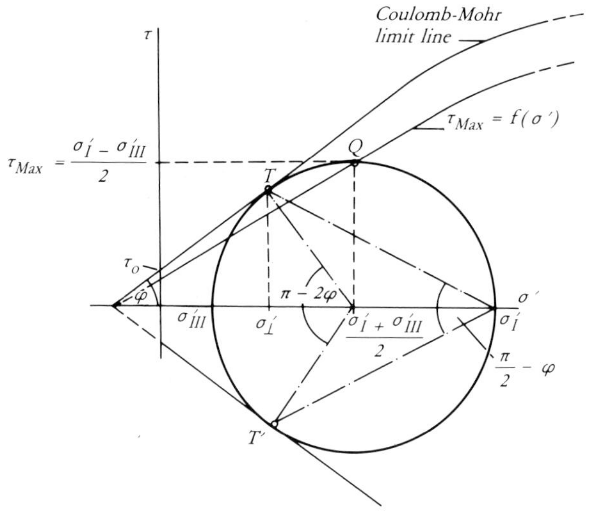

Fig. I.2-3 LIMIT STATES OF STRESS IN MOHR'S STRESS PLANE

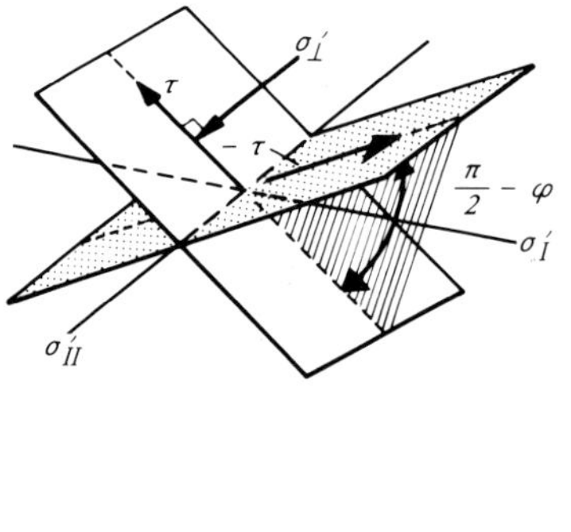

Fig. I.2-4
PLANAR ELEMENTS IN PHYSICAL SPACE ASSOCIATED WITH POINTS OF TANGENCY (T, T′) IN STRESS PLANE OF FIG. I.2-3

Tectonic Limit States. From the existence of the Coulomb-Mohr limit state of stress an important conclusion can be drawn that applies to *any* rock type. Let us consider a stack of horizontal strata, free of horizontal shear stresses and the vertical effective normal stress σ'_V being determined by the overburden weight minus fluid pressure. In the Mohr diagram (Fig. I.2-5) the stresses on a horizontal plane are therefore represented by a point (P) on the $\sigma'_\perp$-axis. In a tectonically undisturbed situation the horizontal effective normal stresses σ'_H will be equal and will represent the smallest principal stress σ'_{III}. The rock will, in general, not yet be in a limit state and tectonic activities are required to lower or raise horizontal stresses. Let us first consider horizontal extension, which leaves the role of the vertical normal stress as the greatest principal stress σ'_I unchanged. Here, then, a modest reduction in σ'_H will induce a limit state in which normal faulting becomes possible and allows the rock mass to extend in a horizontal direction.

However, when horizontal compression is applied, a large increase in the corresponding horizontal stress is required to reach a second type of limit state ($\sigma'_H \rightarrow \sigma'^P_I$). Because of the slope of the limit line, which represents the shear-strength-increasing effect of the effective confining pressure, the differential stress (represented by the diameter of the stress circle) associated with that limit state may be about four times the differential stress required for the extensional limit state. To accommodate shortening, faulting in this limit state will be of the thrust type. Following soil mechanical usage, the compressional limit state is called the *'passive'* limit state, while the extensional limit state is referred to as the *'active'* limit state.

This important *duality of limit states* will still result under more general conditions when the strata are acted upon by shear stresses, as is the case on a sedimentary slope (Example No.2 in Section II.7.3).

Because of the drastic difference in differential stress between the active and passive limit states, in tectonic shortening by thrust faulting the material will generally be much more deformed than in extension by normal faulting. This difference will be particularly pronounced in shaly materials. Thus, very probably, thrust faults will be embedded in a wider zone of pre-faulting deformation than normal faults.

Finally, another implication of the difference between compressional and extensional changes of the horizontal stresses is worth mentioning. It concerns the *switch of fault style* during a single tectonic episode. To start with, in a tectonically undisturbed situation the overburden stress σ'_V is the greatest principal stress. Ideally, it will remain a principal stress when a horizontal compression or extension is imposed upon the rock, but it may no longer

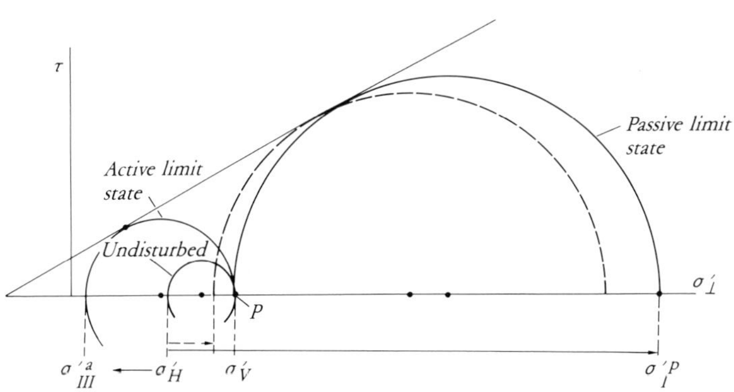

Fig. I.2-5 STRESSES IN HORIZONTAL OVERBURDEN – DUALITY OF TECTONIC LIMIT STATES

keep its role as the greatest principal stress. In compression, a horizontal normal stress exceeds the overburden stress and becomes the maximum principal stress. This does not necessarily mean, however, as was assumed in Fig. I.2-5, that the overburden stress becomes the smallest. The horizontal stress normal to the one that becomes the maximum principal stress ($\sigma_I'^P$, need not increase beyond σ_v', which would make the latter an intermediate principal stress. Critical stress circles such as the dotted one in Fig. I.2-5 then become possible. Obviously, in this case horizontal shortening will not be compensated by thickening of the overburden, as it was in thrusting, but rather by extension in the horizontal direction of lowest relative rock pressure. The type of fault that enables this type of horizontal deformation is the wrench fault. Hence, the stress conditions for *thrusting and wrenching* may lie fairly close together, and relatively small changes in a horizontal compression component can produce a switch between thrust and wrench conditions.

Similarly, in an extensional regime with normal faulting, a relatively small increase in the horizontal stress normal to the direction of extension is needed to make this principal stress slightly greater than the overburden stress, which then becomes the intermediate principal stress. Horizontal extension will then proceed, accompanied by some transverse horizontal shortening instead of thinning of the overburden by normal faulting. Hence, the style of faulting will change from *normal to wrench faulting.*

3. ORIENTATION OF FAULT PLANES

In the last section, when discussing the duality of extensional and compressional limit states, we touched upon the question of fault plane orientation with respect to the directions of principal stresses. As stated in the Introductory Section I.1.2, there is indeed strong evidence from laboratory and field observations that the orientation of fault planes is primarily controlled by the directions of the principal stresses. This has also been corroborated by modern continuum-mechanical theories on the spontaneous shear band formation in brittle, i.e. elastic/frictional-plastic materials, which were advanced in particular by Rice and Rudnicki (1975), Rice (1976), Vardoulakis (1977, 1980) and Vardoulakis et al. (1978). These theories also present a formalism for determining the orientation of an 'embryonic' shear band, but the number of unknown material moduli involved seem to make actual predictions of shear band orientation impossible[3)].

Obviously, however, quantification of the fault orientation with respect to the principal stress directions would be of considerable value in structural analyses. What can be said about it? First, we recall from Section I.2.1, that the dissipation of mechanical work in the fault zone proper requires that the direction of the maximum compressive stress includes an acute, but otherwise still undetermined, angle with the fault plane. In addition to this angle θ (Fig. I.2-6) we need a second parameter to define the orientation of a fault plane or the orientations of the tangential planes of a curved fault. As second parameter we take the orientation of the axis of the intermediate principal stress σ_{II}', the stress which had only a negligible effect on the limit state. One might argue that the actual shear direction on a fault plane element will be parallel to the σ_I', σ_{III}'-plane, since it can be shown that the tangential stress component would be smaller in any other direction on the fault plane; therefore, the whole fault plane element should orthogonally intersect the σ_I', σ_{III}'-plane. In other words, it appears plausible that the fault plane elements pass through the σ_{II}'-axis.

Next we consider the angle θ (Fig. I.2-6). To determine it more closely we resort to the hypothesis advanced by O. Mohr around 1900 for granular materials, that the orientation of the discrete slip elements, along which the material deforms in the limit state, is determined by the interaction of shearing and normal stresses. Whereas the shear stress is the driving agency, the effective normal stress $\sigma_\perp'$ on a potential slip element increases the resistance to slippage. When applying this concept to rocks with a cohesive component τ_0 of the shear strength, we assume that the part τ_0 of the total shear stress τ operates in breaking cohesive bonds, without inducing a notable strength anisotropy in the material. It is then only the excess shear stress $\tau - \tau_0$, which interacts with the normal stress $\sigma_\perp'$ in controlling the orientation of actual slip elements. It is an almost obvious conclusion that actual slip will take place along elements for which the dimensionless quantity $|\tau - \tau_0| / \sigma_\perp'$ attains its maximum. As stated earlier, this maximum value is attained at the points of tangency of a limiting stress circle, as a glance at Fig. I.2-3 will show. Note that this also holds for curved limit lines, since these lines never steepen with increasing effective pressure. *Therefore actual slip will occur only along material elements that in the stress plane are associated with the points of tangency between the critical stress circle and the limit line.* The slip elements, therefore, make with the σ_I'-axis the well-defined angles (Fig. I.2-4)

$$\theta = \pm (45° - \varphi/2) \qquad \text{(I.2-5)}$$

Conjugates. In an isotropic material there would be no preference for the orientation angle θ, at which slip planes (faults) initiate, to be positive or negative, and one therefore has a pair of *conjugate orientations* for the incipience of faults. If, moreover, the directions of the principal stresses remain fixed during the deformation process, the two conjugates have an equal chance to develop and to contribute to the deformation. In addition to the examples of conjugate faults in Figs. I.1-6, II.7-16 and II.7-20, Fig. I.2-7 shows a well-known example of sets of conjugate faults that operate together in accommodating the extension caused by a rising salt dome.

The simultaneous development of two conjugates requires fairly ideal conditions. Usually, rocks exhibit some strength anisotropy that may be associated with closely spaced bedding planes, cleavage, foliation or parallel tension fractures. An equal chance for the formation of the two conjugates will then still exist when the principal stress axes coincide with the symmetry axes of the material, as was obviously the case in Fig. I.1-6.

The preferential development of one fault type is more common than the simultaneous development of conjugate faults, with faults of the conjugate type playing a subsidiary role in the deformation of the rock mass. There are various reasons for this, some of which will

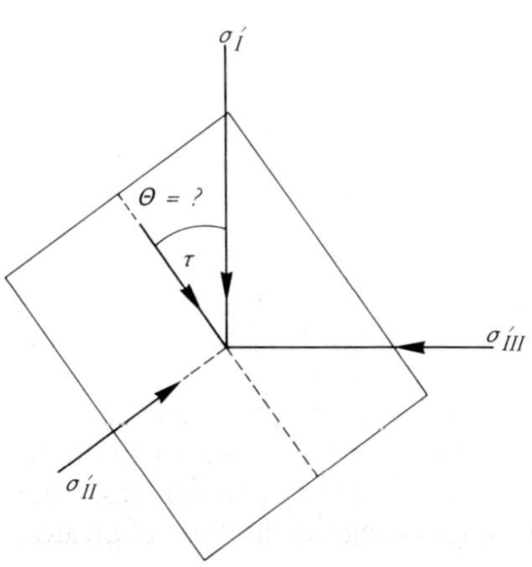

Fig. I.2-6
THE ORIENTATION OF SLIP ELEMENTS

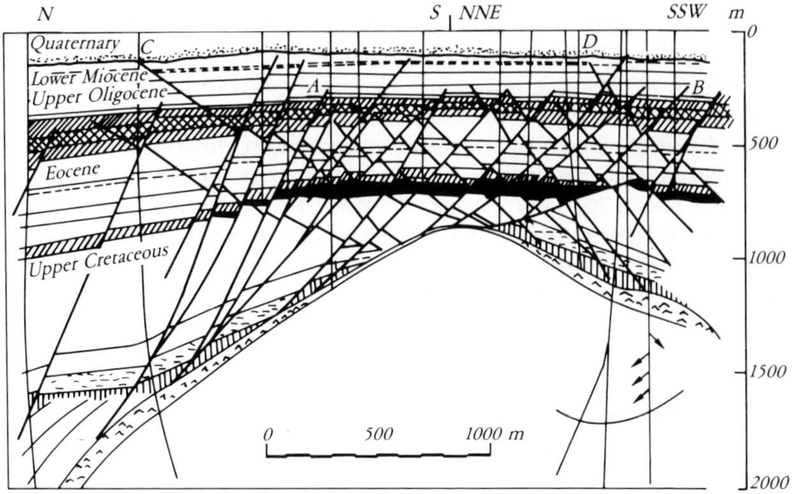

Fig. I.2-7 REITBROOK SALT DOME, NORTH GERMANY
(after Behrmann 1949 simplified by Horsfield, 1980; reprinted by permission of Pergamon Journals Ltd.)
Examples of crossing conjugate normal faults over an anticlinal crest, illustrating a typical problem of subsurface reconstruction. Oil wells are shown as vertical lines, oil accumulations in black.

be discussed later in some detail when antithetic faults (Section I.2.6) and specific styles of fault tectonics are dealt with. The simplest reason for the preferential development and operation of one conjugate fault type is the presence of 'rigid' boundaries which may prevent the second conjugates from developing beyond the incipient state. Secondly, if the principal stress axes are asymmetrically oriented with respect to planes of strength anisotropy in the rock, slip along only one of the conjugate slip directions may take place because of lower shear resistance, in spite of the stresses acting on the two conjugate planes being equal.

Even in the absence of any asymmetry between principal stress axes and anisotropy planes, or in the absence of strength anisotropy itself, preferential development of one conjugate may be caused by a difference in shear softening along conjugate slip planes. Such a difference would result from a difference in shearing activity on incipient slip planes. Such asymmetry in shear strain is indeed possible, even though the stresses are exactly the same on the two potential conjugates. To understand this, one has first to recall that the angle between the directions of maximum shearing is 90°, while the angle between incipient conjugates is only $90° - \varphi$. The 90° angle is bisected by the direction of the greatest (incremental) shortening $\dot{e}_I$, while the angle between the conjugate slip directions is bisected by the σ_I-axis. The two axes do not have to coincide, but may deviate by as much as $\pm \varphi/2$ (see 'Non-coaxiality of Stress and Strain' in Section II.7.4, and Fig. II.7-25). One potential slip direction will then be closer to or even coincide with a direction of maximum shearing, which should result in a preferential reduction in shear strength (shear softening). We shall revert to this mechanism in Chapter I.3 when discussing parallel faults on a slope.

Non-plastic processes, such as compaction or pressure solution, may play a role in enlarging the initial angle between slip planes and σ_I-axis, as is indicated schematically in Fig. I.2-8A. This may make it difficult to determine the original σ_I-direction from the present-day appearance of conjugate faults. An obtuse angle facing the direction of greatest shortening must certainly be expected if the faults were initiated under high pressure and temperature conditions when the shear strength of the rock became practically independent of the effective confining pressure (i.e. $\varphi \rightarrow 0$ in eq. I.2-5) and the incipient conjugates therefore included an angle of 90° (Ramsay, 1980). In contrast, shortening by brittle processes will hardly change the angle between conjugate shear zones, as is explained in Fig. I.2-8B.

Coulomb-Mohr's slip concept may also be considered the basis of *Anderson's* well-known fault classification, which is outlined in Fig. I.2-9. It applies to tectonic stress fields with uniform principal stress directions and with the overburden stress as a principal stress. The latter implies, of course, that there are no horizontal shear stresses operating. Because of these restrictions, various fault types fall outside Anderson's scheme.

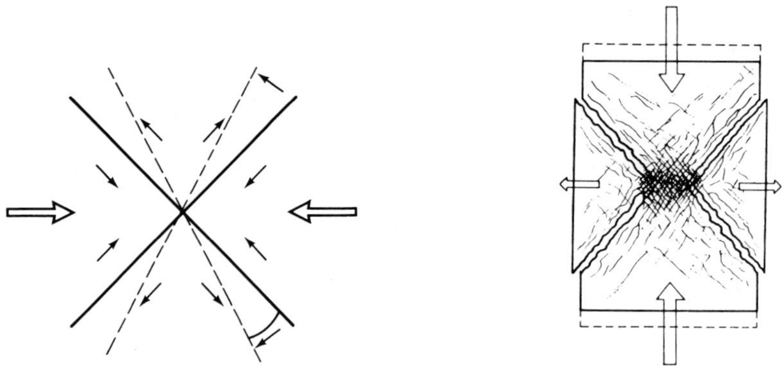

A Conjugate plastic slips (after B. Sander, 1948). Angle to maximum compressive stress may be enlarged by internal rotation.

B Conjugate shear fractures in 'brittle' rock (after L. Müller, 1963;reprinted by permission of Ferdinand Enke Verlag). Angle between rupture surfaces remains preserved during shortening; accommodation of shortening by brecciation and dilatation.

Fig. I.2-8 APPEARANCE OF CONJUGATE SLIP PLANES

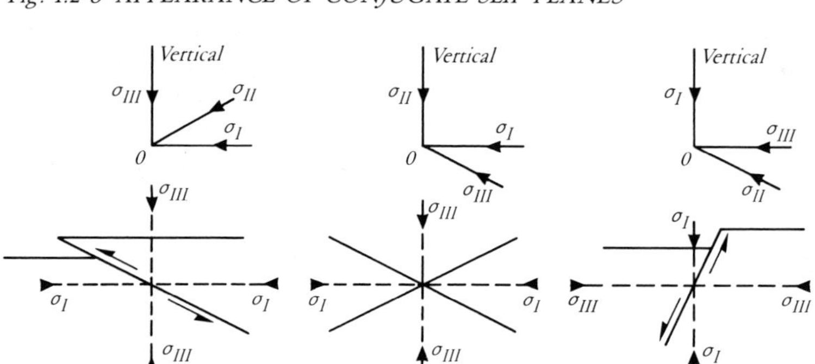

A Thrust faulting *B Strike-slip faulting* *C Normal faulting*

Fig. I.2-9 ANDERSON'S FAULT CLASSIFICATION

The Double Gliding Model. A rock mass in the limit state of a tectonic stress field can deform by contemporaneous or sequential sliding or shearing along many surfaces of both conjugate sets. This mechanism of 'double gliding' (or 'double shearing') is illustrated in Fig. II.7-18 of Section II.7.4. Close spacing of the slip surfaces allows any plane, quasi-continuous deformation to be produced. This generalisation of Coulomb-Mohr's concept will prove extremely useful in describing and understanding the deformation of rock masses by faulting, since many tectonic deformations are of the plane type – at least approximately and in parts of the deformed region – as is discussed and illustrated in more detail in Section II.5.3.

To avoid misunderstanding, it should be noted that Coulomb-Mohr's slip concept is not restricted to plane strain. Although slip takes place only in directions perpendicular to the σ_{II}-axis, in a truly three-dimensional and non-uniform stress field the orientation of this axis may vary in space, and the field of associated slip directions will then allow for three-dimensional, albeit non-uniform, straining. It is true, however, that in a field of uniform principal stress directions and unequal principal stresses the Coulomb-Mohr slips cannot constitute non-planar strain.

This is not incompatible with the observation, frequently made in the field and in experiments, that three-dimensional deformations, even of homogeneous character, have been accommodated by *more than two fault sets.* Often, these multiple sets can be shown to be Coulomb-Mohr slips associated with changes in the stress fields during different deformation phases. Even when, in the field or in experiments, more than two fault sets have formed contemporaneously to accommodate uniform three-dimensional strain (Aydin and Reches, 1982; Reches and Dieterich, 1983), the faults may still be of the Coulomb-Mohr type. This point is pursued a little further in Section II.7.4 ('Two-dimensional vs. Three-dimensional Strain') where it is shown that uniform biaxial extension may be accomplished by slip along multiple sets of Coulomb-Mohr faults, the formation of which is merely associated with the value of the intermediate principal stress changing into the value of the minimum principal stress, while the principal axes of the uniform stress field remain unchanged [4)].

Limitations of Mohr-Coulomb's Fault Orientation. In the literature, Mohr-Coulomb's slip condition I.2.5 is usually considered as part of the limit condition. We have dealt with the slip condition separately from the limit condition proper (eq. I.2-4), because the slip condition, which relates deformation and stresses, is based on more restrictive assumptions and is therefore somewhat less generally applicable than the purely static limit condition which determines the onset of faulting.

Essentially, this limitation stems from a possible *direction dependence of the shear strength* in rocks. Such anisotropy of the shear strength may be associated with an anisotropy of the rock fabric, which already existed before a faulting episode, or it may have been induced by inelastic deformation processes during the tectonic build-up of a limit state of stress. Unfortunately, the amount of experimental work performed on anisotropic rocks is not yet sufficient to provide an overall picture of the range of the strength anomalies involved [5,6)]. It may suffice here to refer to a highly schematic representation in Fig. I.2-10A,B of results obtained for some consolidated sedimentary rocks (Donath, 1967). In case (A) a core is loaded by σ_I parallel to the plane of anisotropy, which results in a reduced angle between the shear fractures produced and the σ_I-direction. When σ_I is oriented at a high angle to the plane of anisotropy, the shear fractures are expected to make higher angles with the σ_I-direction. In Fig. I.2-11 the possible effect of large-scale jointing on fault orientation is illustrated. The two normal faults were formed in a sandstone that is heavily fractured (extension fractures) in the vertical direction. The resulting strength anisotropy might very well be responsible for the strikingly small angle between the two conjugate faults.

But even if loading tests of the type shown in Fig. I.1-4 do not reveal any direction dependence of the maximum differential stress that a rock can withstand under the applied effective confining pressure $(\sigma_I' + \sigma_{III}')/2$, and therefore the slope angle φ of the limit lines in

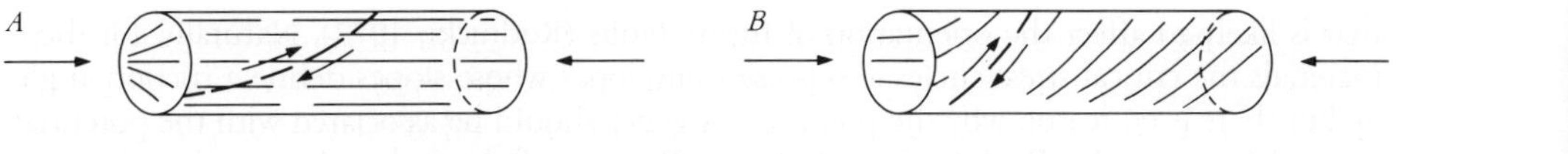

Fig. I.2-10 EFFECTS OF ANISOTROPY ON GEOMETRY OF SHEAR FRACTURES IN TEST SPECIMENS (after Donath, 1967)

Fig. I.2-11 CONJUGATE NORMAL FAULTS IN BUNTSANDSTEIN, HELGOLAND (courtesy of W.T. Horsfield and R.T.C. Frost)

Mohr's stress plane is the same for any direction in the material, the planar elements associated with the points of tangency of limiting stress circles with the limit lines do not necessarily represent the correct orientation of potential faults. It was assumed in the argument we used to arrive at Mohr's formula I.2.5 for the slip plane orientation, that the inelastic deformations that precede slip formation do not induce an anisotropic change in shear strength. As long as these pre-faulting strains consist mainly of a loosening of aggregates of grains, one may expect this to induce only little anisotropy in the microstructure. This is in accord with soil mechanical experience, which has demonstrated the applicability of Mohr-Coulomb's slip concept to many kinds of sandy, silty and clayey materials. In this connection, it is of note that load-induced small anisotropies in sand and clays, although they hardly affect the orientation of slip planes, may greatly affect the amount of continuous straining that precedes the actual formation of faults [7)]. Most likely, the orientation formula I.2.5 will also apply to many kinds of moderately consolidated sediments, i.e. to sedimentary rocks in which intergranular friction and interlocking are considerably stronger than the cohesive bonds between the grains, and to heavily jointed or brecciated rocks.

Quite differently, the cohesive shear strength of heavily consolidated, intact rocks (outside the crystal-plastic deformation range) will be reduced by the growth of numerous brittle microcracks before the onset of faulting. Commonly, these subcritical cracks align along the direction of the greatest compressive stress and most probably induce a strength anisotropy that is likely to affect the orientation of future faults (Rudnicki, 1977). Naturally, for these materials the critical stress circles also possess envelopes whose slopes define a friction angle φ, but there is no reason why the points of tangency should be associated with the potential fault planes (see also Paul & Gangal, 1966). Because of the induced strength anisotropy somewhat different orientations might be more favourable for the development of a fault. In fact, one should expect the angle between incipient fault planes and the σ_I'-axis to be *smaller* than that predicted by eq. I.2-5, since the general alignment of the subcritical cracks is along this axis and the damaged rock may dilate under shear more easily in the direction of the smallest principal effective stress. It is also fairly obvious that these effects will be enhanced by small effective confining pressures and may be suppressed by a sufficiently high effective confining pressure.

Finally, faulting will attain a hybrid form between extension fracturing (Fig. I.2-12A) and slip under effective compression (Fig. I.2-12B), which combines shear movement with extensional separation (Fig. I.2-12C), when the effective normal stress on the fault plane vanishes or attains negative, i.e. tensile, values. Such ***dilational faults*** form under critical stress states which, in the Mohr diagram (Fig. I.2-13), are represented by circles that neither touch the limit lines in the compressive range, nor reach far enough into the tensile half plane to pass through the point $-T_0$, 0, which indicates the stress condition for pure extension fracturing. ($-T_0$ is the effective tensile stress that in uniaxial extension produces separation of the material perpendicular to the direction of the applied tension.) All these critical circles lie to the left of a limiting circle $\tilde{C}$ for dilational shear faulting, whose principal stress values (Fig. I.2-13) can be easily determined when the limit line is straight inside the compression range. Naturally, the rock must possess some cohesive strength τ_0 to enable stress circles to reach into the tensile range. Such critical states of effective stress can be produced, under certain conditions, by an increase in pore pressure which shifts the effective stress circle into

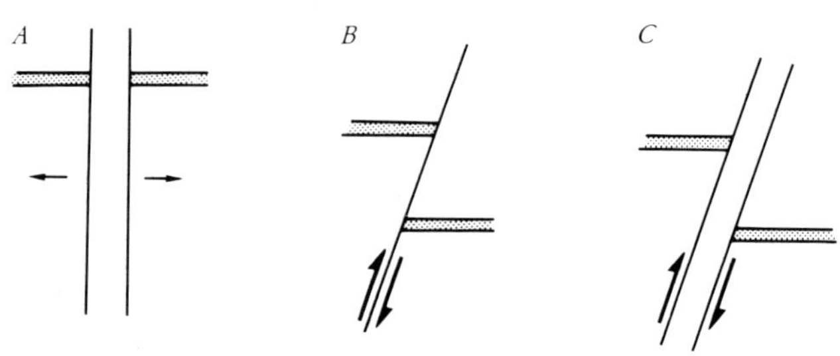

Fig. I.2-12 KINEMATIC FRACTURE TYPES

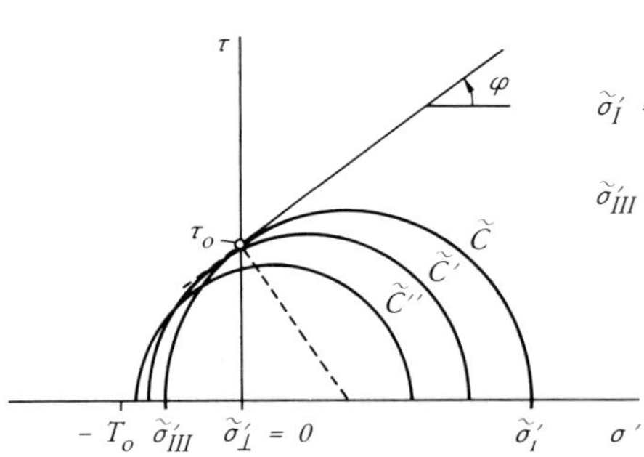

$$\tilde{\sigma}_I' = \tau_o \frac{\sin\varphi + 1}{\cos\varphi}$$

$$\tilde{\sigma}_{III}' = \tau_o \frac{\sin\varphi - 1}{\cos\varphi}$$

Fig. I.2-13
STRESS STATES WHICH PRODUCE DILATIONAL FAULTS

the region of effective tension (see Section II.4.3, Example No. 2 and Fig. II.4-6A,B).

Dilational faults, being a compromise between extension fractures (perpendicular to σ'_{III}) and shear fractures under effective compression, will include an angle θ with the σ'_I-direction, which will be smaller than the corresponding angle when faulting in the same material takes place under effective compression, and may sometimes amount to only a few degrees. This has been observed in experiments (Roš and Eichinger, 1949) and in the field (see, for example, Muehlberger, 1961). In particular, normal faults usually steepen as they approach the surface, even when there is no evident surface topography that might have deviated the σ'_I-direction from the vertical. Such drastic steepening is very likely of dilational origin [8]. It should be noted that less pronounced steepening of normal faults may also occur in loose or only slightly consolidated sediments, where the effective horizontal stresses cannot become tensile, and therefore faulting cannot be of the dilational type. Here it is the dilatancy of the material itself, i.e. the loosening of the granular aggregate in overcoming the interlocking resistance of the grains, which may cause steepening of a normal fault. This dilatancy precedes the formation of a fault, provided the effective confining pressure is small enough (say, of the order of a few MPa). According to soil mechanical experience under these conditions the angle of internal friction is noticeably greater (say, by 10 to 15° for medium fine sand) – and the angle θ correspondingly smaller – than under higher effective confining pressures which suppress dilantancy.

Critical stress conditions in the transition region between shear faulting under effective compression and pure extension fracturing are very difficult to determine experimentally, and hypothetical extrapolations of the limit line from the compressional into the tensile range seem to be hardly more than convenient camouflages of our ignorance in this field. Because of this and the strength anisotropy induced by microcracks we refrain from associating fault plane orientations with stress points on hypothetical limit envelopes in the tensile region.

From these remarks, it should be apparent that some judgement has to be exercised in predicting fault orientations relative to principal stress directions, or conversely, in deducing from observed fault plane orientations the principal directions of the stress field at the time of faulting. This applies in particular to the angular relations in Anderson's fault classification (Fig. I.2-9).

However, this does not impair the extreme usefulness of Coulomb-Mohr's slip concept in the field of tectonic faulting, in which, incidentally, less accuracy in the determination of slip orientations is required than in engineering applications. It should be emphasised that, independent of the question of an accurate determination of the slip orientation, the kinematical content of the concept – deformation by two sets of slip elements – has obtained overwhelming confirmation in experiments on geological materials and in field observations on any scale. Characteristically for the slip model, *rates of slip* are not controlled by the stresses but rather by the movement of the boundaries of the rock mass in the limit state. Hence, the material immediately follows displacements of its boundaries, and internal deformations stop the instant the boundary movements come to a halt. As stated earlier (Section I.1.3), this is a characteristic feature of any ideal plastic process. In Coulomb-Mohr's concept, deformation is accomplished exclusively by sliding, and elastic deformations of the material are neglected. It has therefore become an essential part of the theory of 'rigid/frictional plastic' materials (presented in Part II (Section II.7.4) and already referred to by means of the analogue model of Fig. I.1-7A). In the theory of an 'elastic/frictional plastic' material (Section II.7.5) the slip model will serve to delineate potential faults a posteriori, i.e. after the stress field has been determined by a formalism that does not invoke the slip model itself. The analogue of this idealised material was shown in Fig. I.1-7B.

Although many tectonic faulting processes can be explained in terms of Coulomb-Mohr's slip concept or its double-gliding generalisation, it would be naive to expect such a relatively simple model to cover all cases of faulting in the brittle crust. But we shall see in later sections (I.2.5 and I.2.8) that even the formation of faults that are not of the Coulomb-Mohr type is the result of Coulomb-Mohr-type faulting on a smaller scale.

4. CURVED FAULTS

Although many fault surfaces may be considered as approximately planar in large parts, most faults show some degree of curvature in dip sections or in plan view. Outstanding examples of curved faults are the *'listric'* (shovel-shaped) normal and thrust faults, which have been recognised for a long time in areas where faulting was restricted to all or part of the sedimentary overburden without participation of the substratum ('thin-skinned tectonics'). Particularly well-documented by seismic data are the synsedimentary listric faults in deltaic sediments, which have been described by many authors, mainly from the oil industry. But listric faulting is not restricted to this environment and may also involve crustal basement (cf. the resumé of listric normal faulting by Bally et al., 1981, and the illustrated summary by Shelton, 1984). A.W. Bally is to be credited for compiling, in a genetically differentiated form, reflection seismic evidence for listric faulting in extensional and compressive regimes, and for presenting it to a wider geological audience. For good examples of listric normal and thrust faults, the reader is referred to Bally's pictorial atlas (Bally, 1983). Here, our concern is with the mechanical factors that cause a fault surface to be curved. Our discussion will concentrate mainly on fault curvatures in dip sections since, at present, the causes of changes in dip angle are better understood than the factors that control changes in strike direction.

Since most of the evidence for fault curvatures in dip sections is found on reflection seismic records, it is first of all important to be aware of the effect that *changes in seismic velocity* have on the appearance of dip angles of seismic reflectors. Because seismic velocity usually increases with depth, real dip angles of reflectors at depth will appear smaller on seismic time-sections. Hence, the seismic expressions of normal and thrust faults indicate or exaggerate a decrease in dip with depth which is not there in reality. Even a perfectly plane fault will appear as somewhat concave-upwards, and flattening faults may appear to merge into a bedding plane while in reality still transecting the strata.

To illustrate this and to obtain an idea of the amount of distortion involved, the time-section (Fig. I.2-14) of typical synsedimentary normal faults in the Niger delta is converted into depth sections (Fig. I.2-15), assuming a linear increase in the seismic velocity with depth and using different velocity gradients and an initial velocity typical of the area.

Changes in fault shape may be caused by tectonic events and non-tectonic factors at the time a fault is being formed or at a later stage. Typical non-tectonic factors are non-uniformities and anisotropies in shear strength of the rock mass, compaction caused by overburden weight, and certain types of abnormally high pore pressures. Tectonic events affect the shape of an incipient fault by means of changes in the tectonic stress field that precede the actual formation of the fault. Subsequent changes in the tectonic stress field may change the shape of a fully developed fault. A fully developed fault may also passively change its shape as part of a tectonically or non-tectonically deforming rock mass.

Among the *non-tectonic* factors that may cause flattening or steepening of a fault, we consider first the *variation in shearing strength.* In a uniform stress field a fault cutting across the interface between a sedimentary rock (e.g. sandstone) with high friction angle φ and one with a low frictional resistance (e.g. clay) may change its orientation with respect to a

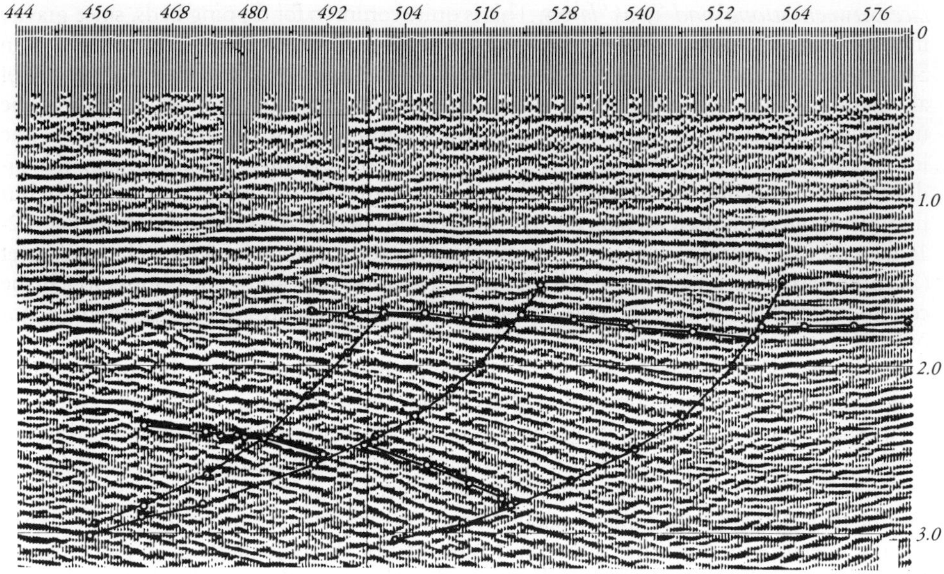

Fig. I.2-14 NIGERIAN GROWTH FAULT (horizontal scale 1:25.000, time section)

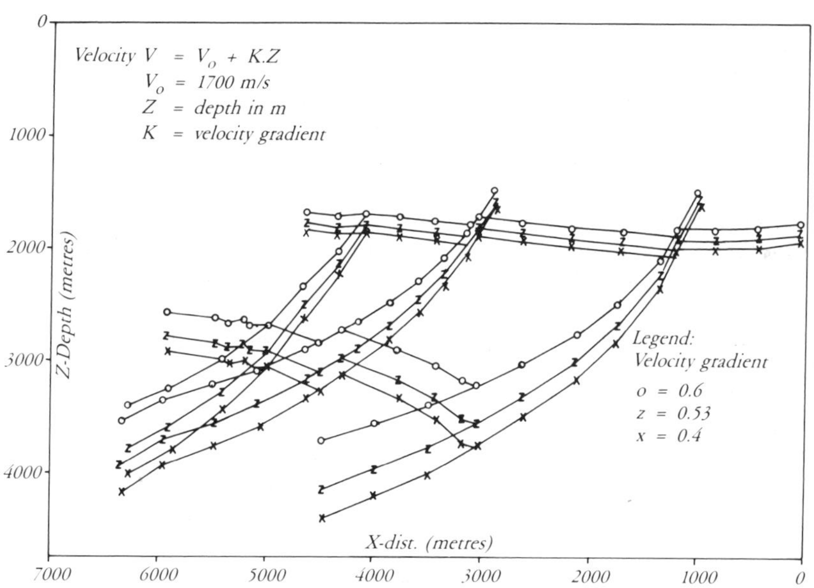

Fig. I.2-15 DEPTH CONVERSION OF FAULTS AND HORIZONS OF FIG. I.2-14 (Courtesy of W. Peet)

uniform σ_I-direction by as much as, say, 15°, in accordance with Coulomb-Mohr's concept. This is easily demontrated in experiments (Wunderlich, 1957).

The effect of a variation in shear strength is increased when a *shear stress acts on the inter-*

face between 'strong' and 'weak' layers. This is quite common for dipping beds, since gravity induces a shear stress on dipping bedding planes (see Example No. 2 and Fig. II.2-7 in Section II.2.1). Interstratal shear stresses are also induced by the flexure of stacks of alternating strong and weak strata, when superincumbent strata are displaced over the underlying ones towards the location of highest curvature (e.g. an anticlinal crest). The important point here is that an incipient fault not only changes its angle with the σ_I-direction when cutting through an interface between layers of contrasting shear strength, but that the σ_I-direction itself is changed. This is demonstrated by the stress diagram of Fig. I.2-16 for a normal fault that cuts across the interface between a strong layer (1) and a weak layer (2). Both layers have to be in the limit state; therefore, each stress circle must touch the corresponding limit line. If the pore pressure is the same on both sides of the interface, not only the tangential stress but also the normal effective stress $\sigma'_\perp$ must have the same value on both sides of the interface. The two limit circles therefore intersect at the corresponding stress point (A). The direction of σ_I is then found easily by means of the stress pole – a method explained in Section II.4.2 ('The pole of the stress circle'). One simply draws through the known stress point A a straight line to represent the planar element that is acted upon in real space by the stress components of A. More precisely, this line represents the intersection of that element with a plane parallel to the σ_I- and σ_{III}-directions in space. It serves then as reference for any other direction determined by the pole method; in particular for the direction of the σ_I-axis, which is determined in the stress plane by the simple construction in Fig. I.2-16 (see also the explanatory Figure II.4-5B) [9].

One will immediately notice from this figure that upon entry into the weaker layer the σ_I-direction flattens somewhat with respect to the interface of the two beds, which implies an additional flattening of the slip element that is synthetic with the shear stress direction on the interface. Under the conditions in the figure (with $\varphi_1 = 42°$, $\varphi_2 = 15°$, $\tau_o^{(1)} = 20$ MPa, $\tau_o^{(2)} = 5$ MPa and an effective normal stress of 100 MPa acting upon the bedding plane) the associated σ_I-directions would differ by 8° and the fault would therefore flatten by 21.5°. Realistic changes in the assumed strength contrast and in the effective normal stress will only slightly affect the change in σ_I-directions which, however, strongly depends on the value of the interfacial shear stress and will be zero for $\tau = 0$.

A more dramatic change in σ_I-direction at an interface between two horizontal layers of a brittle overburden can be caused by the *shearing action of a ductile underflow.* Consider first a diverging underflow such as occurs when shale or salt substrata are squeezed out from underneath accreting sediments into marginal shale or salt ridges or, on a mega scale, when the ascending branch of a convective mantle current separates into two opposite horizontal flow branches. The situation is sketched schematically in Fig. I.2-17A for an idealised two-layer overburden, the upper layer (1) of which consists of brittle Coulomb-Mohr-type material with internal friction φ_1 and cohesive stress $\tau_0^{(1)}$. In contrast with this, and essential for the process considered, it is assumed that the shear strength of the material of the lower overburden layer (2) is practically independent of the effective confining pressure under the prevailing pressure and temperature conditions. The limit line of this material therefore flattens in Mohr's stress plane with increasing effective pressures and eventually degenerates into a line parallel to the $\sigma'_\perp$-axis, as shown in Fig. I.2-17C. Note that the shear strength of this pressure-insensitive material may still be high and that a high shear stress has to be imposed by the ductile underflow to induce a limit state of stress.

In the region of divergent underflow the original horizontal effective overburden stresses are reduced by the imposed basal shear stress, and may be brought into the active limit or rifting state when the shear stress is of sufficient magnitude. It should be noted, however, that in the frictional layer (1) this limit state will remain restricted to an interval above the

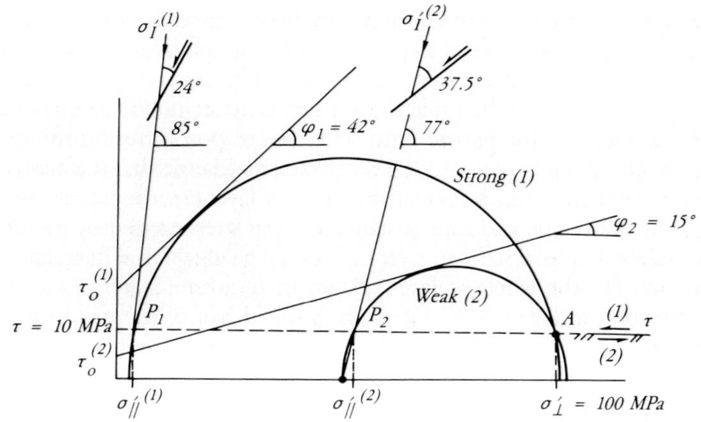

Fig. I.2-16 FLATTENING OF NORMAL FAULT UPON ENTERING WEAKER LAYER
Change in directions of maximum compressive stress and associated slip elements graphically determined by pole-method in Mohr's stress plane (P_1, P_2 'Poles')

A Listric boundary faults in extension region: low-angle normal faults in frictional layer (1) merging with horizontal slip plane in purely cohesive base layer (2).

B Listric thrust fault in compression region: low-angle thrust fault in frictional layer (1) branching off horizontal slip plane in purely cohesive base layer (2).

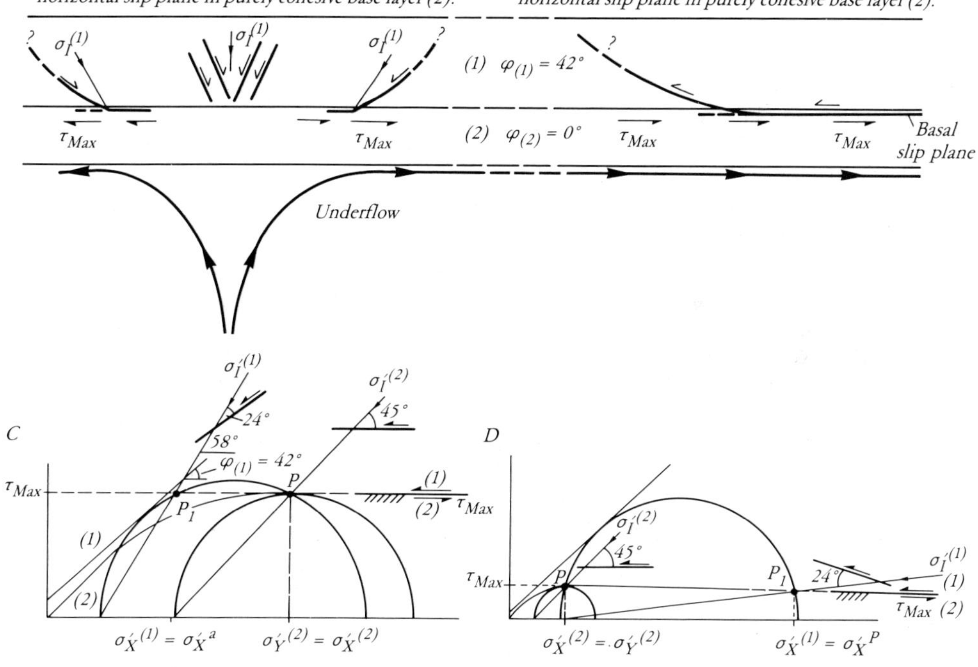

C, D Orientations of maximum compressive stress and associated fault dip (synthetic with basal shear stress) determined by pole method; in Fig. C for the extensional limit state of Fig. A, and in Fig. D for the passive limit state of Fig. B.

Fig. I.2-17 DUCTILE UNDERFLOW OF TWO-LAYER OVERBURDEN.
Material of upper layer has frictional shear strength; material of base layer is purely cohesive (pressure-independent shear strength). Underflow causes elastic/plastic simple-shearing of base layer.

branching of the underflow. The reason for this is simply that, in static equilibrium, the drag force exerted upon the base of the layer has to be fully balanced by the horizontal normal stresses. Hence, outside a central extension region the horizontal normal stress increases monotonically in the direction of the underflow. It can therefore no longer satisfy the limit condition [10] unless it reaches the passive limit state, whereupon thrusting commences – the situation to be considered in Fig. I.2-17B. When we also assume that the elastic stiffness of the brittle upper layer (1) is much greater than that of layer (2), the upper layer will act as a 'rigid' lid on the lower one and fully sustain the shear stress exerted by the underflow. Therefore, the horizontal normal stress in layer (2) will not change in the flow direction and, in contrast with layer (1), the whole of layer (2) can be in a limit state.

The pole construction in Fig. I.2-17C for the right-hand half of the two-layer system of Fig. I.2-17A indicates that a normal fault, induced by basal shear stresses at the margin of the active plastic region of layer (1), will approach layer (2) at a low dip angle (34°) and continue horizontally inside layer (2), as is shown schematically in Fig. I.2-17A. Figure I.2-17C also shows that the horizontal normal stress σ_x' changes discontinuously across the horizontal slip plane, a phenomenon that will be discussed further in Section I.2.9. The listric faults, situated at the margin of the active plastic region, may remain active as long as the enclosed area is extending. In the central part of the graben the σ_I'-direction remains vertical, or nearly so, since there it is not or hardly affected by basal shear stresses. The normal faults, which develop in this part, will therefore be planar.

In the region where the drag force of the ductile underflow has induced the passive limit state, a low-angle thrust fault will branch off a horizontal slip plane in layer (2), as is derived by the pole construction of Fig. I.2-17D and indicated schematically in Fig. I.2-17B.

Similar to the change in dip angle produced by a contrast in friction angle between adjacent layers, and perhaps even more pronounced, is the effect of *abnormally high pore pressures,* i.e. pore pressures that are considerably in excess of the hydrostatic reference pressure. Following Crans et al. (1980), we consider a sediment sequence with overpressured sediments overlain by layers with normal pore pressure; the layers are somewhat inclined, and so the overburden weight induces some shear stress along the bedding planes. The stresses and their changes across the interface between normally and abnormally pressured sediments are indicated in Fig. I.2-18A. Only the shear stress does not change across the interface.

If a normal fault is to transect the interface (Fig. I.2-18B), the stress circles associated with normally and abnormally pressured layers must both touch the limit line. This again implies a difference in the σ_I'-directions and a corresponding change in the slip direction, as is shown by the pole construction in Fig. I.2-18C, where for simplicity the same limit line is applied to both strata. Hence, the normal fault flattens abruptly when it enters the overpressured region (Fig. I.2-18B). By a similar argument, one may show that low-angle thrusts should also suffer an abrupt change in dip at the interface between abnormally and normally pressured rocks; they should steepen when leaving the overpressured region. The reader should be reminded, however, that this dip change will occur only when a shear stress acts along the bedding; otherwise, the fault will intersect the top of the overpressure region undeflected.

It is also obvious from Fig. I.2-18C that a change in dip angle requires a change in the difference between the effective normal stresses. This is probably the reason for the little recognition that this pore-pressure effect has so far received in the literature: At first sight, any change in pore pressure will affect the two normal stresses in the same way and will therefore leave the difference between the two effective stresses unchanged. In reality, however, this difference (represented by the size of the limit stress circles) will normally change. This is simply because of the tendency of strata to expand laterally when the pore

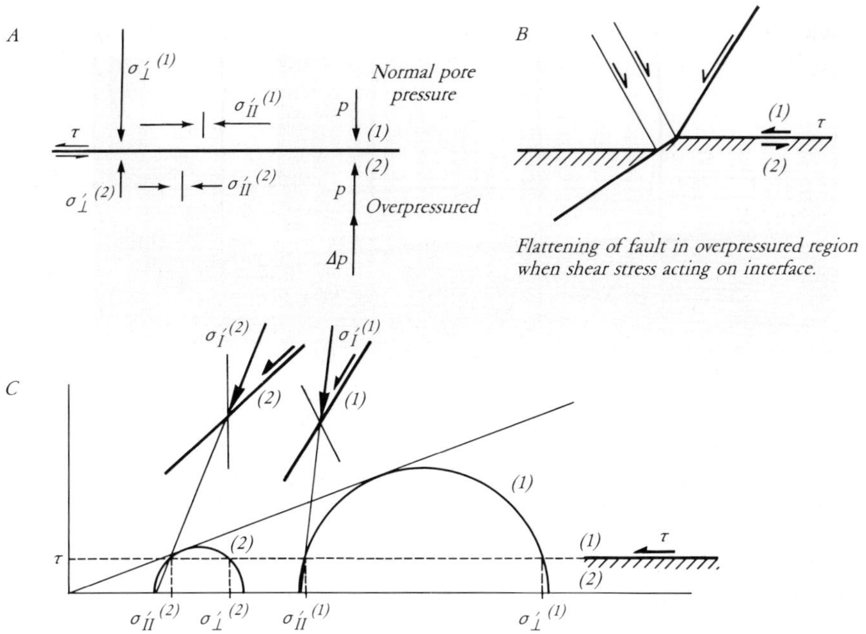

Fig. I.2-18 NORMAL FAULT ENTERING OVERPRESSURED LAYER UNDER SHEAR STRESS

pressure is raised. Such expansion is, however, partly or wholly prevented by interstratal friction forces or other lateral constraints. In response to these constraints, the total stress $\sigma_{\parallel}$ increases, whereas in the case of small dip angles the 'overburden' stress $\sigma_{\perp}$ may remain almost unchanged. (This is discussed further for horizontal strata in Example No. 2 of Section II.4.3.) The effective stress $\sigma'_{\parallel}$ is therefore reduced less than $\sigma'_{\perp}$ by an increase in pore pressure.

The mechanism described is, in particular, operative on *overpressured delta slopes,* where it produces a remarkable flattening of synsedimentary ('growth') faults. There is abundant seismic and well evidence for 'listric' fault shapes associated with overpressured slopes. A particularly illustrative example – Bruce's (1973) seismic section with pore pressure data – is reproduced in Fig. I.2-19. We shall return to this topic in the next chapter where we deal with gravitational gliding in deltaic sediments. In compressional regimes, abnormal gradients of pore pressures – separately or in combination with gradients of shear strength – may be instrumental in the process of 'stepping-up' of bedding plane thrusts to higher gliding horizons, as will be discussed in Chapter I.5.

A kink produced in the slide path by a contrast in internal friction angle or pore pressure will give rise to a series of antithetic faults (Fig. I.2-18B) to accommodate the changes in shape of the gliding rock mass – a phenomenon that we shall return to in Section I.2.6. It is obvious that a more gradual change in pore pressure or friction angle will produce a continuous change in dip angle ('listric' faults) rather than a kink in the fault surface.

A second mechanism by which abnormally high pore pressures may affect the dip of a fault has been already pointed out in connection with shear strength anisotropy. In cohesive rocks a rise in pore pressure may produce conditions for the formation of *dilational* normal

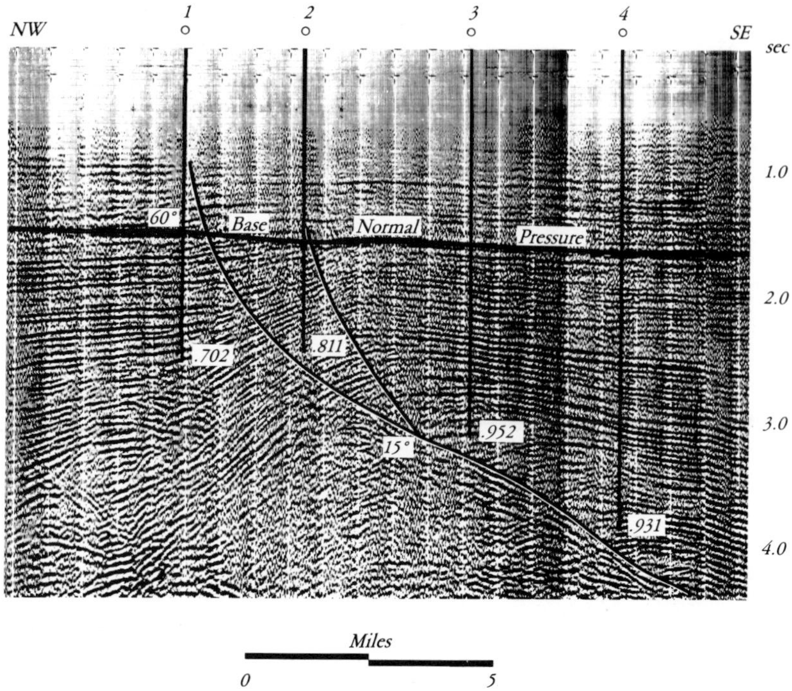

Fig. I.2-19 SEISMIC SECTION WITH WELL LOCATIONS (TEXAS COASTAL AREA) SHOWING FLATTENING OF CONTEMPORANEOUS FAULTS IN OVERPRESSURED SHALE (After C.H. Bruce, 1973; reprinted by permission of "American Association of Petroleum Geologists") bottomhole fluid pressure gradients are recorded for each well.

faults. Such faults will steepen as they approach the surface. But even when all effective stresses remain compressive, the reduction of effective rock stresses by a rise in pore pressure may sufficiently facilitate *dilatant deformation* and the concomitant weakening of the rock, by the loosening of grains or the opening of microcracks or macroscopic joints, to affect the orientation of a fault. Since the rock tends to dilate most in the direction of the smallest compressive stress, dilatancy is likely to steepen a normal fault and to flatten a low-angle thrust fault, particularly at shallow depth.

So far we have considered changes in fault dip which were primarily of non-tectonic origin. We shall now turn to the *tectonic factors,* i.e. the *tectonically induced changes in the stress field that precede or accompany the formation of a fault. Before* a fault is formed the stress field has to be brought – at least in some parts – into a limit state. An initial stress state that is solely due to the overburden weight would not yet be in a limit state. Hence, some increase in differential stress and associated straining is required to initiate faulting. Although later we shall encounter situations in which these changes of the stress field are not accompanied by changes in principal stress directions, it is more common for such changes to be induced by frictional reaction stresses along the base or lateral boundaries of a deforming rock mass. We may reasonably expect these effects to be most pronounced where large straining precedes the actual formation of a fault. This is the case in *low-angle thrusting* where considerable straining will accompany the build-up of the large differential stresses associated with the passive limit state (Fig. I.2-5). Hence, the formation of a low-angle thrust in horizontally shortened strata that rest in frictional contact on undeformed basement will be preceded by the formation of basal shear stresses, which deflect the σ_I'-trajectories and enable the thrust fault to attain a listric shape, as qualitatively sketched in Fig. I.2-20. In comparison with this, the strains preceding *normal faulting* in brittle rocks on a 'rigid' basement plate that is rifted apart, or on a substratum that is locally stretched, will remain small and the associated basal shear stresses may have little effect on the σ_I-trajectories. Therefore, without the interference of strong contrasts in shear strength or pore pressure inside the overburden, the incipient normal faults will be essentially planar. These statements, which at this point may appear somewhat speculative, will be supported by numerical simulations in later chapters.

Strong changes in the field of principal stress directions may also be induced in a horizontal overburden by vertical movements along vertical faults in the basement. As indicated in Fig. I.2-21A, for initial stress isotropy ($\sigma_I^\circ = \sigma_{III}^\circ$), such movements render in the over-

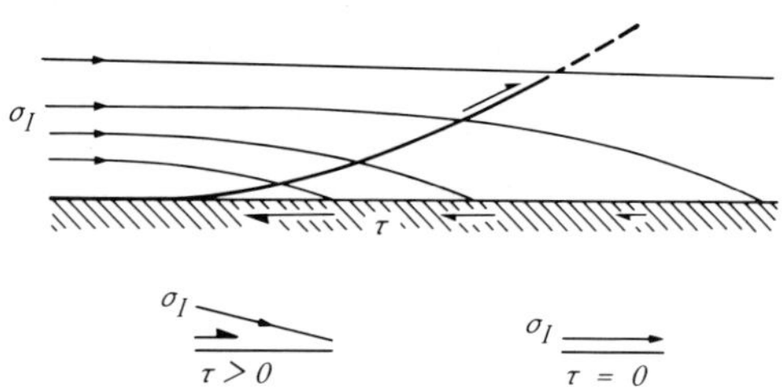

Fig. I.2-20
CURVATURE OF LOW-ANGLE THRUST

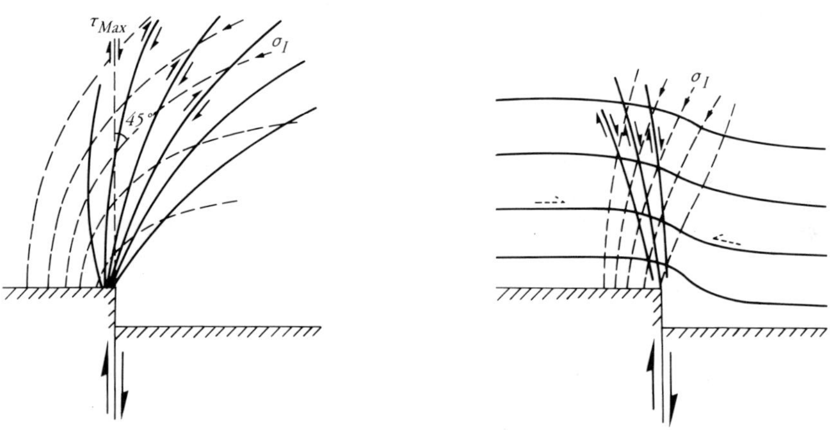

A ***Uniform*** *brittle overburden; maximum shear stress induced along vertical plane which, consequently, is transected by σ_I-trajectories at 45° (see also finite element simulation Fig. II.7-32). Potential Coulomb slips agree with precursor faults observed in experiments with sand (Figs. II.5-11, II.7-33).*

B ***Multilayered*** *overburden with flexuring and interlayer slip preceding faulting. Assuming dense spacing of bedding planes with negligible frictional resistance, the σ_I-trajectories inside the region of greatest bed extension may be inferred as shown in the Figure. Accordingly, normal graben boundary faults should be expected to form first, instead of the steep reverse faults of Fig. A.*

Fig. I.2-21 VERTICAL MOVEMENT OF BASEMENT BLOCKS; POTENTIAL SLIP LINES AND TRAJECTORIES OF MAXIMUM COMPRESSIVE STRESS IN SEDIMENTARY OVERBURDEN

burden the vertical geometrical extension of a basement fault into a plane of maximum shear strain and shear stress, forcing the σ_I-trajectories to intersect that plane at 45°. First faults in the overburden should therefore at least locally have the character of downward-concave reverse faults. This is in agreement with the results of computational fault simulations (see 'Differential uplift', Example No. 5 and Figs. II.7-31,32,33 in Section II.7.5 and Example No. 7 and Fig. II.7-46 in Section II.7.7). It is important to note the partial unloading of the subsiding basement block. The σ_I-trajectories form a 'stress arch' that transfers overburden load to the higher basement block.

While in this case we have assumed a mechanically uniform and isotropic overburden material, it is of interest to speculate on the effect that the presence of *weak bedding planes* of a multilayered overburden would have upon the directions of the principal stresses and associated potential faults. We assume that, before the onset of faulting, the layers are flexed, as sketched in Fig. I.2-21B, and that a set of densely spaced bedding planes exist, the frictional resistance of which may be neglected. The σ_I-axes are then perpendicular to the bedding. Obviously, the limit state will be attained first in the regions where the flexed beds have suffered the greatest extension. Although the σ_I-trajectories still show some arching and overburden load is transferred from the subsiding block to the high basement block, the potential faults now have the character of normal faults which dip towards the subsiding block and steepen downward. It is not difficult to envisage, at least qualitatively, how stress trajectories and potential fault surfaces will be changed when shear stresses are induced by the flexural slip of beds. Since a higher bed slides over the underlying bed towards the point of greatest curvature, the shear stresses will point in the direction indicated by the two dotted arrows in Fig. I.2-21B. The σ_I-trajectories, now no longer perpendicular to the beds, will make an acute angle with the beds which opens against the relative slip direction. The fan of σ_I-trajectories will therefore widen and the upward-convex candidate faults will be even more strongly curved.

This may explain the results of experiments reported by Belousov (1961) in which multilayers of wet clay were deformed by the vertical upwards or downwards movement of a 'basement' block. The pattern and shape of the faults that cut across the flexed layers are shown in Fig. I.2-22. In both the uplift and subsidence phase, the main faults are convex upward and dip towards the relatively downthrown side. In the uplift phase (Fig. I.2-22A) a clay wedge, attached to the uplifted block and bounded by the main faults, is pushed into the overburden. Notice also the secondary reverse faults which release the local compression

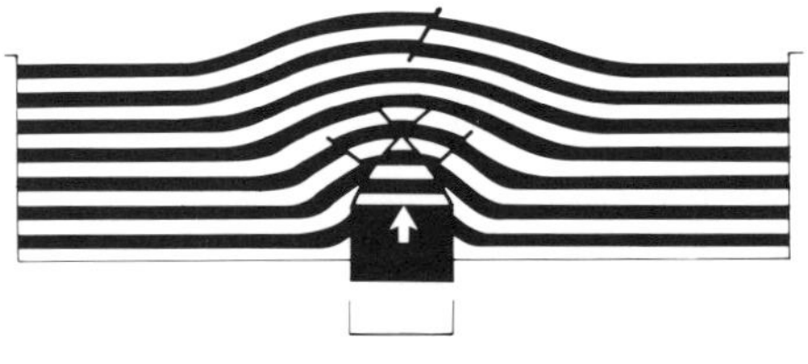

A Uplift phase: Prismatic clay wedge, bounded by slightly upward convex normal faults, is pushed into the overburden.

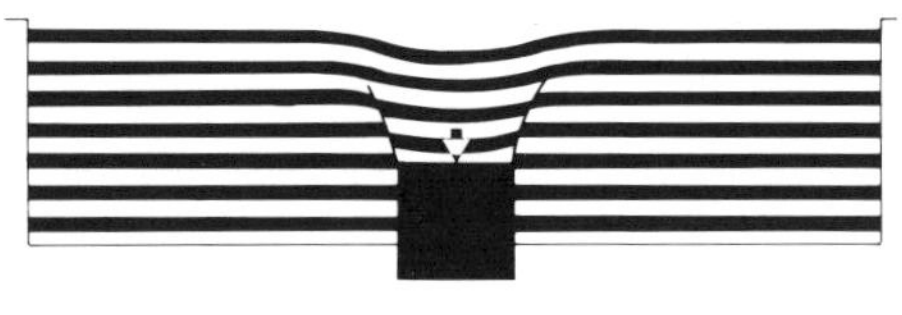

B Subsidence phase: Normal (graben) boundary faults develop.

Fig. I.2-22 VERTICAL MOVEMENT OF 'BASEMENT' BLOCK CAUSING FLEXURAL DRAPE FOLDING AND FAULTING IN MULTILAYER PACKAGE OF WET CLAY (after Belousov, 1961; Scientific American 204: 97-106)

caused by the intruding fault wedge. In the subsidence phase the two faults appear as normal graben boundary faults. The entire fault configuration is contrary to that which, on theoretical grounds, should develop under similar conditions in a uniform overburden (Fig. I.2-21A, Figs. II.7-32F,46) and which is actually observed in model experiments with an homogeneous overburden (Fig. II.7-33). The contrast between the two fault configurations may be sufficient to illustrate the important effect that the presence of weak bed separations may have upon the directions of principal stresses and associated fault elements.

However, tectonic reorientations of principal stress directions, which result in a convex upward curvature of incipient faults, are not restricted to multilayer systems with weak bedding planes. The graben faults produced in the sandbox experiment of Fig. I.2-23 show such convexity in the tilted part, and the kinematic advantages of such a curvature, which allows rigid body rotation of the downthrown block, is illustrated by the transparent model of Fig. I.2-24. Figure I.2-25A illustrates schematically the principal stress orientations responsible for this type of curvature. At further bending of the strata the curved normal faults can rotate into a flatter position (Fig. I.2-25B) as is also demonstrated in Fig. I.2-24. However, bending of strata does not necessarily cause an upward convex curvature of normal faults; the faults may be curved in just the opposite (listric) sense when bending of the strata and fault growth takes place during deposition. Older elements of the synsedimentary faults in the diagrammatic Fig. I.2-25C will be flattened by rotation. The same process may also occur in the central part of 'withdrawal basins'. As sediment is deposited locally on top of ductile shales or salt, ductile material is squeezed out from the central part towards the margin of the basin, i.e. in the direction of decreasing overburden. While subsiding, the basin widens and its floor flattens. This causes older deposits with an originally horizontal surface to be gradually bent into a 'turtle-shaped' anticline. The concomitant extension and faulting of new sediments in the crestal region will result in downward-flattening normal faults.

These fault curvatures were produced by the rotation of fault elements *after* slip had been initiated. Naturally, there are many ways in which an existing and still active fault can be deformed, for example by refolding. An important non-tectonic factor, well-known to soft-sediment geologists, is the synsedimentary or postsedimentary flattening of normal

A Photo showing intermediate stage of experiment with left-hand 'basement' block tilting towards left.

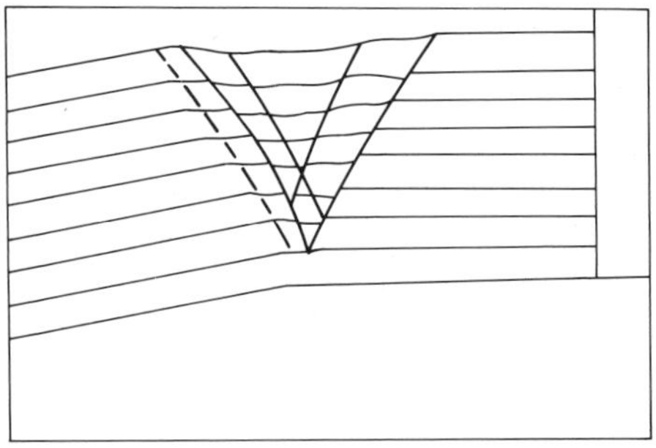

B Sketch based on stereo viewing of photos showing active faults. Note curvature of fault at left which is associated with rotation of beds (also note stippled inactive fault at left).

Fig. I.2-23 GRABEN CAUSED BY POSTSEDIMENTARY TILTING OF 'BASEMENT' BLOCK IN SANDBOX EXPERIMENT (courtesy of W.T. Horsfield, unpubl. work)

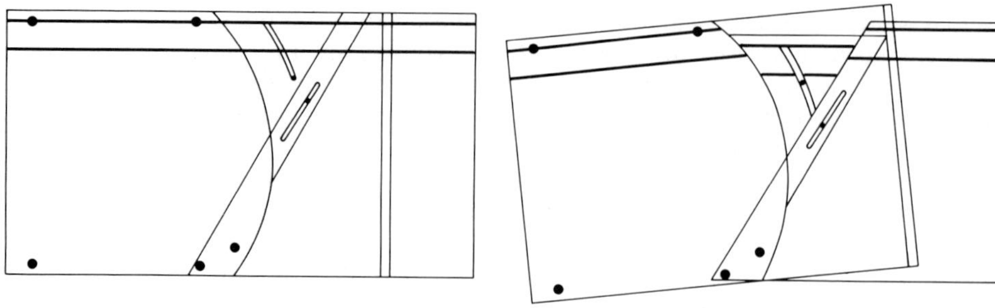

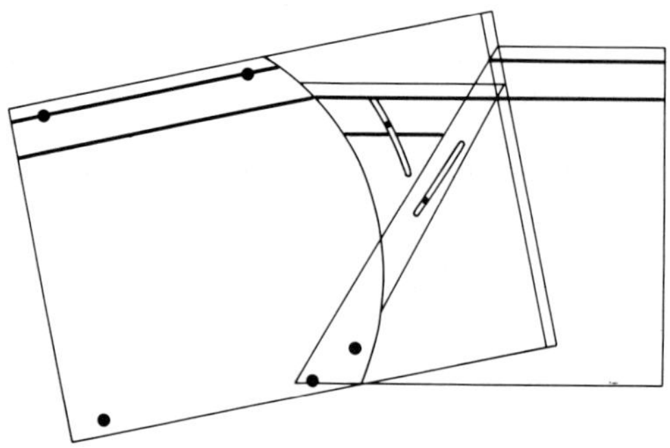

Fig. I.2-24
PERSPEX MODEL DEMONSTRATING OPERATION OF CURVED GRABEN FAULT IN POSTSEDIMENTARY TILTING

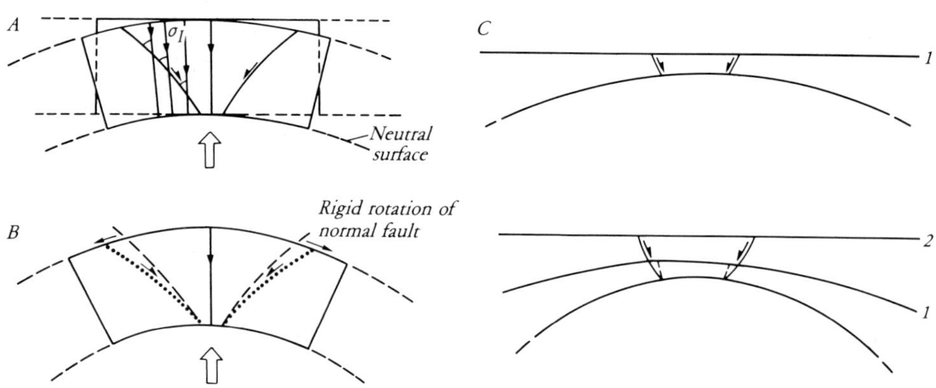

A, B Postdepositional bending of strata on non-extending ('neutral') surface;

C Syndepositional bending; rotational flattening of older fault elements causing concave shape of growing fault.

Fig. I.2-25 CURVATURE OF NORMAL FAULTS ASSOCIATED WITH THE BENDING OF HORIZONTAL STRATA BY VERTICAL FORCES (schematic)

faults by the compaction of dewatering shales (Roux, 1979). Figure I.2-26 presents an example of a seismic profile that clearly shows compaction flattening of 'growth' faults (i.e. faults with stratigraphic thickening on their downthrown side and increased throw with depth) in Pliocene shale. It should be noted, however, that compaction flattening, although increasing listricity, cannot account for the smooth merging of growth faults with sub-horizontal bedding planes which has been unambiguously observed in certain cases.

It is easily seen, for instance, that concomitant with a 35% reduction in the thickness of a horizontal shale bed an original dip angle of a fault plane of 60° would be reduced by 12°. However, flattening of a fault plane is more effectively established in an interval that is not only reduced in thickness, but is also extended at the same time. Uniform horizontal extension of ductile crust without change in volume has been proposed by Jackson and McKenzie (1983) to explain the listricity of normal faults that cut across the whole of the brittle crust. The essence of the concept is shown in Fig. I.2-27 together with the elementary relationship between stretch factor and initial and reduced dip angles. Again, it should be noted that this model accounts for the deformation of a fault plane *after* formation of the fault.

The possible causes of fault curvature dealt with so far are either already effective prior to the actual formation of a fault – as lithological and fluid pressure effects or tectonically induced shear stresses – or they operate after a fault plane has been formed. A third and, unfortunately, least understood factor, is associated directly with the propagation of an active fault inside the rock; this is the *stress field around the leading edge of an active fault plane.* This 'near fault edge' field, which is the strong local disturbance of the overall tectonic stress field by the dislocation of material along the fault, acts as a 'scout' in controlling the direction of any incremental extension of the fault in the dip or strike direction.

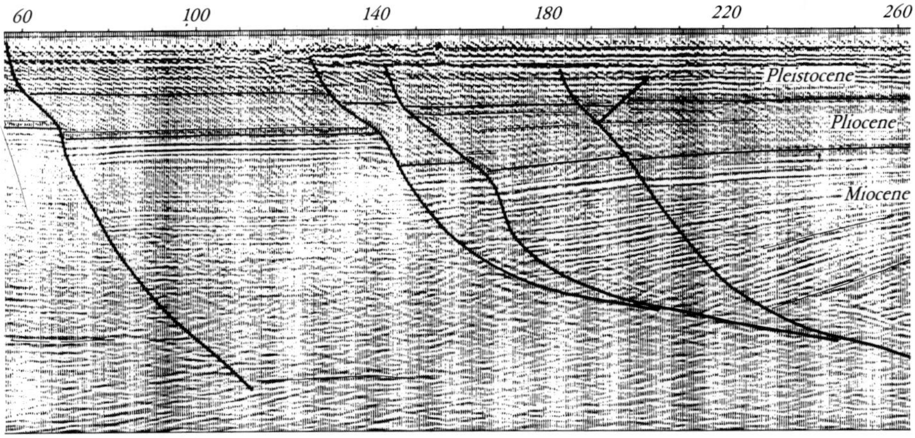

Fig. I.2-26 SEISMIC PROFILE OFFSHORE TEXAS SHOWING COMPACTION FLATTENING OF GROWTH FAULTS IN PLIOCENE SHALE (after Roux, 1979)

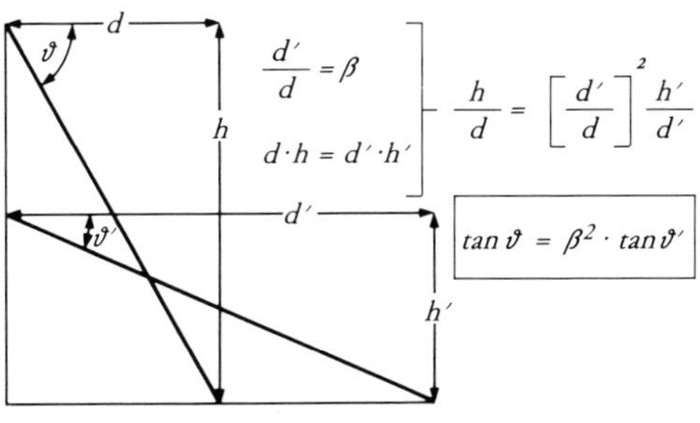

Fig. I.2-27
FLATTENING OF FAULT PLANE BY HORIZONTAL UNIFORM EXTENSION OF DUCTILE CRUST
(proposed by Jackson and McKenzie, 1983)

This stress field is certainly not of a purely elastic nature and the fault can only propagate into rock which near the fault edge is in the plastic limit state. Because of this and the sensitive response of the fault edge field to mechanical heterogeneities and boundaries ahead of the propagating fault, a rigorous analysis of the role of the fault edge field in determining the course of fault propagation entails formidable mathematical complications. Nevertheless, some insight may be gained by the following elementary considerations.

Let us consider a planar fault that has been inactive for some period, the differential tectonic stresses are somewhat relaxed and the fault plane is therefore embedded in an elastic environment. We restrict ourselves to an end part of the fault plane, the trace F of which is shown schematically in Fig. I.2-28A. This may be taken as a dip-parallel section through a normal or thrust fault, or as the plan view of a vertical strike-slip fault. The axes of two coordinates x_1 and x_3 are chosen as indicated, parallel and perpendicular to the fault plane. The sub-critical state of the surrounding tectonic stress field is represented by the stippled stress circle in the stress plane of Fig. I.2-28B. We also assume that σ_1' acts at 45° to the inactive fault plane. Now the fault is reactivated, which is indicated in Fig. I.2-28A as a fault-parallel relative displacement imposed upon the wall rock at some distance from the leading edge of the fault plane. Obviously, on the advancing (+) side of the fault the material will be compressed in the x_1-direction, while on the receding (−) side the material is extended. The material around the fault is assumed to behave linearly elastic up to the moment when the limit state of stress is re-established locally. Now consider the change in elastic stresses in two points Q^+ and Q^-, which are symmetrical with respect to the x_1 axis. Before reactivation of the fault the normal effective stresses $\sigma_1'^{\circ}$ and $\sigma_3'^{\circ}$ which act parallel to the x_1- and x_3-directions, are equal as indicated in the diagram of Fig. I.2-28B, because of the 45° inclination of σ_1°, and the maximum shear stress τ_{13}° acts along the x_1- (and x_3-) axis (see, for example, Fig. II.4-4A,B). The imposed fault displacement will change this state. At the advancing side in Q^+ the normal stress components will be increased by $\Delta\sigma_1'$ and $\Delta\sigma_3'$, while at the receding side in Q^- they will be decreased by the same amounts. Therefore, in the stress plane of Fig. I.2-28B the centres of the corresponding Mohr circles are shifted in opposite direction by the same amount. Moreover, for symmetry reasons, exactly the same increase in shear stress $\Delta\tau_{13}$ is induced at the two points, raising the value of the total shear stress acting parallel to the x_1-axis to $\tau_{13}^{\circ} + \Delta\tau_{13}$. In the stress diagram the Mohr circles for the two points are now fully determined by the position of the centre and the stress point with the coordinates $\sigma_1'^{\circ} + \Delta\sigma_1'$, $\tau_{13}^{\circ} + \Delta\tau_{13}$. Since both circles are the same size, it is evident that, with increasing deformation, the circle associated with point Q^- will first touch the limit line. In other words, the plastic limit state will be attained at the receding side of the fault that advances into a mechanically uniform and unbounded material. After temporary inactivity the propagating fault will therefore *curve towards the receding side.* The phenomenon is observed in particular on basement-controlled wrench faults, which are the subject of Chapter I.4. Note, however, that the asymmetric development of a plastic region at the leading edge of the propagating fault is due to the pressure sensitivity of the shear strength and would not take place in cohesive rocks with $\varphi = 0$ as may be expected to exist in the lower crust. Moreover, our argument applies to the incipience of a renewed fault propagation; as the fault continues to propagate, its deviation will generally give rise to reactive stresses that tend to restore the former slip direction.

In reality, the symmetry in the ideal situation considered above may be disturbed in various ways. First, a different $\sigma_1'^{\circ}$-orientation with respect to the fault plane may exist before the fault resumes propagation. The plane of maximum shear stress will then no longer coincide with the fault plane and its straight extension along the x_1-axis. In Fig. I.2-

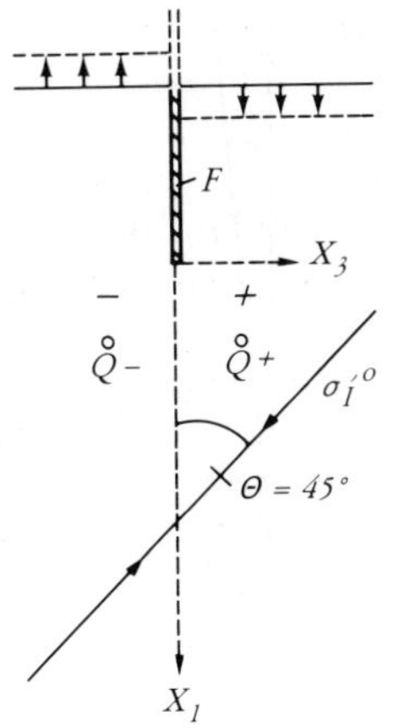

A Strike-or dip section through near-edge region of fault plane with imposed fault-parallel displacement and 45°- orientation of maximum compressive stress prior to reiteration of fault slip.

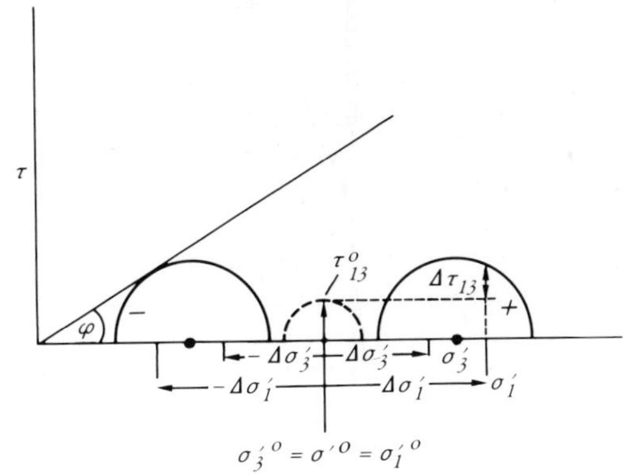

B Stress circles representing stress changes on advancing (+) and receding (−) side of fault plane (in points Q+ and Q− of Fig. A). Stippled circle indicates initial state of stress. Notice that limit state is first attained at receding side.

Fig. I.2-28 LIMIT STATE OF STRESS INDUCED NEAR LEADING FAULT EDGE BY REITERATION OF FAULT MOVEMENT (see text for discussion)

29 we consider the case in which σ'°_1 makes only a small angle θ with the fault plane. A subcritical state of stress, accordant with this, but otherwise arbitrary, is represented by the Mohr circle in Fig. I.2-29B. In the construction the stress pole P was used to relate the stress components, which act parallel and normal to the fault plane F, with the angle θ. A small angle between the σ°_1-direction and the fault plane may be caused, for instance, by extreme shear softening of the fault material, the emplacement of 'lubricating' material inside the fault, or the intrusion of highly pressurised fluids. We consider again the elastic stress changes, induced by a fault-parallel displacement in two points that are symmetrically positioned with respect to the fault plane (Fig. I.2-29A). Again the normal stresses σ_1' and σ_3' increase in Q^+ by the same amount as they decrease in Q^-. The corresponding Mohr circles are constructed in Fig. I.2-29C in the same way as in the previous case, the only difference being that the initial stresses σ'°_1 and σ'°_3 are not equal and therefore do not define the centre of the initial (stippled) stress circle. The result is an asymmetry in the stress circles associated with the points Q^+ and Q^- which, as illustrated by the figure, may now enable the stress circle of Q^+ to touch the limit line first. Hence, the plastic state will be induced at the *advancing* side and the fault will now tend to deviate from the original plane towards the advancing side, possibly developing synthetic splay faults. Without going into further details, we note that this result can only be obtained for values of the inclination angle θ that are considerably smaller than $45° - \varphi/2$ [11)].

But there are more effective conditions to suppress the growth of the stress circle associated with a point Q^- at the receding side and thus to force the propagating fault to curve or branch off towards the *advancing* side. When the propagation of a fault is driven by only one wall advancing as sketched in Fig. I.2-30, while the other wall is at rest and, at the same time, a barrier of stiffer rock ahead of the propagating fault effectively confines the region of elastic deformation, the plastic zone is bound to develop preferentially at the advancing side. We shall return to this situation in Section I.2.8 when discussing the termination of faults [12)].

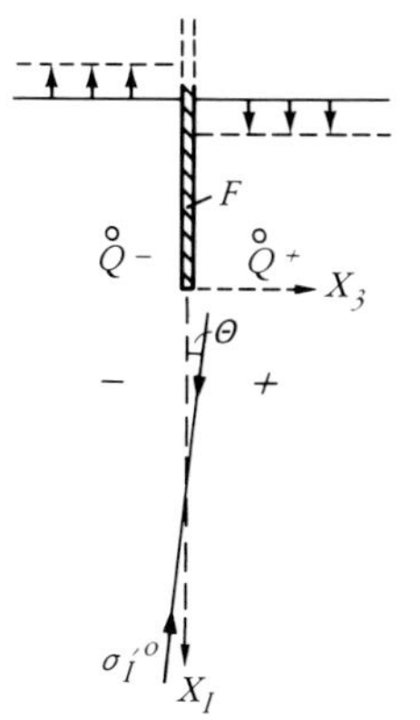

A Orientation of maximum compressive stress near-parallel to fault prior to reiteration of slip. Same magnitude of imposed displacements on advancing and receding side of fault F.

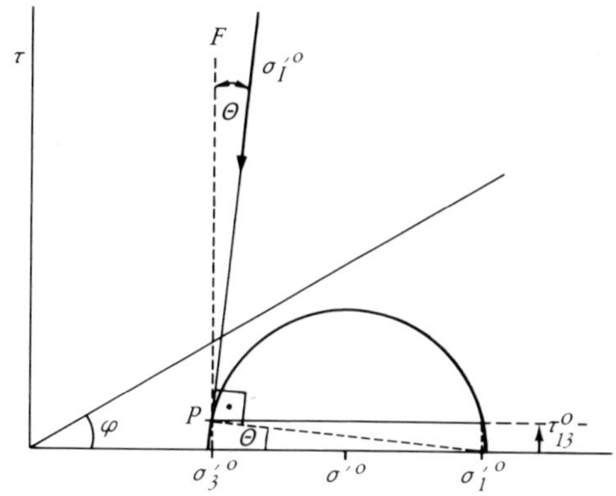

B Stress circle representing initial stress state in accord with prescribed principal stress orientation.

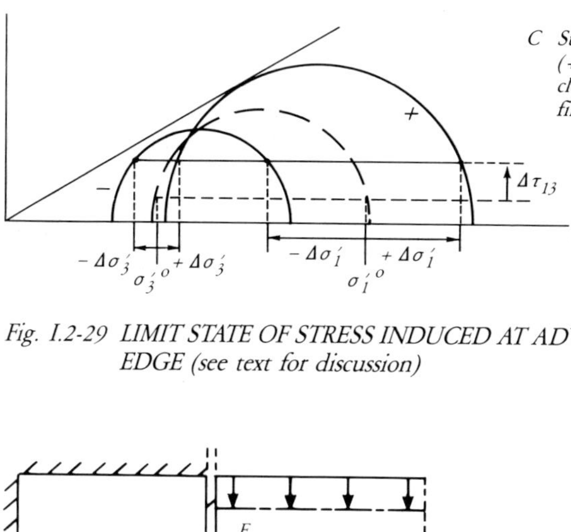

C Stress circles representing stress changes on advancing (+) and receding (−) side of fault plane. Stippled circle indicates initial state of stress. Limit state may be first attained at advancing side.

Fig. I.2-29 LIMIT STATE OF STRESS INDUCED AT ADVANCING SIDE NEAR LEADING FAULT EDGE (see text for discussion)

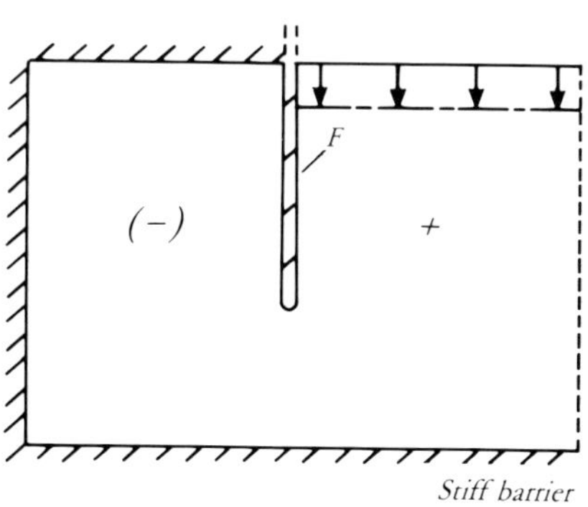

Fig. I.2-30
ONE-SIDE REITERATION OF FAULT SLIP IN PRESENCE OF LITHOLOGICAL BARRIER
Strike or dip section through near-edge region of fault plane. No displacement imposed on (−) side and extensional straining suppressed by barrier ahead of propagating fault.

Model Experiments. Various authors have attempted to model listric normal or thrust faulting in sand – the prototype of a frictional plastic material – or in clayey materials resting on a horizontal base and confined between transparent parallel walls through which the development of faults could be observed. These experiments need some critical comment because of a serious pitfall involved. Consider, for instance, the modelling of rifting in which the basement moves horizontally apart and new basement fills the gap. In experiments this may be simulated by the opposite movement of two base plates that partly overlap or are connected by a rubber strip. The overburden then responds to the local extension of its base by the formation of graben-type normal faults, which may show good listricity when viewed through the transparent side walls. One might be inclined to ascribe this to the local change in σ_I-direction by basal shear stresses that were induced by the 'rifting' of the basement. This would be contrary to our expectation, stated earlier, that the basal shear stresses that are induced by such a rifting process will hardly affect the shape of normal faults in a brittle overburden. This expectation is indeed confirmed by an inspection of the fault traces on dip sections *inside* the overburden of the experiments. Such an inspection can be done elegantly and in a non-destructive way by X-ray computed tomography [13]. In Fig. I.2-31 our experimental set-up is shown schematically in plan view. Two base plates with firmly attached vertical side walls W_1 are connected by a rubber strip which is stretched when the side walls are moved apart by a gear mechanism. In addition, two walls W_2 are installed in a fixed position parallel to the extension direction, allowing the base plates to slide underneath. The whole box is filled with a uniform pack of dry sand and X-ray scans are taken of vertical sections running parallel to the walls W_2 at various stages of basal extension. The formation of shear bands becomes clearly visible because of the shear dilatation of the sand which causes contrasts in X-ray absorption. Figure I.2-32 shows the development of faults in successive stages of the rifting process. Each picture shows two overburden sections parallel to the direction of the imposed basal stretch. The section on the left in each picture is taken midway between the frictional boundary walls W_2, while the section to the right is close to

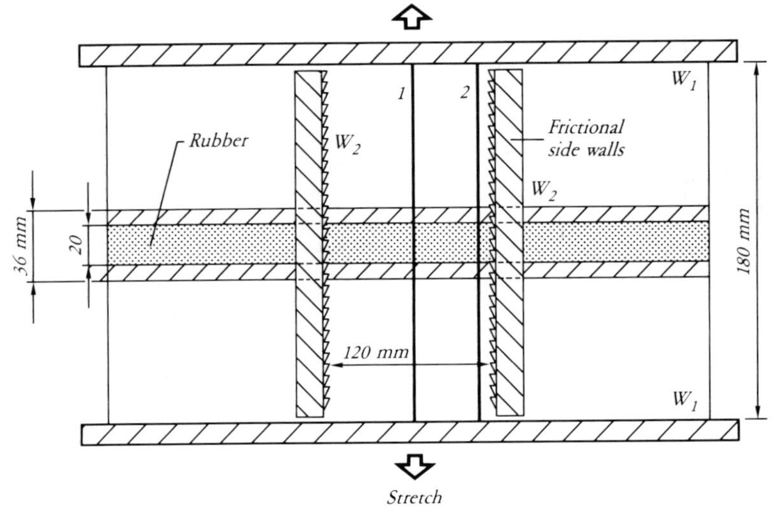

Fig. I.2-31 SANDBOX MODELLING OF HORIZONTAL OVERBURDEN EXTENSION IN TECTONIC RIFTING, PLAN VIEW OF SANDBOX EQUIPMENT

Two base plates are connected by rubber sheet the middle part of which can be stretched from 20 to 36 mm. Two frictional side walls (W_2), covered with sand-paper, are in fixed position allowing base plates to slide underneath. Cross-sections 1 and 2 indicate positions of the X-ray scans shown in Fig. 1.2-32.

a boundary wall. The extent of rifting is shown by the width of the gap between the two base plates. The earliest faults in the central part of the sand pack show only faint listricity, while those in the boundary section are strongly curved – a clear indication that shear stresses between the extending pack and the walls at rest are responsible for the listricity. Note that later infill faults are planar. The result is also corroborated by similar rifting experiments on a somewhat larger scale, in which a horizontal layer of dry sand was again used; this time, however, it was not confined by side walls. All faults were found to be perfectly planar.

The flattening of the fault dip observed near fixed boundary walls is clearly caused by the friction between the sand and the fixed side walls. This friction has a horizontal component as a result of extension of the sand pack. Hence, a shear stress acts upon the sand pack in

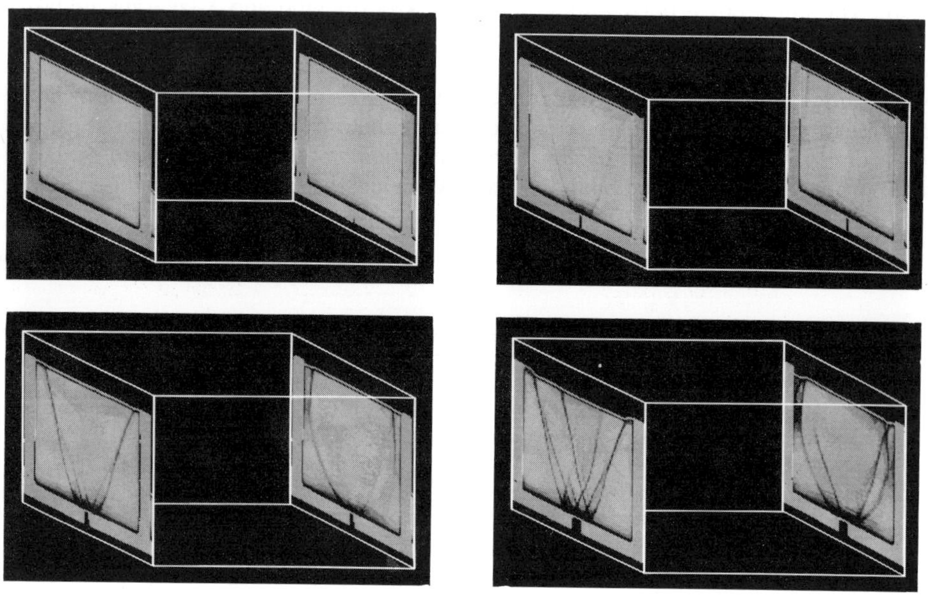

Fig. I.2-32 *SANDBOX MODELLING OF HORIZONTAL OVERBURDEN EXTENSION IN TECTONIC RIFTING. X-RAY COMPUTER TOMOGRAPHY IMAGES OF NORMAL FAULTS SHOWING THE EFFECT OF FRICTIONAL SIDE WALLS ON FAULT SHAPE.* *Outlines represent part of sandbox which is bounded by the scanned cross-sections indicated by 1 and 2 Fig. I.2-31. Mean grain diameter of the dry sand used was 0.25mm. The shear-induced reduction in sand density and the corresponding reduction in X – ray absorption allows observation of fault development already at the very state of incipience of shear concentration. Figure shows the history of faulting at subsequent states of base plate separation (visible as gap in the base). Faulting started from the stretched base interval upward (exept for the two subvertical branches in the upper part of the pack in the section at the right-hand side, which are associated with the outward movement of the adjacent walls). Whereas the first pair of faults show very little listricity in the middle section of the pack, it is strongly curved in the section near a frictional wall. Note that later 'infill' faults are practically straight*

the direction opposite to the movement of the sand. To understand, at least in qualitative terms, how this shear stress affects the orientation of the principal stresses, we think of the shear stress as being distributed across some boundary slice of the sand pack, which makes it equivalent to a horizontal body force. When this friction force is vectorially added to the vertical gravity force, the resultant body force will be inclined and pointing against the displacement of the basement. In the absence of wall friction the principal stress σ_I' is caused only by gravity and therefore oriented vertically. Near the frictional walls it will be inclined to the same side as the resultant body force. Accordingly, one of the two conjugate slip elements will attain a flatter orientation near the wall than it will at some distance from it [14)].

Although confinement of basement extension (or shortening) and normal (or thrust) faults between fixed parallel side walls may occasionally exist in nature (e.g. when the material between parallel wrench faults is extended or shortened), it is much more common for normal or thrust faults to terminate along strike by a rather gradual decrease in fault displacement. Usually, the strike line of these faults shows some curvature, as is well-known for normal faults, many of which have an *arcuate shape in plan* with the concave side facing the downthrown block (Fig. I.2-33). Unfortunately, very little is known about this type of fault curvature and the factors that control it. Therefore, the following comment on the origin of the arcuate shape of normal faults cannot pretend to give a full explanation of this phenomenon.

We consider a normal fault at an initial stage of its development and assume that the fault still intersects the horizontal plane along a straight line segment as sketched in Fig. I.2-34A. The fault grows by increments both in the dip and strike directions. Why would the outward migration of the fault hinges deviate from the initial strike direction? Again, we have to consider the stress field near the hinge line S of the fault plane (Fig. I.2-34A). It is reasonable to assume that the latter is parallel to the dip-slip direction. The material in the vicinity of the migrating hinge line is therefore sheared parallel to this line by the dip-parallel displacement on the fault. Two shear-stress components, both parallel to the hinge line S, but no normal stresses, are induced by the shearing as indicated in the figure. This mode of hinge propagation is referred to in fracture mechanics as Mode III (or 'tear') fracture. Naturally, the same shear stresses also act upon elements that intersect S at right angles, on which they determine, by vectorial addition, the vector of the maximum shear stress as shown in the figure. The important point here is the complete symmetry of the value of the maximum shear stress with respect to the x_1, x_3-plane, which is the extension of the original fault plane [15)]. As long as the near-edge field is in a subcritical elastic state, the superposition of an homogeneous stress field with σ_I' parallel to the x_2, x_3-plane and directed at $45° - \varphi/2$ with respect to the fault plane would not disturb this symmetry. The mean normal stress associated with this stress field is the same in points that are symmetrical with respect to the x_1, x_3-plane. Hence, exactly the same stress circles are associated with these points, and thus no reason for a systematic deviation of the outward-migrating hinge line to deviate from the original fault plane is apparent.

Fig. I.2-33 THE ARCUATE SHAPE OF NORMAL FAULTS IN PLAN

The reason for such a deviation may be sought in the change of stresses that is induced on the downthrown side of the fault by *shear softening* of the fault zone material and the concomitant reduction in shear resistance of the active fault plane. One way of maintaining static equilibrium on the fault is to increase the support of the hanging wall by means of an increase in the fault-parallel normal stress $\sigma_{\|}$. This increase will be restricted to the hanging wall since a discontinuous change of $\sigma_{\|}$ across the active fault plane is perfectly admissible (see Section II.2.1, Example No. 1). However, this causes a gradient of $\sigma_{\|}$ in the x_1-direction (Fig. I.2-34B), which is accompanied by a corresponding gradient in fault-parallel contraction of the material. Since the material is not interrupted in the x_1-direction

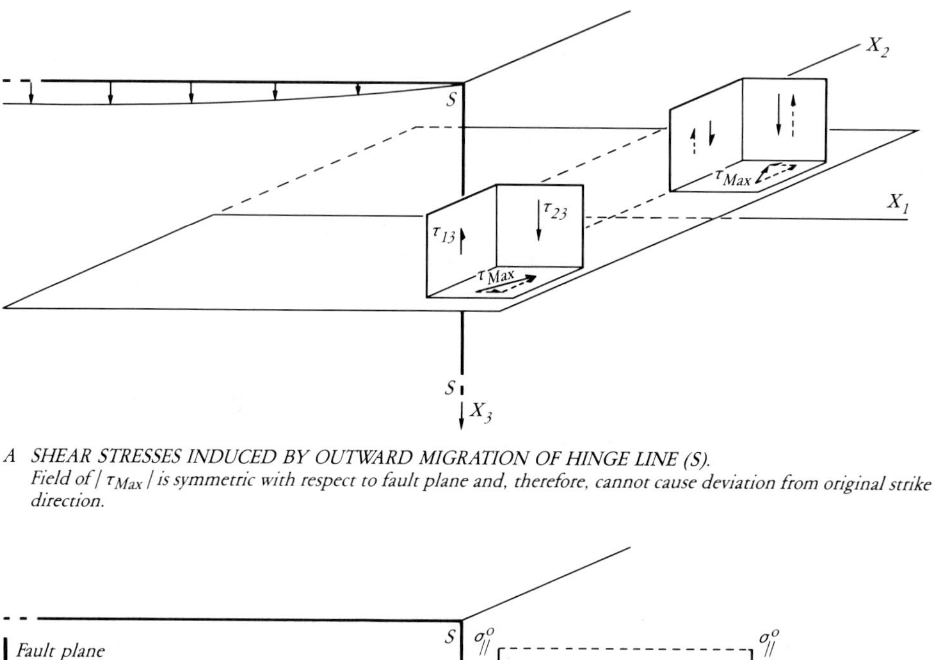

A SHEAR STRESSES INDUCED BY OUTWARD MIGRATION OF HINGE LINE (S).
Field of $|\tau_{Max}|$ is symmetric with respect to fault plane and, therefore, cannot cause deviation from original strike direction.

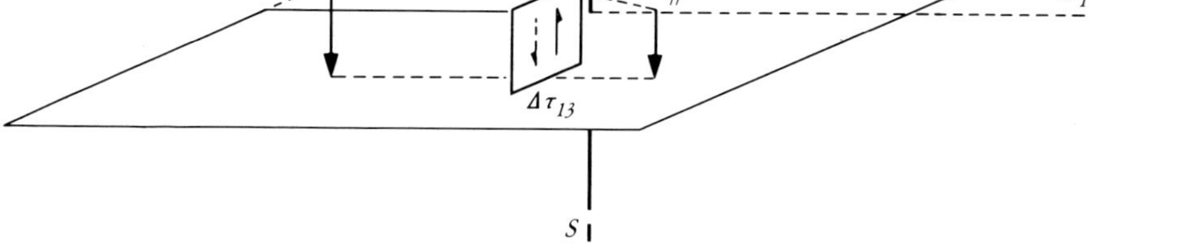

B SHEAR SOFTENING OF FAULT ZONE MATERIAL DISTURBS SHEAR STRESS SYMMETRY.
When loss in shear strength of fault plane is balanced by an increase in fault-parallel normal stress ($\sigma_{//}$) in hanging wall, concomitant differential straining will give rise to an extra shear stress $\Delta\tau_{13}$ at the side of the subsiding block.

Fig. I.2-34 NEAR-FAULT EDGE STRESSES INDUCED BY OUTWARD MIGRATION OF HINGE LINE OF NORMAL FAULT

by a slip plane, the difference in contraction causes vertical shear stresses on the side of the subsiding block in the vicinity of the hinge, as indicated in the figure. When added to the shear stress τ_{13} in Fig. I.2-34A, the additional shear stress disturbs the symmetry in the distribution of maximum shear stress values. The values on the side of the subsiding block are increased and the stress circles for points on this side will therefore be systematically greater than for points located symmetrically on the other side. Hence, the plastic limit state will be preferentially attained at the side of the hanging block and the normal fault will have the tendency to propagate laterally along a *strike line that is concavely curved* towards that side.

Another stress change in the subsiding block may be caused by the strike-parallel *extension of downthrown strata* (Fig. I.2-35A), which reduces the strike-parallel intermediate principal stress. At the inflexion of downbent strata of cohesive rocks the reduction may even reverse the roles of the intermediate and smallest principal stresses, and re-induce the active limit state. As sketched in Fig. I.2-35B, this may cause a secondary normal fault to branch off the original fault at a high angle.

Undoubtedly, the stresses near the termination of a normal or thrust fault in the strike direction, and the associated curvature or branching of the fault, depend on various other parameters that we have not even mentioned; these relate to, for instance, the depositional process, the displacements of the underlying basement, the width of a gliding horizon, or a superimposed strike-slip component (oblique slip). Much research remains to be done in this area.

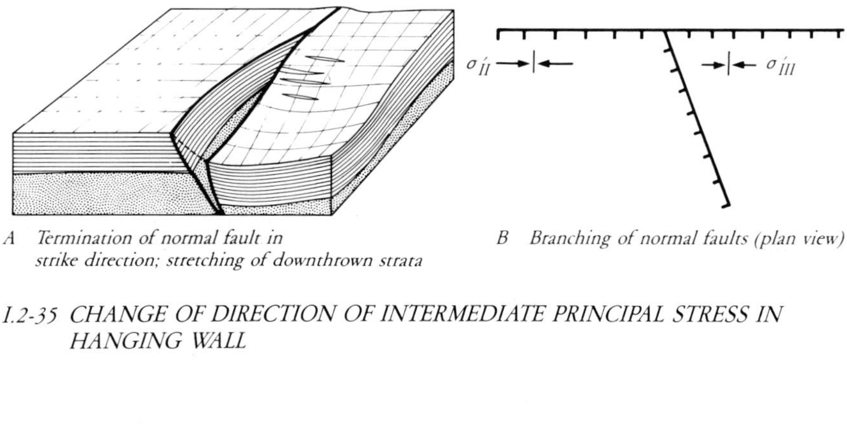

A Termination of normal fault in strike direction; stretching of downthrown strata

B Branching of normal faults (plan view)

Fig. I.2-35 CHANGE OF DIRECTION OF INTERMEDIATE PRINCIPAL STRESS IN HANGING WALL

5. THE SEGMENTARY DEVELOPMENT OF FAULTS

Many, if not most, tectonic faults are generated in separate segments (Segall and Pollard, 1980) which eventually, if shearing continues far enough, interconnect to form continuous faults. The phenomenon is important because it may allow the migration of fluids across faults and it will affect the width and internal structure of a fault zone. The reasons for the discontinuous character of incipient faults, although still insufficiently investigated, may be related to two major factors: variations in lithology and non-uniformities in the tectonic stress field.

Variations in Lithology may be accompanied by variations in the deformational response of the rock. In the bending of an alternating sequence of competent and incompetent (i.e. more ductile) beds, faulting will start in the competent, brittle beds, while the incompetent beds will accommodate a certain amount of strain by creep, or numerous small slips, before being offset by faulting. Connection of the fault segments will then result in faults that – at least in an early state – have a tortuous and splintered shape.

If the bending of a thick incompetent bed of shale, for instance, has not yet led to the formation of a fault cutting across the bed, sets of small parallel faults may form, as shown schematically in Fig. I.2-36. The faults strike transversely to the imposed shear, which they accommodate by rotation and concomitant slip – a process that will be discussed in detail in the following section. The bending of a shale bed may also cause lateral offset of faults in adjacent brittle rock layers. This is illustrated in Fig. I.2-37, in which a normal fault, progressing upwards through a sandstone bed, is laterally offset by the bending of a shale bed. As faulting continues (Fig. I.2-37B) a slice is cut from the shale bed and emplaced as clay gouge on the fault plane.

Instead of concentrating in fairly narrow zones of fault segments, tectonic shearing can – at least during a certain period – pervade a wide area by producing a great number of separate smaller shear bands. This will be the case when over a wider area of reasonably homogeneous rock a uniform stress field approaches the limit state, for instance, by uniform shortening of a thick bed of uniform lithology. Slight differences in shear strength will, however, always be present and at various places will allow the rock to reach peak strength before the main part of the rock mass. Hence, softening and shear concentration will commence at these localities which act as nuclei for shear bands. One would therefore have to expect that in such situations large-scale faulting – if it occurs at all – will be preceded by the pervasive formation of small shear bands.

Incidentally, neighbouring parallel shear bands will generally be arranged *en-échelon*. In other words, closest neighbours in an array of parallel shears will only partly overlap, often in the tip regions only. We think the latter will occur when parallel shear bands form in succession. Without anticipating a rigorous elastic/frictional plastic analysis, one may expect that the formation of a shear band will reduce the shear-band-parallel shear stress in a certain neighbourhood of the newly formed shear band, and will change the critical state of stress in that region into subcritical, in which no shear band can form. Limit stress conditions will, however, be maintained around the leading edge of the shear band and hence allow there a close coexistence of another shear band [16)].

Another reason for the en-échelon grouping of parallel shears is the configuration of the plastic region itself. In the bending layer of Fig. I.2-36, for instance, the limit state of stress will be restricted to an axial zone where extension has been sufficient to induce the limit state. Inside this plastic zone, which is contained within elastic material, potential Coulomb slips will not normally be parallel to the lateral boundaries of the zone itself. Incipient shear faults may therefore form arrays confined between the boundaries of the plastic zone. Later we shall discuss how and under what conditions such an array of parallel faults can operate (Section I.2.6).

Non-Uniform Stress Fields. With the last remark we have already proposed a non-lithological cause of segmentary development of faults, namely the inhomogeneities of a tectonic stress field. While in the foregoing such an inhomogeneity was considered as causing the containment of a plastic region as a whole, variations in the stress field may even be the cause of segmentary fault formation inside the plastic region of a lithologically uniform rock mass. Beach (1975) for instance, mentioned that faults, when propagating into a region with

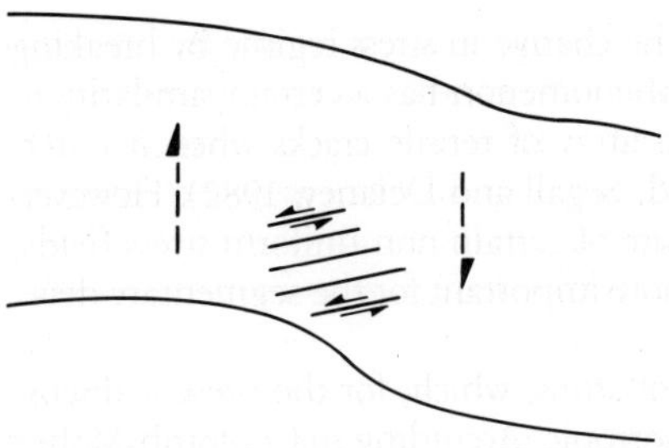

Fig. I.2-36
ZONE OF MONOCLINAL WARPING WITH EN-ECHELON SET OF SMALL FAULTS (highly schematized)

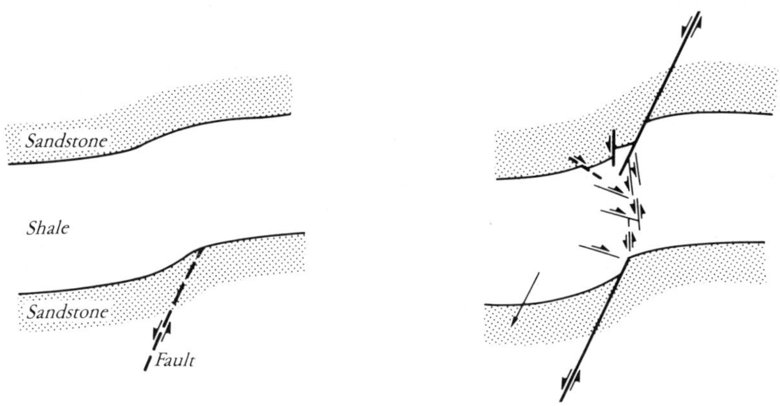

A Continuous flexuring of shale bed in response to faulting of substratum.

B Upward progressing fault experiences lateral offset in dip direction.

Fig. I.2-37 OFFSETTING OF NORMAL FAULT BY FLEXURING OF SHALE BED IN SANDSTONE/SHALE SEQUENCE (simplified after F. Lehner & W.F. Pilaar; 1974, unpubl. report)

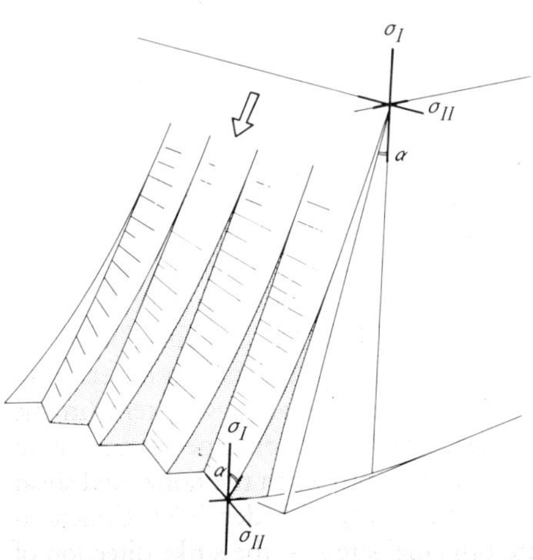

Fig. I.2-38
SEGMENTATION OF NORMAL FAULT IN STRESS FIELD WITH HORIZONTAL PRINCIPAL DIRECTIONS

differently oriented principal stresses, will adjust to the change in stress regime by breaking up into an en-échelon array of smaller shears. The phenomenon has a certain similarity to the breakdown of a tensile crack into an en-échelon array of tensile cracks when it enters a region of changed principal stress directions (Pollard, Segall and Delaney, 1982). However, this is only one manifestation of a more general feature of certain non-uniform stress fields, which, in large-scale tectonic faulting, may be even more important for the segmentary development of faults than variations in lithology.

Let us therefore consider a rock in the limit state of stress, which, for the present discussion, is idealised as mechanically homogeneous and isotropic. According to Coulomb-Mohr's theory, potential slip elements will then everywhere include the same angle with the σ_I-axis. Therefore, in a non-uniform stress field with changing orientations of principal stresses the orientation of slip elements will also change (see, for instance, Fig. II.7-19). Will it then be possible to fit smooth surfaces into the three-dimensional field of potential slip elements in such a way that the tangential plane at any surface point coincides with a slip element? Obviously, only then can a coherent fault be initiated. This essential problem in the application of Coulomb-Mohr's theory to tectonic faulting is rigorously treated in Section II.7.6. The result may be summarised as follows: *A non-uniform field of principal stress directions will not impede the formation of continuous, smooth fault surfaces if the field is associated with a state of plane stress or plane strain* (cf. for example, 'Plane problems' in Sections II.1.2 and II.2.1). *However, in general, it will not allow the incipient formation of a continuous fault surface when it is of a non-planar, truly three-dimensional type* (Mandl, 1987a).

Illustrative examples of non-uniform, non-planar fields of principal stress directions that do not allow the formation of a continuous fault surface, composed of Coulomb-slip elements, are dealt with in Section II.7.6. There it is also made quite clear that such fields need not be of a complex type; for instance, a stress field with two different horizontal principal stresses will become incompatible with the incipience of continuous faults when the *directions of the horizontal principal stresses vary with depth.* Such variation in horizontal principal stress directions may result from a variety of tectonic situations. In the case depicted in Fig. II.7-40 of Section II.7.6 two horizontal principal stresses are rotated by the superposition of a horizontal compression [17]. To consider here a somewhat different case, let us assume a tectonic stress field which is already close to the active limit state, with σ_I vertical, and σ_{II} and σ_{III} different in magnitude but having constant directions. At some later stage the rock is subject to a small horizontal simple shear along the vertical plane through the σ_{III}-axis. While this has no effect on the vertical maximum principal stress, magnitudes and orientations of the horizontal principal stresses will be changed. We assume that the applied shear has reduced the minimum principal stress sufficiently to induce the limit state and to initiate normal faulting. The orientations of the horizontal principal stresses have changed, because the plane perpendicular to the original direction of σ_{II} is no longer free of shear stresses. If this change in orientation is the same throughout the depth interval considered, the field of principal stress directions is still uniform and allows the formation of smooth normal faults. Usually, however, the change in horizontal orientation of the intermediate principal stress will vary with depth for the following reason. The direction of the maximum horizontal shear stress of the imposed simple shearing deformation differs from the direction of the maximum horizontal shear stress of the original stress system (in the present case by 45°) and the maximum horizontal shear stress of the combined stress system has still another direction. This last direction will vary with depth when the horizontal shear stresses of the original and/or superimposed stress system vary with depth [18]. Concomitantly, the horizontal direction of the intermediate principal stress – the strike direction of an incipient normal fault – will change with depth, since it makes an angle of 45° with the direction of the maximum shear stress. Incipient faults therefore cannot have a coherent smooth shape [19], but will consist of segments, that have the same dip angle but vary in strike and slip direction. A normal fault propagating downwards into a depth interval with rotated horizontal principal stress directions should therefore break up, as sketched in Fig. I.2-38, into twisted slip segments, the incoherence of which increases with depth. As faulting continues, the individual segments may be linked by transverse shears and incorporated in a throughgoing shear zone, the width of which should increase with depth, and which still reflects the size of the original steps between the segments.

Conversely, when a normal fault grows upwards into a region in which the horizontal principal stress directions have been rotated (consider Fig. I.2-38 in upside-down position), for instance by some shearing action parallel to the strike of the fault at depth, the fault will also disintegrate into en-échelon segments. It would seem that at shallow depth the eventual interconnection of individual segments will produce a set of faults that are arcuate in plan (Fig. I.2-33). The segmentary development of normal faults, for instance under a superimposed horizontal shear, may thus be another cause of *arcuate* faulting, in addition to those discussed at the end of the preceding section.

As a conclusion to our discussion of the segmentary development of faults, it is important to note that the strike of a fully developed fault zone, composed of incoherent, twisted and later-connected Coulomb-type segments, deviates from the direction of the tectonic σ_{II}-axis and, therefore, *cannot be considered as a Coulomb-type shear zone* – a point we shall return to later (Section I.2.8).

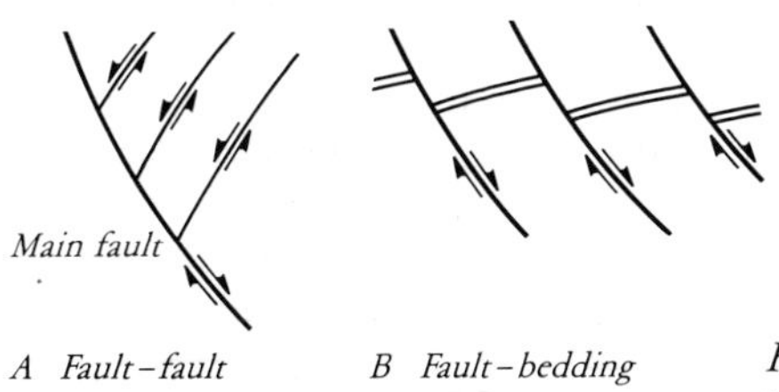

A Fault–fault relationship　*B Fault–bedding relationship*

Fig. I.2-39
TWO DEFINITIONS OF ANTITHETIC FAULTING

6. ANTITHETIC FAULTS

This section concerns the mechanical genesis and operation of antithetic faults in the deformation of a rock mass. The term 'antithetic' is used here in a kinematical sense to indicate minor faults with a direction of shearing that is opposite to the shearing along a major fault as sketched in Fig. I.2-39A, or, in a somewhat wider sense, to indicate faults with a sense of shearing that is opposite to the general direction of an externally imposed shear. To avoid confusion, it should be mentioned that in the literature the term 'antithetic' is also used in a quite different sense for faults that dip opposite to the strata they intersect, as shown in Fig. I.2-39B.

Antithetic Adjustment Faults. A first function of antithetic faults is the accommodation of a gliding rock mass to changes in the geometry of its gliding path. As pointed out earlier (Section I.2.3), any plane deformation of a rock mass can be accomplished by the interplay of conjugate Coulomb-type faults – a process we have referred to as 'double gliding'. Often, one of the conjugate fault types predominates in accommodating the tectonic deformation, while faults of the other type play the subordinate role of adjustment faults. As conjugates, the adjustment faults are of course antithetic to the main faults.

If a rock mass is sliding along a perfectly plane fault whose orientation does not change in time, there is no reason for any adjustment along that fault. Often, however, the dip or strike of the main fault may change quite abruptly and give rise to antithetic adjustment faults. This kinematical consequence of a *'kink' in the sliding path* along a main fault is demonstrated by the transparent model shown in Fig. I.2-40, which can be used on an overhead projector. The successive stages of movement along the main fault show how at the 'kink' antithetic faults are mobilised and later immobilised as the next antithetic fault is formed. The whole process allows a change in the bodily translatory motion. The saw-toothed misfits decrease in size as the spacing of the antithetics becomes smaller. In nature they will be accommodated by small-scale slips, compaction, etc.

An application of this antithetic adjustment mechanism is shown by another transparent model in Fig. I.2-41, which approximates a *gravity slide* with segments of circles and a straight line. Again antithetics are generated at the points where curved faults merge with or depart from the slope-parallel basal slip plane. To allow for the transition between a bodily rotation and a purely translatory gliding motion, the antithetic faults have to be *curved* in the sense shown. The sense of the curvature is the only possible one, also from a mechanical point of view, if the main fault and antithetics are conjugates in the same field of σ_1'-directions. To account for the flattening of the main normal or reverse fault with depth, the σ_1'-axis has to rotate towards the fault with increasing depth; the antithetics must, consequently, steepen with depth to fit into the field of σ_1'-directions (see also Fig. II.7-19 in Section II.7.4).

It should be noted, however, that this conclusion applies only to antithetics that are truly conjugates of the main fault, i.e. those generated in the same stress field. It is not difficult to envisage situations in which the stress field in the rock moving along a curved fault is

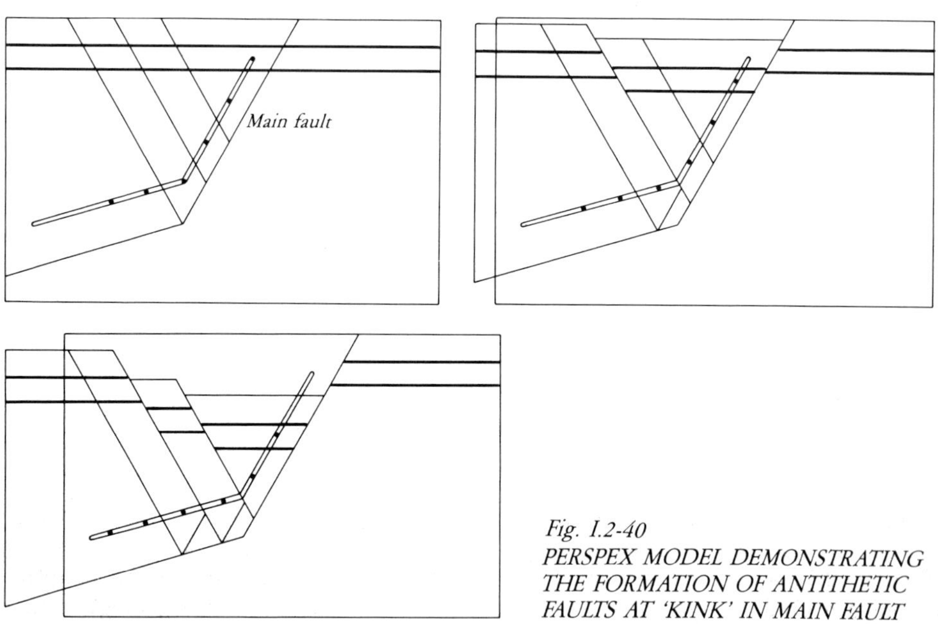

Fig. I.2-40
PERSPEX MODEL DEMONSTRATING THE FORMATION OF ANTITHETIC FAULTS AT 'KINK' IN MAIN FAULT

changed prior to the formation of the antithetics. For example, strata in the hanging wall of a listric normal fault may experience some stretching in the drag zone of the main fault, which reorients the σ_I-directions so that they become closer to the normal of the dragged strata, and give rise to straighter or even 'wrongly' curved antithethics[20].

The seismic section in Fig. I.2-42 shows a major synsedimentary normal fault with a set of antithetic faults that may have been generated at a kink in the upper part of the main fault. Note that the depth scale is about three times the horizontal scale; in reality the faults are therefore much flatter than shown on the seismic record. Note also that the antithetics increase in length updip along the main fault, which is indicative of the synsedimentary origin of the antithetic faults.

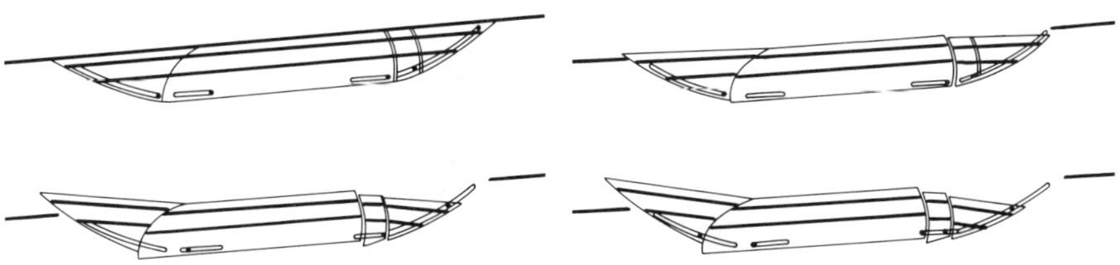

Fig. I.2-41 PERSPEX MODEL APPROXIMATING COMPLETE GRAVITY SLIDE ON DELTA SLOPE BY COMBINATION OF RIGID-BODY MOTIONS
Antithetic faults allow transition between rotational and translatoric gliding. Note that antithetic normal faults become inactive, when entering region of purely translatoric gliding.

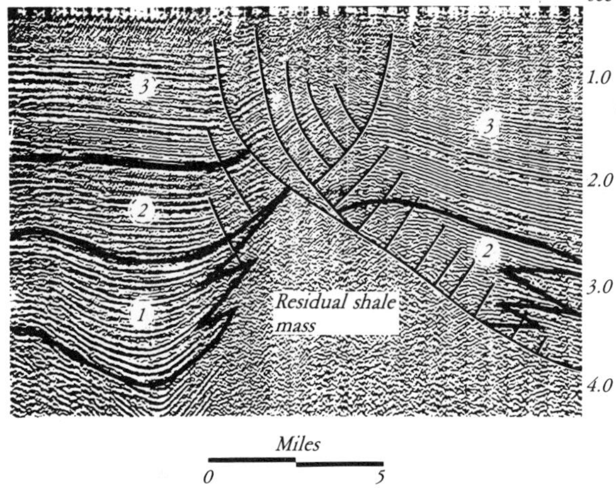

Fig. I.2-42 SEISMIC SECTION ACROSS MAJOR SYNSEDIMENTARY FAULT SYSTEM (After C.H. Bruce, 1973; reprinted by permission of "American Association of Petroleum Geologists")
Set of antithetic faults along SE flank of shale may have been generated at 'kink' in gliding path (numbers on either side of fault system indicate depositional sequence of sand and shale units).

In Fig. I.2-43 it is analysed how, as a result of the activity of an antithetic fault, the dip-slip vector $\vec{V}$ along the main fault is changed at a kink into $\vec{V}'$. The smaller magnitude of the new dip-slip follows from an elementary formula derived in the figure [21].

A second cause of antithetic adjustment faults is the *tilting* of the main fault plane. There are several mechanisms that bring about such a change in fault dip. The fault may be downthrown or upthrown and rotated along a curved neighbouring fault (Fig. I.2-44). Reclining or steepening of the main fault may also be caused by the tilting of underlying strata in regional updoming. In sediments that are still compacting, a fault plane will recline; when the footwall material (e.g. clay) compacts more strongly than the material of the hanging wall (e.g. sand), the hanging wall material has to adjust by subsidiary faulting to prevent the formation of a wedge-shaped gap between the two sides of the reclining fault plane. In Fig. I.2-45 it is shown schematically how the reclination of a normal fault s_1 into a new position s_1' can be adjusted in the hanging block by a set of conjugate antithetic slides s_2. The operation of these antithetics is again demonstrated by a transparent model in Fig. I.2-46.

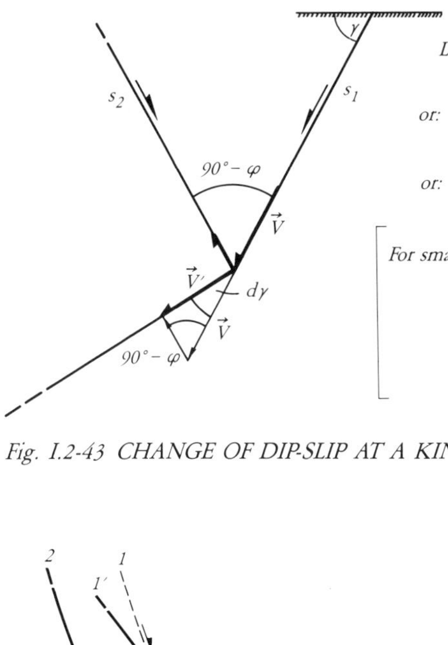

Law of sines: $$\frac{|\vec{V}'|}{\sin(90° - \varphi)} = \frac{|\vec{V}|}{\sin(90° + \varphi - d\gamma)}$$

or: $$\boxed{|\vec{V}'| = |\vec{V}| \frac{\cos\varphi}{\cos(\varphi - d\gamma)}} \qquad (1)$$

or: $$\frac{|\vec{V}| - |\vec{V}'|}{|\vec{V}'|} = \frac{\cos(\varphi - d\gamma)}{\cos\varphi} - 1 \qquad (2)$$

For small dip changes dγ:
$$\frac{|\vec{V}| - |\vec{V}'|}{|\vec{V}'|} \approx d\gamma \cdot \tan\varphi \qquad (2a)$$
$$(\cos(\varphi - d\gamma) \approx \cos\varphi + d\gamma \cdot \sin\varphi)$$

Fig. I.2-43 CHANGE OF DIP-SLIP AT A KINK IN MAIN FAULT (s_1)

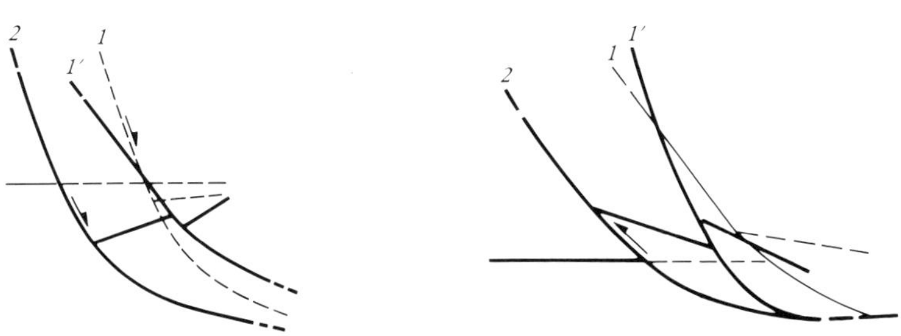

A Reclining of normal fault by rotational downthrow along second fault

B Steepening of thrust fault by rotational upthrow along second fault

Fig. I.2-44 ROTATIONAL TILTING OF FAULTS

The adjustment slips along the antithetics (s_2) and the main fault (s_1') in Fig. I.2-45 can easily be expressed in quantitative terms. In Fig. I.2-47 it is derived by elementary trigonometry that the adjustment slip (δ) along the antithetics is proportional to their spacing (ds) and to the sine of the angle of fault rotation, and is inversely proportional to the sine of

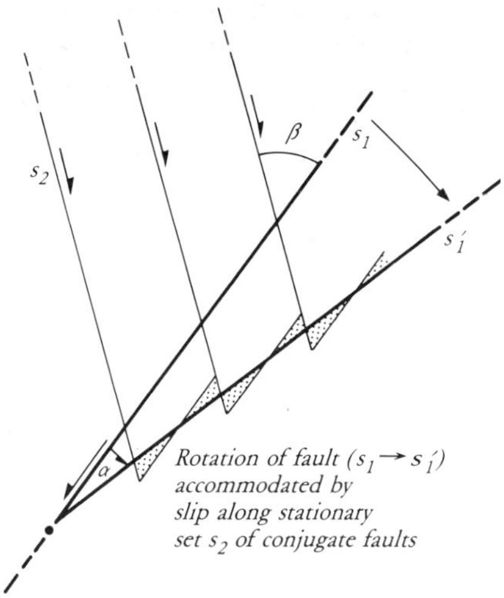

Fig. I.2-45
'RECLINING' NORMAL FAULT WITH DOWN THROWN BLOCK ACCOMMODATING BY GLIDING MECHANISM

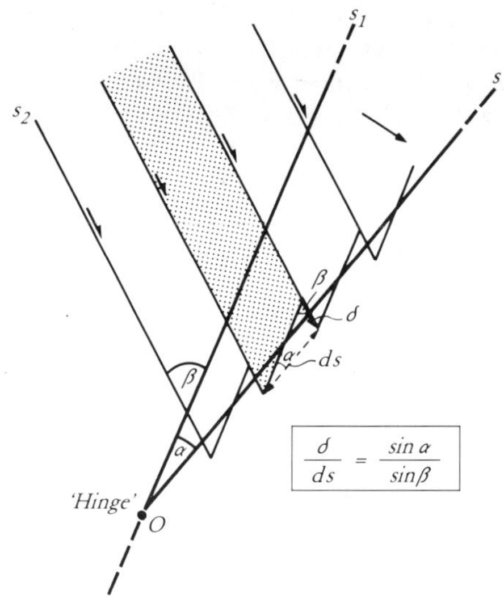

Fig. I.2-47
SLIP (δ) ALONG 'ANTITHETICS' (s_2)

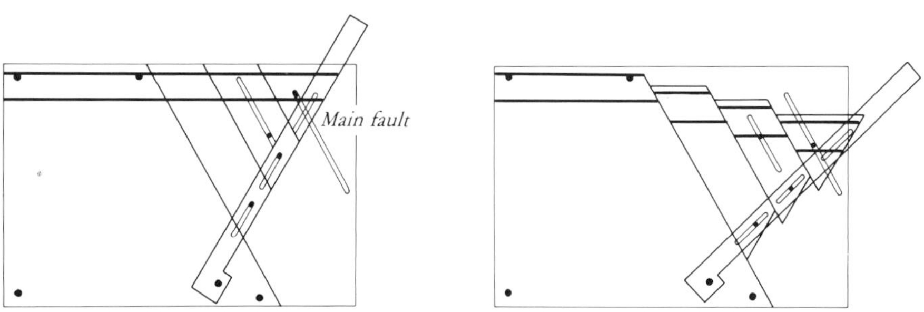

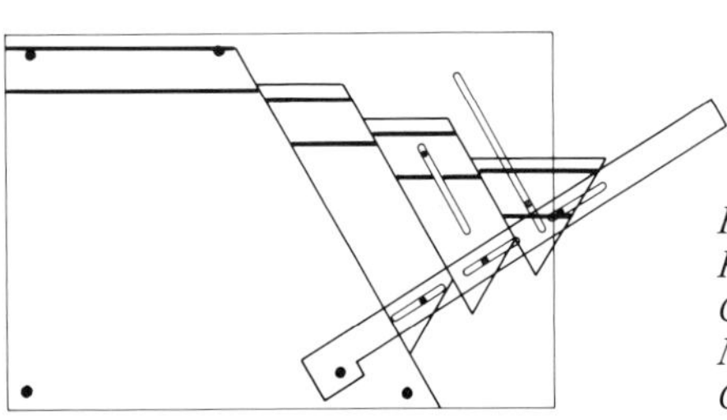

Fig. I.2-46
PERSPEX MODEL DEMONSTRATING OPERATION OF ANTITHETIC ADJUSTMENT FAULTS WHEN MAIN FAULT 'RECLINES' UNDER COMPACTION OR ROTATION OF FOOTWALL MATERIAL

the original angle between the main fault and the antithetics. The adjustment slips along the antithetics also induce slip (Δ) along the main fault, even after its activity has ceased. The calculation in Fig. I.2-48 shows that this slip increases with the distance from the 'hinge' around which the main fault rotated [22]. It is of note that the adjustment mechanism becomes locked when $\sin(\alpha + \beta) < \sin\beta$, since the sense of the displacement Δ along the reclining main fault would then reverse, which is energetical incompatible with a subvertical direction of σ_I (see Section I.2.1). For $\beta = 65°$ the locking position would be reached when $\alpha = 50°$.

Next we shall consider a few field examples of antithetic adjustment faults caused by the reclination of main faults. Figure I.2-49 shows the exposure of the asymmetrically collapsed crest of an anticline in the Zagros Mountains (Iran). A set of subparallel antithetics accommodate the rotation of the normal fault that separates the faulted hinge region from the steepening limb of the fold. Typically, the *spacing* of the parallel antithetic faults decreases towards the surface. What is the reason for this? In the schematic of Fig. I.2-45 and the model in Fig. I.2-46 adjustment of the rock mass was achieved by equidistant antithetic faults, each of which caused the same misfit at the reclining main fault. The misfits were accommodated by deformation of the fault slices without extra faulting. Although suffering the same misfits, equidistant fault slices would not experience the same straining because of their difference in length, and shorter slices would have to sustain greater straining before the formation of a new fault. If we assume that the rock can everywhere withstand the same pre-faulting strain, an equal state of maximum adaptation strain would exist in geometrically similar fault blocks, which implies a decrease of fault spacing in proportion to fault length. It should be said, however, that this similarity argument does not answer the more difficult question of how fault spacing depends on the rock mechanical properties – a problem to which we shall revert later.

The interpreted seismic dip section from the Texas coastal area presented in Fig. I.2-50 shows a family of synsedimentary normal faults with numerous antithetic adjustment faults which are caused by the reclination of the downslope hading normal faults. The reclination

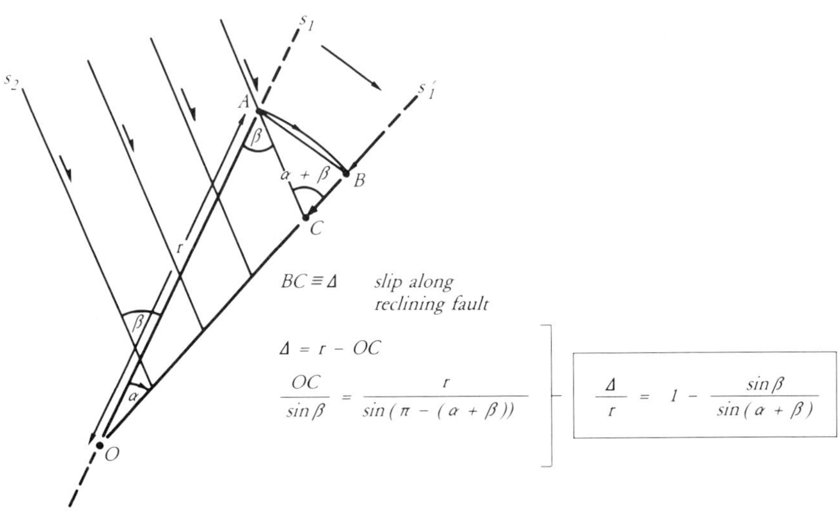

Fig. I.2-48 SLIP (Δ) ALONG ROTATING FAULT PLANE

is the result of differential compaction and the rotation in 'roll-over' structures of neighbouring faults.

The seismic section of Fig. I.2-51 is from the Niger Delta and shows normal fault structures in the roll-over anticline of a boundary growth fault. The interpretation presented at the bottom of the figure emphasises the role of the downdip-heading growth faults (W. Peet, 1977, private communication). Note the pronounced drag zone along the boundary fault (BF). The stretching of beds in this zone might be accommodated by a fan of

A B

Fig. I.2-49 RECLINING FAULT WITH ANTITHETICS IN POST-SEDIMENTARY COLLAPSED CREST, ZAGROS MTS, IRAN
Photo (A), and fault interpretation (B) of NW cliff face of Tang-e Tacob, Kuh-e Khaiz-Badil. The faults are straight or steepen slightly downwards, and show a regular decrease in spacing towards the structural axis. (Courtesy J.F. Lapré)

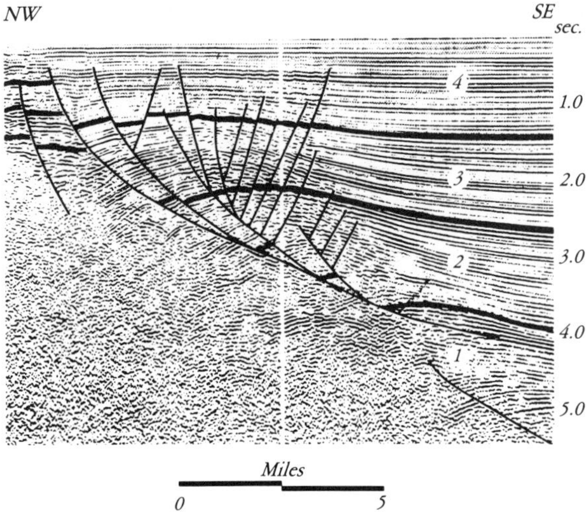

Fig. I.2-50 SEISMIC ILLUSTRATION OF SYNSEDIMENTARY BEDDING-PLANE FAULT SYSTEM, TEXAS COASTAL AREA (after C.H. Bruce, 1973 and 1983; reprinting by permission of "American Association of Petroleum Geologists".) Reclining of SE-dipping faults by rotation in roll-over structure of neighbour faults and differential compaction (Bruce) caused numerous antithetic adjustment faults.

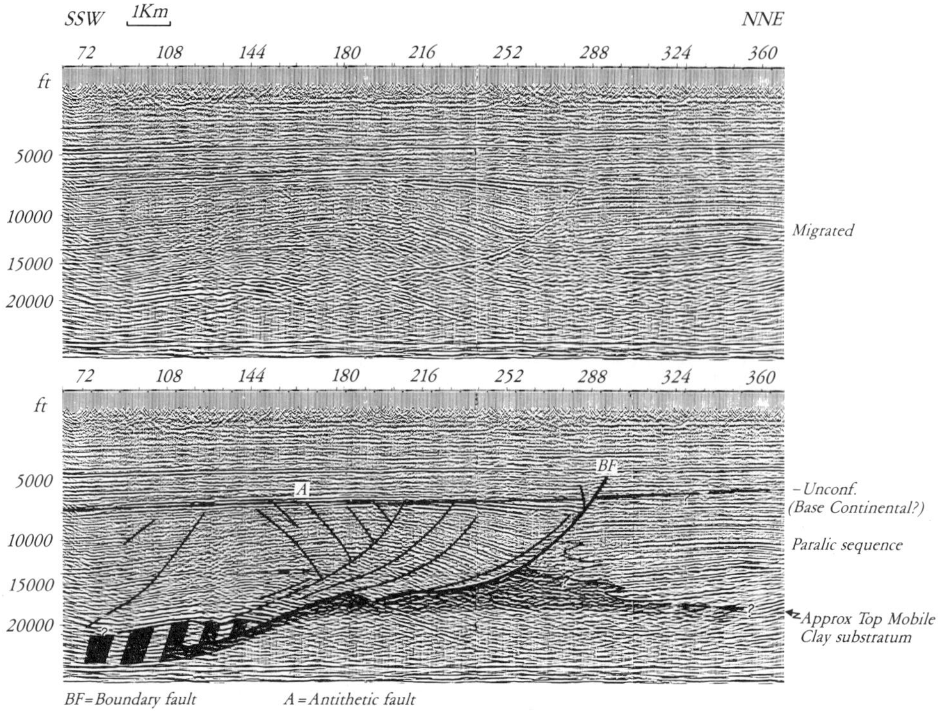

Fig. I.2-51 CRESTAL COLLAPSE STRUCTURE, NIGER DELTA (Nun River field)

synthetic faults. The antithetic faults (A) are given a slightly curved and upward-convex shape in accord with what has to be assumed for conjugates of upward-concave faults.

Figure I.2-52 shows another seismic dip section from the Niger Delta with an interpretation at the bottom. The major growth faults flatten rapidly with depth and eventually merge into bedding planes. Apparently, movement of the underlying shale has affected major slip paths, causing some major faults to recline and the basal slip planes to be displaced downwards. Both the curvature of the faults – most probably associated with overpressuring – and the reclination of major faults strongly suggest the presence of antithetic adjustment faults. This is accounted for in the interpretation, although in reality there may be many more antithetic faults that remain unnoticed on the seismic record because of their small throw.

Accommodation of Crustal Shearing by the "Bookshelf" Mechanism. The second function of antithetic faults is the accommodation of shearing imposed by parallel faults or parallel and predominantly parallel moving boundaries of stronger rocks. The process may perhaps be best illustrated by a row of books on a shelf with a plank resting on top (Fig. I.2-53). The plank may be shifted by tilting the books. The rotation of the books is accompanied by slip between the individual books, which is *antithetic to the externally imposed shear displacement* (movement of the plank). The contacts between the individual books therefore represent parallel and collectively rotating antithetic faults, the displacement of which is caused only by the rotational movement. A finite rotation of the books is accom-

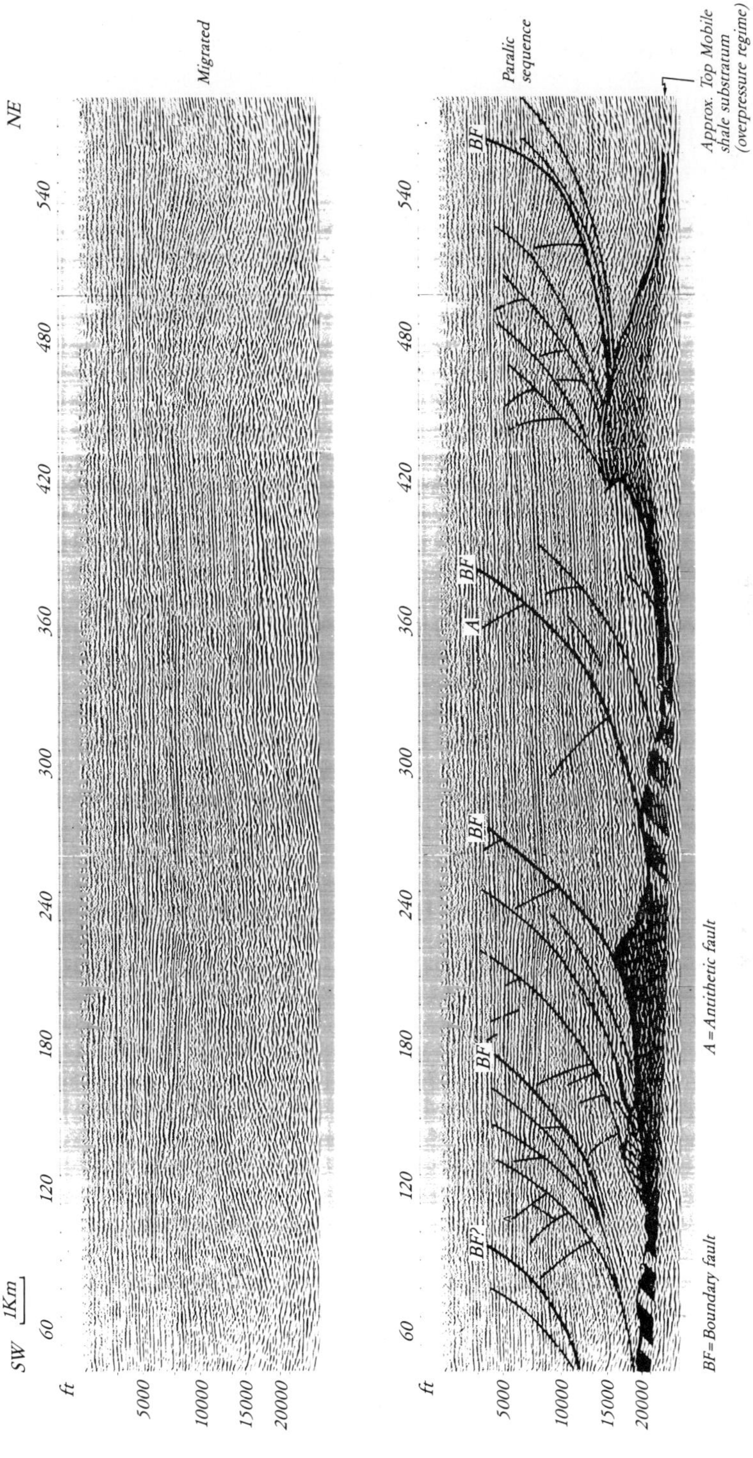

Fig. I.2-52 'FAMILIES' OF GROWTH FAULTS, NIGER DELTA (Greater Forcados area)

panied by a change in the height and length of the row of books: 'domino'-style tilting of the books is associated with some extension of the row and decrease in height, while steepening of tilted books is associated with shortening of the row and increase in height. Accordingly, the shearing deformation accommodated by a set of rotating parallel antithetic faults is that of the type produced between parallel plates which, in addition to movement parallel to their planes, are free to move perpendicularly to their planes. Hence, in contrast with simple shearing *sensu stricto*, there will be some strain perpendicular to the driving plates. In principle, the situation is that of a 'direct' shear test in which only the plate-parallel displacement component of the driving plate and the normal stress acting upon the plates are prescribed. For simplicity of presentation, in the following we shall use the term 'shear' or 'direct shear' to convey this deformation plan.

The bookshelf analogy falls short in one important aspect: it would seem to suggest that planes of weakness, faults or cleavage planes have to exist prior to the onset of rotation, as is indeed required for the formation of kink bands. This is, however, not required for the 'bookshelf'-type process considered here, in which sets of parallel antithetic faults may form and rotate 'in one go' – even in the absence of any strength anisotropy. Since displacement along the faults is then entirely the result of fault rotation, it may often remain small and difficult to detect on seismic records.

There is abundant evidence (Roering, 1968; Beach, 1977) of small-scale arrays of rotated mineral-filled fractures occurring as localised zones of deformation in the interior of an otherwise undeformed area. Similar en-échelon sets of shear fractures have also been observed to precede and localise the formation of larger shear fractures in some biaxial loading experiments on samples of pyrophyllite lava (Sobolev et al., 1978).

Sets of rotating en-échelon shears may be confined to individual beds and associated with interstratal shearing in flexural slip folding, as described by Wilson (1961) and illustrated in Fig. I.2-54. The mechanism, which implies some stretching of beds, may operate near the crestal extension region. In shears of this type it is often difficult to decide whether the fractures have been initiated as tension fractures (i.e. parallel to the direction of the maximum compressive stress) or as shear fractures.

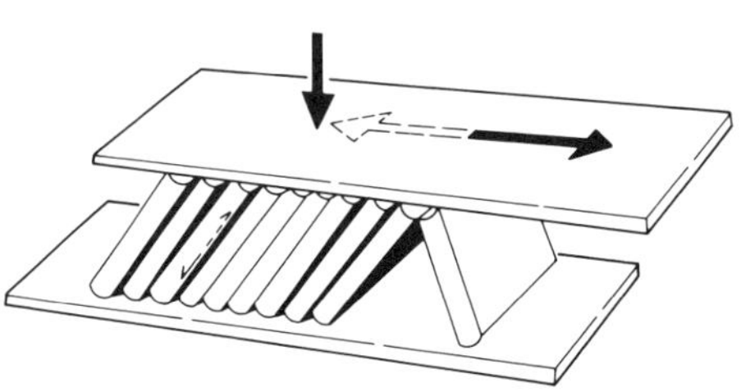

The bookshelf analogy illustrates the two-fold kinematic function of rotating parallel faults: accommodation of an externally imposed shearing and accommodation of crustal extension or shortening. The direction of slip between the individual books, which accompanies the rotation of the books, is opposite to the externally imposed overall shear displacement and represents the action of rotating antithetic faults.

Fig. I.2-53 BOOKSHELF KINEMATICS

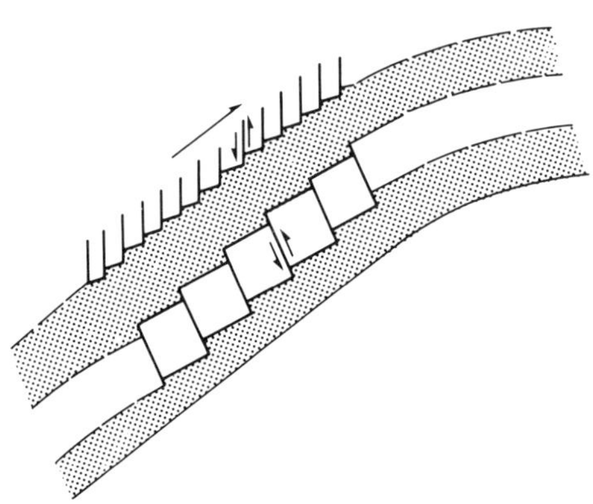

Fig. I.2-54
'BOOKSHELF' MECHANISM ASSOCIATED WITH UPWARD SLIP OF BEDS IN COMPRESSIONAL FOLDING (The mechanism, implying some stretching of beds, may operate near the crestal extension region.)

An example of rotating faults confined between parallel normal faults is shown in Fig. I.2-55. Deformation and tilting of the soft marl and brittle dolomite beds provide clear evidence of the anticlockwise rotation of the cross faults. It seems reasonable to assume that the cross faults have been initiated as conjugates of the steeply dipping normal faults.

Large-scale tectonic examples of rotating faults may be found in tectonic wrench zones, as for instance in the left-lateral Icotea fault zone of West Venezuela (Fig. I.2-56A). The map in Fig. I.2-56B shows a segment of this zone in the Lake Maracaibo district where two sub-parallel branches of the Icotea fault are linked by rotated cross faults with a predominant component of dip slip. The western margin of the wrench zone segment is formed by another set of rotated antithetic faults, the rotated rock slices still being 'anchored' at one side in undeformed rock.

A second, well-documented example of a wrench zone is a coal basin along the great tectonic furrow which traverses the Massif Central of France (Fig. I.2-57A). The very narrow basin (Fig. I.2-57B) is about 1 km deep and bounded to the NW by a major, steeply dipping fault and to the NE by subvertical normal contact with crystalline basement rock. In addition, there is evidence of detachment horizons at depth. After a period of transverse compression, which caused strong folding, the carboniferous basin fill was horizontally sheared in the left-lateral sense by the relative movement of the lateral boundaries – all in all a tectonic setting strongly reminiscent of conventional shear testing. In its very latest phase the imposed shear was accommodated by an en-échelon set of rotating cross faults which indeed show small strike-slip displacement (Letourneur, 1953). Further examples of rotated antithetic wrench faults in the field and in model experiments will be discussed in Chapter I.4.

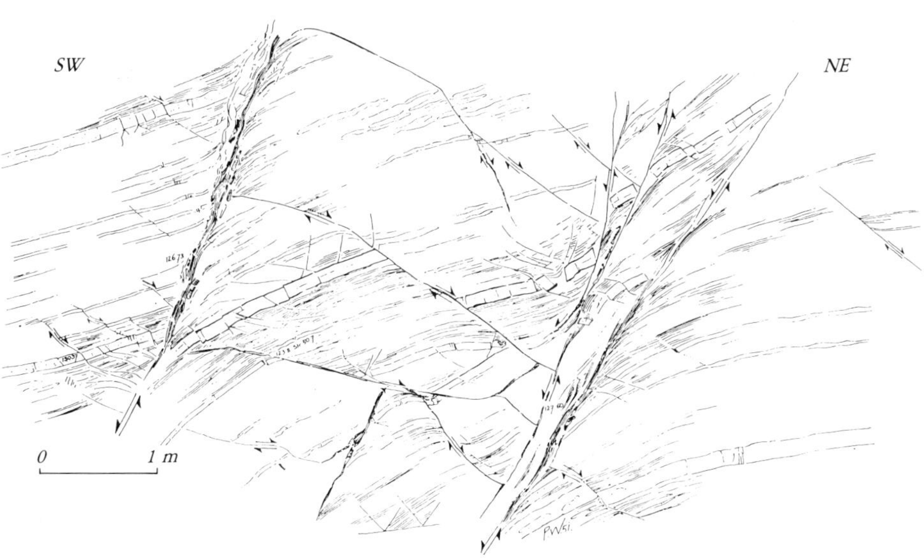

Fig. I.2-55 NORMAL FAULTS WITH ANTICLOCKWISE-ROTATED CROSS FAULTS IN SOFT MARL WITH INTERCALATED BRITTLE DOLOMITE BEDS; HALLWANGEN, NORTHERN BLACK FOREST, GERMANY.
(Courtesy of Prof. P. Wurster, Bonn.)

A Location map

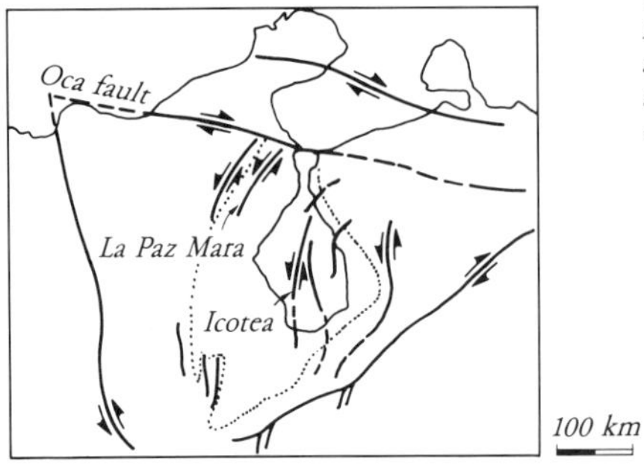

B Map of fault zone segment with sets of rotated parallel faults (with predominant dip-slip) associated with shearing along two subparallel branches of the Icotea fault.

Fig. I.2-56
ICOTEA FAULT WITH CROSS FAULTS, LAKE MARACAIBO AREA, W. VENEZUELA
(Courtesy of Maraven and Shell Internationale Petroleum Maatschappij 'Venezuela Study Team')

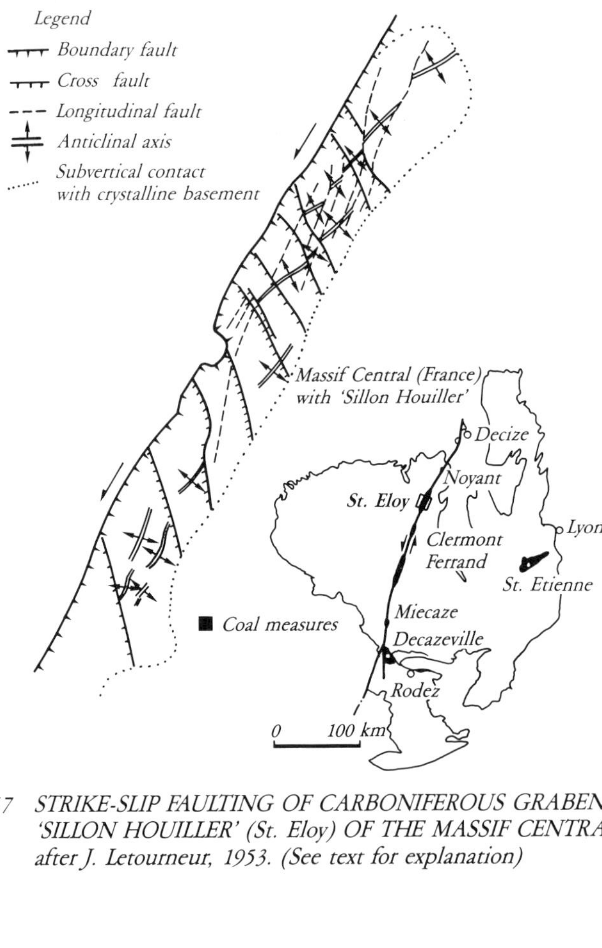

Fig. I.2-57 STRIKE-SLIP FAULTING OF CARBONIFEROUS GRABEN FILL IN THE 'SILLON HOUILLER' (St. Eloy) OF THE MASSIF CENTRAL, FRANCE, after J. Letourneur, 1953. (See text for explanation)

Considering these examples of 'bookshelf'-type accommodation of tectonic shearing, one may wonder why the deformation process, schematically depicted as simple shearing in Fig. I.2-58A, is not accomplished by faulting parallel to the imposed shear direction (Fig. I.2-58B), but by the more complex kinematics of rotating antithetic faults in en-échelon position (Fig. I.2-58C). This question is to be answered next (Mandl, 1987c).

Let us consider an isotropic, cohesive-frictional and as yet unfaulted rock and impose a shear displacement on it, for example, in the horizontal direction. At the same time we apply a certain normal stress $\sigma'_{\perp}$ perpendicular to the (horizontal) shear direction as shown in the insert of Fig. I.2-59A. What is the 'driving' shear stress τ that acts in the direction of the externally imposed shear displacement? The Mohr diagram of Fig. I.2-59A represents the stress state required to produce slips parallel to the external shear direction; $\sigma'_{\perp}$ and $\tau^{(1)}$ are the normal and tangential stresses acting on these slip planes. The stress circles in Figs. I.2-59B and C represent limiting states that have in common that the normal stress $\sigma'_{\perp}$ on horizontal planes is the same as in the case of Fig. I.2-59A, but that, in contrast with this case, the horizontal normal stress $\sigma'_{\parallel}$ is *smaller* than $\sigma'_{\perp}$. Using the concept of the stress pole (Section II.4.2), one can easily determine the orientation of potential slip planes associated with these limit circles. In Fig. I.2-59B $\sigma'_{\parallel}$ has been reduced to $\sigma'^{(2)}_{\parallel}$ to allow slip along vertical planes, i.e. in a direction perpendicular to the direction of external shear. A succession of infinitesimal slips along such planes in combination with infinitesimal rotations would accommodate horizontal simple shearing *sensu stricto*. A further reduction of $\sigma'_{\parallel}$ in Fig. I.2-59C would allow slip along planes inclined as indicated. Slip along these planes, combined with finite clockwise rotation, would accommodate a direct horizontal shear, accompanied by some thinning of the shear zone in the vertical direction.

Although the normal stress $\sigma'_{\perp}$ on horizontal planes is the same in all three cases, the shear stresses on these planes (indicated by vertical arrows in the Mohr diagrams) are different. Somewhat surprisingly, the shear stress $\tau^{(2)}$ that is required to set in operation an ideal (i.e. simple shear accommodating) 'bookshelf' mechanism with reiterated infinitesimal sliding perpendicular to the direction of imposed shear is *smaller* than the shear stress $\tau^{(1)}$ required to produce the 'normal' 'card-deck'-type shearing. Still smaller shear stresses $\tau^{(3)}$

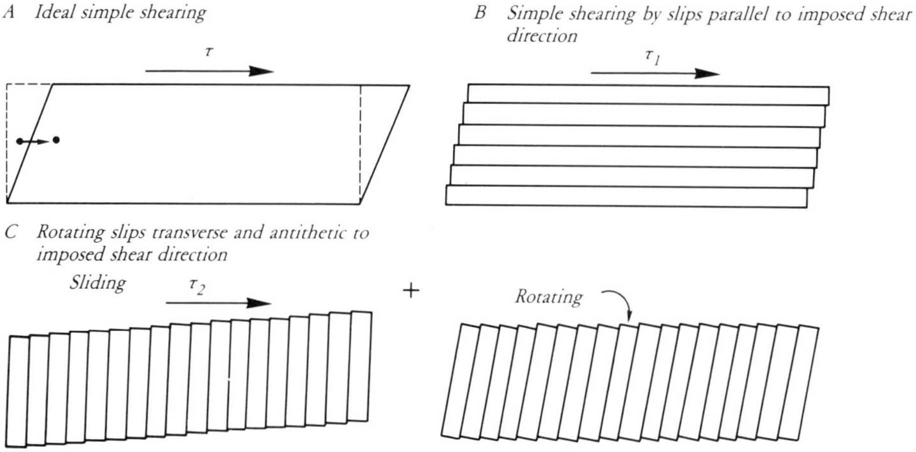

Fig. I.2-58 SHEAR ACCOMMODATION BY DISCRETE SLIDING

are sufficient to drive direct shearing when the activated slip planes are inclined as shown in Fig. I.2-59C.

It should be emphasised that antithetic slip planes rotating in the *'domino' style* shown in the diagram of Fig. I.2-59C can form only when the effective normal stress $\sigma'_{\parallel}$ parallel to the externally imposed general shear direction (right-lateral in the figure) is considerably smaller than the effective normal stress $\sigma'_{\perp}$ perpendicular to that direction. This situation can also be obtained by drastically increasing the load $\sigma'_{\perp}$ while leaving $\sigma'_{\parallel}$ unchanged, as is illustrated in Fig. I.2-60. Again the rotating antithetics are initiated at a high angle with respect to the general shear direction.

It is of note that antithetic slips of this type, termed X-shears, have been observed in high-pressure shear tests on quartz gouges (Logan et al., 1979) and in similar tests on layers of initially intact Indiana limestone (Bartlett et al., 1981) 23).

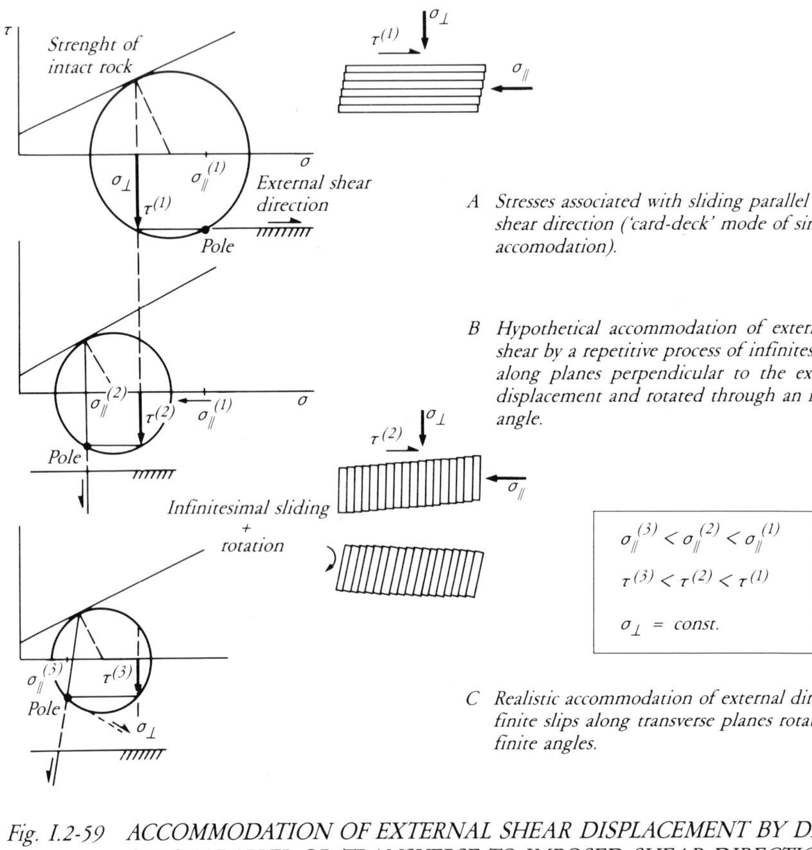

A Stresses associated with sliding parallel to external shear direction ('card-deck' mode of simple shear accomodation).

B Hypothetical accommodation of external (simple) shear by a repetitive process of infinitesimal sliding along planes perpendicular to the external shear displacement and rotated through an infinitesimal angle.

C Realistic accommodation of external direct shear by finite slips along transverse planes rotated through finite angles.

Fig. I.2-59 *ACCOMMODATION OF EXTERNAL SHEAR DISPLACEMENT BY DISCRETE SLIPS PARALLEL OR TRANSVERSE TO IMPOSED SHEAR DIRECTION.*

MOHR DIAGRAMS DEMONSTRATE THE INTERDEPENDENCE OF NORMAL AND SHEAR STRESS COMPONENTS.
At fixed load perpendicular to the external shear direction, the 'driving' shear stress – represented by the vertical arrow in the Mohr circles and acting in the direction of the external shear – decreases concomitantly with a decrease in the normal stress that acts in the same direction.

In 'domino'-style rotation of antithetics the thickness of the shear band will be reduced and, at the same time, the length of the band will increase as, a glance at Fig. I.2-2 may show. Without going into the kinematical details (Lovering, 1928), it is nevertheless obvious that an effective operation of the bookshelf mechanism in 'domino' style requires that the rotational stretching is somehow accommodated by slip of the rotating slices along band-parallel planes of weakness (bedding planes or faults) or by stretching of the adjacent rock parallel to the direction of the imposed boundary shear. Such an extension may be imposed by up- or down-bending of the sheared strata or by the extensional thinning of interlayers

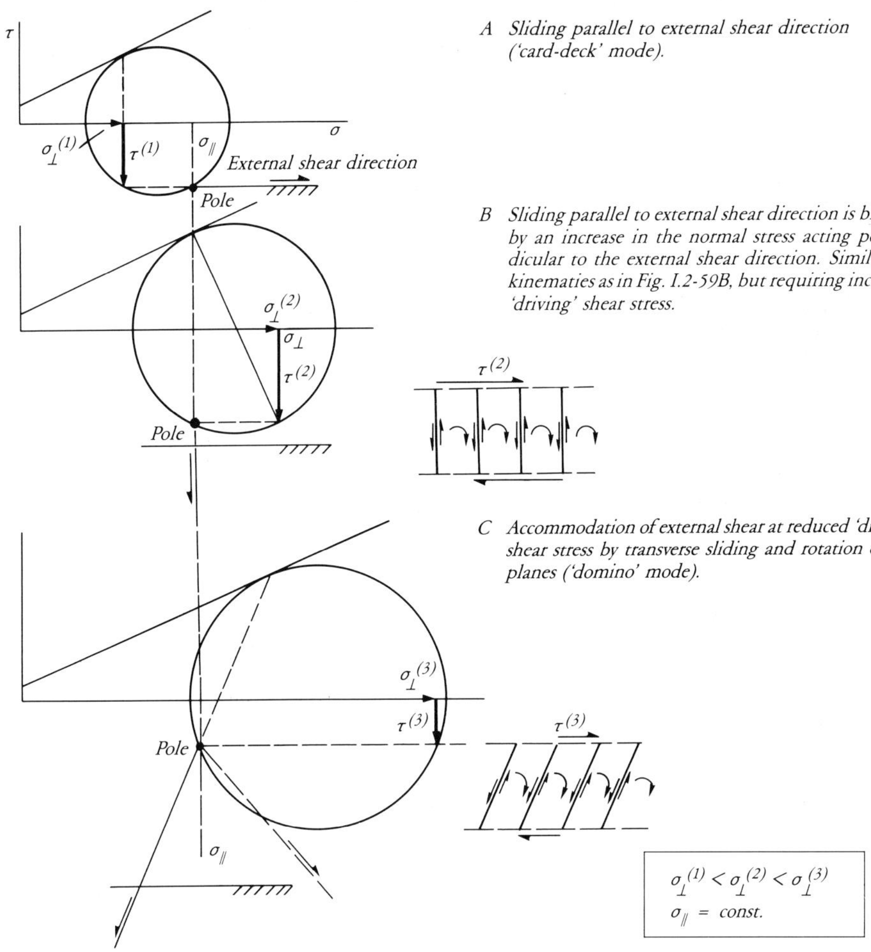

Fig. I.2-60 ACCOMMODATION OF EXTERNAL SHEAR DISPLACEMENT BY DISCRETE SLIPS PARALLEL OR TRANSVERSE TO IMPOSED SHEAR DIRECTION (Similar to Fig. I.2-59)

Comparison of 'driving' shear stresses for different values of the normal stress perpendicular to the external shear direction; normal stress parallel to that direction is kept constant.

and substrata. A favourable combination of extension and boundary slip may exist, for instance, near the hinges of a growing anticline with bedding plane slip, as was assumed in the example of Fig. I.2-54.

When such accommodation mechanisms are not available, parallel antithetics of the type shown in Figs. I.2-59C and I.2-60C may still be initiated under the applied external shear, but rotation will be stopped almost immediately by the concomitant rise in the band-parallel normal stress, and shear displacement along these fractures will remain very small. The process will therefore result in the formation of sets of parallel *shear joints* rather than faults.

Far more important than 'domino'-style rotation of antithetics in accommodating tectonic shearing is the *'dilational' mode*, in which parallel antithetics rotate into a steeper position with respect to the general shear direction. As sketched in Fig. I.2-2, this operation tends to increase the thickness of the band of rotating rock slices and to decrease its length. When the rotating slices are still connected to the surrounding rock, as indicated in the figure, and are not allowed to slide along the boundary of the shear band, the rotating fractures tend to open up.

The mechanical conditions for the initiation of this bookshelf mode are explained in Fig. I.2-61. In contrast with the 'domino' mode, incipience of the 'dilational' bookshelf mechanism requires a $\sigma'_{\parallel}$ that greatly exceeds $\sigma'_{\perp}$. This is accomplished in the construction of the figure by increasing $\sigma'_{\parallel}$ and leaving $\sigma'_{\perp}$ unchanged. Again a limit state of stress (Fig. I.2-61B) is established where the 'driving' external shear stress τ^*, which acts parallel to the direction of the imposed shear displacement, is smaller than the shear stress τ that would

A Stresses associated with sliding parallel to external shear direction ('card-deck' mode).

B Reduction of 'driving' shear stress by transverse sliding and rotation of slip planes under increased normal stress acting parallel to external shear direction; normal stress perpendicular to that direction is kept constant. Rotation of slip planes is characteristically accompanied by thickening of the shear band ('dilational' mode).

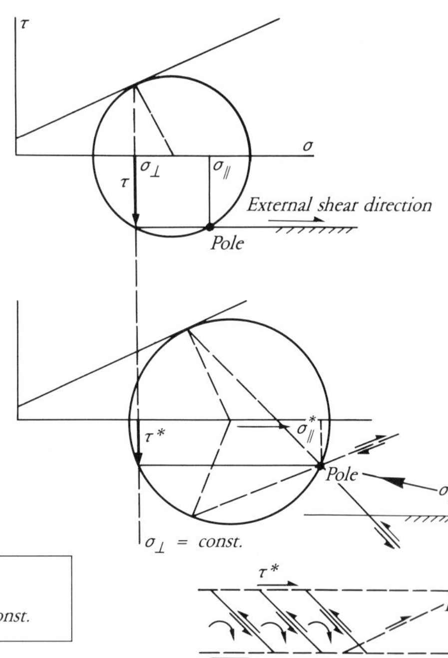

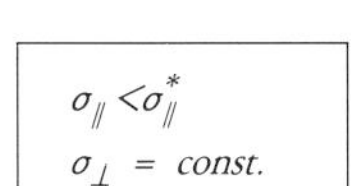

Fig. I.2-61 ACCOMMODATION OF EXTERNAL SHEAR DISPLACEMENT BY DISCRETE SLIPS PARALLEL OR TRANSVERSE TO IMPOSED SHEAR DIRECTION

be required in 'card deck'-mode shearing. The reader may easily verify that a similar result can be obtained by considerably decreasing $\sigma'_{\perp}$ while keeping $\sigma'_{\parallel}$ constant. The 'dilational' bookshelf mode can therefore form only under direct shear when an extra compression is provided parallel to the direction of external shear, or an extra extension perpendicular to it.

The cross faults that interlink the parallel normal faults in Fig. I.2-55 are an example of a dilational bookshelf operation. The sigmoidal deformation of the rotated cross faults may have aided in accommodating the rotational dilatation of the cross-faulted segment. Tectonically far more important than these cross faults are those that operate as dilational bookshelf mechanisms in wrench zones, as the examples of Figs. I.2-56 and 57 illustrate. We shall deal with this wrenching mechanism in more detail in Chapter I.4.

It should be added here that cross faults rotating between subparallel main faults need not be conjugate to these. In fact, the stress states that lead to the formation of rotating antithetics in the Figs. I.2-59C, 60C and 61B would also allow the formation of a conjugate set of shears (indicated in the figures by stippled lines) which are transverse to the direction of external shear. The sense of displacement on these conjugate shears, however, is synthetic with the direction of external shear, and it is important to note that active faults of this conjugate type cannot co-rotate with the external shear, since this would reverse the direction of motion and shear stress along these faults, and rotate the σ_I-axis into a direction that is incompatible with the direction of the external shear stress – the very driving agency of the 'bookshelf' operation.

Further, it should be emphasised that incipience and operation of *rotating* antithetics require quite considerable differences between the normal stresses $\sigma'_{\perp}$ and $\sigma'_{\parallel}$, and these need not be attained in arbitrary limit states of simple or direct tectonic shearing. In such cases antithetic faults are *non-rotational*. This may be illustrated by considering the simple shearing (*sensu stricto*) imposed upon a rock mass by the horizontal movement of parallel 'walls' (e.g. fault planes or contacts with stronger rocks). We assume that the rock mass has a horizontal surface and rests on a horizontal base of very low shear resistance. We also assume that prior to shearing the horizontal normal stresses are equal and that the rock responds to the imposed shearing in a predominantly linear elastic (i.e. very brittle) manner. The simple shearing will produce a difference between the greatest and smallest horizontal principal stresses, which increases until the limit state is reached, without changing the mean stress $(\sigma'_I + \sigma'_{III})/2$. This is expressed in the Mohr diagram of Fig. I.2-62 as concentric growth of the stress circle. In the limit state the σ_I-axis makes an angle of 45° with the imposed shear

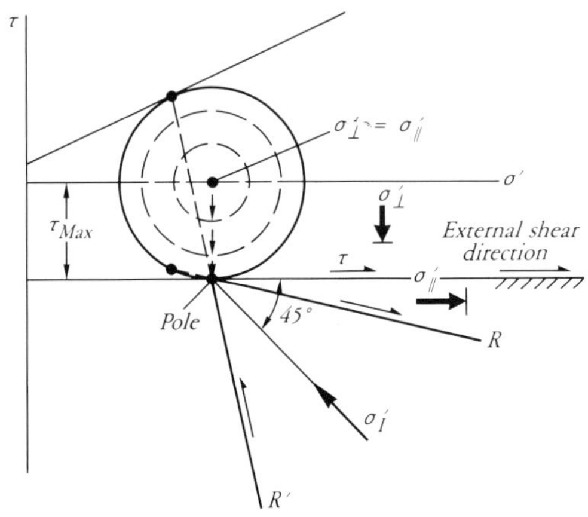

Fig. I.2-62
STRESS DIAGRAM OF SIMPLE SHEAR IMPOSED ON HORIZONTAL LAYER IN STATE OF HORIZONTAL STRESS ISOTROPY.
Mohr circle grows concentrically as external shear stress (τ) increases towards limit value τ_{Max}. Pole construction of conjugate horizontal slip directions R and R' relative to external shear direction.

direction, since it coincides with the direction of greatest shortening, which in small elastic simple shearing makes an angle of 45° with the imposed shear direction. The associated directions of potential conjugate slips R and R′, which are commonly referred to as 'Riedel' shears (after W. Riedel (1924) who studied the formation of such conjugate shears in thin, wet clay cakes on a 'shear table'), are both transverse to the imposed shear direction. At this limit state of stress the antithetics R′ do not constitute a bookshelf mechanism – which would have to be of the dilational type – since rotation would not result in a reduction of the 'driving' shear stress. In fact, it would even decrease $\sigma'_{\parallel}$ and increase $\sigma'_{\perp}$ – just the opposite of what would be required for the operation of the dilational bookshelf mode. However, this does not exclude the possibility that *later* changes in the limit state of stress may allow the same R′-shears to operate in a dilational bookshelf mode. We shall encounter this phenomenon in the structural development of wrench zones (Section I.4.3).

In summary, our analysis has asserted that simple or, more generally, direct tectonic shearing can be brought about by the formation and collective rotation of parallel antithetic faults, at an external shear stress lower than that required for faulting parallel to the general shear direction. The crucial mechanical factor in this mechanism is the component of normal stress that acts in the direction of the externally imposed displacement component parallel to the 'driving' boundary 'plates'. This stress component can affect the magnitude of the 'driving' shear stress – in contrast with the usual assumption which attributes such an effect only to the effective normal stress that acts upon the boundary plates.

7. THE TECTONIC SIGNIFICANCE OF SHEAR SOFTENING

In the enumeration of the basic phenomena of tectonic faulting in Section I.1.2, we mentioned the process of shear softening, which accompanies the formation of a tectonic fault and involves some reduction of shear strength, resulting from structural changes caused by inelastic shearing of the rock. On the 'micro'-scale the inelastic softening mechanisms may have a predominantly brittle character or may be associated with ductile processes, some of which are still insufficiently understood or not yet known even (see 'Shear Softening', Section II.7.7). Here we shall not be concerned with the micro-mechanical and physico-chemical aspects of shear softening, but with its tectonic implications, the importance of which is not always recognised.

The significance of shear softening in processes of tectonic faulting is twofold: firstly, it creates conditions for the spontaneous formation of a fault by *localising the shear deformation* and, secondly, it makes an active fault into a *'plane' of weakness*.

As to the first statement, it should be mentioned that many rock-mechanical experiments (see, for example, Hallbauer et al., 1973; Sobolev et al., 1978) have established that faults do not form by the propagation of a single flaw or small crack, but rather form inside some diffuse zone of tectonically strained material, as may also be inferred from the study of faults in the field. As deformation increases inside such a zone, at a certain stage the resistance of the material will begin to decrease in some place – as a results of inhomogeneity of the material or the stress field – and shear deformation will start concentrating there by causing further localised softening of the material [24)].

The second role played by shear softening in tectonic faulting lies in the fact that shear displacement along a newly formed fault is accompanied by a reduction in shear strength of the material in the shear band from the peak value to a residual value, as shown, for instance, by the curves of Fig. I.1-5. In the brittle regime, full reduction in shear strength may be achieved by a shear displacement of only a few millimetres or centimetres, resulting in a zone or plane of weakness embedded in material that still possesses peak strength. For the ductile regime there is, unfortunately, very little known about the – presumably large – amount of shear strain required to produce, at a given rate of shearing, some structural weakening of the material.

The important consequence of fault zone weakening is that the fault will remain active even when – within certain limits (see Section II.4.3, Example No. 3) – the stress field and its principal directions may change and the fault no longer 'fits' into the changed field of stress. This may be an explanation for the surprising fact that, in many tectonic processes in the brittle crust which involved faulting as the main deformation mechanism, the operation of a few major faults was sufficient to accommodate very large rock-mass deformations. In view of the stress control of faulting this is indeed astonishing, since one can hardly think of a tectonic process that does not impose changes on the original stress field. In the absence of shear softening one would, therefore, rather have to expect a successive generation of new fault sets with minor displacement and differing geometries, and the inactivation of their predecessors.

Although shear softening may be considered the main reason for the *localisation* and relative permanence of tectonic faults, *other factors* may contribute too: faults may form in pre-existing weak zones of bedding planes, or the localisation may be controlled by geometrical factors. We consider a simple illustration for the latter in Fig. I.2-63, in which an homogeneous brittle overburden rests horizontally on a ductile substratum without shear stresses acting along the overburden base. Therefore, the same horizontal force must act on every vertical cross-section. We also assume an elongated depression of the free overburden surface, as indicated in Fig. I.2-63A. While the mean overburden stress at the position A of the smallest vertical cross-section is smaller than that at a position B outside the depression region, the mean horizontal normal stress at A is greater than that at B, as follows directly from the simple force balance in the figure. The mean stresses at the two locations may therefore be represented by Mohr circles of the type shown as solid circles in Fig. I.2-63B. Next, we consider this original state of stress being disturbed by a horizontal extension of

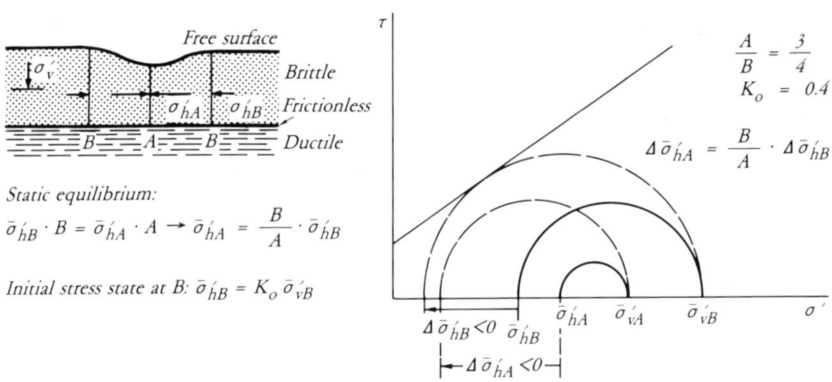

A Schematic cross-section with effective mean normal stresses before extension.

B Mohr diagram of mean stresses before (solid circles) and after (dotted circles) extension indicating that limit state is only attained outside depression region.

Fig. I.2-63 EFFECT OF BOUNDARY GEOMETRY ON FAULT LOCALISATION. Brittle overburden with cylindrical surface depression subject to horizontal extension.

the brittle overburden. We assume that the rate of extension is small enough so that shear stresses induced at the overburden base by adjustment flow of the ductile substratum can be neglected. Static equilibrium of the horizontal force components then requires a greater reduction in horizontal mean stress at A than at B. In spite of this, the active limit state will not be reached in the depression regime but outside it, as indicated by the dotted circles in the Mohr diagram. Hence, somewhat surprisingly, the first graben faults in this idealised situation would develop outside the depression region. In contrast, horizontal compression would induce the (passive) limit state and associated thrusting inside the depression region, as can easily be shown by the same line of argument.

Returning to shear softening of fault zone material, we have to mention next that the reduction in shear strength and the concomitant drop in the shear stress on the fault plane require a *rearrangement of the stress field* to maintain static equilibrium. No general rules can be given for these stress changes, since they are also controlled by the reaction of adjacent rock masses. Without specific conditions being considered, some of which will be dealt with in the next section, it can only be stated that the new stress state is represented in the Mohr diagram (Fig. I.2-64) by any of the infinite set of stress circles that pass through the common stress point P*, which represents the new tangential and normal stresses acting on the fault plane, and that do not cross the limit line which still represents the shear strength of the intact material outside the fault. Two limit circles (active and passive) can be drawn through P* in Fig. I.2-64 but, normally, the stress state attained directly after reduction of the shear strength will be *sub-critical*, as indicated, for instance, by the dotted circles. In other words, shear softening of the fault material has the striking implication that the material outside the fault zone proper 'drops back' into a sub-critical (elastic) state and fault blocks will therefore tend to move as *quasi-rigid* masses.

Further, shear softening makes it possible for the fault-parallel effective normal stress $\sigma'_{\parallel}$ to change discontinuously across the fault plane, while a jump in the shear stress τ and the (total) normal stress $\sigma_{\perp}$ is prohibited by the conditions of static equilibrium (Section II.2.1, Example No. 1; Section II.7.3, Example No. 3). A 'jump' in $\sigma'_{\parallel}$ is kinematically possible since the corresponding difference in fault-parallel shortening or extension of the adjacent fault blocks may be accommodated by slip along the active fault. The stress jump is limited

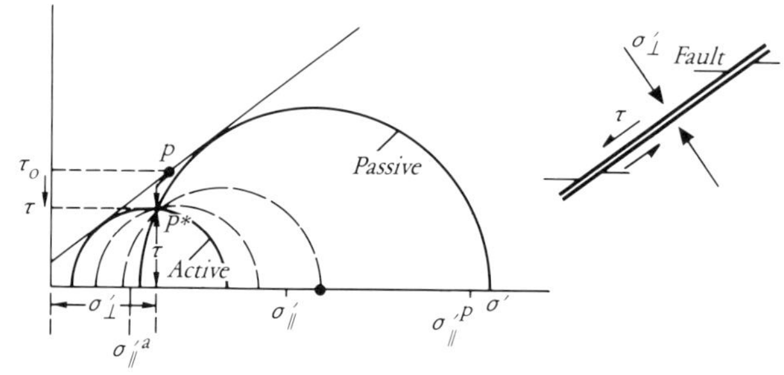

Fig. I.2-64 MOHR DIAGRAM OF POSSIBLE STRESS CHANGES IN THE VICINITY OF SHEAR SOFTENING FAULT.

Dotted stress circles represent infinity of possible sub-critical stress states in adjacent fault blocks. Solid circles represent the two possible limit states.

Fault-parallel effective normal stress $\sigma'_{\parallel}$ may discontinuously change across fault and attain values $\sigma'^{a}_{\parallel} \leqslant \sigma'_{\parallel} \leqslant \sigma'^{P}_{\parallel}$.

by the values $\sigma_{\parallel}'^{a}$ and $\sigma_{\parallel}'^{P}$ associated with the two limit circles that are drawn through P* in Fig. I.2-64 tangential to the limit line of the intact material. We may conclude that a critical and a sub-critical stress state or, in the extreme, the active and the passive limit state, may coexist on the opposite sides of a shear-softened fault.

We may recall that discontinuous changes in $\sigma_{\parallel}'$ may also be associated with an abrupt change in rock mechanical properties or pore pressure (Section I.2.4; Figs. I.2-16 and 18). But shear softening of the fault material allows such stress jumps to take place even when the material properties and the pore pressure are the same on both sides of the fault. This offers an explanation for the frequently observed phenomenon of adjacent fault blocks differing quite drastically in the degree of deformation they have suffered, even when the blocks do not differ in lithology or pore pressure. In particular, one block may have remained comparatively undisturbed, whereas the other is severely faulted, folded or fractured by extension cracks. The translatory movements of huge thrust sheets of competent rocks over large distances (Chapter I.5), or the gravitational gliding of sediment sheets on overpressured delta slopes (Chapter I.3), are impressive examples of 'block gliding' on major fault planes.

If shearing along a fault inside a rock mass of uniform strength did not reduce the shear strength of the fault material, only one limit circle could pass the common stress point in Fig. I.2-64, which would then lie on the limit line (P*⇒ P), and no stress jump across the fault plane could arise. This is in complete accord with the concepts of the plasticity theory for rigid/plastic and elastic/plastic materials (represented by the analogue models of Figs. I.2-7A,B), which do not allow slip and stress discontinuities to occur simultaneously on a slip surface inside the plastic material [25]). Hence, if a Coulomb-type slip surface is to represent a fault that intersects a rock mass in the limit state and, at the same time, carries a discontinuity in the fault-parallel normal stress, the slip surface has to be considered as a thin zone of somewhat weaker material.

A change in $\sigma_{\parallel}'$ across a fault plane is associated with a change in the orientation of the principal stress σ_I'. Again, as in the case of abrupt changes in shear strength or pore pressure across a bedding plane (Figs. I.2-16 and 18C), the pole method (explained in Section II.4.2) may be used to determine in a simple graphical way how the direction of σ_I' with respect to a fault plane varies in dependence on $\sigma_{\parallel}'$. This is illustrated in Fig. 1.2-65, which also shows that the extreme variation is obtained for the active and passive limit states.

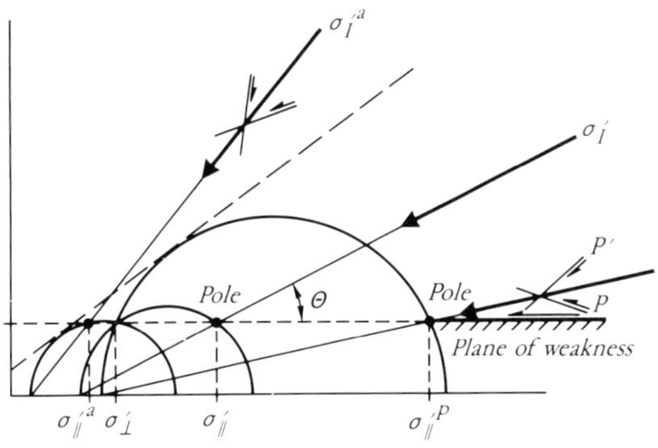

Fig. I.2-65 ORIENTATION OF σ_I' WITH REFERENCE TO SHEAR-SOFTENED FAULT PLANE. A jump in the fault-parallel normal stress ($\sigma_{\parallel}'$) across the fault plane is accompanied by a discontinous change in σ_I'-direction.

Mathematical Modelling of fault formation as a concentration of shearing in narrow zones has to take shear softening into account. It therefore has to employ the mechanical theory of an 'elastic/frictional plastic material with strain hardening and softening', the analogue model of which was shown in Fig. I.1-7C. The theory is presented in Section II.7.7, together with the results of a numerical application in modelling the development of faults in the overburden above a vertically uplifted basement block (Example No. 7, Differential uplift). The calculations were carried out with a finite-element program developed by J.N. Thomas and J.V. Walters [26)], which has been applied successfully to model the incipience of various fault structures. We shall conclude this section by presenting one of these computer simulations, which impressively demonstrates the role of shear softening in faulting.

We consider a horizontal layer of mechanically homogeneous and isotropic rock resting on a base that is extensible in the horizontal direction, but does not deform vertically. The rock layer and base are subject to a gradually increasing, uniform horizontal strain. The configuration of the system and the imposed boundary displacements are indicated in Fig. I.2-66A. The horizontal normal stress in the undisturbed state is assumed to be 60% of the overburden stress. The linear elastic behaviour of the rock is specified by a Young's modulus of 4.10^4 MPa and a Poisson's ratio of 0.25. The strain hardening and softening of the rock, which starts when the elastic limit is reached, is characterised by an initial friction angle of 30° which rises with increasing plastic strain to a peak value of 45°, decreasing thereafter in the softening range – somewhat arbitrarily chosen – to its initial value. The cohesion is assumed to decrease monotonically from an initial value of 5 MPa to zero. (The whole process of plastic hardening and softening is of the type depicted in Fig. II.7-45 of Section II.7.7.) Since some type of non-uniformity in strength, configuration or boundary conditions is needed to trigger the first fault in the uniformly strained layer, a small vertical incision is added in the middle of the free surface. (The orientation of the triggered fault was found to be insensitive to the type of triggering non-uniformity.)

Because of the symmetry of the configuration, the finite-element analysis is performed for the right-hand side of the rock layer only. The strain is built up in small increments. Fig. I.2-67A shows part of the layer covered with the net of eight-noded isoparametric elements with four integration points, with a finer mesh near the incision. The figure shows the extent of the plastic region after a few strain increments. Blank elements are still in the elastic state, small circles represent plastic hardening, while crosses indicate that the peak stress has been reached and plastic softening is taking place. The plastic zone is seen to start from the surface [27)]. Figure I.2-67B shows the plastic zone at a somewhat later stage when a thin softening band has formed from the surface incision downward, while its surrounding has 'dropped back' into an elastic state. After some further straining (Fig. I.2-67C) the softened band has reached the base and a second softening zone has begun to develop. Figure I.2-67D shows the state at which the first two shear bands (faults) have been formed. In the deformation state shown in Fig. I.2-66B all shear bands have been developed, including bands of both conjugate types [28)]. In Fig. I.2-66C the same shear bands also show in the contour plot of the field of total vertical displacement components, in which the bundling of parallel contours indicates the simple shear character of the deformation bands. Sequences of development, spacing and dip angles of the 'faults' are indicated in Fig. I.2-66D.

The computation clearly demonstrates how the shear strength reduction inside a shear band is accompanied by a drop in differential stress outside the band, and this causes the surrounding material to return to an elastic state of stress. Inside this zone of elastic unloading a second shear band cannot form. Hence, it is evident that shear softening is the most important material property in determining the ***spacing of parallel faults***. The more brittle the rock, the greater the reduction in strength and the more drastic the accompanying unloading will be. However, when the tectonic deformation continues after the formation of a first fault, as in the present case, the continuing straining tends to maintain the plastic limit state. Obviously, as a result of these competing processes, the subcritical, elastic state will be restricted to a certain surrounding area of the first fault. The width of this elastic fault margin will determine the smallest fault spacing possible.

A Configuration of layer with incision at surface to trigger first fault

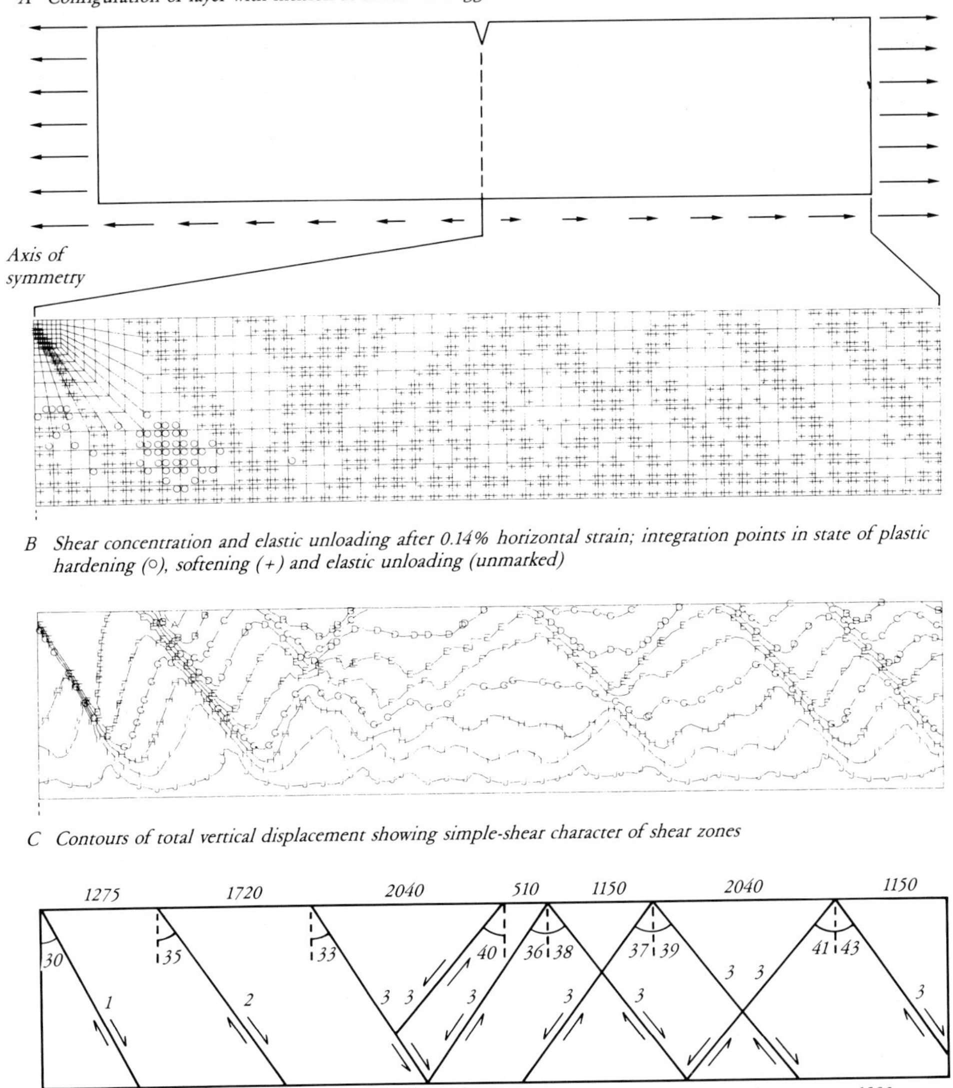

D Sequence of development, spacing (metres) and dip angle of shear bands

Fig. I.2-66 FINITE ELEMENT SIMULATION OF NORMAL FAULTING IN UNIFORM LAYER UNDER UNIFORM HORIZONTAL STRAINING OF ENTIRE BASE (after Witlox, 1986.)

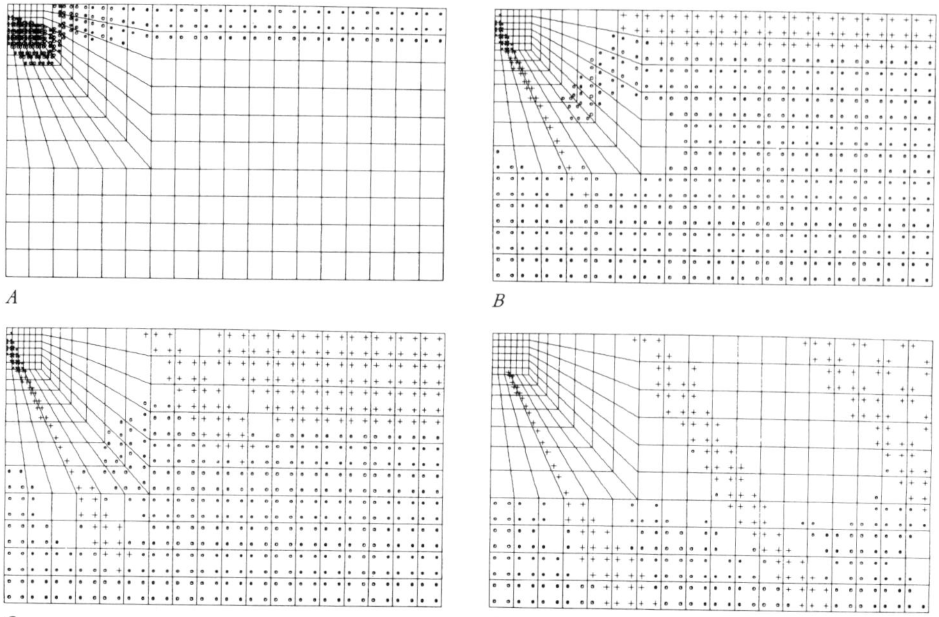

A B C D

Fig. I.2-67 SUCCESSIVE STATES IN THE DEVELOPMENT OF PLASTIC SHEAR ZONES IN THE UNIFORMLY STRETCHED LAYER OF Fig.I.2-66. Enlargement of left-hand part of right half layer.

In view of the competition between unloading by shear softening and reloading by continuing straining, one must also expect that the width of the elastic separation zone of parallel faults not only increases with the relative amount of total strength reduction, usually referred to as the 'brittleness index', but also with the *rate of plastic softening* with respect to the shear strain imposed by the continuing straining of the rock layer (i.e. the drop in shear strength per increment of shear strain). This has been confirmed by numerical parameter analyses (Witlox, 1986).

It should also be noted that in our example the first fault was triggered by an incision at the overburden surface, whereas in reality material or stress heterogeneities may represent several simultaneously active fault 'nuclei', and these may cause a fairly irregular spacing of the first-generation faults. But as straining continues the intervals between these faults will also be faulted and these 'infill' faults will usually be parallel to one of the adjacent first-generation faults and regularly spaced.

Essentially the same results as for the extended layer were also obtained in computer simulations of thrust faulting in uniformly shortened horizontal layers.

Finally, it should be noted that the spacing of parallel faults also depends strongly on geometrical parameters. Numerical parameter analyses by Witlox indicate that an increase in *thickness* of the stretched rock layer – all other parameters remaining unchanged – results in a proportional increase in fault spacing [29).

8. REDISTRIBUTION OF STRESSES, PRECURSORY AND SECONDARY FAULTS

In this section we shall be concerned with ***subsidiary fault structures*** which are caused by changes in the original tectonic stresses during the process of faulting. Such redistribution of stresses during the formation and further operation of tectonic faults is quite a common phenomenon, a special case of which are the stress changes induced by shear softening of the fault material, discussed in the preceding section. Redistribution of stresses in the process of tectonic faulting presents a central problem in tectono-mechanics, since it causes the formation of subsidiary fault structures, the branching of faults, the displacement transfer from one fault to another, the formation of fractured zones, joint sets and other features, which greatly affect the movement of fluids in the crust, the strength of the rock and the way rock responds to mining and rock engineering operations.

Despite their practical importance, until now relatively few of the problems of stress redistribution in faulting have been tackled by mechanical analysis. In fact, most of the theoretical mechanical work has been concentrated on the changes in stress and strain that take place around the leading edge of the slip motion along an existing plane of weakness inside an intact linearly elastic material ('elastic dislocation model'; see, for example, the survey article by J.C. Savage, 1980). Elastic-frictional plastic analyses of faulting processes, which have been carried out in recent years by means of finite element calculations, are restricted to two-dimensional problems of incipient normal and reverse faulting.

In view of this situation, the account we wish to give of the relationship between subsidiary fault structures and the stress changes that take place during the formation or continued operation of a main fault will partly indicate tendencies rather than rigorous solutions of mathematical analyses.

The changes in the tectonic stress field in the process of tectonic faulting may be subdivided into two classes: primary stress changes, i.e. changes that precede and lead up to the actual formation of a fault, and secondary changes that postdate the incipience of a fault. Although there may be continuous transitions between the two types, we shall deal with each one separately.

Primary Stress Changes. When a tectonically undisturbed horizontal rock layer, resting on a 'lubricating' (i.e. practically frictionless) base is subject to a horizontal extension, the principal stresses will have horizontal and vertical directions that remain unchanged during the loading period prior to faulting. Similarly, horizontal compression will proceed without change of principal stress directions, as long as pre-fault folding can be neglected. A typical example is the formation of Anderson-type vertical wrench faults in a horizontally compressed layer on a horizontal detachment horizon, as is shown schematically in Fig. I.2-68, or the gliding of a detached thrust block on a lubricating substratum consisting of, for instance, evaporites or highly overpressured shales. In both cases, the principal stress directions are still the same when the limit state is reached and determine the orientation of the incipient fault. This *stress control* is most common in normal faulting, where the invariable direction of gravity tends to maintain a constant σ_1'-direction. An example of this is also presented in the chapter on rock plasticity (Section II.7.4; Fig. II.7-26A).

In contrast with the development of Anderson-type faults in a uniformly stressed overburden with stationary principal stress directions, the directions of the principal stresses may be changed drastically by competent boundary rocks, moving or at rest, which oppose the tectonic deformation by frictional and normal reaction forces.

Thrust sheet. A simple example of this situation has been mentioned in connection with the problem of fault curvature and concerned the deflection of the σ_1'-direction near the

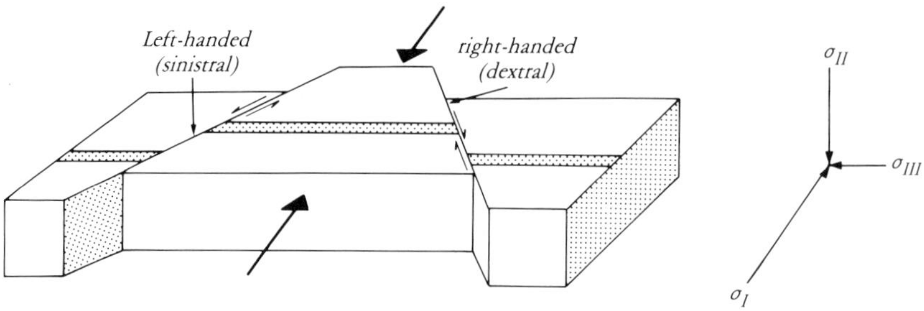

Fig. I.2-68 CONJUGATE WRENCH FAULTS ('Anderson' type)

base of a thrust block by basal friction (Fig. I.2-20). We elaborate on this in Fig. I.2-69 which presents the result of a two-dimensional finite-element analysis of a rectangular thrust sheet of uniform mechanical properties, which rests in frictional-cohesive contact on a thin substratum of homogeneous weaker material attached to rigid basement. The configuration is shown in Fig. I.2-69A together with the assumed material properties and the stress/strain response of the elastic/frictional plastic materials. Figure I.2-69B presents an end section of the finite-element net with the plastic hardening (o) and softening (+) zones which developed in the competent thrust sheet and the weak substratum after an incrementally imposed horizontal end displacement of 7.5 m. Note that the plastic hardening and softening zones in the weaker substratum develop ahead of the corresponding zones in the thrust sheet. The σ_I-directions at this stage are plotted in Fig. I.2-69C, and this clearly shows the deflecting effect of the shear stresses that have developed at the base.

Differential uplift. A typical example of the principal stress directions being disturbed by the displacement of competent boundary rocks is the uplifting or subsidence of a 'rigid' basement block, which was discussed briefly in Section I.2.4. We refer again to Fig. I.2-21A, which illustrates how the pattern of principal stress directions in a uniform overburden is distorted by differential vertical displacement along a vertical basement fault (see also the principal stress pattern in Fig. II.7-32E, which was calculated for the uplift situation in Fig. II.7-31 assuming elastic/frictional plastic material behaviour). The stresses in the overburden are redistributed by the transfer of overburden load from the subsiding to the uplifted basement block via a zone of vertical shearing that extends from the basement fault into the overburden. A small displacement along the basement fault is sufficient to cause on 'arch'-like deformation of the originally vertical σ_I-lines [30]. An important aspect of these stress changes is that they do not usually lead directly to the formation of a main fault cutting through the whole overburden, as might be expected on purely kinematical grounds, but instead produce *precursory faults*. As pointed out earlier (Section I.2.4), these faults should be upward-convex high-angle reverse faults above the 'subsiding' basement block, in accordance with the pattern of potential Coulomb slips in Fig. I.2-21A (see also Figs. II.7-32F and II.7-46).

These precursors can easily be produced in model experiments with dry sand (Fig. II.7-33) or clayey material (Sanford, 1959; Stearns et al., 1978) [31], and have also been observed and studied in detail in high-pressure experiments on layered samples of sedimentary rocks (Logan et al., 1978). It is obvious that a precursory fault can only accommodate a small displacement step in the basement, since the subsidence of the lower block poses a space problem along the precursor fault as indicated in Fig. I.2-70A. In sandbox experiments and in sediments at shallow depth the separation across an active precursor fault is accommodated by strong dilation of the material along the fault. This is demonstrated in Fig. I.2-71 by the plot of accumulated dilation along the precursor faults observed in a sandbox experiment. Obviously, dilatancy of fault zone material can only allow for a short period of activity on a single precursor, after which the activity will be transferred to steeper precursors

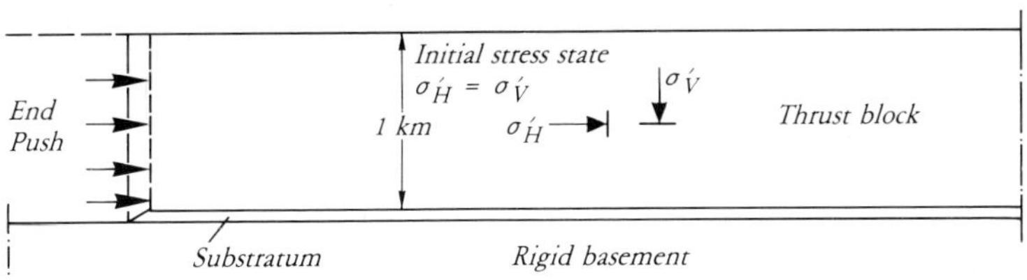

Material parameters	*Thrust block*	*Substratum*
Young's modulus E	*4.0 10^5*	*2.0 10^5 bar*
Poisson's ratio ν	*0.25*	*0.25*
Initial cohesion	*100*	*50*
Final cohesion	*0.0*	*0.0*
Initial friction angle	*30°*	*30°*
Peak friction angle	*45°*	*45°*
Residual friction angle	*30°*	*30°*
Amount of plastic strain to reach peak friction angle	*0.5%*	*1%*
Amount of plastic strain to reach max. rate of cohesive softening	*1%*	*1%*
Rate of plastic dilation reaches maximum at peak stress and decreases to zero in softening range.		

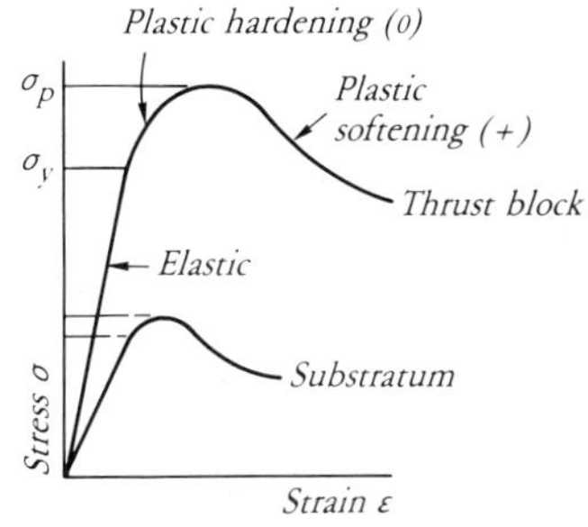

Assumed stress/strain response

A Thrust block on weak substratum (schematically) and assumed elasto-plastic material behaviour

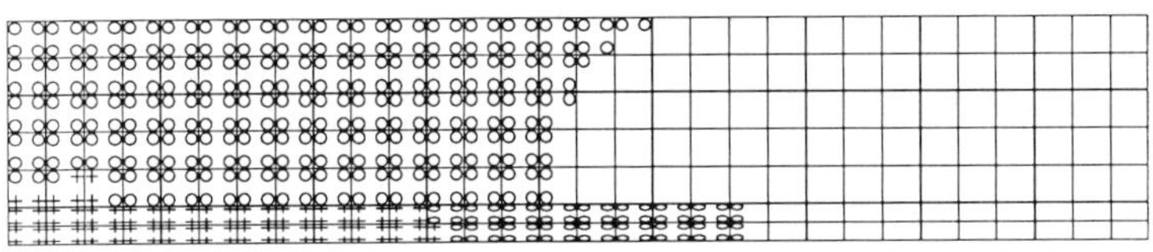

B Finite element net of end part with growing plastic hardening (o) and softening (+) zones in thrust sheet and weak substratum

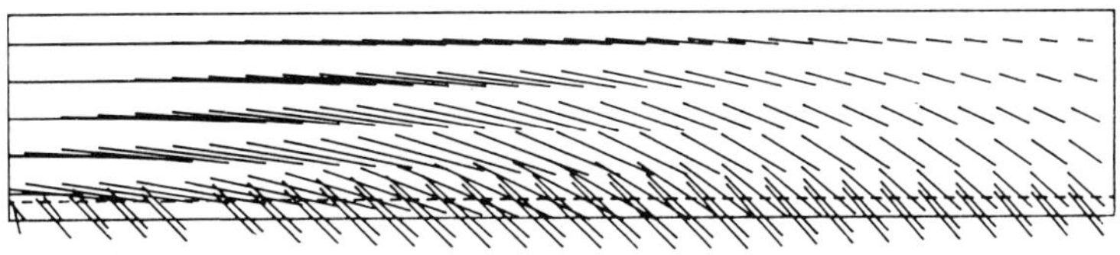

C Directions and relative magnitudes of σ_I

Fig. I.2-69 THRUST BLOCK ON WEAK SUBSTRATUM. Conditions after 7.5m of total end displacement (after G.H. Mäkel & J.V. Walters).

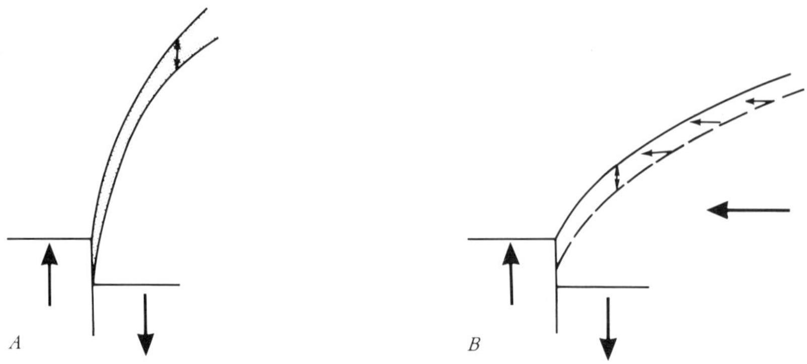

Fig. I.2-70 PRECURSOR FAULT CAUSED BY DIP SLIP ALONG VERTICAL BASEMENT FAULT. Accommodation of fault separation
A) By shear dilatancy of fault material
B) By supply of material under lateral compression

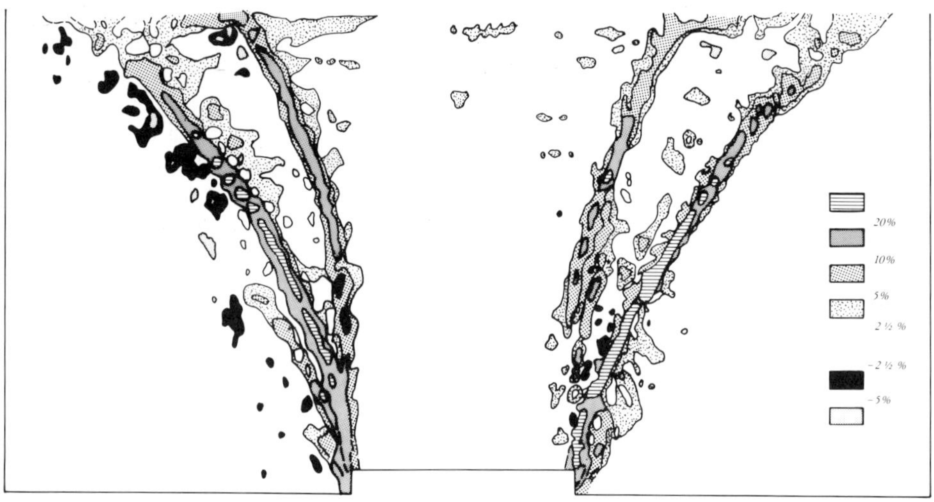

Plot of accumulated dilation during 1 cm uplift of a 10 cm wide basement block bounded by vertical faults. The plot was made using a stereophotogrammetric technique (Butterfield et al., 1970.) Numbers give volume change in percents of original specific volume. Negative numbers indicate zones of compaction. Zones of positive dilation are closely associated with the curved precursor shears.

Fig. I.2-71 SAND DILATION ALONG SHEAR ZONES IN UPLIFT EXPERIMENT (after W.T. Horsfield and R.M. Harkness; unpubl. work)

until finally a nearly vertically dipping main fault is formed. It is important to note, however, that under higher effective overburden pressures dilatancy will be practically suppressed and precursor faults can only develop beyond some embryonic state when the function of dilatancy is taken over by another mechanism. The only suitable mechanism we can think of is the supply of material by lateral compression, as is depicted schematically in Fig. I.2-

70B. In accordance with this, experiments with sandpacks under a horizontal compression perpendicular to the vertical sides of the uplifted basement block have shown an increased development of precursors, and these have flatter dips and extend farther over the subsiding block than precursors in a sandpack without extra horizontal compression (J.V. Walters and F.J. Nieuwland; unpublished). We therefore conclude that, under conditions of suppressed dilatancy, the uplifting of a vertically sided basement block will only produce prominent precursor faults when the overburden is under extra lateral compression at the uplift site. Such compression would be imposed when the basement is locally shortened by the uplift of a block with upwards-diverging sides, while the local extension caused by the subsidence of such a block would suppress the development of precursory reverse faults.

Simple-Shear Tectonincs. We consider next a class of tectonic deformations in which the *kinematic control* is even more severe and in which weaker rocks are sheared by stronger boundary rocks. The separation between the two rock types should be approximately planar and parallel, with the boundary rocks moving predominantly parallel to the planar interfaces. The conditions are mechanically very similar to the shearing between the rigid platens of a conventional shear apparatus, and will be referred to as *'simple-shear tectonics'* even if the distance between the boundary rocks does not remain strictly constant or the boundaries are not perfectly parallel. The three main types of simple shear tectonics are reviewed schematically in Fig. I.2-72. Type A is shearing of incompetent strata sandwiched between an overriding competent thrust sheet and competent substrata, or between two competent beds, for instance in the straight limbs of a flexural slip fold. Type B is horizontal shearing of detached overburden strata between horizontally moving competent boundary rocks. Type C, which is the most important type, represents horizontal shearing of an overburden by strike-slip faulting in the basement. Types A and B can be compared directly with simple shearing in a shear apparatus. The deformation is of the plane type, and the directions of the shear-induced greatest and smallest principal stresses lie in the deformation plane, which is the plane perpendicular to the 'driving' walls and parallel to the imposed shear direction.

Basement-induced wrenching. The conditions in type C are more complex. In comparing type C with a shear apparatus, one has to note first that the overburden at some lateral distance from the basement fault is passively dragged along by the moving basement 'plates' in the undeformed state, and secondly that the horizontal shear stress imposed by the moving basement plates decreases in the upward direction and vanishes at the overburden surface. Hence, close to the overburden surface the material may be considered under simple shear, imposed by the undeformed outer parts of the overburden. This further implies that, in type C, close to the horizontal surface two principal stress directions are parallel to the surface. Naturally, at depth the directions deviate from the horizontal under the increasing

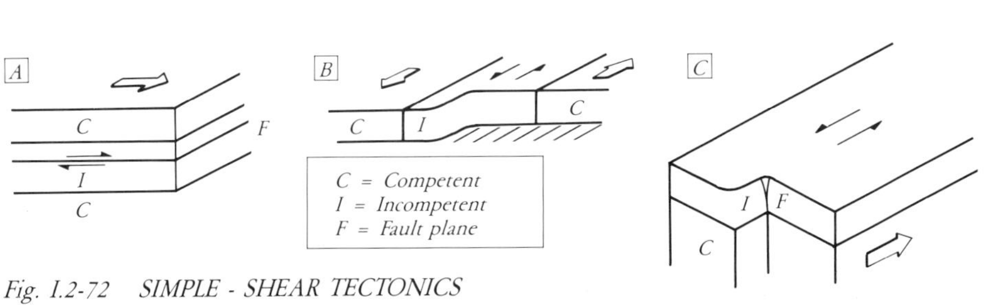

Fig. I.2-72 SIMPLE - SHEAR TECTONICS

influence of the basal shear stresses, which has a profound influence on the development of subsurface structures. We shall deal with these effects in Chapter I.4 and here restrict ourselves, with respect to type C, to stress changes near the surface.

To understand the changes in principal stress directions and the formation of precursory shears under tectonic simple shearing, we first recall what was said at the end of Section I.2.6, with reference to Fig. I.2-62, on simple shearing of an isotropic elastic material. In small simple shearing the direction of the maximum compressive stress, which is induced by the shear, makes an angle of 45° with the simple shear direction. However, this does not have to be the σ_I-direction of the total state of stress that results from the superposition of the shear-induced state and the state of stress that existed before shearing. Only when the state of stress in the future deformation plane was isotropic before shearing, i.e. all normal stresses parallel to that plane were equal in magnitude, will σ_I under simple elastic shear make the angle of 45° with the simple shear direction. As shearing proceeds in this case, the Mohr circle grows concentrically (Fig. I.2-62), while the 45° orientation of σ_I is maintained. Therefore, when the limit state is reached, the potential Coulomb slips (Riedel shears) will be obliquely orientated with respect to the shearing boundaries, as indicated in Fig. I.2-73A. As mentioned earlier, H. Cloos (1928) and W. Riedel (1929) have studied the en-échelon formation of these precursory shears in shearing a thin, wet clay cake on a 'shear table'. Their experiment, which has been repeated in its original or some modified version by geologists all around the world, is shown schematically in Fig. I.2-74.

It should be recalled that we have inferred the 45° orientation of σ_I from the assumption that the material deforms linear elastically up to the peak stress. Geological materials will hardly exhibit such extreme brittleness, but will reach the elasticity limit (yield stress) at stresses considerably below peak stress. Why should the 45° orientation, which was attained in the elastic range, be maintained in the following range of frictional-plastic strain hardening? It seems that this question has not yet been satisfactorily answered for frictional plastic materials [32)].

It should also be emphasised that the occurrence of precursory shears in simple shearing is typical for a frictional plastic material, i.e. a material whose shear strength increases with the effective confining pressure ($\varphi \neq 0$). A glance at Fig. I.2-73A will show that for $\varphi = 0$ the synthetic Riedel shears will coincide with the imposed shear direction, and a throughgoing fault will develop directly parallel to the moving walls.

In many cases of simple shear tectonics it has to be surmised that the normal stresses in the future plane of deformation would not have been equal before the onset of simple shearing. We therefore consider two extreme cases of *initially anisotropic states of stress*, in which both initial principal stresses σ°_I and σ°_{III} act parallel to the future plane of deformation. In the first case (Fig. I.2-73B), which represents exactly the initial condition in a shear box ('direct shear test'), σ°_I is oriented perpendicular to the moving walls; in the second case (Fig. I.2-73C) it acts parallel to these walls.

In the first case (Fig. I.2-73B) the shear motion will rotate the σ_I-axis from its initial position to make an acute angle with the shear direction. At some stage of the rotation the limit state will be reached and a first set of Riedel shears may form. The greater the initial difference in principal stresses, the earlier this state is reached and the more obliquely oriented are the conjugate Riedel shears relative to the imposed shear direction. Examples of such Riedel shears generated by simple shear under 'transpression' are given in Section II.7.4 ('Simple-shear kinematics vs stress control'; Figs. II.7-27,28).

Since a first set of synthetic Riedel shears, which was formed under transpression, can accommodate only very little movement of the boundary walls, σ_I has to rotate further to generate new Riedel shears that are less steeply inclined towards the moving walls. If this

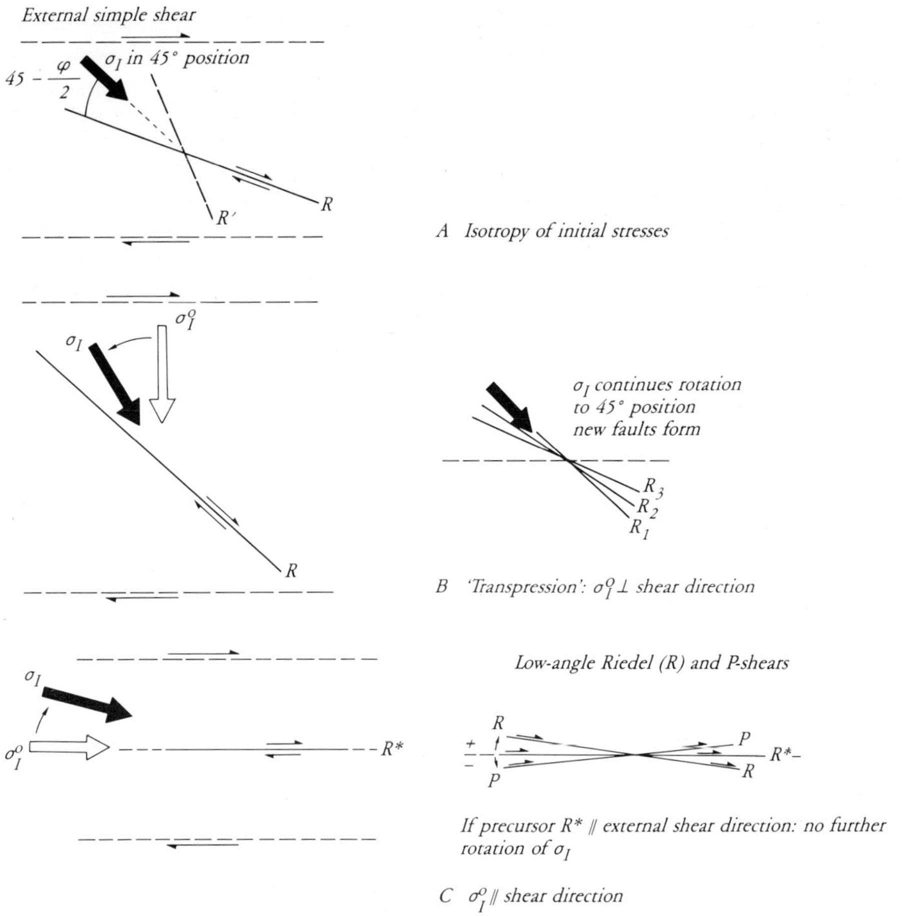

Fig. I.2-73 *HISTORY OF STRESS ROTATION AND ORIENTATION OF PRECUSOR FAULTS IN SIMPLE-SHEAR TECTONICS.*
Three states of initial principal stresses σ_I^o, σ_{III}^o in plane of deformation.
Further development: as anastomosing network of faults.

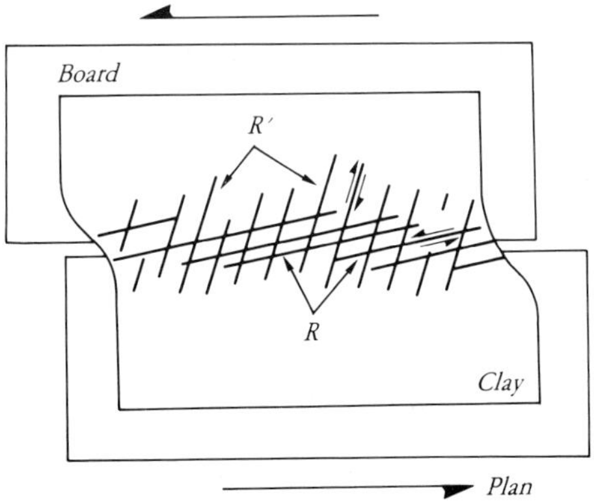

Fig. I.2-74
DIAGRAM OF THE RIEDEL EXPERIMENT

rotation were taking place in the elastic range, the superimposed transpression would certainly not cause the rotation to proceed beyond the 45° position of elastic simple shearing in the absence of anisotropic pre-stressing. Again the question arises as to how far the stress rotation will proceed in the plastic deformation range, which is obviously reached when the first Riedels form. Would it not be reasonable to expect the rotation to continue until new synthetic Riedels can form parallel to the moving wall and can allow a smooth throughgoing fault plane to form? This expectation is contradicted by the experimental evidence discussed in Section II.7.4 ('Simple shear kinematics vs stress control'), according to which the σ_I-rotation stops at 45°.

When σ°_I acts parallel to the moving walls, σ_I will obviously be rotated in the opposite direction until a first set of synthetic shears is formed, these having an orientation close to that of the imposed shear. The greater the difference in principal pre-stresses is, the better aligned the precursors will be with the direction of the applied simple shear. As indicated in Fig. I.2-73C, synthetic shears may even make a small negative angle with the simple shear direction; in soil mechanics such shears are referred to as P-shears. Obviously, in cases of wall-parallel pre-stress, little σ_I-rotation may be required and the synthetic shears may easily combine into a throughgoing wall-parallel fault, with respect to which the final σ_I-orientation may approximately satisfy Coulomb's criterion.

In contrast with this pre-stress case, the first generation of synthetic Riedel shears (R) formed under the initial stress situations of Figs. I.2-73A and B can only accommodate little displacement of the moving boundary rocks. One might indeed wonder why Coulomb shears, obliquely oriented with respect to 'rigid' walls, are not completely suppressed. But numerous experimental and field observations have given overwhelming evidence that the Coulomb-slip tendency of rocks in the frictional plastic range can hardly be completely suppressed by obstructing boundaries, since slight local compaction, reorientation of particles, some local weakness of the boundaries, or in particular the presence of a free surface in Type C of Fig. I.2-72, will allow minor Coulomb shears of the oblique type to develop. To accommodate further movement of the boundary walls synthetic shears making a smaller angle with the imposed shear direction have to develop. This, however, would require the σ_I-direction to rotate beyond the 45° position, where it was found to stop, as we have said above. The apparent dilemma is resolved by considering the stresses between two adjacent synthetic Riedel shears. Movement along these Riedels causes the material to be compressed in the region of overlapping (Fig. I.2-75A). This causes both an increase in the magnitude of σ_I and its closer alignment with the direction of the Riedel shears, as is indicated somewhat exaggeratedly in Fig. I.2-75B. Once a sufficiently large principal stress difference is built up, new faults will develop at a small angle to the external shear direction, these connecting the parallel 'master' Riedels. These new faults may be synthetic low-angle Riedels or synthetic P-shears, as indicated in the figure. Depending on the precise orientation of the σ_I-axis in the overlap zone of the earlier Riedels, the new low-angle shears will vary in orientation, and may include shears that are parallel to the external shear direction. As a result of this process, a series of shear lenses is formed, and the shearing is concentrated in a narrow band of anastomosing Coulomb-type shears, which ultimately develops into a throughgoing fault parallel to the externally imposed shearing motion (Skempton, 1966). The phase of shear lens formation is illustrated in Fig. I.2-76 by experimental results. Figure I.2-76A shows a result of microscopic studies of shear zones in clay by Morgenstern and Tchalenko (1967). Whereas these shear zones were produced in a shearbox, the stage in the development of shear lenses shown in Figs. I.2-76B,C has been observed at the surface of a pack of dry sand, which was sheared on a shear table. The first experiment corresponds with the classes A and B of simple-shear tectonics in Fig. I.2-72, and the second experiment models

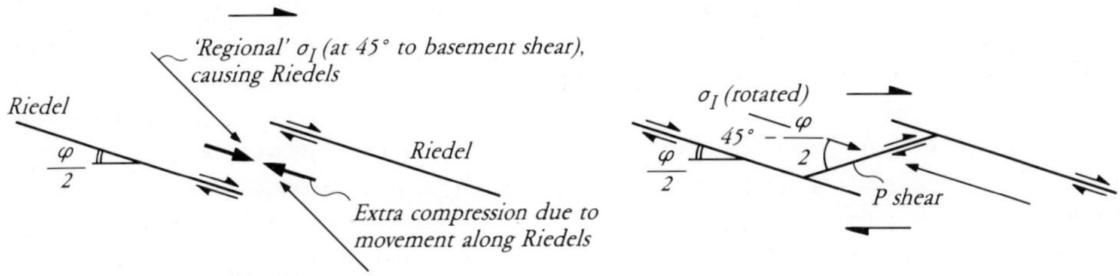

A Extra component of compressive stress due to displacement along Riedel shears

B Maximum rotation of σ_I (under most favourable conditions) between two Riedel shears, and orientation of associated P shear

Fig. I.2-75 THE INTERLINKING OF SYNTHETIC RIEDEL SHEARS

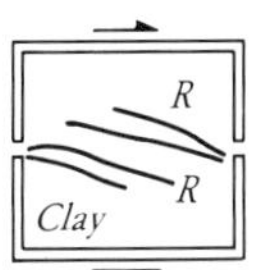

A Development of P shears in a shear box (after Tchalenko, 1970; reprinted from GSA Bulletin 81, p1626, by permission of GSA.)

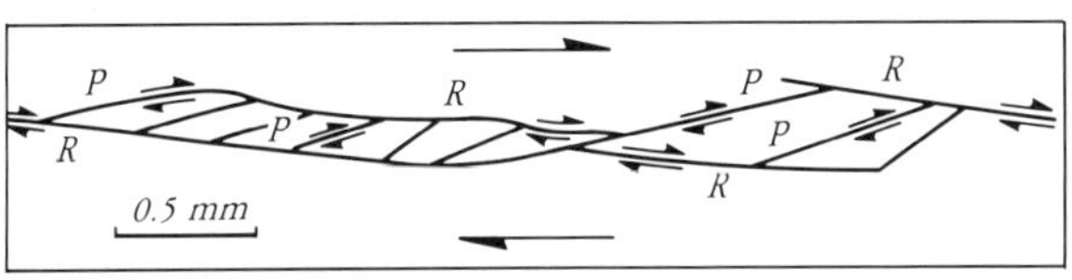

B

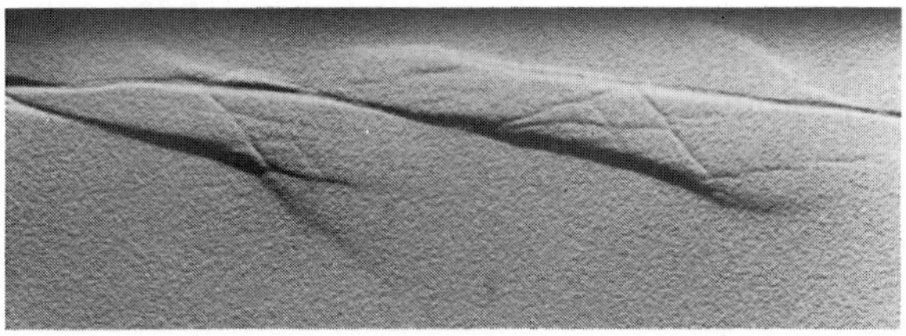

C

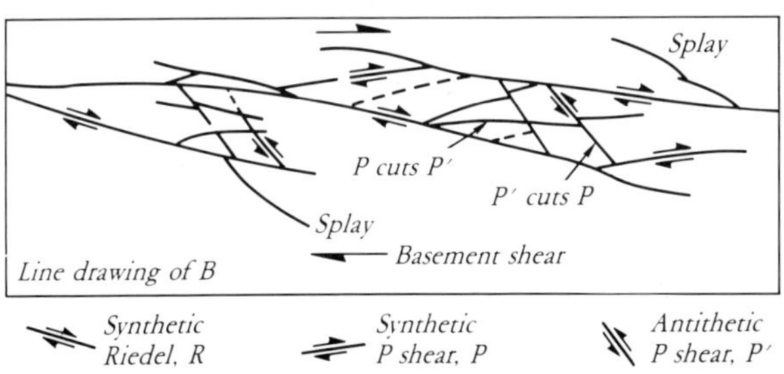

B, C Close-up of overlap between Riedel shears at surface of sandpack on shear table (after Naylor et al., 1986).

Fig. I.2-76 THE FORMATION OF SHEAR LENSES

an example of the basement-driven type C. In spite of the differences in set-up, the two experiments show essentially the same process of shear concentration. Similar results were obtained in high-pressure shearing experiments on consolidated material (limestone) by Bartlett et al. (1981) [33].

Antithetic shears make a high angle with the imposed shear direction, and their activity is therefore practically restricted to situations in which they are allowed to rotate in 'bookshelf' mode. Further details of the shear zone development of class C and related field evidence will be presented in Chapter I.4.

While the σ_I-direction is drastically changed inside the zone of anastomosing faults, outside this region it has remained as it was when the first Riedel shears were formed, or it has been slightly changed by the elastic unloading that is caused by shear softening inside the central shear zone. We conclude this mainly from the stress data obtained from numerous shearing experiments with a ring-shear apparatus (Mandl et al., 1977), which agree with the results of in-situ stress measurements at depth in wells near the San Andreas fault in California (Zobak et al., 1980) and also with the average orientation of tensional microfractures observed in high-pressure shearing experiments (Bartlett et al., 1981). We also recall from Section I.2.7 (Fig. I.2-65) that the abrupt change in σ_I-orientation at the boundary of the anastomosing shear band must be accompanied by an abrupt change in the shear-band parallel normal stress.

Hence, we arrive at the conclusion that *a final throughgoing fault, which develops under tectonic 'simple shearing' in rocks that initially were at a state of isotropic stress or transpression in the deformation plane, is not of the Coulomb-type, albeit that it is formed by the cooperation of Coulomb-type precursory shears.* The non-Coulomb-type fault zone, which was forced to develop parallel to moving boundary rocks, is bounded by planes of maximum shear stress (or nearly so) and is therefore referred to as *'simple-shear fault'.*

Secondary Stress Changes. The modifications of the stress field after the formation or reactivation of a fault may be attributed to four main causes. First, it is obvious that the *geometry of the fault surface* itself may oppose continued movement along the fault (Fig. I.2-77) and thus greatly disturb the local stress field. Since the formation of a fault reduces the principal stress difference in some neighbourhood of the fault, the material adjacent to the fault is, at least temporarily, brought into an elastic state of stress. The development of stresses around an irregularity of the fault surface during a first phase of fault movement may, therefore, be modelled with some confidence by elastic blocks in frictional contact along a pre-cut 'fault'. Figure I.2-78 presents the result of a photoelastic experiment in terms of principal stress trajectories corresponding to an early stage of displacement along a thrust fault with a 'ramp'-type irregularity. The originally horizontal maximum compressive stress is strongly deflected and is now almost perpendicular to the fault in the ramp region. At the ends of the ramp the principal stress difference attains extreme values. These are therefore natural localities where a plastic limit state will be re-induced and secondary fault structures will form to adapt the moving block to the shape of the fault.

Such adaptation faults can be quite puzzling to the interpreter since they may at first sight appear as conjugates of the main fault, but have the 'wrong' sense of motion. Figure I.2-79 shows schematically the post-depositional rotation of a hanging-wall block along a listric fault. When the rotation is not accompanied by a translatory component of motion on the downthrown side, local compression will induce a limit state in the rotating block and rotate the σ_I-direction as sketched in Fig. I.2-79C. In this locally altered field of σ_I-directions adaptation faults of the thrust type will be formed. A rather similar situation may exist when the curvature of the main fault is upwards convex, as may be readily seen by turning Fig.

I.2-79 upside-down. (The reader may notice a similar situation above the uplifted basement block in the experiment of Fig. I.2-22A.) Analogously, a block upthrusted along a steep curved thrust fault may be locally adjusted by secondary normal faults.

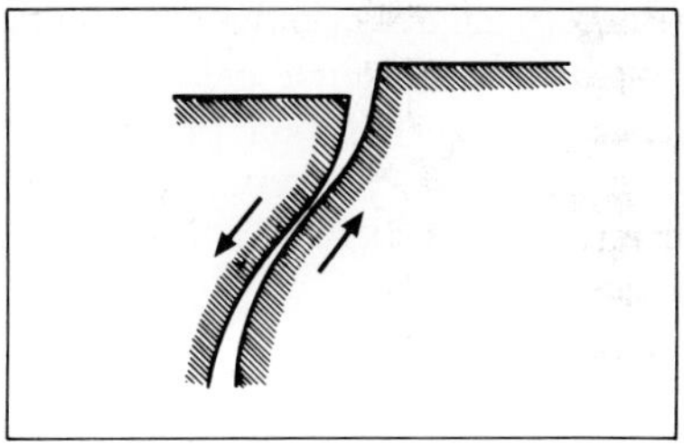

Fig. I.2-77
FAULT GEOMETRY OPPOSING MOTION
(after E.M. Anderson, 1951)

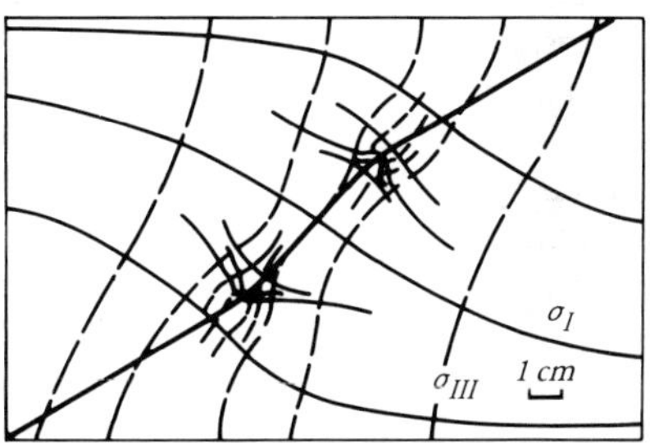

Fig. I.2-78
REDISTRIBUTION OF PRINCIPAL STRESSES IN HORIZONTALLY COMPRESSED BLOCK OF PHOTO-ELASTIC MATERIAL WITH PRE-CUT THRUST 'FAULT'
(after D.W. Barber and G.M. Sowers, 1974; by permission of U.S. National Academy of Sciences).
'Ramp' in thrust fault experiences extra loading (σ_I is maximum compressive stress) with stress concentrations at each end. Fault is still closed.

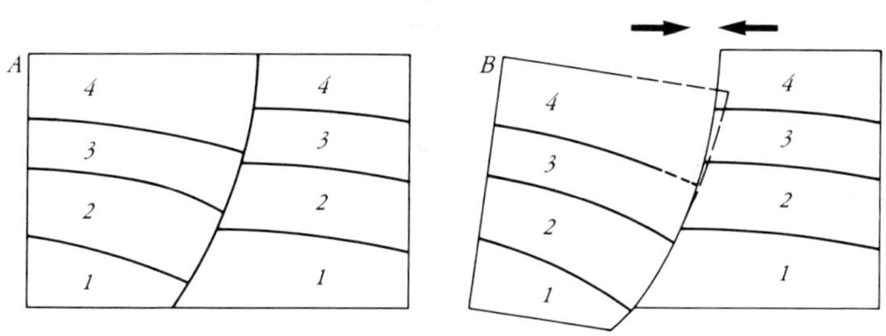

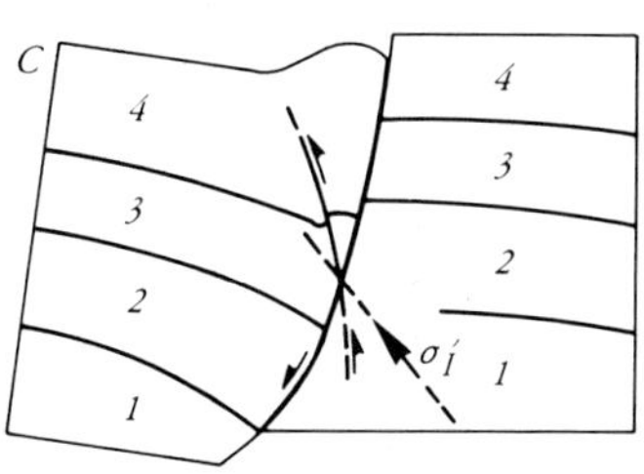

The overlap in B indicates the volume problem and implies localised additional horizontal stress concentration, which may be realised by bending of the strata and/or reverse faulting. Note that the volume problem will occur only when the rotation is not associated with movement along a basal slip plane; i.e. there is not enough room for the block to move outwards to the left (after A. Koopman, unpubl.).

Fig. I.2-79 SCHEMATIC MODEL SHOWING THE EFFECT OF POST-DEPOSITIONAL ROTATION OF A HANGING WALL BLOCK ON A PRE-EXISTING FAULT OF A CERTAIN CONCAVE GEOMETRY.

A second, also quite obvious cause of stress changes in faulting is that the *outer configuration of a rock mass is changed* by motion along faults and the loading conditions are thereby altered. One example may suffice to illustrate this. Figure I.2-80 shows schematically a thrust sheet overriding its foreland. The overridden block experiences an extra overburden and shear load. The action of the extra overburden is to raise the average compressive stress and to decrease the principal stress difference. These changes impede faulting of the overridden block and any new thrust fault will have to form ahead of the 'strengthened' part. In addition, the emplacement of extra overburden may induce abnormally high pore pressures in less competent substrata, which may reduce the effective overburden stress sufficiently, without affecting the shear stress, to allow the formation of a bedding-plane fault along which the stack of competent sheets can be pushed (Gretener, 1972).

A third type of stress redistribution takes place at the *edges of active fault planes.* This is illustrated in Fig. I.2-81, which in a schematic way shows the leading edge of a fault under simple shear conditions. There is not much theory required to conceive that compression on the advancing side of the fault and stretching on the receding side of the fault produce a deflection of the σ_I-trajectories as sketched in the figure. As far as the edge region is in

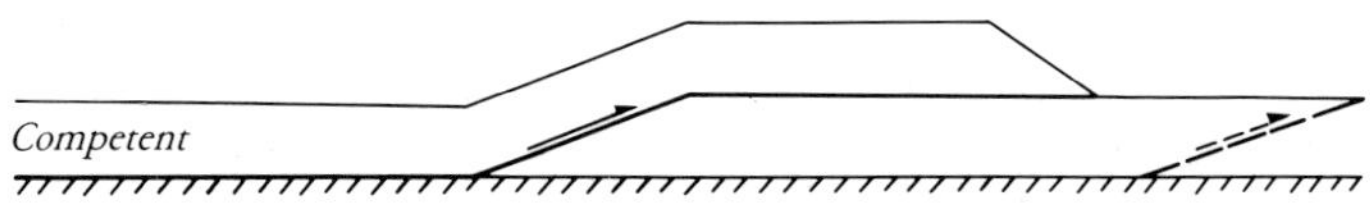

Fig. I.2-80 OVERTHRUSTING (change of boundary conditions during faulting)

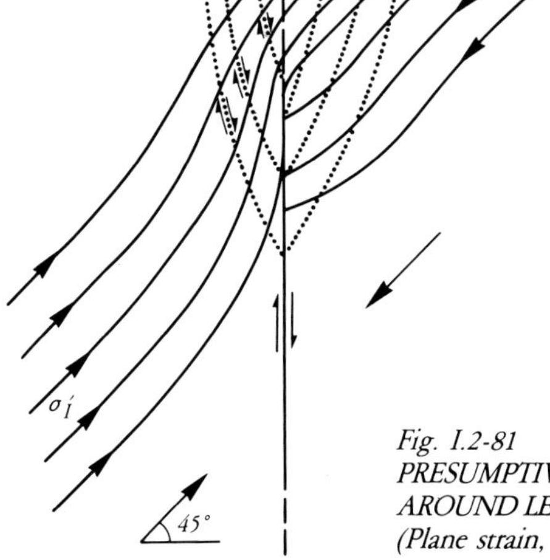

Fig. I.2-81
PRESUMPTIVE MODIFICATION OF STRESS FIELD AROUND LEADING EDGE OF PROPAGATING FAULT. (Plane strain, simple shearing)

the plastic state, subsidiary faults – the 'type A faults' of Chinnery's (1966) elastic analysis – may splay off the main fault as indicated in the figure, or the main fault itself may be deflected. We have already discussed (Section I.2.4; Figs. I.2-28,29) that, because of the pressure-sensitive shear strength of the rock, the plastic edge region develops asymmetrically, even if the fault propagates into a uniform and unbounded foreland. We have also noted (Fig. I.2-30) that, in addition, the splay faults or the shape of the propagating main fault may be affected by boundaries, in particular barriers of stiffer rock ahead of the propagating fault. Arrested by such barriers the fault may terminate by fanning out into numerous synthetic faults, which distribute the slip of the main fault into an almost continuous deformation of the rock ahead. Such splay-type fault termination is often encountered in wrench zones (Anderson, 1951; Chinnery, 1966; Freund, 1977; Arthaud and Matte, 1977; see also the small-scale illustration in Fig. II.7-22), where it shows a strikingly asymmetric development towards either the advancing or receding side of the main fault.

Figures I.2-82 A,B summarise diagrammatically the results of finite element calculations that have been carried out to explore the effect of a strong lithological barrier upon the development of splay faults under plane strain conditions. Only a part of the main fault adjacent to the leading edge is considered and is represented by a narrow band of weakened material. The fault zone and surrounding rocks are again treated as elastic/frictional plastic material with plastic strain hardening and softening. The fault has been temporarily at rest and the stress field is assumed to be uniform with σ°_I inclined at 45° towards the fault plane (σ°_I = 30 MPa, τ_{max} = 5 MPa). The fault is then reactivated by prescribing a relative displacement parallel to the fault as indicated in the figures. In the left-hand figure the region is bounded in the lateral direction by fault-parallel planes (stippled) on which the

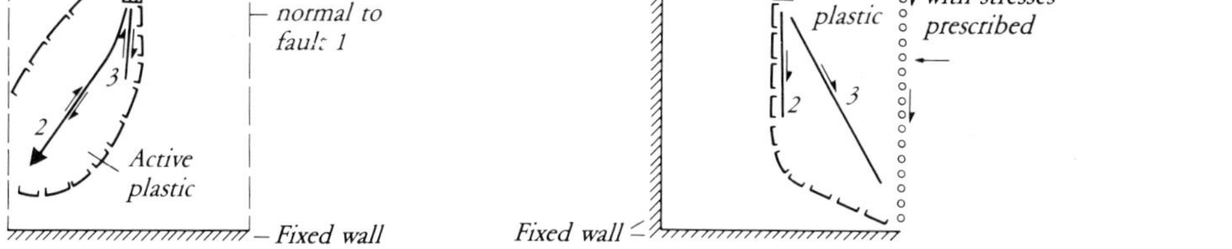

A Two-sided displacements allow the preferential development of a plastic zone and 'Riedel'-type splaying at the receding side of the main fault.

B One-sided displacement and rigid confinements cause plastic zone and secondary faults to develop at the advancing side of the main fault.

Fig. I.2-82 ARREST OF ADVANCING FAULT TIP BY 'RIGID' BARRIER (after H.W.M. Witlox, unpubl. work.) Diagrammatic representation of numerical simulation of the development of secondary fault zones in elastic/frictional plastic material in plane strain; softened zone 1 represents fault tip in regional stress field with maximum compressive stress σ^o_I acting at 45° to fault (σ^o_I = 30 MPa, τ_{Max} = 5 MPa). Reactivation of fault is accompanied by fault-parallel displacement behind the fault tip.

initial shear stress τ°_{max} is maintained and only displacement parallel to the fault is allowed. The propagation of the main fault is impeded by the fixed barrier ahead. In spite of the small displacements that can be accommodated under these boundary conditions, the calculation clearly shows that a plastic edge zone with splay faults develops towards the receding side of the main fault. It should be noted that it is irrelevant for the process whether the relative displacement along the fault is imposed by actually prescribing opposite movements as shown in the figure, or by moving only the advancing side. A quite different development is observed (Fig. I.2-82B) when the (dotted) fault-parallel boundary plane on the advancing side is free to deform, but remains loaded by the initial (regional) stresses, while no receding displacement is allowed to take place at the upper boundary. The whole advancing side of the region considered becomes plastic and splay faults develop in the indicated sequence. Thereby it is practically immaterial whether, on the fault-parallel boundary plane at the opposite side, all displacement is assumed to vanish, as shown in the figure, or the same conditions as in Fig. I.2-82A are prescribed on this plane. It is essential, however, that the fault-parallel boundary at the advancing side remains free, allowing the material to move out laterally. Otherwise the situation would be reduced to that of Fig. I.2-82A, in which only the relative fault displacement matters.

In spite of the rather special character of the imposed boundary conditions, these results strongly suggest that a fault that is terminated by a lithological barrier will splay towards the (relatively) receding side when the lateral confinement at the advancing side is not relaxed by allowing material to move laterally away from the near-edge region of the fault. Conversely, if at the advancing side such lateral escape of material is possible and, at the same time, only the advancing fault block is moved, the terminating fault will splay towards the advancing side. One may envisage of such situations in strike-slip faulting, or one may consider the branching of a bedding plane fault into synthetic and antithetic toe thrusts, as further bedding plane gliding becomes impeded by an increase in friction. (The latter may be immediately seen by turning Fig. I.2-82B into a horizontal position.) It should be further noted that the barrier considered is a boundary on which all displacement components vanish. But faults may also terminate against a barrier along which material can slip, for instance when a fault terminates against another active fault or hits a bedding plane of low frictional resistance. The slip direction along such a barrier will introduce an additional factor of mechanical asymmetry. Regrettably, it seems that this problem has not yet been analysed.

Finally, the fourth cause of stress redistribution in faulting is *shear softening* of the fault material. The tectonic significance of this phenomenon has been discussed in Section I.2.7, where it has been pointed out that, by reducing the principal stress difference, shear softening allows the material adjacent to the fault to drop back into a subcritical state of stress and also makes it possible for the σ_I-direction and the fault-parallel normal stress to vary discontinuously across the faults.

As a first example let us schematically consider in Fig. I.2-83A a *symmetric graben*. When the shear stresses on the two graben faults are reduced, the hanging wedge will become seized between the weakened fault planes and increase its effective pressure upon the fault, and so recover frictional strength. At the same time the fault-parallel effective normal stress $\sigma'_{\parallel}$ will drop somewhat. One may expect that these stress changes are accompanied by a reorientation of the σ'_I-direction, as sketched in the figure. This is indeed confirmed by finite element simulations of symmetric graben formations, the very simplest example of which is presented in Fig. I.2-84. The graben is formed by the extension of a uniform layer on a horizontal rigid base. In contrast with the simulation of normal faulting which has been considered before (Fig. I.2-66), the base is here assumed to be frictionless, which allows a small extension, imposed from outside upon the whole layer (Fig. I.2-84A), to be accommodated by a single graben without deformation of the material outside the graben. Again, the first fault is triggered by a weak element at the surface. Figure I.2-84B shows the fully

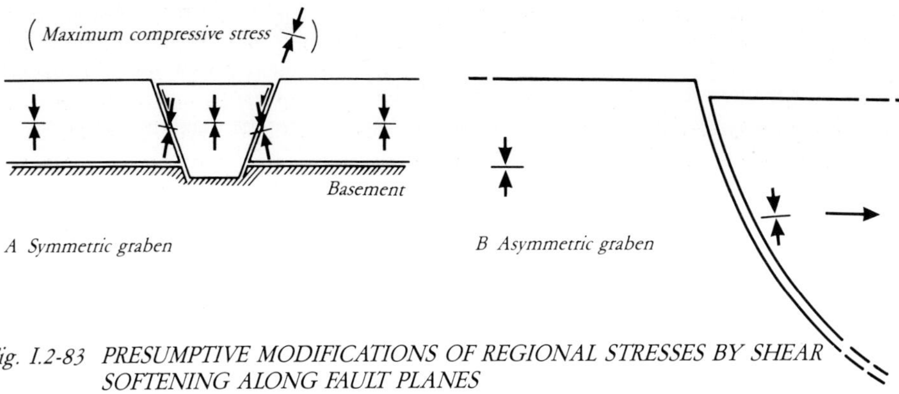

A Symmetric graben *B Asymmetric graben*

Fig. I.2-83 PRESUMPTIVE MODIFICATIONS OF REGIONAL STRESSES BY SHEAR SOFTENING ALONG FAULT PLANES

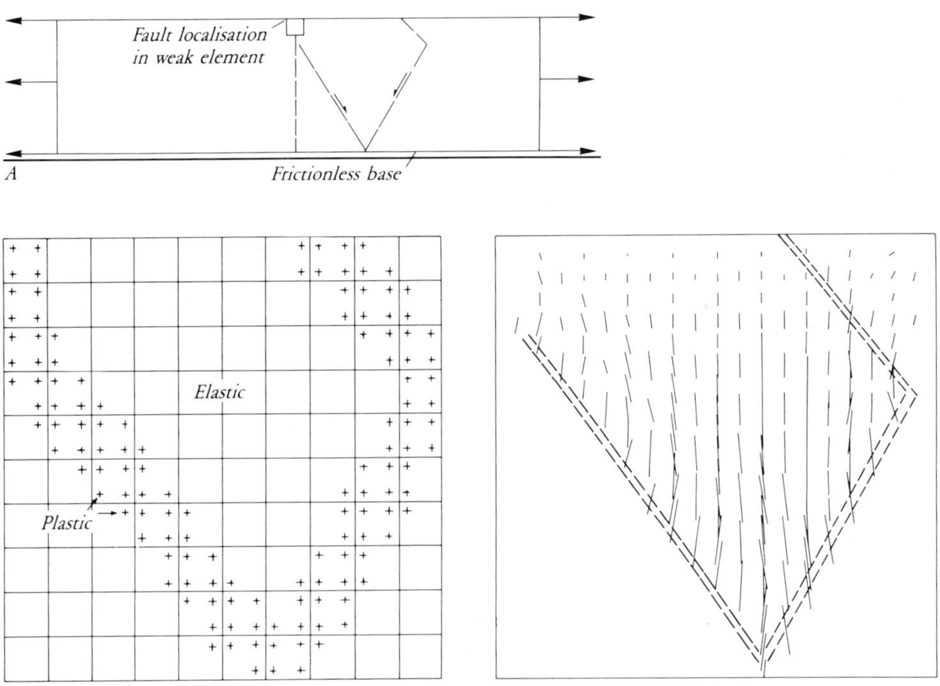

B Fully developed softened shear zones

C Direction and relative magnitude of maximum compressive stress inside graben

Fig. I.2-84 FINITE ELEMENT SIMULATION OF GRABEN FORMATION UNDER UNIFORM HORIZONTAL EXTENSION
(after H.W.M. Witlox; unpubl work, 1986)

developed softened shear bands, which are surrounded by elastic material. Except for the small fault in the right-hand upper corner, the graben is fairly symmetric. Figure I.2-84C shows the direction field of σ_I inside the graben. While in the central part σ_I still has the direction and magnitude of the original overburden stress, its direction is strongly rotated near the graben boundaries in the expected sense.

The opposite type of stress reorientation may be expected in the case of an *asymmetric graben*, as sketched in Fig. I.2-83B. The subsiding block is moving to one side and, instead of the fault pressure increasing in response to shear softening, it is now the fault-parallel stress $\sigma'_{\|}$ which increases. This may become plausible when one imagines a complete loss of shear strength at the fault. Obviously, $\sigma'_{\|}$ would then become the maximum compressive stress.

The state of stress in a moving fault block may, in principle, remain sub-critical when the block glides along a planar fault contact. Such a sub-critical state was represented by the dotted circle in the Mohr diagram of Fig. I.2-64. The reader may observe, however, that it requires only modest stress changes to bring the 'detached' stress circle again in contact with the limit line of the non-softened material of the fault blocks, thereby producing an active limit state. In this state secondary structures of the normal fault type could form. As a first example, we consider a *gravitationally gliding rock sheet* on a lubricating (e.g. overpressured) substratum. Large parts of the sheet show very little deformation and are in a sub-critical state of stress, represented by circle 1 in the stress diagram of Fig. I.2-85. The stresses in the weak substratum underneath the gliding sheet must be in the limit state to make gliding possible. Accordingly, stress circle 2 touches the limit line for the weak material. In addition, we assume, while not strictly necessary for the following, that the substratum is overpressured, the pore pressure difference between sheet and adjacent substratum being Δp. It is now not difficult to envisage a local steepening of the basal slip plane, or a local change in basal friction, causing a reduction in the base-parallel normal stress $\sigma'_{\|}$ in the gliding sheet, which is sufficient to induce locally the active limit state represented by circle 3 in Fig. I.2-85. The associated σ'_I-direction, again determined by the pole method, deviates only a little from the vertical and is fairly insensitive to the somewhat arbitrary choice of the stress circles. This enables the formation of conjugate normal faults that are slightly asymmetrically oriented with respect to the vertical.

This type of faulting, with minor throws, may take place contemporaneously with the gravitational nappe gliding, and may leave the substrata unaffected. It may have been responsible for the striking shape of the mountain peaks ('Tschingelhörner') of the 'Glarus' nappe shown in the introductory Figure I.1-1, since erosion has taken place preferentially along the normal faults in the Permian sandstone of the nappe. (Note that the direction of transport in Fig. I.1-1 is opposite to that of Fig. I.2-85.)

Simultaneously with a sub-critical (elastic) state of stress, or even with an active limit state in the gliding sheet, a passive limit state may exist in the incompetent strata that are overridden by the competent sheet. We refer again to the Mohr diagram of Fig. I.2-65, which demonstrates the possible coexistence of these states on opposite sides of a softened fault plane. Also shown in that figure are the secondary Coulomb faults (P, P′) that would be associated with the orientation of the maximum compressive stress σ'^{P}_{I}. It is tempting to interpret the lineaments that transect the Malm limestone underneath the 'Glarus' nappe in Fig. I.1-1 as synthetic secondary thrust faults (P-shears) associated with the passive limit state.

Similarly, extensional and compressional regimes may be active simultaneously on the opposite sides of wrench faults. But *decoupling of different tectonic stress regimes* may also be accomplished without the intervention of an active fault, when creep flow of a weak interlayer allows for different straining on either side of the layer, as is summarised in

Fig. I.2-86. On a large scale, the decoupling of extensional and compressional stress regimes is known to take place in active convergent plate margins where, on top of a compressional imbrication zone, gravitational faults dip towards the plate margin. It should be noted that decoupling of tectonic stress regimes does not necessarily imply that the associated tectonic events have occurred at exactly the same time and place. Decoupling interlayers may allow different tectonic stress states to coexist after the stacking of the differently stressed masses and to relax independently in the course of time. This may have been the case, for instance, at the northern front of the Eastern Alps, where extensional faulting accommodated the extensional strain in the downbending foreland plate prior to the superpositioning of the imbricated thrust mass of the Northern Calcareous Alps and flysch belt (Kollmann, 1977).

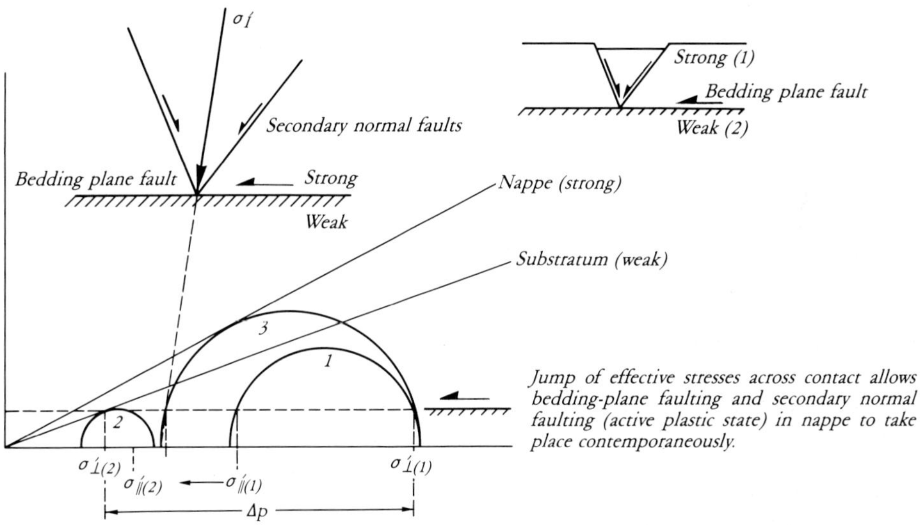

Fig. I.2-85 GRAVITATIONAL GLIDING OF STRONG ROCK-SHEET ON OVERPRESSURED SUBSTRATUM

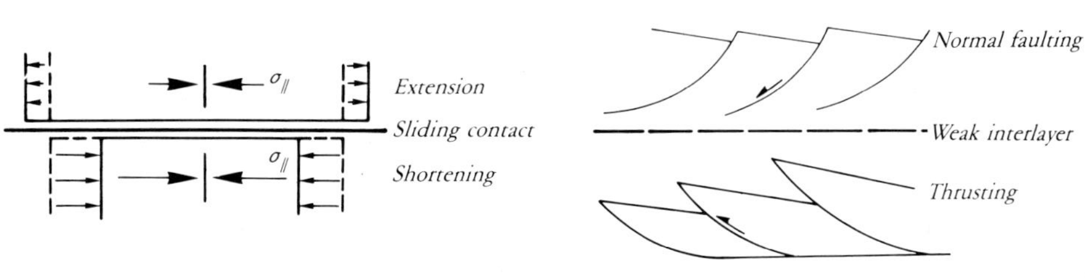

Fig. I.2-86 DECOUPLING OF TECTONIC REGIMES BY WEAK INTERLAYER

9. MODEL EXPERIMENTS.

The experimental situation of tectonic processes appears particularly attractive in cases in which the mathematical analysis becomes exceedingly complex. In problems of tectonic faulting such complexities immediately arise when a problem is formulated as a three-dimensional one rather than the usual restriction to two dimensions being imposed.

To ensure that an experiment will yield qualitatively and quantitatively useful information, a one-to-one correspondence which applies to any stage of the deformation process has to be established between model and prototype. In other words, the experiment has to be properly scaled. This can only be achieved by referring both model (M) and prototype (P) to the same mechanical theory. Since we are concerned with tectonic faulting (under quasi-static conditions), the mechanical theory comprises the boundary conditions, i.e. the geometry of the rock body considered and a specification of the mechanical interaction with its surroundings, the conditions of stress equilibrium, a limit condition for the shear stress the rock can withstand, and stress/strain relationships that characterise the deformational behaviour of the material before and after a limit state of stress has been reached.

The first exposition of scaling requirements for tectonic model experiments was presented in 1914 by Koenigsberger and Morath, who gave a clear presentation of the rules for the proper scaling of rock strength. The paper received little attention, and it is M.K. Hubbert (1937) who has to be given credit for bringing model theory to the attention of a wide geological community. In his introductory paper, Hubbert also pointed out the requirements for an adequate modelling of rate-depending deformation in its simplest form of a (linear) viscous flow. While scaling rules for viscous material behaviour are required in modelling processes of tectonic folding, the experimental or mathematical modelling of faulting processes in the brittle range is based on the theory of frictional plastic material behaviour. As has been stated repeatedly, this automatically excludes all phenomena of creep and stress relaxation and makes our model experiments rate-independent, enabling the same structure to be produced at quite different rates of experimentation (for further discussion see Sections II.7.1 and II.7.4)

Let us briefly discuss the essential points in modelling faulting. Let us assume that in model M the linear dimensions are 10^{-4} times those in the prototype P. The specific weight in M and P being about the same, the overburden stress in M will – in the normal gravity field – also be 10^{-4} times the corresponding overburden stress in P. Will that enable, for instance, thrust faults to develop both in M and P when the horizontal compressive stress raised in M is 10^{-4} times the corresponding stress in P? This will obviously depend on whether in both cases the critical stress condition (I.2-4) is satisfied. Let us first consider that in both cases the cohesive strength τ_0 is of negligible magnitude in comparison with the pressure-depending friction term, and that the angle of internal friction has about the same value (say 40°) in both materials. Then the linear equation I.2-4 will be satisfied simultaneously for P and M enabling thrust faults to develop in both.

Usually, rocks have a cohesive strength τ_0 and accordingly, the cohesive strength of the material in M, in the normal gravity field, would have to be scaled down by a factor of 10^4. However, in sedimentary rocks in a depth range from, say, 1 to 10 km, the frictional (effective stress dependent) component of the shear strength usually predominates by far and τ_0 may be neglected. Moreover, in relatively densely jointed rocks we have to consider the cohesive strength of the jointed rock mass which may be considerably lower than that measured on an intact sample in the laboratory. Note also that scaling down a cohesive strength of say 5 MPa (50 bar) in the material of P by a factor of 10^4 would require a cohesive shear strength in M of only 5 g/cm^2, which is difficult to maintain in the experimental material.

As far as shear strength scaling in fault modelling under normal gravity conditions is concerned, it is therefore expedient and in accord with the underlying mechanical model to use dry, densely packed sand as model material. It should also be mentioned at this point that faulting experiments carried out with slightly cohesive clays, gypsum, barytes, etc. usually correspond with extremely high cohesive strengths of the rock mass in the prototype.

In using sand or any other granular material to simulate tectonic faulting one is not interested in the individual grains but in the deformation behaviour of the whole aggregate – its frictional-plastic character. Therefore, when faulting in a heavily jointed rock mass is being simulated by sandbox experiments, there is no comparison between grain size and the joint block dimensions involved.

Unfortunately, however, the size, distribution, packing etc. of the grains in the model material may have some unscaled effect on the development of faulting in the experiment. This is mainly a result of the grain-dependent phenomenon of dilatancy. To illustrate this point, the reader is reminded of the precursor faults above a subsiding basement block, which have been discussed in the preceding section (Figs. I.2-70, II.7-33). In sandbox experiments, such faults operate in the absence of extra horizontal compression because of the high shear dilatancy of the sand (Fig. I.2-71), but would not, or to a much smaller extent, operate under the corresponding prototype conditions, since then dilatancy is suppressed. Also grain-size-dependent, and therefore unscaled, is the width of a shear band.

Finally, one has to consider the elastic/frictional plastic stress strain relationships. If we disregard, for the moment, the elastic component and plastic hardening and softening, the limit stresses in such an 'ideal plastic' material are independent of the total strain since, like a viscous liquid, the ideal plastic material has no 'memory' of past plastic deformation. Hence, if both M and P are in the limit state, geometrically similar displacements and, therefore, identical strains (e.g. the same relative shortening) can be imposed without affecting the state of stress. In other words, the flow rules of ideal frictional plastic materials do not interfere with proper scaling.

In reality, both the M and P materials do exhibit some (linear) elastic response which, unfortunately, is in conflict with the scaling based on ideal frictional plastic behaviour. Assume the same elastic moduli for P and M and apply scaled, i.e. geometrically similar, displacements. This implies equal elastic straining in M and P and, consequently, equal elastic stress changes. However, since the down-scaled shear strength of M is 10^{-4} times that of P, M only requires an elastic strain that is 10^{-4} times that required by P to reach the limit state under elastic deformation. Since the geometry of M was already scaled before elastic straining, this smaller strain implies a displacement in M that is 10^{-8} times that in P. Thus, as far as the elastic part of the process is concerned, scaling rules would be severely violated. (For our quite realistic scaling factor, 1 mm horizontal shortening in M would correspond to 100 km shortening in P!)

This conclusion concerning the elastic deformation is, however, based on the assumption of equal elastic stiffness of M and P. If Young's modulus of M were 10^{-4} times that of P, elastic scaling would be perfect. Fortunately, Young's modulus for dry sand under the very low effective pressures in a sandbox is very much lower than that for sand under a geological overburden. Although the elastic stiffness of sand under very low static loads is very difficult to determine, a ratio of 1000 between the Young's modulus of P and M would seem a conservative estimate. This reduces the deviation from proper scaling in the elastic range, for sand models under normal gravity conditions, to a tolerable degree (1 mm displacement in the model corresponding now to 100 m or less in the prototype). Nevertheless, considerable uncertainty remains with regard to the exact relationship between the pre-faulting displacements in sand model and prototype. Additional uncertainties in displacement scaling arise in the plastic range for the unscaled hardening and softening processes in model and prototype.

Although these uncertainties may allow no direct comparison of displacement magnitudes, or at least call for great caution, qualitative aspects such as sequence, shape or pattern of model faults may be interpreted in terms of the prototype with greater confidence. Nevertheless, it remains highly advisable to consider model studies and computational simulations as two complementary and mutually corrective approaches.

NOTES

1) Note that the *full* pore-fluid pressure had to be subtracted and *not,* as is sometimes done, a pore pressure reduced by the factor $1 - K/K_S$ where K and K_S are, respectively, the compression moduli of the bulk material and the material of the skeleton of a porous rock. This factor was introduced to simplify formally the laws of *poro-elasticity* as is explained in detail in Section II.6.2, but has no relation to the shear strength of the material. Since some confusion has arisen about this point, the reader is referred to the discussion in Section II.6.3.

2) It should be understood that, at present, elastic/frictional plastic modelling of brittle deformations with shear zones (faults) developing is restricted to processes which, on the macro-scale applied, are not self-accelerating, but quasi-static in the sense that they may be considered as sequences of static equilibrium states.

3) These theories, originally developed for homogeneous and isotropic materials in uniform stress fields, have been extended recently by various authors to include material inhomogeneities, anisotropy, shear dilatancy, pore fluids, non-uniformity of the stress field, etc. and to elucidate how these factors affect the formation of shear bands. Unfortunately, it is still rather difficult to present the essence of these highly formalistic theories in a form that would suit the purpose of this book.

4) This mechanical interpretation complements the purely kinematical statement that three fault sets, each of them with two independent shear contributions, are necessary and sufficient to accommodate general three-dimensional straining without volumetric changes and externally imposed rotations (Reches, 1978).

5) From soil mechanical testing (at pressures of a few bar) we may conclude that shear strength anisotropy due to particle orientation will hardly amount to more than 2° in the orientation of slip planes (Arthur and Menzies, 1972; Oda, 1976). Since from other studies one may conclude that the anisotropy effects decrease with increasing confining pressures, one may probably neglect strength anisotropy in sandy material.

6) In two other investigations (Chenevert and Gatlin, 1965; McLeamore and Gray, 1967) of sedimentary rocks it was stated that the lowest shear strength is encountered when σ_1' acts at about 30° to the bedding plane.

7) Although the peak strength of *sand* seems to be hardly affected by inherent (bedding) anisotropy or by load-induced anisotropy (formation of load-carrying grain structures), these anisotropies do have an effect on the mechanical behaviour of sand, which might be of tectonic importance. If a sand pack is loaded with σ_I' oriented *parallel* to the bedding and a certain stress ratio σ_I'/σ_{III}' below the critical state is gradually built up in a triaxial test (Fig. I.1-4B), the sample may experience more than twice as much straining as in a comparable test in which σ_I' is oriented *perpendicular* to the bedding (Arthur and Menzies, 1972).

It was also found (Arthur et al., 1977) that first loading of sand induces anisotropy along a pair of directions that are about $\pm 30°$ off the σ_I'-axis. When, in a second loading procedure, σ_I' is applied at about 60° from the former direction, the sample shows a drastic reduction in 'stiffness' and experiences much more straining before a pre-set stress ratio (below critical state) is reached.

In *clayey* material, anisotropy may have a greater effect. When loaded under a confining pressure with σ_I' perpendicular to the bedding, the sample will deform continuously to a considerable extent, assuming 'barrel' shape, before it fails. When σ_I' is applied parallel to or at an acute angle to the bedding, shear bands will form with little preceding deformation of the sample (Prof. A.W. Bishop, London; private communication). This phenomenon might be of interest in connection with whether a *fault entering a thick overpressured shale bed* will proceed as a fault or will terminate in a diffuse region of continuous shale deformation.

8) In the standard type of normal faulting the initiating stress field is characterised by the vertical overburden stress σ_V being the maximum principal stress σ_I. From the formula for $\tilde{\sigma}_I'$ in Fig. I.2-13 we can then immediately determine the greatest effective overburden that would still allow normal faulting of the dilational type. For example, a weak, slightly lithified clay with an average bulk density of 2.4, normal hydrostatic water pressure, a friction angle of 30° and a cohesive strength $\tau_0 = 1$ MPa, would allow dilational normal faulting down to a depth of 120 m.

9) Of course, this change in σ_I'-direction can also be determined analytically by means of formula II.4-4, which relates the orientation of the σ_I-axis to a Cartesian coordinate axis parallel to the bedding and to the plane in which σ_I' and σ_{III}' are acting, and by inserting the difference of normal stresses from the limit condition in Cartesian stress components (eq. II.7-5).

10) A glance at the limit condition in Cartesian stress components (eq. II.7-5), with subscripts 2 and 3 indicating the horizontal and vertical coordinate axes, will be sufficient to show that the plastic limit state can exist only very locally, since an increase in a horizontal normal stress σ_2 in the horizontal x_2-direction is incompatible with an overburden stress σ_3 and a horizontal shear stress that do *not* vary in the horizontal direction.

11) This disagrees with Chinnery's (1966) conjecture, which was based on a purely elastic analysis for $\theta = 45° - \varphi/2$, that faults would tend to deviate to the advancing side when propagation is resumed.

12) Note that in the diagrams of Fig. I.2-28,29 identical stress states were assumed at Q^+ and Q^- prior to the resumption of fault propagation. In low-angle normal and thrust faults, a difference in stress states caused by gravity might also slightly favour the growth of a plastic zone in the hanging wall.

13) In computer tomography an X-ray source moves on a circular track and emits a fan of rays which penetrate a thin slice of the object. The intensity of the transmitted rays is measured by an array of detectors positioned on a circular arc, and recorded for any position of the moving source. From these data the X-ray absorption in any object element is calculated by computer and the final image is generated for any cross-section of the object.

14) Further, it is also obvious that the intermediate principal stress cannot be oriented normal to the frictional wall. Instead it will make an acute angle (with the wall) that opens against the direction of basement movement. The normal fault, although dipping in the opposite direction to the movement of the basement, should therefore have a curved strike line that is *convex towards the side of the downthrown block.* This may also be inferred from Fig. I.2-32. It seems, however, that normal faults with concave dip curvature and convex strike curvature are less common than listric faults that are concave towards the downthrown block in both dip and strike.

15) Expressed in polar coordinates centred at the fault edge, and referred to Fig. I.2-34A, the shear stresses induced by the 'tearing' mode fracture (mode III fracture) are (cf. Lawn & Wilshaw, 1975; p. 54): $\tau_{13} = \tau_{31} = - K_{III}(2\pi r)^{-1/2} \sin \theta/2$ and $\tau_{23} = \tau_{32} = K_{III}(2\pi r)^{-1/2} \cos \theta/2$, where θ is the angle (counted anticlockwise) between the positive x_1-axis and the position vector, and K_{III} is the 'stress intensity factor' – a parameter determined by the geometry of the elastic body containing the fracture and by the external load system. The maximum shear stress $\tau_{max} = [\tau_{13}^2 + \tau_{23}^2)]^{1/2} \approx r^{-1/2}$ acts in the x_1, x_3-plane and on an orthogonal plane the normal of which includes the angle $\theta/2$ with the positive x_1-axis.

16) This qualitative statement is in agreement with the results of Chinnery's (1966) two-dimensional and purely elastic analyses of the stress changes associated with the growth of a planar shear crack in an homogeneous elastic medium subject to uniform far-field stresses.

17) The example of Fig. II.7-40 may stand for many, more complex tectonic situations in which the horizontal stresses are anisotropic and their differential stress varies with depth. The compressive second phase in Fig. II.7-40 may, for instance, be produced by a (sinistral) horizontal simple shear that does not affect the role of the overburden stress as *intermediate* principal stress, nor introduces a dependence of the stresses on x and y. When the resultant stresses are in the limit state conducive to wrench faulting, potential fault elements will be vertical and their strike line will be straight. The strike direction, however, will change with depth because of the changes in the direction of horizontal principal stresses. (This is elaborated in the text immediately following.) It is obvious that the twisted elements cannot combine into a coherent smooth vertical fault surface, the incoherence becoming more noticeable the wider the fault zone extends along strike. Note that this incoherence of wrench faults in the vertical direction should not be confused with the well-known en-échelon array of such faults ('Riedel' shears) in plan, which is discussed in Chapter I.4.

18) The maximum of the horizontal shear stress is half the difference in magnitude of the horizontal principal stresses (see Chapter II.1, eq. II.1-4). The value of this shear stress will therefore vary with depth when the principal stress difference varies with depth (as is the case for the stresses of the first tectonic phase in the example of Fig. II.7-40).

A variation of the horizontal principal stress difference with depth may be caused by variations in lithology and effective confining pressure. Such a change may, in particular, take place across the *interface of beds* that differ in stress response to the same regional straining. A second tectonic phase, which changes the direction of bed-parallel intermediate principal stresses, may then also cause this principal direction to change discontinuously across the interface. Faulting across such an interface will therefore be complicated by an abrupt change in strike direction.

19) Mathematically, this is easily shown by means of the integrability condition II.7-19 in Section II.7.6. Only the unit vectors $\vec{e}_{II}$ and $\vec{e}_{III}$ along the axes of σ_{II} and σ_{III} are functions of the depth coordinate z, while all quantities are independent of x and y. Therefore, only the x- and y-components of the curls of $\vec{e}_{II}$ and $\vec{e}_{III}$ do not vanish. Since, moreover, the vertical unit vector $\vec{e}_I$ has zero x- and y-components, all inner products of vectors in eq. II.7-19 vanish except for the second one, which means that the condition is not satisfied.

20) Displacement along antithetic adjustment faults is usually small and the limits of seismic resolution make it difficult to decide on the curvature of these faults. Moreover, antithetics, which appear straight on reflection seismic time-sections, are convex-upward on a depth-converted section because of the increase in seismic velocity with depth.

21) If instead of a 'kink' the dip of a normal fault changes gradually, the rock mass sliding along the fault may adjust to the shape of the fault partly by rotation, partly by antithetic faulting. The rigid body rotation would not contribute to a change in the magnitude of dip-slip. The combination of 'double gliding' deformation with bodily rotation has not yet been resolved satisfactorily.

22) When the reclination of the main fault is caused by compaction of the footwall and at the same time the material of the hanging wall is compacting too, although to a lesser degree, the dip-slip formulae in Figs. I.2-47 and 48 have to be modified sligthly. The reference position s_1 of the main

fault will then be flatter, since the gap between s_1 and s_1' is reduced by footwall compaction. Accordingly, the angle α in the formulae is to be replaced by a smaller angle and β by a somewhat greater angle.

23) The question here is: What may have allowed $\sigma_{\parallel}$ to attain the required low value in the experiments? In the experiments by Logan et al. glass developed at the gouge/host-rock interface, which may have allowed horizontal extension of the gouge.

It should not be too difficult to contrive a shear apparatus that would allow the driving plates to be extended during the shearing process. We would expect the enforced decrease in $\sigma_{\parallel}$ to promote the predominant formation of X-shears.

24 In theoretical work on shear band localisation (Rice, 1973; Rudnicki & Rice, 1975; Vardoulakis, 1977; and others) it is a special point of discussion whether the formation takes place at peak stress or somewhat before or after. Attempts to compare theoretical results predicting pre-peak formation of shear bands in homogeneous, uniformly strained materials with experimental results appear to be ineffectual, because the experimentally determined load curve is not recording the switch from uniform deformation to shear band localisation in an *homogeneous* material, but represents the load displacement behaviour of an *inhomogeneous system* (i.e. the sample containing one or several shear bands which may or may not intersect the test sample).

25) Slip and stress discontinuities could only coexist on slip surfaces which are part of the boundary of the plastic region. See also note 22 of Chapter II.7.

26) See also note 17 of Chapter II.7.

27) This is not a general phenomenon, but results from Young's modulus being assumed constant. Uniform straining therefore produces a uniform reduction in horizontal compressive stress throughout the whole layer. This makes Mohr circles that are associated with shallow depth grow faster proportionally and touch the limit line earlier than the large circles that correspond with greater depths.

28) The width of the calculated shear bands will to some degree depend on the mesh size used.

29) As to the observed proportionality, it is still an open question how general the conditions for strict proportionality are.

30) Naturally, the arching of σ_I-trajectories is also found in purely elastic analyses (Hafner, 1951; Sanford, 1959; Couples, 1977; Gangi et al., 1977).

31) Experiments of this type have a long history in soil mechanics ('trap-door problem'): Terzaghi, K., 1936: "Stress distribution in dry and saturated sands above a yielding trap-door", Proc. Int. Conf. Soil Mech., Cambridge, Mass.; Vol. I, 307-311; Fiebinger, K., 1936: "Ein Beitrag zur Bestimmung des Druckes von kohäsionslosem, trockenem Sand auf einer beweglichen Bodenklappe", Thesis, Techn. Univ. Vienna; Loos, W. & Breth, H., 1949: "Kritische Betrachtung des Tunnel und Stollenbaues und der Berechnung des Gebirgsdruckes", Bauingenieur H.5, 29-135.

32) In frictional plasticity, stresses depend on infinitesimal strain increments rather than on total strains. The direction of maximum compressive incremental strain will always make an angle of 45° with the direction of simple shearing. One might therefore expect the plastic deformation to leave the elastically established 45° orientation of σ_I in a mechanically isotropic material unchanged. However, owing to the pressure sensitivity ($\varphi \neq 0$) of a frictional plastic material, the directions of the maximum compressive stress and the greatest incremental shortening do not have to coincide (see "Non-coaxiality of stress and incremental strain" in Section II.7.4).

33) The stiffer the rock and the narrower the confinement of the precursor shears, the faster the pressure will build up between Riedel shears and will induce P-shears. This may explain why in the experiments by Bartlett et al. P shears were found after a displacement along the 'basement' fault of about one-third of the thickness (0.5 cm) of the sheared limestone layer.

I.3. EXTENSIONAL FAULTING

Most of the extension in the brittle crust is accommodated by normal faulting. In comparison to this mechanism, ductile stretching and extension by open tension fractures are of minor importance, since in the brittle crust ductile stretching is restricted to a few rock types, such as evaporites or shales, and open extension fractures can only exist under small effective overburden loads. Extension may be imposed upon a crustal segment, or some part of an overburden, in various ways, giving rise to different styles of normal faulting, which are the subject of this chapter. For the purpose of a discussion on the mechanics involved, the styles of extensional faulting may be roughly grouped in three classes (Fig. I.3-1). In the first one, extension is confined to some upper strata ('thin-skinned' extension); in the second class, the extension is imposed upon brittle layers by the ductile flow of substrata, and in the third class a brittle overburden is extended by the rifting apart of a more competent basement. (Note again that we do not restrict the term 'basement' to a highly brittle crystalline basement underlying the sedimentary column, but apply it to any mechanically stronger ('competent') strata that control the deformation of overlying weaker rocks.)

Extension may be confined to upper strata in essentially two ways. As sketched in Fig. I.3-1A, a detachment plane or layer may allow sheets of strong rocks (nappes), or of weak deltaic sediments, to glide or creep down a slope, driven by their own weight. This process may take place without the interference of substrata. The second mode of thin-skinned extension is associated with the bending of a thick layer (Fig. I.3-1B) by lateral compression or by local subsidence of substrata. Faulting will be restricted to the actually extending (upper) part of the bending layer and will be most intensive in the region of greatest curvature ('crestal faulting').

Extensional fault structures, which are generated by the flow of a ductile base, may attain quite different appearances, and these seem to depend strongly on the details of the basal flow. The Figures I.3-1C,D,E,F,G indicate schematically parallel normal faulting caused on a delta slope by deep-seated creep flow (C), the widening of a sediment trough by the lateral extrusion of salt or shale into marginal ridges (D), crestal grabens above rising domes or ridges (E), and fault blocks accommodating the horizontal extension of a ductile base by 'domino'-style rotation (F) or by movements in 'horst and graben' style (G).

Extensional overburden structures are often directly caused by faulting in a competent basement. The basement may be involved by means of the rifting and subsidence of basement blocks along normal faults, as is sketched in Fig. I.3-1H. A horizontal extension of the overburden is also imposed by an extensional offset of a wrench fault in the basement (Fig. I.3-1I), generating 'pull-apart' structures (discussed in the chapter on wrench faulting

(Section I.4.7)). Finally, two-dimensional horizontal extension of the basement will generate a mosaic-type configuration of normal and oblique-slip faults, as sketched in plan view in Fig. I.3-1J, which will, in turn, induce overburden structures, the complexity of which may thwart any mechanical analysis.

In general, a number of the extensional processes outlined above will operate together in an extensional province. This holds in particular for the gravitational tectonics on a submarine delta slope, which comprises extensional processes of the type sketched in Figs. I.3-1A,B,C,D,E. For this reason, and because of the importance of deltaic sediment complexes as prolific hydrocarbon provinces, we shall turn first to the mechanics of gravitational delta tectonics.

1. GRAVITATIONAL FAULTING ON DELTA SLOPES

In Chapter I.2 (Sections 4 and 6), when dealing with fault curvature and antithetic subsidiary faults, we have repeatedly referred to normal faults on delta slopes. While we were primarily concerned there with features of individual faults, we shall consider here extensional faulting in the context of mechanical models of outbuilding river deltas. At present such models are still restricted to two dimensions, dealing with stresses and deformations under plane strain conditions in a vertical dip section of the deltaic sediment body. Such a model applies to parts of a delta, where geometry and stratigraphy of the sediment body, sedimentation rate and fluid pressure vary little or not at all in strike direction. When parameter variations along strike cannot be neglected, the model may still be applied (providing the variations are symmetrical with respect to a certain dip section), albeit restricted to the plane of symmetry (see 'Plane Strain' in Chapter II.5).

The basic gravitational processes that operate on a prograding delta slope are outlined in the dip section diagrams of Fig. I.3-2. The diagram of Fig. I.3-2A represents the 'thin-skinned' process of a sediment sheet gravitationally gliding along a path that is not affected by substratal deformations and which extends over a considerable distance subparallel to the delta slope ('thin-skinned' process). Under the weight of the outbuilding delta wedge the basal substrata will bend downward as sketched in Fig. I.3-2B. It is, however, unlikely that this isostatic subsidence can interfere with the deformation processes in the upper layers of the slope, because the flexural deformation is 'smeared out' over much larger distances than the actual slope (Walcott, 1972) and may develop at slower rates than that of the delta progradation. In contrast, mobile substrata of salt or overpressured shales beneath a prograding or stationary delta slope (Fig. I.3-2C) may deform as a result of differential loading, gravitational instability of the lighter material, or divergence of the underlying crustal basement (not shown in the figure), and thus form synclinal troughs and diapiric structures. The overlying sediments may then slide along normal faults into the troughs, while extension above rising diapirs will be accommodated by roof grabens (Fig. I.3-1E). Since these extensional processes on the slope are controlled by substratal deformations, they are of the 'thick-skinned' type. The reader can find various examples of thin- and thick-skinned extensional delta structures in Bally's seismic atlas (Bally, 1983). The two types of processes may operate in different parts of a delta, but also in the same part. An indication of the latter is seen in Fig. I.2-52, where thin-skinned slide structures have been affected by the flow of the underlying shale.

In the following we consider 'thin-skinned' and 'thick-skinned' models separately and compare the conditions for their application.

The 'Thin-Skinned' Gravity Slide Model (Crans et al., 1980; Crans and Mandl, 1980, 1981; Mandl and Crans, 1981). In this model, which is essentially based on Terzaghi's theories of land and submarine slides (Terzaghi, 1950, 1956), submarine sediment masses glide along paths that are not affected by substratal deformations (Fig. I.3-3). On the gentle delta slopes,

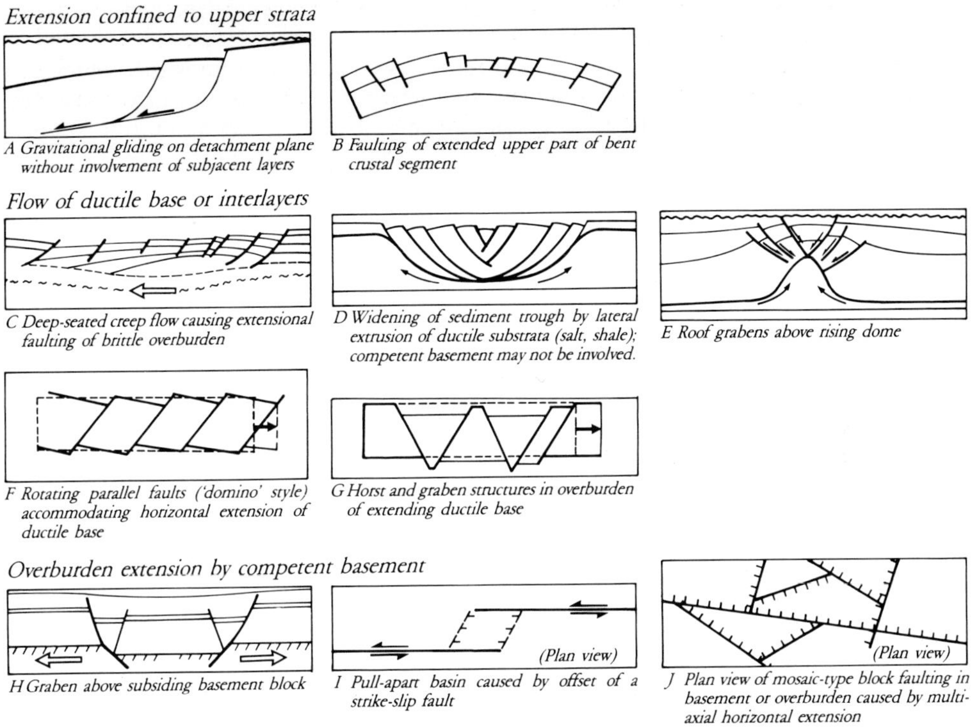

Fig. I.3-1 SCHEMATISED MODES OF EXTENSION THROUGH NORMAL FAULTING

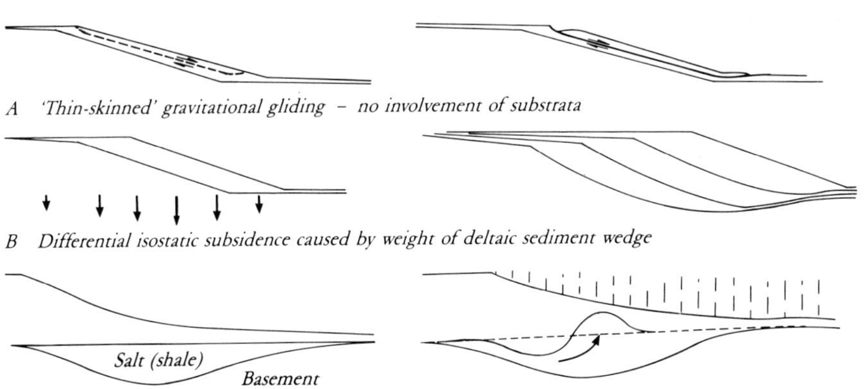

Fig. I.3-2 DIAGRAM OF GRAVITATIONAL PROCESSES IN OUTBUILDING DELTAS

with typical slope angles (δ) between 0.5 and 3°, sliding subparallel to the delta slope requires pore pressures well in excess of the corresponding pressures of hydrostatic fluid columns to reduce the 'effective' weight of the overburden, which is carried by the intergranular contact pressures – the source of the sediments frictional resistance (see Sections II.3.2 and II.7.2, "The effective stress principle"). Such abnormally high pore pressures are common in large deltas where low-permeability (clayey) sediments are deposited at rates that prevent complete drainage and compaction, and therefore allow excess pressures to build up (see, for example, the pore pressure data in Fig. I.2-19). The overpressuring may start at very shallow depth, but excess pressures of magnitudes measurable in bore holes are usually first encountered at depths below a few hundred metres. This overpressure top runs approximately parallel to the slope, as indicated in Fig. I.3-4 for an idealised prograding slope, and only drops off upslope, where coarser sediments favour drainage, and at the base of the slope, where the sedimentation rate is extremely low. The facies boundary remains essentially horizontal during a normal progradation phase.

In Fig. I.3-5A it is shown schematically how the pore pressure p, after an approximately hydrostatic increase with depth in a top layer of comparatively high permeability, rises drastically at the transition to very low-permeability sediments in the state of delayed compaction. Note that in the figure the 'depth' coordinate z is measured along an axis perpendicular to the slope surface (Fig. I.3-3) and, therefore, differs slightly from the actual depth below the sediment surface. It is often observed that the gradient of the abnormal pore pressure quite closely approximates the gradient of the total 'overburden' stress $\sigma_\perp$ that acts upon slope-parallel planes. Hence, in the depth interval where this takes place, the 'effective overburden' stress $\sigma'_\perp$ $(=\sigma_\perp - p)$ will maintain the value it attained at a shallower depth, represented by Z_{ig} in the idealised pore pressure profile of Fig. I.3-5A. In the same interval the frictional shear strength on slope-parallel planes will remain constant (except for changes in the angle of internal friction φ), while the shear stress τ on slope-parallel planes increases with depth independently of the excess fluid pressure.

In view of the 'thin-skinned' character of the model and of the small slope angle, it seems justified to treat the delta slope as infinitely long. We may further assume that the upward drainage flow is perpendicular to the slope. This is strictly the case when the compaction behaviour of the sediment does not change in dip direction and the axes of any permeability anisotropy are parallel and perpendicular to the slope. The upward drainage flow is then driven by the negative gradient of the excess pore pressure and, according to Darcy's law (Chapter II.10, Section 2), exerts the upward drag force $\eta V/k_z$ (eq. II.10-2) upon the sediment in a unit cube, where η is the viscosity of the fluid, V (<0) the seepage rate in the negative z-direction and k_z the permeability in the flow direction. The stress components $\sigma'_\perp$ and τ are then obtained as rigorous solutions of the equations for static equilibrium (see Example No. 5 in Section II.2.2, reading z for x_3):

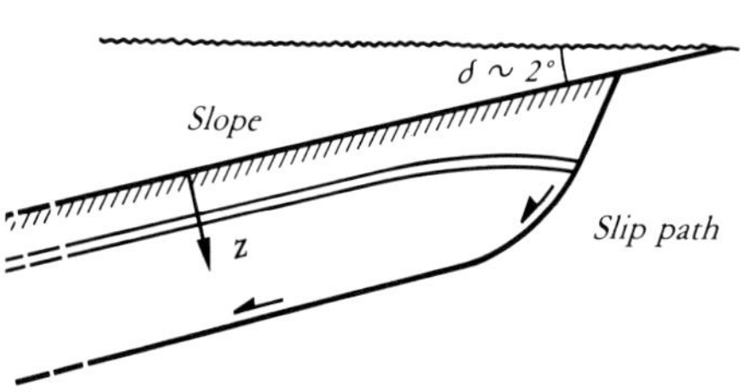

Fig. I.3-3 'THIN-SKINNED' SUBMARINE SLIDING (Schematic dip section.)

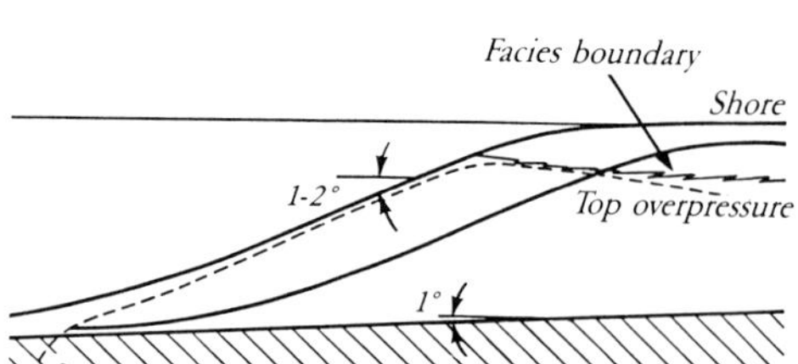

Fig. I.3-4 PROGRADING DELTA SLOPE (idealised dip section)

are then obtained as rigorous solutions of the equations for static equilibrium (see Example No. 5 in Section II.2.2, reading z for x_3):

$$\tau = \gamma' z \sin \delta$$
$$\sigma'_\perp = \gamma' z \cos \delta + F_z \qquad \text{(I.3-1)}$$

where $\gamma' = (1-\Phi)(\gamma_s - \gamma_w)$ (I.3-2)

is the 'submerged' weight, with Φ being the porosity and γ_w, γ_s the specific weights of the pore water and the solid component, respectively, and

$$F_z(z) = \eta \int_0^z (V/k_z)dz < 0 \qquad \text{(I.3-3)}$$

is the resultant drag force exerted by the drainage flow upon the solid material inside a column of unit cross-section and extending from the slope surface down to the 'depth' z.

When, as stated above, $\sigma'_\perp$ remains constant in a certain interval $z \geqslant Z_{ig}$, the disappearance of the gradient $d\sigma'_\perp/dz$ implies, in view of the above expressions for $\sigma'_\perp$ and F_z, that

$$\gamma' \cos \delta + \eta V/k_z = 0 \qquad \text{(I.3-4)}$$

This relation states that the z-component of the submerged weight of the sediment in a unit volume of the depth interval considered is completely balanced by the drag force of the upward drainage flow. Hence, in this interval the fluid carries the weight of the sediment it is seeping through, and throughout the whole interval the sediment is only loaded by the effective overburden stress

$$\sigma'_\perp = \gamma' Z_{ig} \cos \delta \qquad \text{(I.3-5)}$$

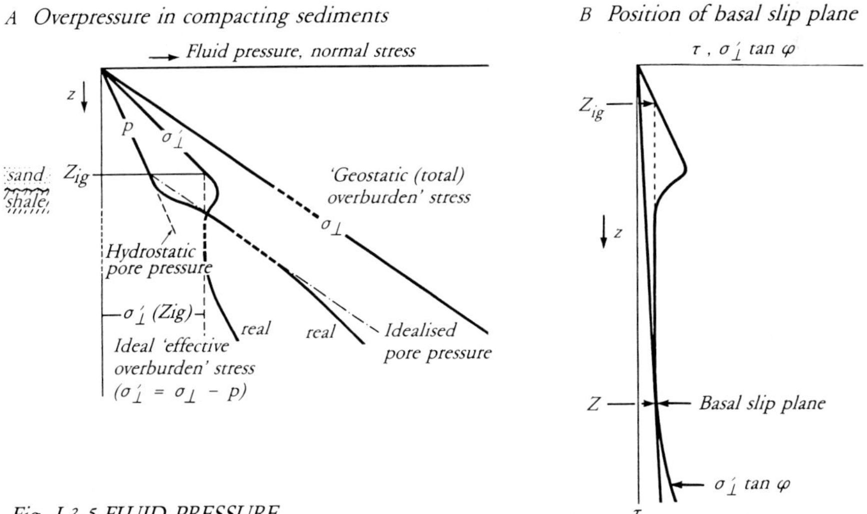

Fig. I.3-5 FLUID PRESSURE

which corresponds to the shallower depth Z_{ig}. This makes good sense in terms of realistic subsidence rates and shale permeabilities, as is shown by a quantitative analysis in Ch. II.10 (Example). Naturally, with increasing depth and age of the overpressured sediment, the local drainage flow and the associated seepage drag will eventually diminish, giving rise to a gradual increase of $\sigma'_\perp$, as shown schematically in Fig. I.3-5A.

While the effective 'overburden' stress $\sigma'_\perp$ remains constant in the overpressured interval considered, the shear stress τ on slope-parallel planes increases linearly with depth (eq. I.3-1). Consequently, when the compacting highly overpressured sediments have reached a certain thickness, the shear stress τ will attain the limiting strength value at the critical 'depth' Z. This is shown schematically in Fig. I.3-5B, in which the friction angle φ is assumed to be constant and the small cohesive strength τ_o of the deltaic sediment is neglected.

At the *critical depth Z*, the sediments will start sliding somewhere on a slope-parallel plane (point A in Fig. I.3-6A). For the idealised pressure profile of Fig. I.3-5A, the critical 'depth' is determined by equating the expression for τ (eq. I.3-1) with the shear strength $\tau_o + \sigma'_\perp \tan\varphi = \tau_o + \gamma' Z_{ig} \cos\delta \tan\varphi$, which yields:

$$Z = \tau_o(\gamma' \sin\delta)^{-1} + Z_{ig} \tan\varphi(\tan\delta)^{-1} \qquad \text{(I.3-6)}$$

For realistic parameters Z becomes about 10 times the depth (Z_{ig}) at which overpressuring started.

Naturally, the exact position of a basal slip plane will also depend on variations of φ and τ_o with depth, which, in reality, will make the $\sigma'_\perp \tan\varphi$ line in Fig. I.3-5B more 'wrinkly'. Obviously, this may even allow for several slip planes being activated at the same time, since the τ-line may touch the strength line in more than one point.

Finally, when drainage is prevented by impermeable strata (e.g. salt), the pore pressure can only increase with depth at the hydrostatic gradient, with the exception of the top of the sealed sequence. There a jump-like increase in pore pressure is accompanied by a corresponding jump-like decrease in $\sigma'_\perp$. If this reduction is sufficient to bring the τ-line and the strength line in Fig. I.3-5B in contact, slip will start at the *top* of the sealed, overpressured sequence.

The Gliding Process. While for the idealised slope considered and a prescribed pore-pressure profile the shear stress and effective normal stress that act on any slope-parallel plane are known (eqs. I.3-1,5) from considering static equilibrium, the third stress component $\sigma'_\parallel$ which acts parallel to the slope (Fig. I.3-6A) is not yet known. Its value depends on the constitutive behaviour of the material and cannot be ascertained from statical considerations alone. We know, however, that the value of $\sigma'_\parallel$, which is associated with given stresses $\sigma'_\perp$, τ, cannot exceed the value of $\sigma'^p_\parallel$ of the passive limit state, nor can it be smaller than the value $\sigma'^a_\parallel$ of the active limit state (see Section I.2.2, 'Tectonic limit states', or Section II.7.3, Example No. 2, Fig. II.7-11). When slope-parallel sliding starts at critical depth in some, as yet undetermined, local area, the material at that location must be in the active limit state with $\sigma'_\parallel = \sigma'^a_\parallel$, as shown in the Mohr diagram of Fig. I.3-6A. The overlying sediments will experience sufficient extension to be also brought into the active limit state, if this state has not already been established in the 'hanging' slope before the onset of slope-parallel gliding. The slide can then extend from the slip base up-dip towards the slope surface (Fig. I.3-6B) along one or more Coulomb-type normal faults. In the overpressured region the normal faults steepen up-dip and attain a listric shape, as has been explained in Chapter I.2 (Section 4, Fig. I.2-18)[1)]. To illustrate this further, we consider in Fig. I.3-7 the limiting stress circles that are associated with the basal slip plane (z=Z) and with two higher levels inside the overpressured interval where $\sigma'_\perp(z) = \sigma'_\perp(Z_{ig})$. Since τ and $\sigma'_\perp$ are known for any z-value in this interval, the associated active limit circles are defined and the direction of the maximum compressive stress can be readily constructed by means of the pole method (see Section II.4.2). As clearly shown in the diagram, these directions and the associated potential slip elements that are synthetic with the slope steepen gradually in an upward direction (decreasing z).

As gliding proceeds (Fig. I.3-6B), shear softening will reduce the shear strength of the activated part of the potential slip plane and, consequently, decrease the value of the basal shear stress ($\tau \rightarrow \tau^*$) which opposes the gliding motion. While the overburden-carrying stress $\sigma'_\perp$ remains constant, the slope-parallel stress $\sigma'_\parallel$ will increase monotonically in downslope direction to compensate for the imbalance in slope-parallel forces caused by the reduction of the basal shear resistance. The stress diagram in Fig. I.3-6B demonstrates that for a whole range of increased $\sigma'_\parallel$-values the stress circle no longer touches the limit line. Therefore, ideally, over a certain length the moving sediment sheet should not be disturbed by faulting and should move as a single block.

Of course, the *length of the 'rigid' block* is limited, since somewhere downslope $\sigma'_\parallel$ may reach the passive limit value for which the associated stress circle again contacts the limit line (Fig. I.3-6C). This marks the *compressive 'toe'* of the slide, where the basal slip plane terminates in one or more low-angle thrusts of the type P indicated in the stress diagram. Hence, extensional head region, compressional 'toe' and an intermediate 'rigid' sheet constitute the ideal, complete gravity slide, the main fault features of which have been illustrated by the kinematic model of Fig. I.2-41 in Section I.2.6. Naturally, the increase in 'slope pressure' $\sigma'_\parallel$ must be balanced somewhere downslope, since otherwise the whole slope would turn into an accelerating slide. This balance may be accomplished, for instance, by

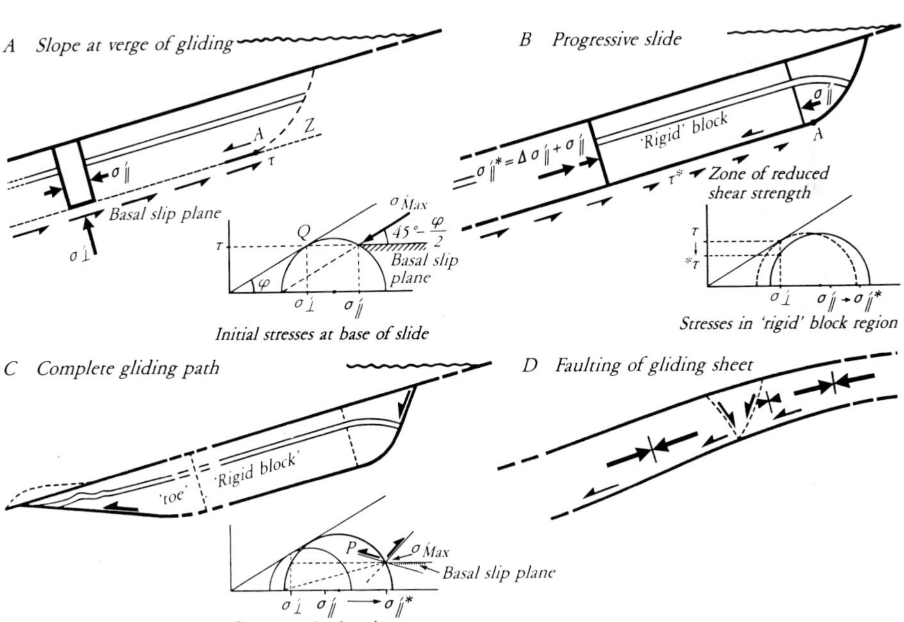

Fig. I.3-6 GRAVITATIONAL GLIDING ON DELTA SLOPE

the 'buttressing' effect of a regional decrease of slope angle, or by a decrease in overpressure near the 'foot' of the delta slope, which implies an increased effective overburden and hence an increased frictional strength.

Although the compressional 'toe' of a deltaic gravity slide is not a topic of extensional tectonics, it is considered here because it limits the maximum length L of the sliding sediment sheet. An average distance between the extensional 'head' region and the 'toe' of the slide can be estimated by equating the unbalanced part of the slope-parallel weight component of the sediment package and the extra support force that is provided by the higher 'slope pressures' $\sigma'_{\parallel}{}^{p}$ in the passive limit state. The balance reads:

$$\overline{[\sigma'_{\parallel}{}^{p} - \sigma'_{\parallel}{}^{a}]} \cdot Z = L \cdot \Delta\tau \qquad \text{(I.3-7)}$$

where the term on the left represents the average difference between the limiting values of the slope-parallel normal stresses times the thickness of the slide, and $\Delta\tau$ on the right-hand side is the average shear reduction of the basal shear strength. Using the expression for the undisturbed shear stress at the potential glide base $\tau = \gamma' Z \sin\delta$, we may rewrite equation (I.3-7) as

$$L/Z = \overline{[\sigma'_{\parallel}{}^{p} - \sigma'_{\parallel}{}^{a}]}(\gamma' Z \sin\delta)^{-1}[\Delta\tau/\tau(Z)]^{-1} \qquad \text{(I.3-7a)}$$

The difference of slope-parallel normal stresses can be expressed in terms of $\sigma'_{\perp}$ and τ by means of Coulomb-Mohr's limit condition for the passive and active limit states, but the actual determination requires assumptions for the fluid pressure, since $\sigma'_{\perp}$ as function of z depends on the overpressure profile. If the idealised overpressure of Fig. I.3-5A is assumed and changes in sediment thickness and Z_{ig}-level after sliding has started are neglected, after some manipulations equation (I.3-7a) can be brought into the simple form [2)]

$$L/Z = \pi (\cos\varphi)^{-1}[\Delta\tau/\tau(Z)]^{-1} \qquad \text{(I.3-8)}$$

where $\Delta\tau/\tau$ is the relative reduction of the basal shear stress by shear softening, i.e. a kind of 'brittleness index'.

Extrapolating from soil mechanical evidence for clays which are normally compacted under effective loads, one would not expect $\Delta\tau/\tau$ to exceed a few percent. Accordingly, the length of the slide could, in general, exceed its thickness by two orders of magnitude. This means that the distance between the extensional head and a potential toe of the slide on clays that are not overconsolidated may easily exceed the available length of the slope. Such a slide would remain *'open-ended'*. In those parts of the Niger Delta, for instance, where the listricity of normal faults and the associated structures strongly suggest the existence of basal slip planes, corresponding toe structures could not be found. (The toe structures observed at the foot of the delta slope appear to be related to deep-seated substratal creep and cannot be explained by the thin-skinned model.) The question then arises of how, in an open-ended slope-parallel slide, the extension and inflow of new sediment at the head of the slide is accommodated. The answer lies in the increase of $\sigma'_{\parallel}$ in the downslope direction (Fig. I.3-6B), which maintains static equilibrium on the slope after the onset of sliding (at least as long as no dramatic drop in basal resistance occurs, for example, by large-scale liquefaction). This extra stress causes compaction of slope sediments parallel to the slope, which allows the basal slip to spread gradually downslope and new sediment to be supplied along normal faults at the slide head. The rate of sliding in open-ended slides will be controlled by the rate of the slope-parallel compaction. When a toe develops, the rate of sediment influx at the slide head may become another slide-rate-controlling parameter.

While, ideally speaking, the slide rate should not increase downslope, one can easily envisage mechanisms that may make it increase in the downslope direction, thereby producing a local stretching of the gliding sediment sheet (Fig. I.3-6D). This, in turn, may locally induce the active limit state and corresponding normal faulting and graben-type structures. Mechanisms producing such variations in downslope slide rate may be associated with gently updoming substrata, non-uniformities in basal shear or compaction behaviour, local rise in overpressures, etc.

The slide model does not provide information as to *where the slide starts* (point A in Fig. I.3-6A). We may, however, argue that at the shelf-edge different compaction regimes exist inside the overpressured sediment (Fig. I.3-8). Onshore, where $\delta=0$, the maximum compressive stress is vertically oriented and compaction is therefore also vertical. On the slope, however, the direction of the maximum compressive stress rotates with depth inside the overpressured zone, as was shown in Fig. I.3-7, and so does the direction of maximum compaction. Therefore, at the shelf edge the compacting sediments experience some stretching and reduction of $\sigma_{\parallel}'$, which promotes the active limit state. Normal faulting is, therefore, most likely to occur first near the shelf edge. Other factors, such as strength inhomogeneities at the facies boundary (Fig. I.3-4), or local overloading by shelf sediments emerging from a regressing sea, may have a similar effect.

The sediments in the active and passive end regions of a gravity slide deform by internal slip along conjugate Coulomb slip surfaces ('Double-gliding model', Section II.7.4). When the variation of the overpressure with depth is known on the infinite slope, the three stress components $\sigma_{\parallel}'$, $\sigma_{\perp}'$ and τ are statically determined by the equilibrium equations, Coulomb-Mohr's limit condition and the condition that the submarine slope surface be free of effective stresses. The fields of principal stress directions and associated Coulomb slips can then be directly calculated. Fig. I.3-9 shows the computed slip line fields for both the active and passive plastic regions in a sediment sheet, which under the ideal overpressure conditions of Fig. I.3-5A has attained the critical thickness Z. [3)] The slip lines, which form the outer boundary of the complete slide, curve smoothly into the basal slip plane. In the case of a stationary delta, where the supply of sediment at the head of the slide is just sufficient to maintain a planar surface without increasing the thickness of the slope, the boundary slip surface at the head of the sliding sediment mass will develop into a *growth fault*.

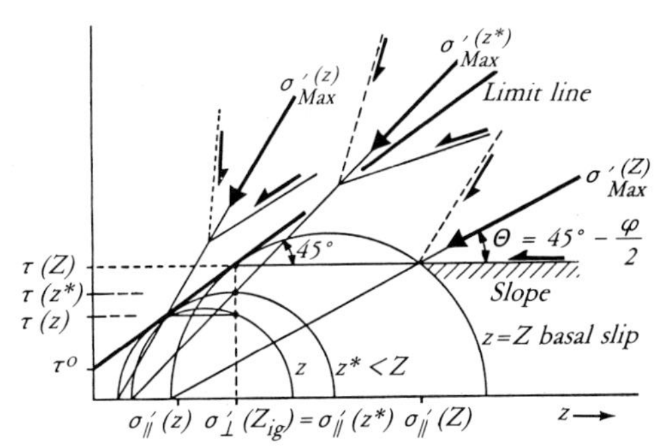

Fig. I.3-7 OVERPRESSURED ZONE IN LIMIT STATE (flattening of maximum compressive stress direction with depth).

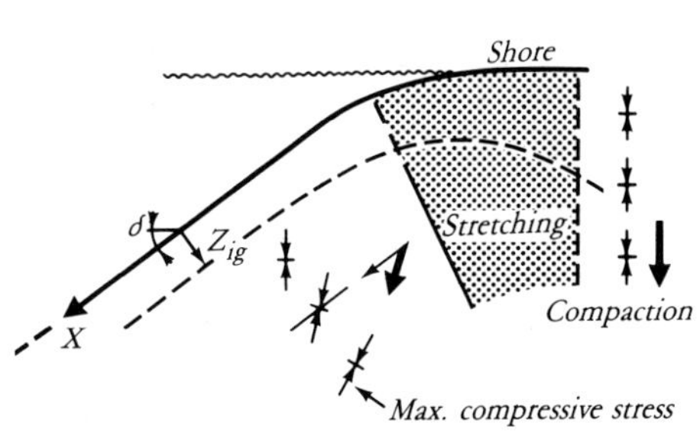

Fig. I.3-8 ONSET OF FAULTING AT SHELF EDGE

(The term 'growth fault' refers to any fault formed contemporaneously with, and more or less continuously active with, deposition, such that its throw increases with depth; strata on its downthrown side are thicker than the correlative strata on the upthrown side.)

The deformation patterns in the head and toe regions can be determined by applying the frictional-plastic flow rules, which are based on the double-gliding model (see Section II.7.4, 'Flow rules'; Fig. II.7-30), to the stationary slip line field and by prescribing the rate of slope-parallel displacement along the interfaces between the plastic and non-plastic regions. The reader is referred to a series of numerical calculations reported by Crans et al. (1980). In the head region, the calculated deformations of the sediment layers have the typical appearance of the 'roll-over' anticlines, which generally develop on the downthrown side of major growth faults (see, for example, Fig. I.2-51). Typically, the beds of these structures dip into the growth fault and show synsedimentary thickening towards the fault. The calculations demonstrate that different roll-over structures can correspond to the same stress state on the slope, the differences being caused by differences in the prescribed velocity profiles (the same holds for toe regions). In open-ended slides, for instance, the rate of downslope displacement may vary with depth because of differences in slope-parallel compaction. Instead of the continuous or quasi-continuous deformations calculated from flow rules, the deformation of the downthrown sides of listric normal faults (and similarly of the upthrown side of listric toe thrusts) may be discontinuous and concentrated in faults along some of the potential slip lines in Fig. I.3-9. Note, however, that these slip lines describe incipient faults, the shape and activity of which may be severely changed as gravitational sliding proceeds along the curved gliding path.

Synsedimentary Growth. As sedimentation continues and the thickness of the sediment package increases, while the delta progrades, the stresses and the fluid pressure change and the growing slide structures may have to adjust to these changes. There are essentially two parameters that control the synsedimentary development of the slide structures: *sedimentation rate* and *rate of overpressure relaxation.* The first determines the increase in the driving weight component on the slope, the second represents the opposing effect of a restrengthening of the basal slip plane by an increase in effective overburden. A comprehensive discussion of the interplay of the two factors and of its structural consequences is beyond the scope of this section. We shall therefore restrict ourselves to a few typical cases.

First we consider a growth fault as it would develop in a growing sediment package when the pore-pressure distribution is of the ideal type of Fig. I.3-5A. To begin with, the sediment layer has reached the critical thickness Z and the effective 'overburden' stress $\sigma'_{\perp}$ has attained the values $\gamma' z \cos \delta$ for $z < Z_{ig}$ and $\gamma' Z_{ig} \cos \delta$ for $Z_{ig} \leqslant z \leqslant Z$, and the shear stress $\tau = \gamma' z \cos \delta$. The head region of the slide is in the active limit state. As to the change in overpressure top, we assume that it rises somewhat less than the sediment surface – the distance between the two levels increasing to such an extent that at any time the growth fault merges tangentially with the active slip base. The results of a computer simulation (Crans et al., 1980) of fault growth and development of roll-over anticlines are shown in Fig. I.3-10. The right-hand side of the diagram presents the change in Z_{ig}. The growth is simulated in 10 sedimentation steps until the sediment package has reached a thickness of 5000 m. The result of the calculation is noteworthy in two respects. It shows the final fault as the envelope of previous ones and therefore bordering a seam of pre-sheared material. Secondly, the regression of the fault causes a drag reversal of the roll-over structures. The deformations of the initially slope-parallel bedding planes have been computed by using the previously mentioned frictional-plastic flow rules and by prescribing a reasonable creep profile downslope, as shown in Fig. I.3-10. The calculated model might approximately apply to synsedimentary fault growth in a *thick, overpressured shale sequence*.

Next, we consider the case in which rapid deposition of very low-permeability sediments forces the overpressure top to rise in step with the sediment surface. Since Z_{ig} now remains constant ($z=0$ representing the rising surface), the effective stress $\sigma'_{\perp}$ will remain unchanged on the slip base. Consequently, the value of the shear stress τ on the slip base cannot rise above the limit value it had when slip started. As the sediment package grows, the shear stress τ will therefore increase less with depth than in the initial sliding state. To balance the increasing downslope weight component of the accreting slope sediment an extra increase in the 'buttressing' pressure $\sigma'_{\parallel}$ is needed. Obviously, this will shift the passive limit state upslope, decreasing the distance L between slide head and toe. In other words, toe thrusts would tend to migrate upslope.

If, however, the *relaxation of overpressure dominates,* as in the case of very low sedimentation rates or the deposition of relatively high permeability sediments, the Z_{ig}-value will in-

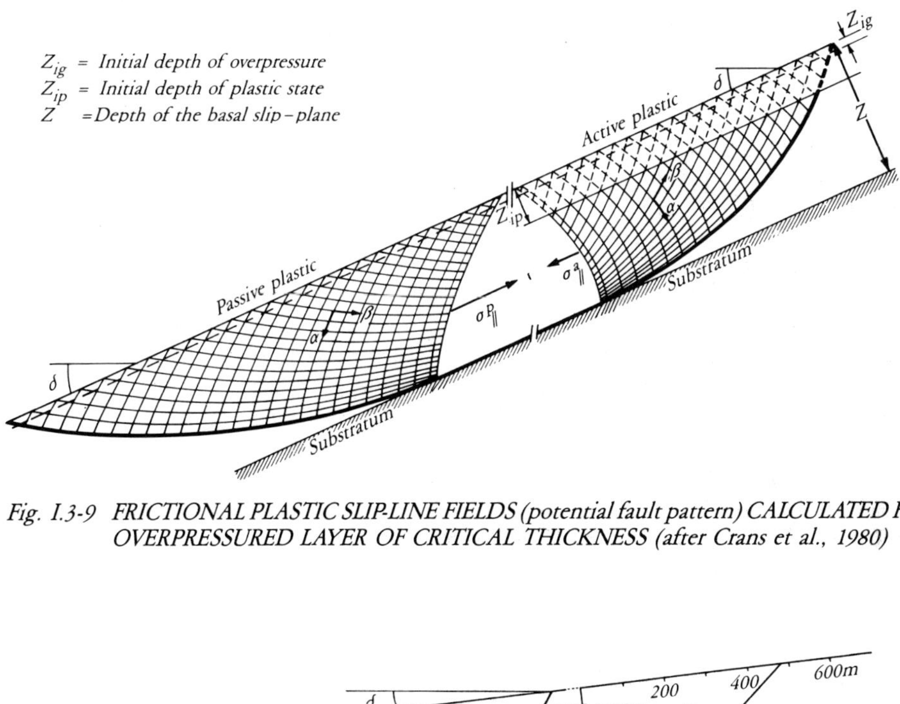

Fig. I.3-9 FRICTIONAL PLASTIC SLIP-LINE FIELDS (potential fault pattern) CALCULATED FOR OVERPRESSURED LAYER OF CRITICAL THICKNESS (after Crans et al., 1980)

Fig. I.3-10 COMPUTER SIMULATION OF GROWTH FAULT AND 'ROLL-OVER' STRUCTURES IN GROWING SHALE SEQUENCE

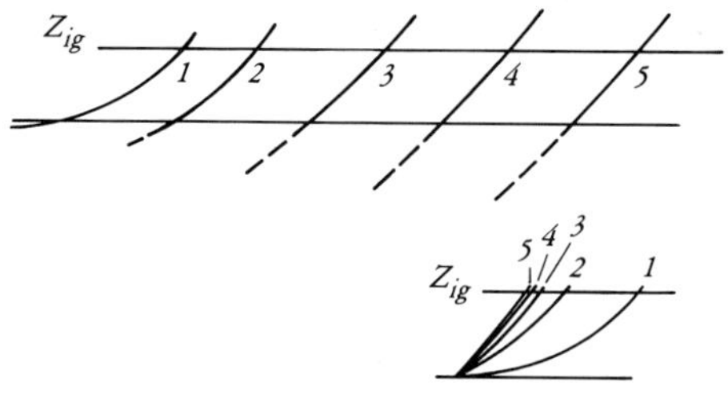

Fig. I.3-11
SYNSEDIMENTARY FAULT GROWTH AT STATIONARY OVERPRESSURE LEVEL ($\hat{Z}_{ig}$) (Good drainage conditions)

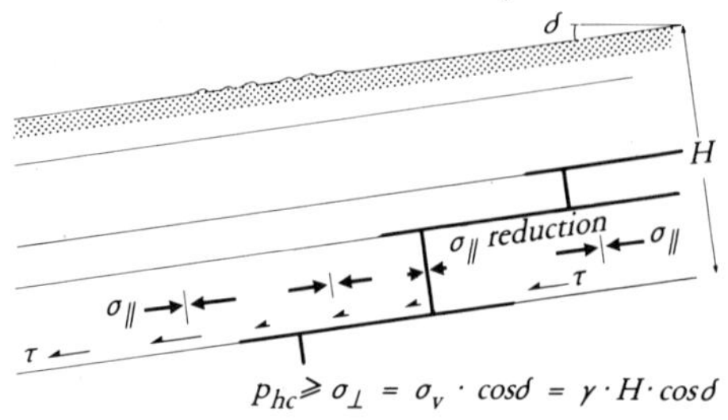

Fig. I.3-12
UPDIP MIGRATION OF HYDROCARBONS ON DELTA SLOPE WHERE $\sigma_{\parallel} \geqslant \sigma_{\perp}$

crease faster than is the case in Fig. I.3-10. The corresponding increase in $\sigma_{\perp}'$ tan φ will then exceed the increase in τ and the slip base will be restrengthened. This will first allow a reduction of the stress increase $\Delta\sigma_{\parallel}'$ (Fig. I.3-6B) that was induced to balance the loss in basal shear resistance. As a consequence, the toe region may first migrate downslope, but eventually the slide will be arrested by the restrengthening of the slip base.

In reality, draw-down of the overpressure top may remain restricted to the slide head near the delta edge where coarser sediment is deposited. This will affect the shape of the growth faults while leaving the slip base practically unaffected. This is illustrated in Fig. I.3-11 by the successive growth fault shapes (1,2,3,..) calculated for the case of an overpressure top that remains at a fixed distance from the slip base. Typically, the growing faults, which were originally curved in the overpressured region, straighten. Although our simple model does not make allowance for downslope variations in Z_{ig}, it seems plausible enough that the straightening growth fault will 'splinter' when joining the still-active slip base and will develop upwards- or downwards-diverging fault bundles ('horse tails') as sketched in Fig. I.3-11.

Hydrocarbon Migration. In concluding this subsection on thin-skinned delta tectonics we turn to aspects of hydrocarbon migration. In general, an important part of the hydrocarbons generated in deltaic complexes has to migrate through overpressured shales to reach the sandy potential reservoirs, which are mainly situated in the roll-over anticlines of major growth faults. By far the most efficient mode of hydrocarbon migration through thick rock barriers of very low permeability is 'hydraulic' fracturing of the rock by the intrusion of high-pressured hydrocarbon fluids. Basically, there are two types of *intrusion fractures:* 'dyke'-type fractures which cross beds and 'sill'-type fractures which follow bedding planes. The role of these natural hydraulic fractures in hydrocarbon migration has been recently re-examined by Mandl and Harkness (1987) and the ways in which they function are also discussed in the chapter on Poro-elasticity (Section II.6.4, Example No. 1). For hydrocarbons to fracture an intact bed of very low permeability, the hydrocarbon pressure has to overcome a component of total normal stress and the tensile strength of the rock. Since the tensile strength across bedding planes in sedimentary rocks is negligible and the hydrocarbon pressure (p_{hc}) in a mature source rock equals the total vertical overburden stress (σ_v), hydrocarbons can penetrate from a maturing source bed into (dipping) basal bedding or slip planes of the overlying shales. However, at locations where the sum of the total bed-parallel rock pressure $\sigma_{\parallel}$ and the bedding-parallel tensile strength $T_{\parallel}$ is smaller than the total stress $\sigma_{\perp}$ on the bedding planes, hydrocarbon dykelets will be preferentially formed. Since we have seen (Fig.

I.3-7) that along the basal slip plane of the slide $\sigma_{\parallel}'$ exceeds $\sigma_{\perp}'$ (and also the vertical overburden stress σ_V'), and increases in the downslope direction in the sliding sediments far beyond the effective overburden stress, and since obviously the same holds for the corresponding total stresses, we envisage a process of upslope sill – dyke migration of hydrocarbons in which, at the lower slope, migration along basal bedding planes predominates, while higher up on the slope in and near the extensional head region of the slide, the shale beds are crossed via a dyke-dominated fracture system.

For the reconstruction of migration paths on delta slopes, it may be useful to note that the formation of short dykelets will not be completely suppressed by the high slope-parallel rock pressures in the lower slope and along the glide base. When a fluid sill forms along a dipping bedding plane, the stress $\sigma_{\perp}'$ will be reduced by the pressure of the injected fluid and, because of the concomitant decrease in shear strength, the beds may no longer be capable of carrying the full original shear stress. As shown in Fig. I.3-12 for gently dipping beds on a long delta slope, the local reduction in shear stress by the intruded high-pressure hydrocarbons will cause a drop in bed-parallel rock pressure $\sigma_{\parallel}$ at the upslope part of the sill, which is compensated for by an increase of this stress in the down-dip direction to maintain slope stability. The reduction in $\sigma_{\parallel}$ will allow dykes to form and interlink sills at different stratigraphic levels. Although for a full understanding and quantification of this step-up mechanism further mechanical analysis is required, one may argue that the unhindered spreading of high-pressure fluids along a bedding plane actually impedes step-up to higher levels, because the location of $\sigma_{\parallel}$-reduction will shift and the subsequent increase in slope-parallel pressure will close an already formed dykelet. Favourable conditions for the 'sill – dyke step-up' through shales under high bed-parallel compression may, therefore, exist where the bedding planes are short or interrupted by lithological barriers.

Normal faults are normally expected to control hydrocarbon migration, firstly because they form barriers against transverse hydrocarbon flow, and secondly because they may serve as preferential migration paths. In deltaic environments the sealing quality of a normal fault will mainly result from the emplacement of a low-permeability clayey fault gouge, but also from fault gouges that have been produced by cataclasis only (Weber et al., 1978; Mandl et al., 1977). In both cases, the sealing capacity of the narrow fault zones proper is associated with the capillary pressure barrier of the water-saturated fault gouge against the entry of the non-wetting hydrocarbon fluids (see also Section II.9.3, "Migration and trapping of hydrocarbons"). Once, however, high-pressure hydrocarbons have found access to the interior of a normal fault, for instance via a permeable sand stringer that is dragged into the fault, the fault can be 'wedged open' by the fluids. The process is similar to the intrusion into bedding planes, but requires less hydrocarbon pressure, since the total normal stress acting on a normal fault amounts to, say, only about 65% of the total overburden stress.

In general, the wedging-open of a (non-vertical) normal fault will be a transient phenomenon, since the loss of frictional resistance will be accompanied by some slip and rearrangement of stresses, which almost immediately causes the fault to close again. In highly brittle rocks the closure may be imperfect, leaving some migration channels open along the 'closed' fault plane. In normally consolidated clayey material this is rather unlikely and we envisage along-fault migration as proceeding in periodic episodes, during which hydrocarbons enter the fault and accumulate in short 'stringers' or thin *'pockets'*, which then travel upwards along the fault plane. The driving force is not the high injection pressure required to emplace the hydrocarbon inside the fault, but a kind of 'buoyancy' force, namely the difference between the gradient of the upwards-decreasing total normal stress that acts from outside upon the fault plane and the smaller gradient of the static fluid pressure in the hydrocarbon pocket. [4)]

The Delta Slope on Mobile Substrata ('Thick-Skinned' Model). As pointed out earlier, the gravitational deformations of deltaic sediments may be strongly affected by the creep flow of deep-seated mobile strata of shales or salt, as has been recognised by many authors (e.g. Quarles, 1952; Merki, 1972; Dailly, 1976; Evamy et al., 1978; Humphris, 1978). The main driving forces of substratal flow are *differential overburden load* and *buoyancy.* A gradient in overburden load may cause the underlying mobile material to flow in the direction of decreasing load and to produce 'waves', 'pillows', ridges or domes of salt or shales, which in turn give rise to faulting of the overburden. This has been demonstrated in early experiments by Rettger (1935), where clay, resting on an undeforming base of sand, is squeezed by the differential load of a stationary sand wedge in the direction of the decreasing load (Fig. I.3-13). The extension of the overburden above the flat synclinal trough is accommodated by a normal fault, which extends into the clay. Notice that inside the clay the normal fault separates into a fan of faults; these have been dragged by the flow into flatter positions (see also Fig. I.2-27). (It seems therefore reasonable to assume that the faults have formed in succession – the flattest one being the oldest.)

The deformations of overburden and mobile substratum are coupled via the deformation of the interface. Any mechanical modelling of the process will therefore have to concentrate first upon the deformation of the interface. This problem is complicated, however, by the fact that the overburden deforms as an elastic/frictional plastic material, while the deformation of the substratum is dominated by viscous flow. It is obvious that some elastic stiffness and, in particular, the frictional shear resistance of the overburden will restrain the deformation of the interface, suppressing strong curvatures and delaying the growth of disturbances of the interface.

In a first approximation, the restraining effects of the overburden are neglected in a model developed by Lehner (1977), in which the overburden is represented by a distribution of vertical 'dead loads' that follow the deformation of the interface without resistance. In other words, the overburden is considered as an assemblage of infinitesimally thin vertical slices in frictionless contact, which adjust independently to movements of the interface. We restrict ourselves here to the case in which the overburden is at rest and has a planar slope whose angle is kept constant (δ = const.) by sedimentation, this compensating any local subsidence.

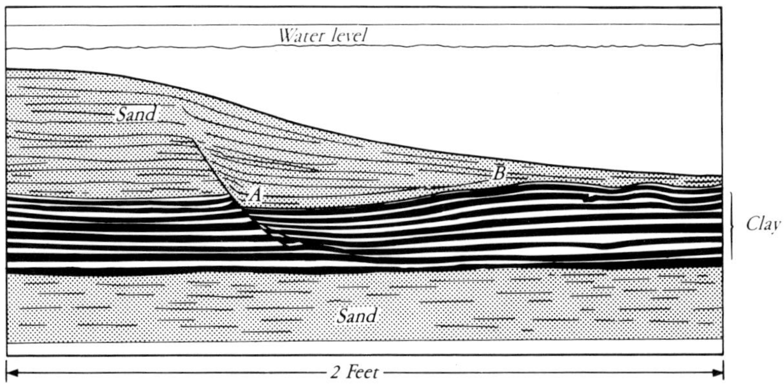

Fig. I.3-13 NORMAL FAULT PATTERN CAUSED BY LOCAL EXCESS LOADING (after Rettger, 1935)

The specific weights of overburden and mobile substratum are assumed to be equal in order to restrict the analysis to the effects of differential loading, excluding additional effects of buoyancy. The configuration of a dip section at the initial time $t=t_0$ is sketched in the diagram of Fig. I.3-14. It is assumed that at the initial time the interface between overburden and mobile substratum is still horizontal, as may approximately be the case when the delta slope has rapidly prograded before coming to a halt. The initial thickness of the mobile substratum is a prescribed function of the horizontal x-direction, corresponding to the relief of the underlying basement. Therefore, the deformation history of the interface is described by the thickness h of the mobile substratum as a function of location x and time t ($h=h(x,t)$).

The theory is based on two fluid laws: a kinematical one, expressing the conservation of volume of an incompressible fluid, and a mechanical one, relating flow rate and pressure gradient. If it is assumed that the material of the mobile substratum is incompressible, the change of volume in a vertical slice of thickness δx during the time interval δt is $(\delta h/\delta t)\cdot\delta t\delta x$, where δh is the change in height. This volume change has to be balanced by the difference δQ between the inflow and outflow of material during the time interval δt. When q is the rate at which material flows through a cross-section of height h at x and δq is the change in rate over a distance δx, the net flow is $\delta Q = (\delta q/\delta x)\cdot\delta t\delta x$. Conservation of volume is therefore expressed, with reference to a cross-sectional element of thickness δx, by $\delta h/\delta t + \delta q/\delta x = 0$, which in the limit for infinitesimal thickness of the element ($\delta x \Rightarrow 0$) and an infinitesimal time interval ($\delta t \Rightarrow 0$) becomes the well-known differential equation

$$\partial h/\partial t + \partial q/\partial x = 0 \qquad \text{(I.3-9)}$$

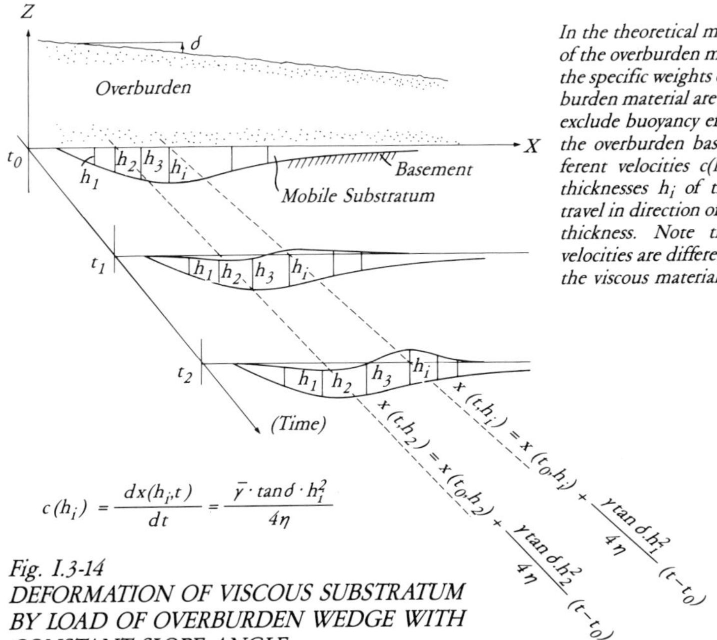

In the theoretical model the shear strength of the overburden material is neglected and the specific weights of substratum and overburden material are assumed to be equal to exclude buoyancy effects. The distortion of the overburden base results from the different velocities $c(h_i)$ at which different thicknesses h_i of the mobile substratum travel in direction of decreasing overburden thickness. Note that these 'kinematic' velocities are different from the flow rate of the viscous material itself.

Fig. I.3-14
DEFORMATION OF VISCOUS SUBSTRATUM BY LOAD OF OVERBURDEN WEDGE WITH CONSTANT SLOPE ANGLE (after F. Lehner, 1973, 1977).

A relationship between the local flow rate q and the pressure gradient dp/dx in the substratum is obtained by assuming that the layer thickness h remains much smaller than both the typical lateral dimensions of the problem and the radius of curvature of the interface, and that the fluid behaves as a (linearly) viscous material with viscosity η. The total flux q through a local cross-section of height h and unit breadth can then be approximated by the corresponding flux between two fixed parallel planes at the distance h [5]:

$$q = -(h^3/12\eta)dp/dx \tag{I.3-10}$$

Because the flow is parallel to the walls, the local fluid pressure may be considered equal to the weight of the local unit cross-sectional material column, consisting of mobile material, overburden and water column. Since we have assumed the same density for overburden and mobile material, it is easily seen that the weight of the material column (of constant length) varies in the x-direction only because of the difference in specific weight of water (γ_w) and overburden (γ_{ov}). The fluid pressure in the mobile substratum therefore decreases in the x-direction (i.e. the direction of increasing water depth) as

$$dp/dx = -\gamma' \tan\delta \quad \text{where } \gamma' = \gamma_{ov} - \gamma_w. \tag{I.3-11}$$

Hence, in the absence of a density difference between overburden and ductile material, the driving force in the mobile substratum is determined solely by the slope angle δ.

Substituting dp/dx from eq. (I.3-11) in eq. (I.3-10) and inserting the resulting expression for q into eq. (1.3-9) yields the differential equation for $h = h(x,t)$:

$$\partial(h^3 \tan\delta)/\partial x + (12\,\eta/\gamma')\,\partial h/\partial t = 0 \tag{I.3-12}$$

which for a constant slope angle δ reduces to the first-order equation

$$h^2\,\partial h/\partial x + (4\eta/\gamma' \tan\delta)\,\partial h/\partial t = 0 \tag{I.3-12a}$$

It can be shown that any solution h(x,t) to eq. (I.3-12a) may be constructed from a known initial distribution $h(x,t_o)$ with the help of the relation [6]

$$(dx/dt)_{h=\text{const.}} = (\gamma' \tan\delta/4\eta)h^2 \tag{I.3-13}$$

which represents the velocity at which a layer thickness h propagates in the x-direction. This velocity should not be confused with the actual flow velocity of the ductile material itself, the mean value of which (obtained by dividing the total flux q in eq. (I.3-10) by h) is only one third of the travel speed of the 'signal' h.[7] The position x(h,t) of a given layer thickness at time t is then obtained from its initial position $x_o(h,t_o)$ by the integral

$$x(h,t) - x_o(h,t_o) = (\gamma' \tan\delta/4\eta)\,h^2\,(t-t_o) \tag{I.3-14}$$

The deformation of the substratum/overburden interface in time can then be graphically constructed as illustrated in Fig. I.3-14, in which each h-value travels along a straight line in the x,t-plane. Since, according to eq. (I.3-14), higher thickness values of h travel faster than smaller ones, the initially flat interface is deformed into a synclinal trough with an outward-migrating ridge. Notice, however, that the mobile layer can nowhere attain a thickness greater than the maximum thickness that the layer had in the initial state; and in the special case of a uniform initial layer thickness the interface would not deform at all. Thus, *without buoyancy, a constant overburden slope cannot produce diapiric structures*. It can do so, however, as we shall see later, when the angle of the overburden slope decreases in the downslope direction.

The velocity at which a layer thickness h (m) propagates in the direction of decreasing overburden is inversely proportional to the viscosity (1 poise = 0.1 Pa.s = 1 dyne cm s^{-1}) and directly proportional to the square of the layer thickness, if the ductile material behaves as a Newtonian (linearly viscous) fluid. The latter assumption may reasonably well approximate the flow behaviour of *rock salt* that is not completely dry but contains a few tenths of a percent (by weight) of brine (Urai, 1983; Spiers, Urai et al., 1986)[8)]. Although a prediction of travel velocities based on eq. (I.3-13) remains speculative in view of the uncertainties regarding the viscosity of salt in steady-state creep, it may be illustrative to insert into eq. (I.3-13) the often quoted value of 10^{18} poise for the viscosity of salt. If, in addition, a slope angle of 3° and a specific submerged sediment weight $\gamma' = 12\ 200$ kg m^{-3} m s^{-2} are assumed, the formula predicts that a layer thickness h = 300 m will travel at a velocity of 4.5 mm/year. This value is considerably lower than the rate at which most deltas advance during the progradation phase. Even if we allow the viscosity of the salt to be reduced by an order of magnitude by the rise in temperature during burial under the sediment wedge, the travel velocity in the 300 m layer would, in general, still remain below the rate at which a delta slope locally progrades. Moreover, we have to recall that the model has neglected any restraint of the interface deformations by the strength and stiffness of the overburden. Therefore, it seems rather *unlikely that salt layers with a thickness of only a few hundred metres will deform fast enough to interfere with thin-skinned gravitational gliding during rapid progradation of a delta.* Since, however, the mobility of the (Newtonian) viscous substratum increases with h^2, one may easily envisage deeply buried salt deposits of sufficient thickness and with viscosities sufficiently reduced by elevated temperatures to form deep synclinal troughs and seaward-migrating ridges while the delta slope is still prograding. The widening troughs will then give rise to *deltaic megastructures*, which are bounded along the synclinal flanks by seaward- and landward-dipping major growth faults which terminate in the mobile substratum. An example of this 'thick-skinned' process will be discussed in the next section (Fig. I.3-26). It is also quite feasible that thin-skinned slide structures are embedded in the megastructures, which are caused by deep-seated flow, or that slope-parallel sliding paths, which formed during a progradation phase, were deformed by substratal flow after the progradation had stagnated.

Instead of salt, the mobile substrata may consist of *shales* (as, for instance, in the Niger delta). Although seismic records leave no doubt that thick shaly substrata under the differential load of a sediment wedge may deform in a similar mode to that for salt, a direct application of Lehner's model becomes somewhat problematic, since the assignation of a viscosity coefficient requires that a stationary creep flow be established by a given differential load. However, soil mechanical evidence strongly suggests that such a stationary ('secondary') creep phase is not established in clays; although under a given load the creep rate of clays decelerates first ('primary' creep), in contrast to salt it does not approach a constant rate ('secondary' creep), but either tends to zero or accelerates again ('tertiary' creep) at a certain stage, the latter rapidly resulting in shear failure (faulting). Obviously, the thicker the shaly substrata, the more continuous deformation they can accommodate by creep before the onset of failure. Hence, we conjecture that, at some stage depending on the thickness of these strata, the substratal shear flow will concentrate into a basal slip zone, undercutting the anticlinal shale structures that have been formed during the primary creep phase.

So far we have assumed that the overburden has a uniform slope angle δ. Accordingly, the driving force of the substratal flow did not vary in the x-direction. When the *slope angle varies* in the x-direction, the pressure gradient (eq. I.3-11) in the mobile substratum will vary, too. In a substratum of uniform initial thickness this will give rise to non-uniform flow rates and, consequently, to changes in layer thickness. Normally the slope angle of a delta slope has its maximum value δ_0 at the shelf edge of the delta and it decreases gradually in the downslope direction. The associated driving force in a ductile substratum of uniform initial thickness will then decrease in the downslope direction and the decrease in flow rate will cause ductile material to pile up. In other words, the downslope flattening of the overburden slope may cause a *diapiric* deformation of an initially flat interface between overburden and mobile substratum. The effect can be evaluated in quantitative terms by means of eq. (I.3-12), if we assume that tan δ decreases exponentially in the x-direction:

$$\tan \delta = \tan \delta_0 \cdot e^{-kx} \tag{I.3-15}$$

where k is a characteristic constant of the delta and δ_0 the maximum slope angle at the delta edge at x = 0. With this formula, which was obtained by D. van den Abeele in 1968 (unpublished study of delta profiles), and with a certain transformation of variables [9]), the problem of a slope with varying slope angle can be formally reduced to that of a constant slope angle.

Several typical cases have been calculated by Beenker (1981). Figure I.3-15 shows the result for the simplest case of a horizontal substratum with a uniform initial thickness $h_0 = 300$ m and a viscosity of 10^{18} poise, overlain by a stationary overburden with a slope angle that

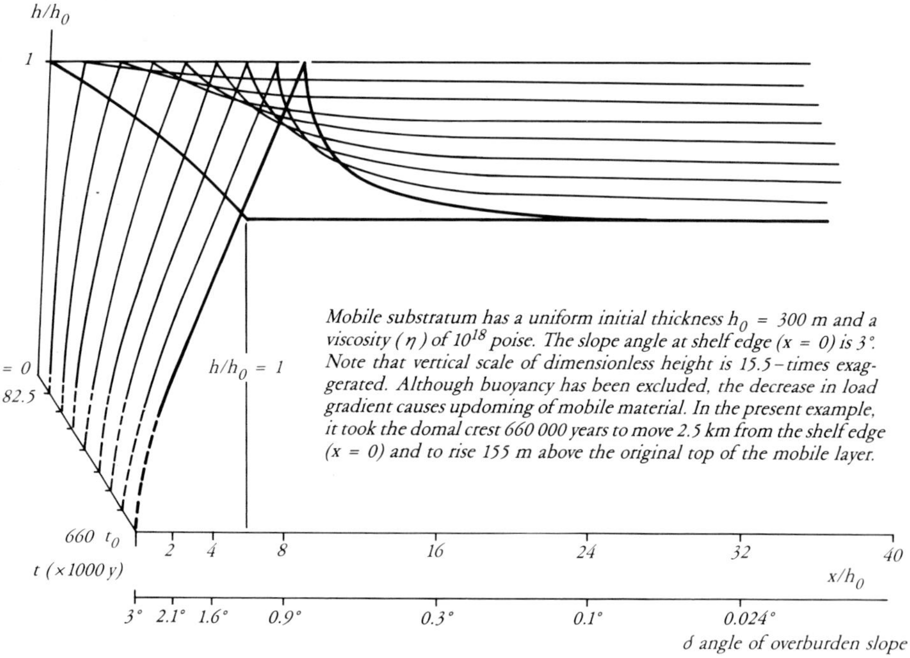

Fig. I.3-15 DEFORMATION OF MOBILE SUBSTRATUM UNDER OVERBURDEN WITH EXPONENTIALLY DECREASING SLOPE ANGLE

decreases exponentially from $\delta_0 = 3^o$. The submerged specific overburden weight is again $\gamma' = 12\ 000\ kg\ m^{-2}s^{-2}$. The deformation history of the mobile layer is presented in a three-dimensional plot. Layer thickness and horizontal distance are expressed in dimensionless form as multiples of h_0 and the unit of time is 1000 years. The decrease of the overburden slope angle δ with distance is indicated at the base line of the figure. Typically, a diapiric deformation is formed, which after 660 000 years will have reached a height of about 155 m above the initial position of the layer top. The growing disturbance steepens in time, but notice that the vertical height scale is 15.5-times exaggerated with respect to the scale of the horizontal distance. As stated earlier, the resistance of the overburden will smoothen the shape of the disturbance, but may hardly extinguish the asymmetry of the disturbance. Because of the decrease in δ, the travel velocity of the growing disturbance decreases as a function of the distance covered. While after 660 000 years the disturbance has travelled a distance of 2.5 km, it will take 3.3 million years for the disturbance to travel a distance of 6 km, at which time the diapir will have reached a height of 670 m. Again, the travel velocity is inversely proportional to the viscosity and proportional to the square of the initial layer thickness. It should be recalled that buoyancy effects have been neglected in the present model. If the overburden has a greater density than the mobile substratum, the growth rate of the diapiric disturbance will obviously be increased.

Finally, it should be noted with respect to the solution in Fig. I.3-15, that only part of the trail of the travelling disturbance is realistic. The dimensionless thickness distribution at the left of the (solid) $h/h_0 = 1$ line should be disregarded for the following reason: behind the shelf edge ($x<0$) slope angle and driving force are zero, and the ductile material should therefore be at rest, while underneath the stationary slope ($x>0$) the ductile material is moving. In reality, this contrast in flow behaviour must cause a synclinal depression of the mobile layer underneath the stationary delta edge. This is not adequately described by the present model, where an ad hoc assumption has to be made to specify the boundary condition at $x = 0$. In the present calculation it is assumed that the layer thickness drops instantaneously to zero at $x = 0$, which obviously vastly exaggerates the formation of the synclinal depression. Fortunately, the mathematical theory of this type of partial differential equation ensures that the choice of the boundary condition at $x = 0$ does not affect the thickness distribution in the region ahead of the $h/h_0 = 1$ line.

From the analytical model of substratal creep deformations we have learned how disturbances of the overburden/substratum interface are formed and propagated by the horizontal gradient in overburden load. If differences in specific overburden weight are neglected, the differential loading is completely characterised by the angle of the delta slope, while flow rates and travel velocities of interfacial deformations are proportional to the square of the layer thickness (for a Newtonian fluid) and inversely proportional to the coefficient of viscosity. The growth rates and travel velocities of interfacial disturbances predicted by the model form upper bounds, since the substratal material in squeeze flow underneath a delta slope at rest moves faster with respect to the stationary overburden than substratal material that is dragged along by a moving delta slope and since the deformational resistance and energy dissipation of the overburden have been neglected. Hence, when the configuration and the (temperature-dependent) viscosity of the underlying ('wet') salt are roughly known, the velocities of interfacial deformations predicted by the analytical model may be compared with the estimated rates and periods of delta progradation to define when and where in the delta thin-skinned gravitational gliding may have taken place without the interference of deep-seated creep flow.

However, when such interference takes place, either during a period of progradation – for instance, on a very thick substratum of hot salt or on shales with overpressures generated by the load of the overlying delta slope – or, more commonly, during a long stagnation period of progradation, a more realistic modelling of overburden deformations would require a more general approach. This should, in particular, account for the elastic/frictional plastic behaviour of the slope sediments, but it should also describe the slow viscous flow of the mobile substratum in more detail than the analytical model, to ensure a correct determination of the shearing interaction between substratum and overburden. In addition, the general model should allow for a seaward shift of the delta slope, which may be caused by the accretion of new slope sediments without slope material being actually displaced in the seaward direction, or may be the result of slope material being carried seaward by substratal flow. Such more general modelling of the two-layer system has been recently undertaken by Last (1986), by adopting and developing further a computer program that was based on the explicit finite-difference method (Wilkins, 1964; Cundall, 1976) and has been used previously to model geomechanical continuum-type problems involving large deformations and non-linear material behaviour (Cundall et al., 1980). We restrict ourselves to a summary of the main conclusions that have been obtained so far for stationary and prograding planar slopes on viscous substrata of uniform initial thickness or of an initial shape similar to that shown in Fig. I.3-14.

Not surprisingly, the computations confirm the effect of layer thickness on the mobility of the substratum. Also in agreement with Lehner's analytical model, a depression of the substratum/overburden interface forms at a location that depends on the position of the edge of the delta slope and of the landward termination of the mobile substratum: when the delta slope is rapidly deposited on the substratum such that the delta edge is located above a mobile substratum, the depression of the interface will form below the delta edge, as may be expected from the abrupt change in driving force associated with the change in slope angle. If the edge of an outbuilding delta has not yet reached the landward termination of the mobile substratum (salt basin), the depression will form underneath the delta slope at the landward end of the substratum. The development of the landward depression of the interface is accompanied by a thickening of the mobile substratum further downslope, and this may develop into anticlinal structures at the toe of the delta slope. When the edge of the delta slope has moved over the synclinal depression and, therefore, differential loading has ceased, further growth of the synclinal trough can only be sustained by buoyancy.

Owing to the flexing of slope sediments, zones of extensional straining develop above the flanks of the interfacial syncline, where eventually the stresses may attain the active limit state. The boundaries of these active plastic regions may be expected to become major growth faults. Hence, these boundary faults, associated with the substratal creep flow, will be either situated at the landward margin of the mobile substratum, or in the interior at locations where the delta edge has stayed long enough to enable a synclinal trough to develop.

The computations have demonstrated that the extent of the plastic extension zones, where normal faulting is to take place, strongly depends on the *thickness of the overburden* and on the *progradation rate* of the delta. It is important to note that, on a stationary delta slope, the time required for the development of an active plastic extension region (or of a passive plastic toe region) that extends vertically from the mobile substratum to the slope surface increases considerably with the thickness of the overburden. In other words, the thicker the sedimentary overburden on a given mobile substratum, the longer the delta progradation has to stagnate to enable major normal faults to develop in response to the substratal creep flow. For example, it was found for the case of a stationary delta slope ($\delta = 2.3^\circ$, $\varphi = 30^\circ$, $\tau_0 = 0$) resting on a layer of salt ($\eta = 10^{18}$ poise) 50 km long and of the shape shown in Fig. I.3-14 (for t_0) with a maximum thickness of 500 m, that the active plastic zone and the passive plastic toe region were fully developed after 0.5 million years, when the overburden had a maximum thickness of 2750 m, but that no more than the onset of such a development had become visible after the same stagnation period when the overburden was 750 m thicker.

In the thicker overburden, displacement of material was almost exclusively in the vertical direction, accompanied by squeeze flow of substratal material (in accord with Lehner's model). In contrast whith this, the extension inside the active plastic region of the thinner overburden was accompanied by a large horizontal displacement component, as a result of which the bulk of the slope which was not in the plastic limit state, was pushed out seaward. Note that this advance of the slope, which was 2 km in the example above, is not caused by sediment accretion. The flow in the substratum is then predominantly simple shear flow; this implies that the moving overburden exerts a shear stress in the down-slope direction upon the mobile substratum, similar to the shear stress at the base of a gravitational slide, but contrary to the direction of the interfacial shear stress exerted by substratal squeeze flow upon an overburden at rest. The reader will note the similarities between a delta slope with fully developed active and passive plastic regions, separated by a 'rigid' wedge 'riding' on substratal simple shear flow, and thin-skinned gravitational gliding.

Finally, Last's computer simulations of delta slopes that prograde by sediment accretion at various rates between 1 and 20 cm/year, have demonstrated that the size of the plastic extension regions depends strongly on the rate of progradation. The faster the delta progrades, the less the plastic region extends from the deforming substratum/overburden interface into newly deposited sediment layers, and the less it therefore interferes with thin-skinned gravitational gliding at shallower depth.

2. OVERBURDEN EXTENSION BY HORIZONTAL STRETCHING OF DUCTILE BASE

The extension of an overburden above a synclinal depression in a ductile substratum, which we considered in the previous section, is a special and complex case, in which a component of horizontal extension is combined with subsidence of the ductile base, and in which faulting in the overlying sediments depends on the shape and deformation history of the subsidence 'bowl'. In this section we shall deal with the extension of an overburden by horizontal stretching of a ductile base without the formation of a synclinal depression into which sediments may subside along normal faults. Such horizontal extension, accompanied by widespread thinning of ductile substrata, may be imposed on individual strata of various dimensions, or on whole parts of the brittle crust. On the largest scale it is associated with the penetrative stretching of the ductile lower crust during rifting, examples of which can be found in the Basin and Range Province of the western USA, the Baja California, the Bay of Biscay, the Gulf of Suez, the Afar depression in Ethiopia, the Viking graben of the North Sea, and the East Greenland coastal flexure (Angelier et al., 1981; Angelier & Coletta, 1983; de Charpal et al., 1978; Stewart, 1980; Wernicke, 1981; Wernicke & Burchfiel, 1982; Morton & Black, 1975; Nielsen & Brooks, 1981; Ziegler, 1982). Although induced by the extension of a ductile base, faulting of the overburden is, in a sense, 'thin-skinned', since no differential movements are imposed by the ductile base upon the overburden in the vertical direction. The surface of the ductile base may be only slightly disturbed by the movement of fault blocks, as is indicated schematically in Figs. I.3-1F,G.

These figures represent the two kinematically possible modes of faulting by which the brittle overburden can accommodate the horizontal extension of the ductile base: the 'domino-style' rotation of parallel normal faults and the formation of 'graben-horst' structures. Both modes can be observed in the field, which raises the question as to the cause of the preference for one of the two modes and the choice of the dip direction when extension is accommodated by the 'domino-style' rotation of parallel faults. In other words, what are the parameters that determine the operational conditions of the two processes? In the following we shall propose some answers, based on model experiments and mechanical arguments (Mandl, 1987c).

Following Hans Cloos (1930), we experimentally simulate processes of ductile stretching of the base of a brittle overburden by the extension of a rubber sheet. Assuming that the frictional shear strength of the overburden by far outweighs its pressure-independent cohesive strength, we model the overburden by a dense pack of dry sand (see Section I.2.9). The table and rubber sheet are schematically shown in Fig. I.3-16. The rubber sheet is either free to enable uniform stretching, or firmly attached to parallel rods fixed to side bars to enable selective stretching of individual segments or a gradual spreading of the stretched region.

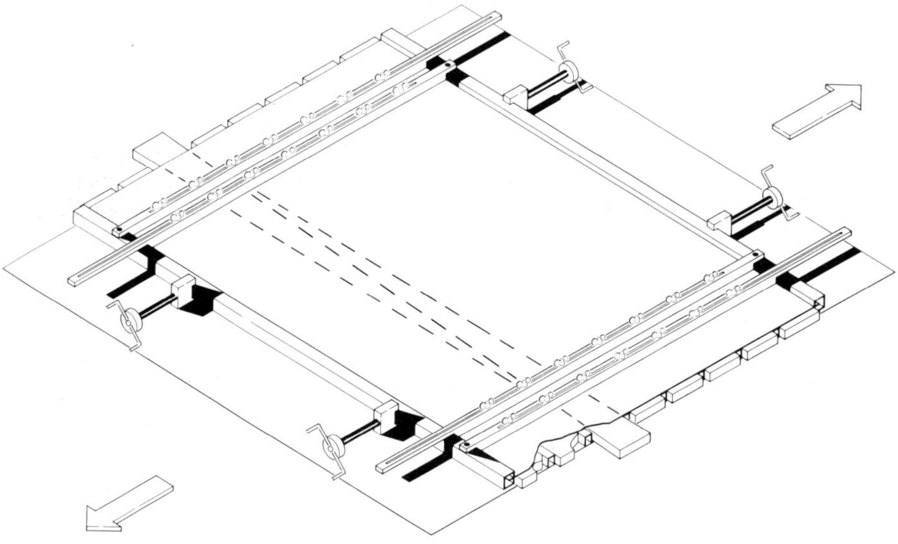

Fig. I.3-16 TABLE WITH RUBBER SHEET (schematic) FOR MODELLING HORIZONTAL EXTENSION STRUCTURES; Sheet is reinforced by steel rods which enables the performance of various modes of non-uniform stretching.

Uniform Extension of the Base. Figure I.3-17 shows a vertical cross-section through a sand pack that has been uniformly stretched by 13% and wetted before sectioning to make the fault planes visible. It is obvious that both types of conjugate faults had the same chance to develop, and this results in a system of small grabens and horsts and short arrays of parallel faults. Terminating experiments at different stages of extension, we observed that the short arrays of parallel faults are mostly 'infill' features which develop as the basal extension continues. The simultaneous operation of conjugate faults is essentially of the type (Fig. I.3-18) already postulated by Freund (1974). The faults are densely spaced since displacements along individual faults remain small because of the planar shape of the extending base, which does not allow graben "keels" to develop. The experiments have been complemented by a computer simulation of the process, which has been discussed in the concluding part

Fig. I.3-17
HORIZONTALLY LAYERED SANDPACK WITH UNIFORM HEIGHT AND VERTICAL BOUNDARY WALLS AFTER 13% UNIFORM STRETCHING OF RUBBER BASE (Vertical cross-section parallel to direction of stretching). Faults become visible after wetting of pack. Both conjugate fault types have the same chance to develop outside boundary region.

('Mathematical modelling') of Section I.2.7 (Figs. I.2-66,67), to demonstrate the decisive role of shear-softening in controlling the spacing of parallel faults.

Strength Anisotropy. The sand pack of Fig. I.3-17 has been built up by the deposition of horizontal layers. When, in contrast, the sediment body with a horizontal surface has been built up by the deposition of "foresets" (Fig. I.3-19A), uniform stretching will produce only one set of parallel normal faults, which accommodate the extension of the base by 'domino'-style rotation (Fig. I.3-19B). This is obviously caused by the anisotropy in shear strength that is associated with this type of deposition. It should be noted that these faults, although syn-

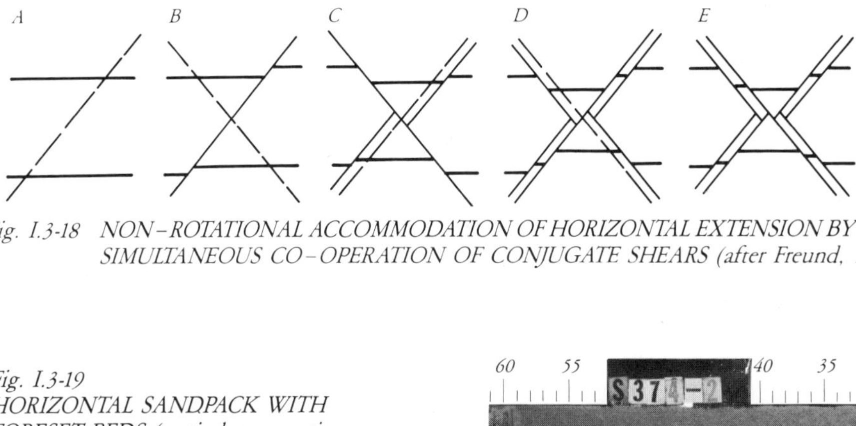

Fig. I.3-18 NON-ROTATIONAL ACCOMMODATION OF HORIZONTAL EXTENSION BY THE SIMULTANEOUS CO-OPERATION OF CONJUGATE SHEARS (after Freund, 1974)

Fig. I.3-19
HORIZONTAL SANDPACK WITH FORESET BEDS (vertical cross-section parallel to applied basal stretching)

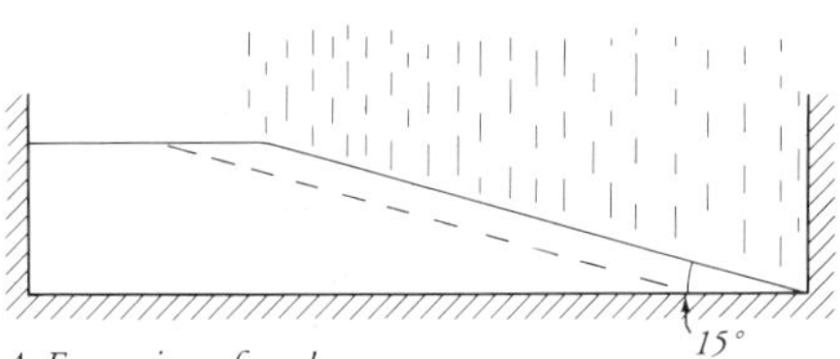

A Formation of pack

B Preferential development of one set of parallel rotating faults

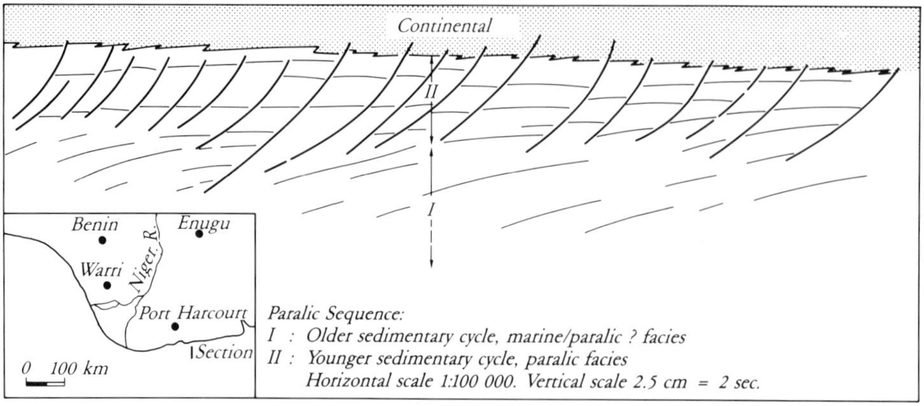

Fig. I.3-20 DIP SECTION OFFSHORE NIGERIA (after Evamy et al.,1987; reprinted by permission of "American Association of Petroleum Geologists"); PARALLEL FAULTS ACCOMMODATING EXTENSION OF MOBILE SUBSTRATA.

thetic with the foresets, dip considerably steeper than the foresets. The parallel normal faults of the section from the Niger Delta shown in Fig. I.3-20 are most likely associated with such foreset-induced strength anisotropy of the overburden, which is extended by the basinward flow of shaly substrata.

Non-Uniform Extension. It is important to emphasise that a strictly uniform stretching of an overburden of uniform thickness does not impose any shear stress upon the base of the overburden. Naturally, the uniform extension will reduce the horizontal normal stress, but this reduction will not vary in horizontal direction and the horizontal normal stresses will remain self-equilibrating. Hence, the equilibrium of horizontal stresses will not involve horizontal shear stresses. This does not contradict the fact that the extension of the rubber base is transmitted to the overlying sand through frictional forces. These forces are in static equilibrium at any basal element (of arbitrary smallness) and therefore cannot result in a net shear stress. It should be noted, however, that shear stresses will be induced along the overburden base *after* the onset of faulting in the overburden. Even if the extension of the ductile substratum is perfectly uniform, the extensional response of the brittle overburden by discrete faulting is less uniform. This is sketched in Fig. I.3-21A, in which the stretching of the overburden is localised in a graben. The unfaulted blocks lag behind or advance ahead of the substratal flow and thus experience basal shear stresses of opposite directions. Naturally, these fault-induced basal shear stresses affect the direction of σ_I. This is illustrated in Fig. I.3-22 by the σ_I-trajectories in the vicinity of two parallel faults, which were generated in the numerical simulation of a uniform layer under uniform basal extension (presented in Fig. I.2-66).

Prior to faulting, an extending ductile substratum may exert shear stresses upon a brittle overburden only if (1) the substratal extension is *not* uniform, or if (2) the stretching of the ductile substratum is concomitant with creep flow at a rate higher than that at which the overburden is displaced (similar to the squeeze flow considered in the preceding section). The first situation arises when the region of basal stretching is spreading out laterally. The experimental set-up to simulate this, by increasing the stretched area of a rubber sheet, is sketched in Fig. I.3-21B, together with an indication of the stress changes caused by the non-uniform stretching. Before the horizontal normal stress σ_h has dropped to the value required for the onset of normal faulting, the difference in extensional straining produces a

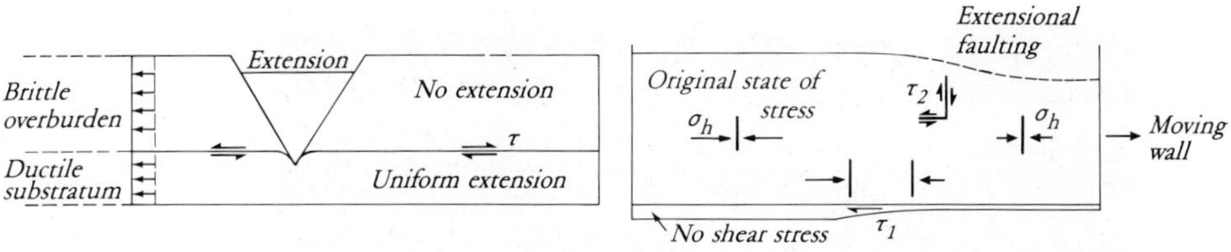

A Uniform extension of ductile substratum non-uniformly accommodated by faulting in brittle overburden

B Shear stresses induced in non-uniformly stretched overburden; experimental set-up.

Fig. I.3-21

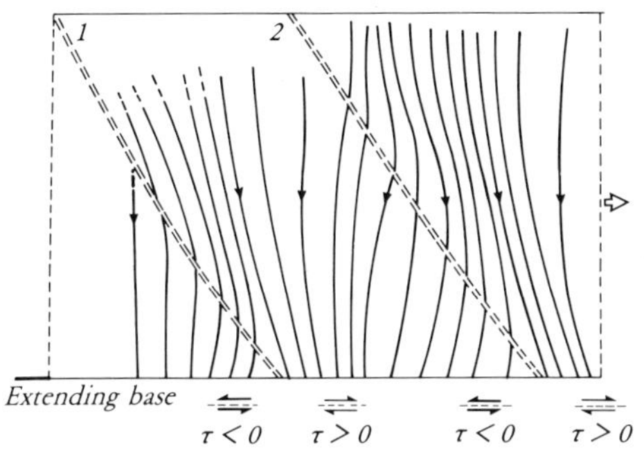

The trajectories have been obtained from the stress field determined by the finite element simulation of normal faulting in Fig. I.2-66. Faults 1 and 2 of Fig. I.2-66D are considered. The stress trajectories are not drawn in the left hand upper corner where the stress field is strongly affected by the notch at the surface that was required to trigger the first fault.

Note that the trajectories which before faulting were vertical, are deflected in opposite directions in foot wall and hanging wall. The inclinations of the σ_I-direction at the stretched base indicate that basal shear stresses of alternating direction have been induced by the process of faulting. The resultant of these stresses vanishes, however, since the uniform extension imposed upon the base of the layer cannot produce a net shear force.

Fig. I.3-22 TWO PARALLEL NORMAL FAULTS WITH σ_I-TRAJECTORIES IN LAYER WITH UNIFORMLY EXTENDED BASE

horizontal gradient of σ_h. The resulting imbalance of forces has to be compensated for by a basal shear stress τ_1 which acts upon the overlying rock in the direction of decreasing stretching (Fig. I.3-21B). Note, however, that this shear stress only exists in some vicinity of the momentary location of the basal extension front and disappears when the front has moved on. The temporary action of this locally induced shear stress can therefore neither favourably nor adversely affect the 'domino'-style rotation of parallel faults in extending overburden, but we shall see that it may determine the preferential dip direction of parallel faults.

A typical experimental result is shown in Fig. I.3-23. The stretching proceeded from left to right as indicated, causing the development of a set of almost equidistantly spaced faults. The fault traces show particularly clearly in samples analysed by computer tomography (see Note 13 in Ch. I.2), as can be seen in Fig. I.3-23. The faults dip *antithetically to the direction of stretch propagation* and, therefore, also antithetically to the shear stress imposed from below upon the overburden (Fig. I.3-21B). (The incongruous faults visible near the fixed side wall at the left were caused when the sample was taken.)

Faugere and Brun (1984) have modelled the (non-uniform) extension of a brittle overburden in a somewhat different way by depositing a sand pack on a ductile substratum of silicone oil (with an additive of powdered lead sulphide to match the specific weight of the sand). The two-layer system was placed in a rectangular Plexiglass container and rested on a thin layer of mercury to eliminate friction along the base of the ductile substratum. Extension was induced by releasing a movable side wall, which was then pushed outwards by the

X-ray computer scan

Fig. I.3-23 UNIFORM SANDPACK AFTER STRETCHING HAS PROGRESSED HALFWAY FROM LEFT TO RIGHT.
Parallel faults dipping oppositely to direction of stretch propagation.

horizontal pressure of the two-layer system. The first extensional feature observed in the brittle layer was a graben adjacent to the moving side wall, followed by the development of a long faulted region where faults were predominantly parallel and dipping away from the moving wall. Accordingly, extension of this region was accommodated by block tilting in 'domino' style, with the fault blocks reclining towards the moving wall. Marker profiles in the ductile substratum clearly indicated that creep flow towards the moving wall had exerted a shear stress upon the overburden in that direction. Thus, again, the preferentially induced parallel faults dipped in the opposite direction to the shear stress imposed from below upon the overburden.

In assessing the results of the two experiments Mandl (1987c) pointed out that, in the equipment used, the results could have been affected by certain boundary effects. It is, therefore, gratifying that numerical simulations by Witlox (1986) corroborate these results. The simulations are of the type shown in Figs. I.2-66,67, but with a non-uniform strain distribution prescribed along a continuously growing interval of the overburden base. When the strain gradients, which in the calculations were maintained unchanged during the expansion of the strained interval, were not too low, the dominant faults that formed in succession at the advancing strain front, were all found to dip in the opposite direction to the expansion of the stretched interval.

It is remarkable that in the two experiments and in the numerical analysis the parallel faults turned out to be planar, not even showing a noticeable listricity near the base, in spite of the fact that some deviation of the σ_I-direction from the vertical must be associated with the locally induced shear stress. We have already drawn attention to this point in Section I.2.4 when discussing tectonically induced stress changes as possible causes of listricity. Although the point is still not completely understood, we recall that only very little extensional straining is required to produce the active limit state, and that, therefore, in the present case, the associated basal shear stress may remain too small to produce a change in the σ_I-direction that would be noticed as curvature of the fault planes.

Although experiments and numerical simulations strongly suggest that basal shear stresses promote the formation of parallel faults and control the dip direction, they do not provide an explanation for this connection. The explanation may be essentially the same as the one we are going to propose next to explain the effect a surface slope has on the fault dips in an extending layer.

Sloping Surface. Consider a long slope that is brought into the active limit state by a *uniform extension* of its base (Fig. I.3-24A). The stresses that act on the conjugate potential slip planes are the same and, from the kinematical point of view, the horizontal extension could be equally well accommodated by a single set of downslope-dipping and upslope-reclining faults as by the conjugate set of upslope-dipping and downslope-reclining parallel faults, or by a combination of both potential fault sets. Our experiments with a uniformly stretched rubber membrane (Fig. I.3-25) clearly show that the slope promotes downslope-dipping parallel faults (type 1 in Fig. I.3-24A) to accommodate the uniform basal extension by reclining in the upslope direction. Although this is as might be expected intuitively, a satisfactory explanation of this phenomenon is by no means trivial.

The explanation we propose for this phenomenon has been already mentioned in Section I.2.3 (see the discussion of "conjugates"). It is based on the phenomenon of shear softening and on the possible non-coaxiality of principal stress axes and principal axes of

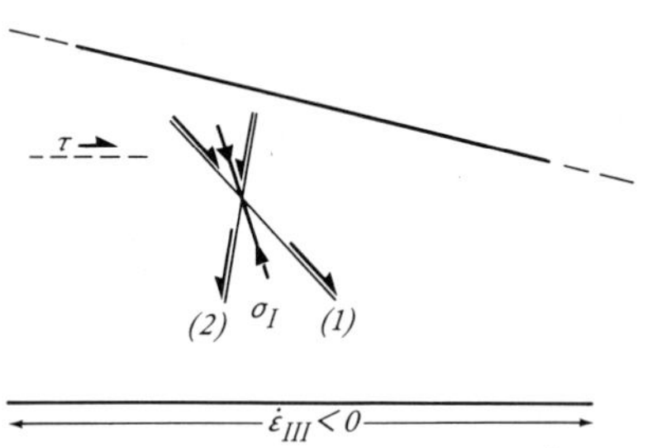

A Stress and Coulomb-type faults on slope brought to active limit state by extension of base.

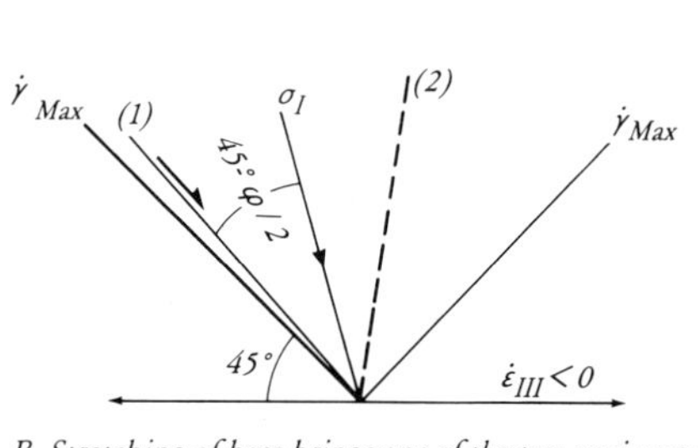

B Stretching of base brings one of the two conjugate slip directions (1) closer to the direction of maximum shear rate. This allows earlier softening of the material along this potential fault.

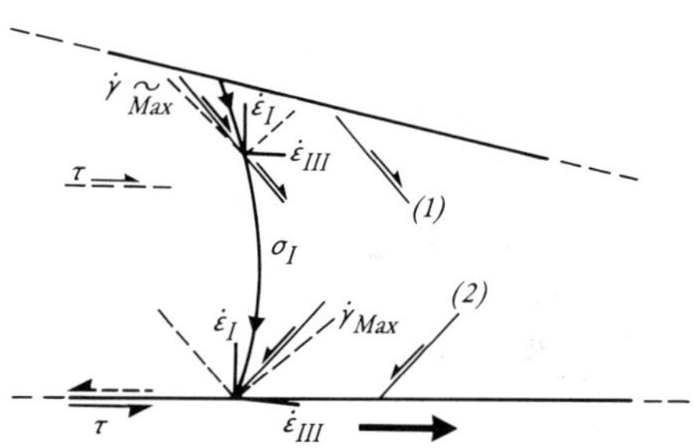

C In addition to extension of base, substratal squeeze flow in direction of decreasing overburden. Associated basal shear stress causes opposite deviation of principal stress direction in deeper part of sediment wedge and exclusive mobilisation of conjugate fault type (2).

Fig. I.3-24
NON-COAXIALITY OF STRAIN INCREMENTS AND STRESS IN FRICTIONAL PLASTIC MATERIAL AS POSSIBLE EXPLANATION OF THE PREFERENTIAL DEVELOPMENT OF ONE FAULT TYPE IN A SEDIMENT WEDGE WITH EXTENDING BASE

A Pack with one inclined side wall, slope angle of pack 20°.

B Pack with two slopes.

Fig. I.3-25 HORIZONTALLY LAYERED SANDPACKS WITH SLOPES AFTER UNIFORM STRETCHING OF BASE

incremental strains (or strain rates) in frictional plastic materials. In the extreme, corresponding stress and incremental strain axes may deviate from each other by the angle $\pm\ \varphi/2$ (see 'Non-coaxiality of Stress and Strain', Section II.7.4 and Fig. II.7-25). Considering now the uniform stretching of the base that brought the slope (Fig. I.3-24A) into the active limit state, we notice that the smallest principal strain rate (extension being counted as negative) will act parallel to the horizontal base. Consequently, the maximum shear rate $\dot{\gamma}_{Max}$ will act along planes inclined at ± 45° to the base as indicated in Fig. I.3-24B. Since the slope causes the σ_I-axis to deviate from the vertical, one of the conjugate slip planes ('1') will lie closer to a plane of maximum shearing rate. We may therefore expect that during a stretching episode the material on that potential slip plane will experience more shearing than the material on the conjugate plane. Consequently, shear softening will begin earlier on planes of type 1 than on its conjugates of type 2 and will allow the preferential development of downslope-dipping faults.

Next, we consider the case when some part of the sediment wedge has been brought into the active limit state by the combined action of basal extension and a shear stress, imposed upon the base of the wedge by some relative movement of underlying strata, such as by the substratal squeeze flow underneath a stationary delta wedge. As indicated in Fig. I.3-24C, this implies a reversal of horizontal shear stress at a certain depth. The slope-induced shear stress decreases with depth until it reaches zero, whereupon it increases with increasing depth, but now acting in the opposite direction. Consequently, the orientation of σ_I will also change with depth, as is qualitatively indicated in Fig. I.3-24C. It becomes fairly obvious that the deviation between stress axes and strain rate axes near the base will be in the opposite direction to the deviation at shallow depth. Applying the same argument of earlier shear softening as in Fig. I.3-24B, we therefore conclude that in the deeper parts of the sediment wedge in the limit state only faults of type 2 may develop, while in the shallower part type 1 will predominate.

If the sediment surface is horizontal, the direction of σ_I will be vertical at the surface and the faults of type 2 in Fig. I.3-24C are likely to extend to the surface. This may account for the dip direction of the parallel faults produced in the experiments of Faugere and Brun, and ourselves (Fig. I.3-23), and in Witlox's computer simulation.

One may also readily envisage slopes or parts of slopes where the shearing effect of a substratal squeeze flow or an outward prograding extension initially outweighs the opposite shear stress associated with the slope angle, and enables the slope to be cut by *upslope-*

A

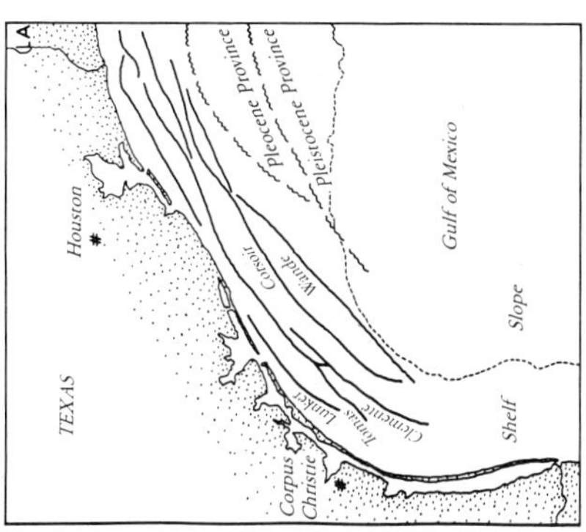

Fig. I.3-26
SEISMIC DIP SECTION THROUGH CORSAIR FAULT SYSTEM OFFSHORE TEXAS (after Christensen, publ. by Bally, 1983; reprinted by permission of "American Association of Petroleum Geologists"); (see text for explanation).

B

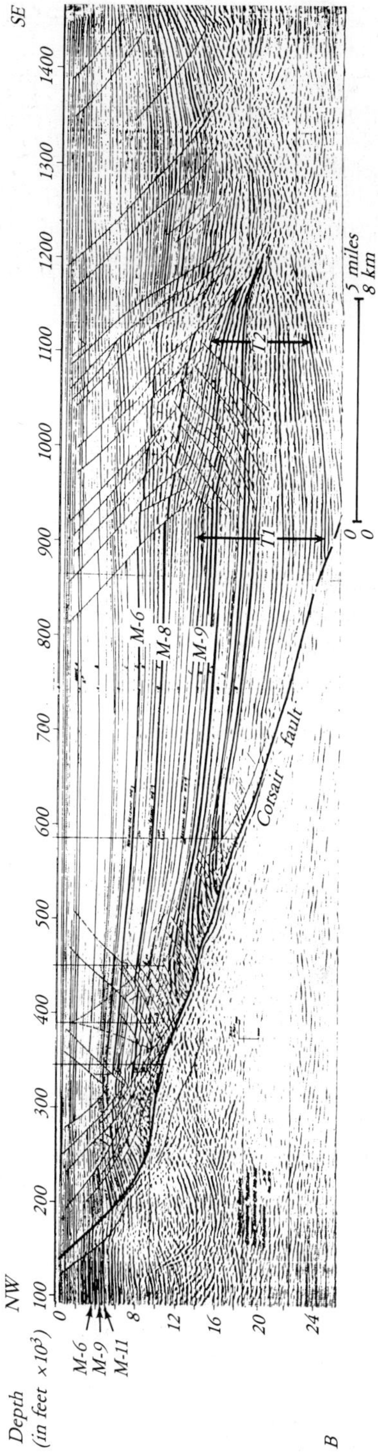

dipping parallel normal faults. Extension of the slope will then be accommodated by the rotation of tilted blocks in the downslope direction. Some superficial structures of this type have been discussed by Judson (1947), but other examples of this mode of slope extension in the geological literature are very scarce. The following example should therefore be of interest.

Figure I.3-26 presents a seismic section from the Texas continental margin whose Tertiary structural development is largely controlled by basinward creep of thick salt, which is driven by differential sedimentary loading and results in the formation of subparallel salt ridges and asymmetric troughs. As these troughs form, the sediment layers experience extension and this is accommodated by normal faulting. The seismic section shown in the figure is a dip section through such a sedimentary trough, here bounded upslope by the regional Corsair growth fault (interpreted by A.F. Christensen; Bally, 1983). The parallel normal faults in the 'toe' region of the Corsair fault between the arrows T_1 and T_2 dip antithetically to the slope and may have accommodated the extension caused by the withdrawal of salt into the neighbouring salt ridge, by rotating in the downslope direction. This set of parallel faults is overlain by a set of basinward-dipping normal faults that, in contrast to the lower set, are synsedimentary and are part of the fault system along which the prograding sediments glide into the next trough that had started to develop when depletion of mobile salt (reduction of the layer thickness) had stopped the growth of the Corsair sediment wedge. The seismic data permit the assumption that the lower fault system was still active during an early phase of activity of the upper set of parallel faults, the dip of which was determined by the delta slope.

In this context we recall an early graben experiment by Hans Cloos (1930). In his experiment (Fig. I.3-27) a clay cake was stretched by means of a base of rubber belts and produced numerous rotating normal faults. In a striking resemblance to the toe structures in the previous field example, a basal fault set is overlain by an oppositely dipping set. Although

Fig. I.3-27 CLAY CAKE EXPERIMENT BY HANS CLOOS, 1930.
Clay layer was horizontally extended by stretching of rubber base

the details of the experimental set-up are not sufficiently known, we presume that the 'necking' of the clay cake and the formation of a central graben is an indication that basal straining was not uniform and gave rise to basal shear stresses, while the development of a pronounced surface slope controlled the dip of the upper faults.

Spacing of Parallel Faults. An intriguing question which arises quite naturally in connection with sets of parallel faults is that of fault spacing. In the experiment of Fig. I.3-25 the spacing of the parallel faults decreases with the thickness of the overburden, and in Fig. I.3-23 the spacing of the parallel faults near the vertical rigid wall decreases towards the surface. This *variation* in spacing is similar to the upward decrease in spacing of the antithetic faults associated with a reclining main fault (Fig. I.2-49), which was explained in Section I.2.6 by means of a similarity argument. This argument may be used here again to interpret the variations in fault spacing in the experiments of Figs. I.3-23,25.

Consider the schematic slope section in Fig. I.3-28A with two geometrically similar quadrilaterals. When a uniform extensional strain (ε_{III}) is applied to the base of the slope, the two quadrilaterals remain geometrically similar. Let us assume, for the moment, that at corresponding points on the two quadrilaterals the same amount of straining is required to bring the stresses into the active limit state, then faulting would start simultaneously at these points. While the deformation as a whole now becomes discontinuous, the deformation of the individual fault blocks remains continuous and has to accommodate the sawtooth-like misfits that rigid fault blocks would generate at the planar base. Since we have assumed that the material at corresponding points of the two similar regions will sustain the same amount of straining before shear failure, corresponding fault blocks in the two regions must also be geometrically similar, which implies a linear decrease in fault spacing in the downslope direction.

The same argument applies to the upward decrease in spacing of the parallel faults near the rigid wall in the experiment of Fig. I.3-23, in which the rigid wall represents, for instance, a boundary fault or the contact with a more competent rock. Kinematically, extension of the wall region could be achieved by equidistant parallel faulting, as is illustrated in Fig. I.3-28B, were it not for the misfits at the rigid wall. Again, the existence of an equal state of maximum adaptation strain in the fault blocks requires geometrical similarity of the blocks and, therefore, a linear decrease in the fault spacing towards the free surface.

At this point we have to recall that, in using the similarity argument above, we have assumed that at geometrically corresponding loci the material sustains the same strain before the onset of faulting. This would indeed be the case were the material properties

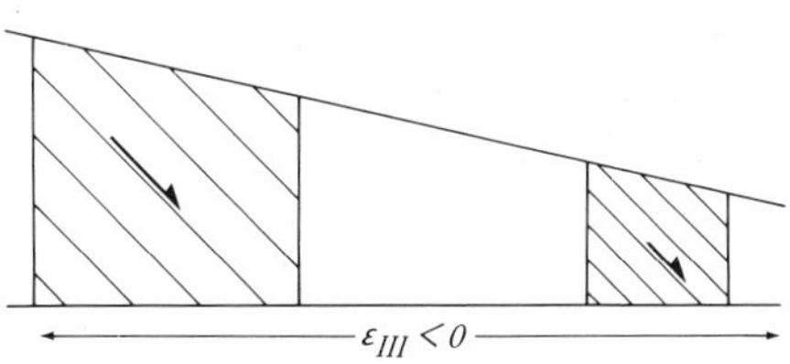

A Linearly varying spacing in sedimentary wedge with uniformly stretched base

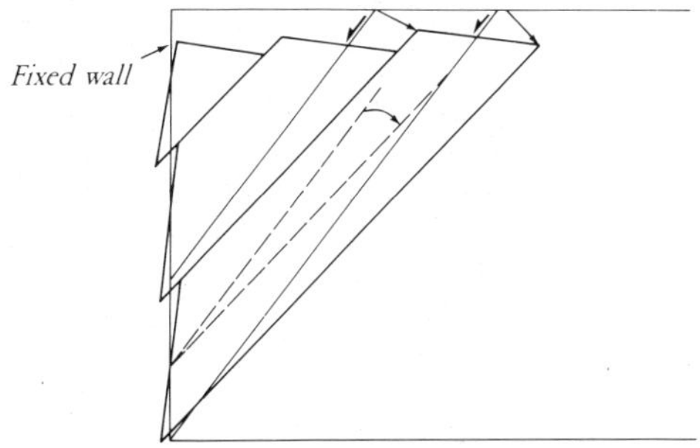

B Unrealistic fault spacing near fixed wall

Fig. I.3-28 FAULT SPACING

and the initial states of stress the same at the corresponding points. Usually, however, the initial states of stress will differ, mainly because of the difference in overburden thickness. Consequently, identical strains that may induce faulting in one of the geometrically similar regions will not yet do so in others. For example, in the case in Fig. I.3-28A, vertical and horizontal normal stresses increase linearly with depth inside the sediment wedge before the base is stretched. The stresses therefore differ between geometrically corresponding loci, for instance, on the base of the sediment wedge. If the material is mechanically homogeneous, which implies uniform elastic moduli and uniform shear strength parameters, a uniform basal stretching will produce the same reduction in horizontal normal stress along the base. It can readily be seen that the active limit state is attained first at the location of smallest overburden load and spreads gradually in the upslope direction along the base as stretching is increased. Pre-fault straining will therefore be greater in upslope parts of the wedge than in downslope parts, this enabling somewhat greater fault spacing in the upslope parts than that predicted by strict geometrical similarity. It may also be shown (Mandl, 1987c) that the similarity argument will strictly apply when, in a material with constant angle of internal friction, both the modulus of elasticity and the cohesive strength increase linearly with depth. This may be the case approximately in young sediments, particularly when they are still in the process of consolidation.

Although this argumentation enables us, at least in certain cases, to understand variations in fault spacing, and even to extrapolate from areas where fault spacing is known to neighbouring areas of unknown spacing, it does not reveal how fault spacing depends on the material properties of the rock. This problem has been addressed in Section I.2.7 (see 'Mathematical modelling') where we concluded that the elastic unloading caused by the shear softening of the fault zone material plays the key role in determining the distance between parallel faults, since inside the region of elastic unloading around a fault no other fault can form. The extent of this region depends on the shear softening characteristics ('brittleness') of the material and on the imposed boundary conditions. In the present case, the relevant boundary condition is the horizontal straining of the base, which tends to maintain an active limit state, while the shear strength reduction generates a subcritical state. Hence, the spacing of the parallel faults is the result of the competition of the two processes.

Slope Failure. Although we consider basal stretching the most prominent process that produces sets of parallel faults, which accommodate extension by forming an array of tilted blocks, it may not be the only mechanism to produce this extensional style. Wernicke and Burchfiel (1982) considered as an alternative a rock sheet that slides down a non-extending gliding base. The sheet is dissected by a suite of downslope-dipping parallel planar faults and upslope is bounded by a major listric fault. While sliding, the sheet extends by 'domino'-style rotation of the faults, made kinematically possible by the presence of the curved boundary fault. Rocks below the detachment plane do not participate in the extension of the gliding sheet. Instead, extension and movement of the completely detached sheet are accommodated by space created at the foot of the slide by rifting and/or erosional processes.

The reader will notice an important difference between this 'thin-skinned' model and the one discussed in the first part of the previous section on gravitational delta tectonics. The deltaic gravity slide was in static equilibrium and the imbalance of frictional resistance along the glide base and the driving weight component was compensated for by a slope-parallel 'buttressing' stress $\sigma_{\parallel}$ that monotonically increased in the downslope direction (Fig. I.3-6B). The bulk of the sliding mass was therefore not in the active limit state and extensional faulting was concentrated in the head region of the slide, which would correspond with Wernicke and Burchfiel's listric boundary fault. Since static equilibrium was maintained, the slide could not accelerate. The space for slow downslope movement was provided by compaction. In contrast, extensional gliding of the sheet in the Wernicke-Burchfiel model requires the removal of a supporting buttress, to induce an active limit state throughout the gliding sheet. The result is an accelerating slide or, more likely, a succession of such 'catastrophic' sliding episodes and the disintegration of the slide mass by normal faults. It seems difficult, however, to find a mechanical argument for the preferred formation of downslope-dipping normal faults in a sliding sheet of this type, and one would expect horst-and-graben structures to be more likely to form, possibly with short suites of parallel infill faults, as is commonly observed in catastrophic landslides.

It would therefore seem more likely that the planar parallel faults have been formed by some substratal extension *before* the slide was induced by a local concentration of ductile extension (rifting) or by the erosive removal of slope support.

3. HORIZONTAL EXTENSION AND SUBSIDENCE

In the preceding section the extending base of a brittle overburden was kept planar. In nature, the extension imposed by ductile substrata and the accompanying faulting of the overburden are often complicated by vertical displacements, in particular by synclinal deformations of the overburden base. These deformations range in scale from continental rifting to the local thinning of an individual incompetent bed, and may be produced by quite different mechanisms, such as thermal contraction, flow (for instance, driven by differential loading as sketched in Fig. I.3-1D), differential compaction, or dissolution. The origin of such synclinal deformations in extensional areas is outside the scope of this book and our concern is only with the effect of local subsidence on the faulting of the overburden. Moreover, we have to content ourselves with the results of a few exploratory numerical simulations, since the mechanical analysis of faulting in this field is still in its beginning and is hampered by insufficient geological data to describe the deformation history of the overburden base with sufficient accuracy to serve as boundary conditions for the mathematical problem.

Let us again consider a uniform rock layer as specified in Fig. I.3-29, together with its dimensions, material parameters and initial stresses. The rock is assumed to behave as elastic/frictional plastic material with strain hardening and softening. Inside an interval, slightly longer than the height of the rock layer, the overburden base is deformed by a component of uniform horizontal extension in combination with a vertical displacement component of sinusoidal distribution. Outside this interval the overburden base is horizontally displaced without extension of the base. The finite element mesh is shown in Fig. I.3-30A for the right-hand half of the overburden block. The displacement of the basal element is built up in small increments for two different cases: (1) the horizontal displacement (u_o) is at any step half of the vertical displacement (v_0) at the centre of the deformed interval; (2) the two displacement components are equal.

In the first case ($u_o = v_0/2$) the trajectories of σ_I form a typical arch (Fig. I.3-30B) and the plastic shear zones (cross-hatched) in Fig. I.3-30C represent steep, 'chimney'-like boundary faults and shallow grabens on either side of the shallow, sinusoidal subsidence trough. A comparison with the calculated stress trajectories and final boundary faults above the relatively subsided sides of a vertically uplifted rigid basement block, which are presented in Ch. II.7 (Section II.7.5, Example No. 5: 'Differential Uplift', Figs. II.7-31, 32E, and Section II.7.7, Example No. 7, Fig. II.7-46D), shows remarkable similarity.

The formation of almost vertical boundary faults is associated with the 'arching' of the σ_I-trajectories. Obviously, the 'stress arching' may be suppressed, by sufficiently increasing the contribution of the horizontal extension component. It was found (Fig. I.3-30D) that this requires at least a horizontal outward displacement of the graben 'shoulders' that is about equal to the maximum vertical displacement (v_0). The associated graben boundary faults then assume the more familiar orientation shown in Fig. I.3-30E. It should be noted that the displacements prescribed at the overburden base were small, although sufficient to enable first faults to develop. Larger displacements have been applied in sandbox experiments with a similar equipment to that sketched in Fig. I.2-31 (without frictional side walls), but with an additional device to simulate subsidence by pulling the rubber base down in proportion to the horizontal stretching. In the experiments the first faults developed in complete accord with the calculations, but with increasing basal deformation new faults were formed. In the experiment corresponding to Fig. I.3-30E these faults were formed between and parallel to the inclined boundary faults. The later faults in the experiment comparable with the analysis in Fig. I.3-30C had about the same dip angle as the 'infill' faults of the first experiment, but were mainly formed outside the inititial fault 'chimney'.

Although these results are of a very preliminary nature, and many more parameter studies are needed to arrive at a satisfactory mechanical model, the results indicate the importance of the ratio between the horizontal movement of the synclinal shoulders and the subsidence of the trough to the configuration and history of overburden faulting.

4. UPDOMING

Overburden extension may be also caused by local or regional updoming or downbending of strata. In Section I.2.4 (Figs. I.2-23,24,25) we have already dealt in a schematic way with the orientation and shape of normal faults generated by post- and syndepositional bending of horizontal strata. From experimental evidence we know that the position of the faults and, in particular, the formation of a first graben (Fig. I.2-23), are highly dependent on variations in the curvature of the bending layers. Naturally, in layers of reasonably uniform strength properties, faults will be initiated at locations of greatest curvature, since these are also the locations of the greatest reduction in σ'_{III}.

Where the extension of strata is caused by the growth of ridges or domes of underlying ductile material (e.g., salt or shale), the history of stress changes and faulting in the overburden is greatly complicated by the changes in the shape of the bodies that intrude into and pierce the overburden. The fault structures generated by a rising dome or ridge of ductile material may therefore be quite complex (see e.g., Fig. I.2-7). To indicate at least the main trends in the development of these structures, we consider in Fig. I.3-31 very schematically the orientations of principal stresses induced by an idealised cylindrical dome in a horizontal overburden with σ_{II} and σ_{III} initially equal and acting in the horizontal plane. The intruded rock will be extended along the circumference of the rising dome. Therefore, in a circular region around the crest of the dome σ_{III} will become different from σ_{II} and oriented tangentially to the dome. When the curvature of the domal surface is greatest at the crestal rim, normal faults will be formed there with a radial strike direction and will turn into radial flank faults upon continued piercement. Above the crest of the dome the pattern of radial faults is likely to change, since the curvature of the top may cause a radial orientation of σ_{III} (Fig. I.3-31). Compatible with this stress orientation,

Interval $-L \leqslant x \leqslant L$ of overburden base is deformed by vertical displacement $v(x)$ (with maximum v_o) and uniform horizontal extension u_o/L. Uniform horizontal boundary displacement u_o outside deformed interval.

Overburden parameters

Total length	:	*10 km*
Height	:	*2 km*
Interval	:	$2L = 2.4$ *km*

$v_o = 7\ m$

Elastic moduli: $E = 4.10^4 MPa,\ \nu = 0.25$

Frictional plastic moduli: φ *increases from 30° to 45° at peak strength,* τ_o *decreases from 5 MPa to 0, no dilatancy.*

Initial state of stress: $\sigma^o_H = 0.6\ \sigma^o_V$

Displacement components: $|x| \leqslant L:\quad u = u_o \dfrac{x}{L}$

$$v = \frac{v_o}{2} \cdot \left(1 + \cos \frac{\pi \cdot x}{L}\right)$$

$$|x| > L: \quad u = u_o, \quad v = o$$

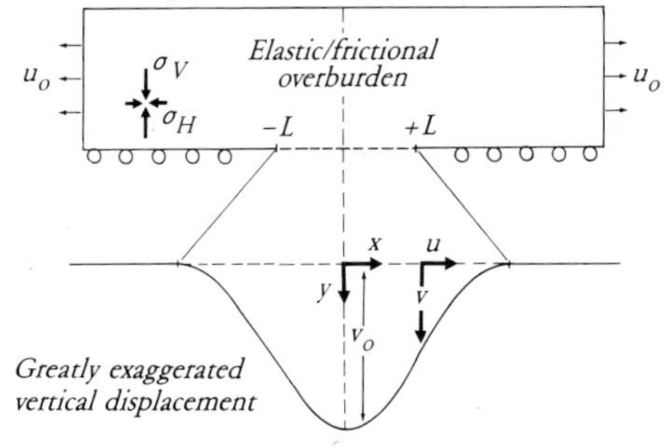

Fig. I.3-29
NUMERICAL MODELLING OF FAULTING INDUCED BY EXTENSION AND SUBSIDENCE OF BASAL SEGMENT; GEOMETRY AND DISPLACEMENTS

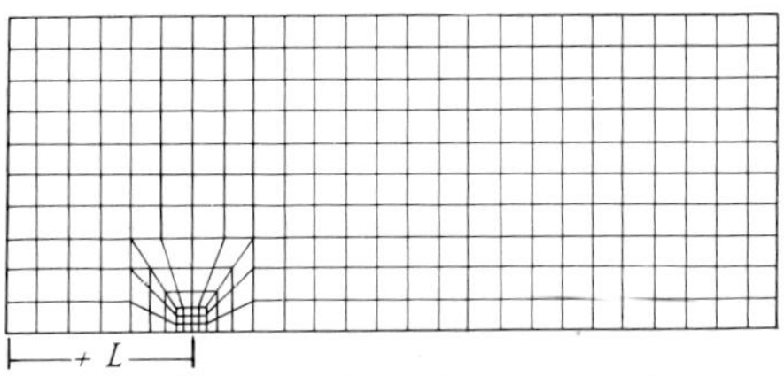

A Finite element mesh of 8 – noded isoparametric elements with four integration points

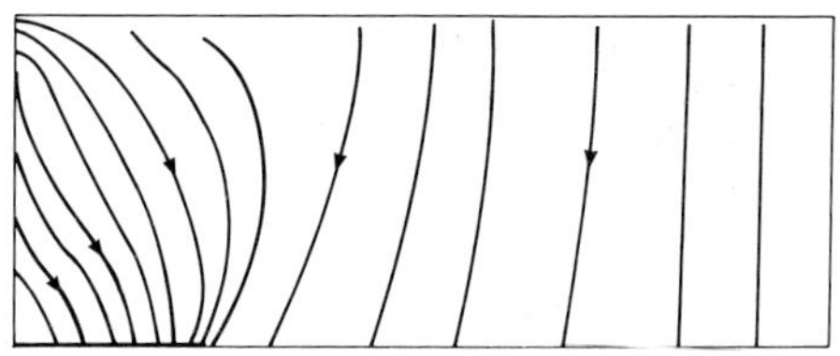

*B Horizontal displacement (u_o) is **half** of vertical displacement (v_o); σ_I-trajectories at earliest stage of through-going softening zone still show the 'arching' which is typical for local subsidence.*

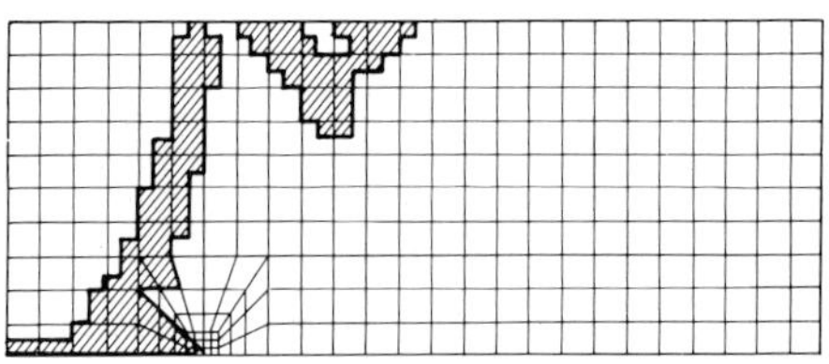

C Fully developed softened shear zones surrounded by elastically unloaded material; displacement conditions as in (B); note the steep precursor fault zone.

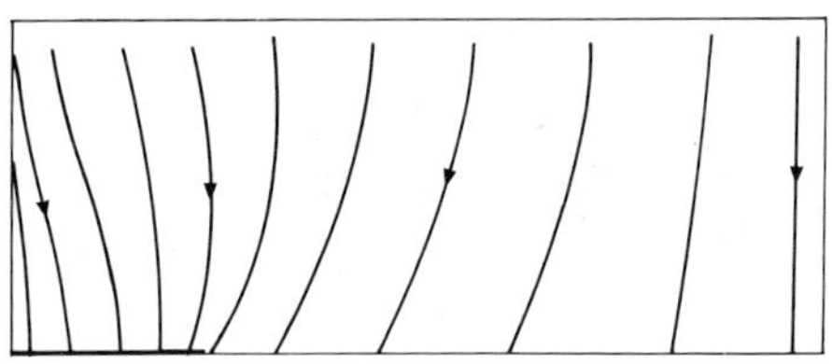

*D Horizontal displacement (u_o) is **equal** to vertical displacement (v_o); σ_I-trajectories at earliest stage of throughgoing softened shear zone; note that 'arching' is almost abolished.*

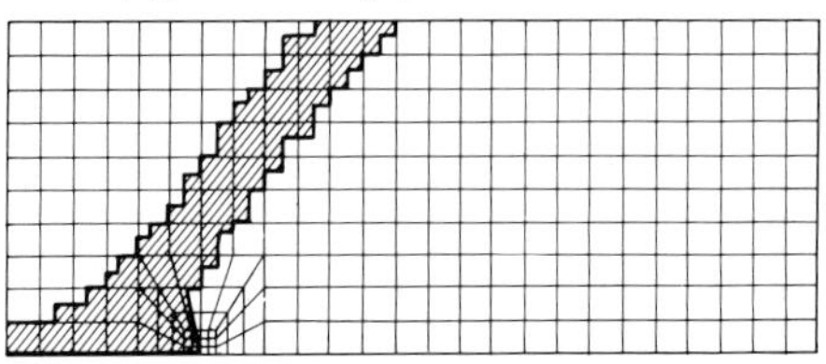

E Fully developed softened shear zone surrounded by elastically unloaded material, representing normal graben boundary fault; displacement conditions as in (D)

Fig. I.3-30 NUMERICAL MODELLING OF FAULTING INDUCED BY EXTENSION AND SUBSIDENCE OF BASAL SEGMENT. Trajectories of maximum compressive stress and softened shear zones in right-hand half overburden region in Fig. I.3-29.
(obtained from unpublished numerical analyses by W. Sassi and H.W.M. Witlox, 1986)

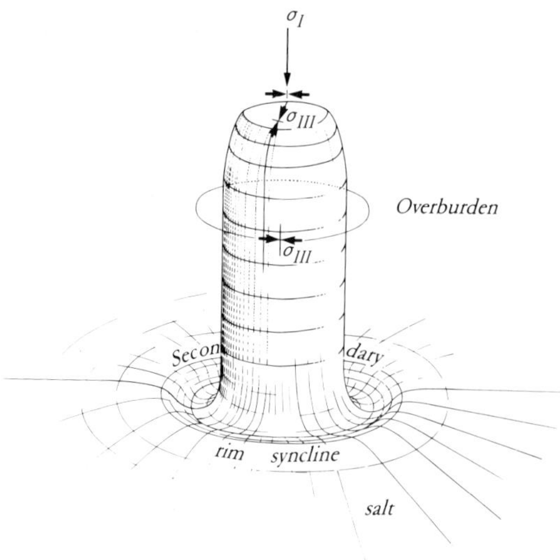

Fig. I.3-31
PRINCIPAL STRESSES
NEAR RISING SALT DOME

radial normal faults may link in the central part of the crest and form one or two roof grabens, as has been observed in the field and in experiments (Parker and McDowell, 1955).

5. OBLIQUE SLIP

In normal faulting σ_I is vertical or subvertical and σ_{II} is parallel to the strike direction of the fault. When, after a period of inactivity, a normal fault is reactivated by a new extensional stress system with changed directions of σ_{II} and σ_{III}, movement along the pre-existing fault plane will no longer be pure dip-slip, but slip will be oblique to the dip direction, consisting of a dip-slip component and a strike-slip component. If it is assumed that the shear strength parameters on the fault plane are independent of the direction, the new slip will be in the direction of the maximum shear stress on the fault plane. A formula for the direction of this shear stress vector in a fault plane of given orientation is derived in Ch. II.1.1, Example No. 5 (eq. II.1-17) for the case that the principal axes of the slip-inducing stress system are vertical and horizontal. We consider here the reactivation of a normal fault with dip angle δ by a new extensional stress system with σ_I vertical and σ_{II} rotated horizontally through the angle ω from the strike direction towards the hanging wall (see Fig. II.1-17). The direction of the vector of maximum shear stress ($\vec{\tau}$) in the fault plane is defined by the angle θ which the vector makes with the strike direction (pointing from right to left when seen from the downthrown side, cf. Fig. II.1-17). The angle θ is considered positive in clockwise direction, and a negative value therefore implies that the shear stress vector points downward on the fault plane causing slip with a normal dip-slip component. The angle θ can then be expressed in terms of the principal stresses of the reactivating stress system (eq. II.1-17b):

$$\tan\theta = -\ [(\Sigma_1 - 1)\cdot(\Sigma_2 - 1)^{-1} - \sin^2\omega]\ 2\cos\delta/\sin 2\omega \qquad \text{(I.3-16)}$$

where

$$\Sigma_1 = \sigma_I/\sigma_{III} > \Sigma_2 = \sigma_{II}/\sigma_{III} > 1 \qquad \text{(I.3-16a)}$$

We see that for vanishing ω, i.e. for σ_{II} being oriented along strike, reactivated movement will indeed be pure dip-slip, since tan θ tends to $-\infty$ and θ therefore becomes $-\pi/2$, which means that the vector of maximum shear stress in the fault plane points downward at a right angle to the strike direction. To obtain an idea of the effect an out-of-strike rotation of σ_{II} will have upon the direction of the induced slip, we consider a fault plane with dip angle $\delta = 60°$ and assume a rotation angle $\omega = 30°$. Formula I.3-16 then simplifies to

$$\tan\theta = -1.15\,(\Sigma_1 - 1)\cdot(\Sigma_2 - 1)^{-1} + 0.29 \qquad \text{(I.3-16b)}$$

The orientation of the maximum shear stress vector will then depend on the stress ratios Σ_1 and Σ_2 of the reactivating stress system. If we choose, for example, $\Sigma_1 = 2.32$, $\Sigma_2 = 1.2$, the angle θ is $-82°$, while for the choice of $\Sigma_1 = 2.29$, $\Sigma_2 = 1.5$, the angle θ is $-70°$. Hence, in the first case the downward slip will deviate by 8°, in the second by 20°, from the dip direction towards the side pointed at by the σ_{II}-axis. These examples may sufficiently illustrate how sensitively the reactivated slip direction depends on the *ratio between the intermediate principal stress and the smallest principal stress.* Thus contrary to the initiation of a fault, which is practically independent of the value of the intermediate principal stress, this stress becomes most important in determining oblique slip on an existing fault plane or any other plane of weakness.

The values for Σ_1 and Σ_2 have not been chosen arbitrarily, but rather to ensure that slippage is actually generated. If it is assumed that cohesion on the fault plane may be neglected with respect to the frictional shear strength, the ratio of the induced maximum shear stress and the effective normal stress on the plane must equal tan φ_{red}, where φ_{red} is the (reduced) friction angle on the fault plane before reactivation. This condition can be expressed in terms of Σ_1 and Σ_2 after some lengthy, but elementary, algebraic manipulations, which we refrain from reproducing here [10]. The stress ratios chosen above satisfy the slip condition for $\varphi_{red} = 20°$.

NOTES

1) These listric faults do not necessarily merge smoothly with the basal slip plane. It should be recalled that the stresses inside the growing sediment layer gradually change during sedimentation before the critical thickness is reached, and that at some depth an extensional limit state may have been induced (for example, by compactional downslope displacement of material or by some ductile extension of deeper strata) before the onset of basal slip. *Parallel precursory faults* may therefore have formed inside the overpressured region, and these would meet the basal slip plane at an acute angle.
   Although these precursor faults will have little throw before the onset of basal slip, some may develop into major faults by joining with the basal slip plane into a gliding path. The sharp change in dip angle at the junction of a normal fault and the basal slip plane will cause the formation of antithetic adjustment faults (see Section I.2.6, Figs. I.2-40,41,42).

2) We refer to Coulomb-Mohr's limit condition in the explicit form of eq. II.7-5 of Section II.7.3 (Example No. 2, "Submarine slope"), where σ_2 and σ_3 are to be replaced by $\sigma_{\|}$ and $\sigma_{\perp}$ of our present notation. Since the plus and minus signs of the square root correspond to the passive and active limit states, respectively, we obtain for the difference between passive and active limit values of the slope-parallel rock stress at depth z:

$$\sigma'^{p}_{\|} - \sigma'^{a}_{\|} = 4(\cos\varphi)^{-1}\sqrt{[\sigma'_{\perp}(z) + \tau_0 \cot\varphi\,]^2 \tan^2\varphi - \tau^2(z)} \qquad (1)$$

where $\tau(z)$ is given by eq. I.3-1 and $\sigma'_{\perp}$ by eq. I.3-5 for $z \geqslant Z_{ig}$. Inserting these expressions in eq. (1) and making use of the critical depth formula I.3-6 for the ideal overpressure profile of Fig. I.3-5A, yields

$$\sigma'^{p}_{\|} - \sigma'^{a}_{\|} = 4(\cos\varphi)^{-1}\sqrt{(Z\gamma'\sin\delta)^2 - (z\gamma'\sin\delta)^2} = 4\gamma'\sin\delta\,(\cos\varphi)^{-1}\,Z\sqrt{1-(z/Z)^2} \qquad (2)$$

Integrating this expression from 0 to Z, thereby neglecting the positive value of Z_{ig} with respect to the much greater depth Z of the basal slip plane, and dividing the result by Z, we obtain for the average difference of the slope-parallel limit stresses:

$$\overline{\sigma'^{p}_{\|} - \sigma'^{a}_{\|}} = 4\gamma'\sin\delta\,(\cos\varphi)^{-1}\cdot Z\cdot\pi/4 \qquad (3)$$

Inserting this expression in eq. I.3-7a gives eq. I.3-8.

3) Note that in Fig. I.3-9 the plastic limit state does not extend from the slope surface downwards, but begins at a certain depth Z_{ip}. At shallower depths the sediment remains in a non-plastic state. This result (Crans and Mandl, 1981, Part II (b)) is based on the (idealising) assumption that the initial ratio $\sigma'_{\|}/\sigma'_{\perp}$ (i.e. a relative 'slope pressure at rest') in sediment deposited on the slope remains unchanged until, with increasing effective overburden load, the plastic limit state is attained. Depending on the pore-pressure profile and the strength parameters τ_0, φ, a certain minimum overburden (Z_{ip}) is required to induce a limit state in the underlying layers. No such minimum overburden would be required in the special case of a cohesionless frictional material under normal hydrostatic pore pressures.

It is of note that in the determination of the thickness of the non-plastic 'crust' on an ideal slope, only the equilibrium equations and the limit condition have been used. The result should therefore still apply when the limit condition defines the onset of a *viscous (Bingham fluid) flow*, instead of plastic deformation. Ideally, therefore, a slope in the state of steady creep flow should carry with it an undeformed crustal layer.

The slip line fields of Fig. I.3-9 can be also represented in analytical form (Crans and Mandl, 1981, Part II(b)), which, however, is too complex to be of much use. For a purely cohesive material the equations reduce to the well-known equations of cycloidal slip lines (Hansen, 1965), which indicate that some slight listricity is caused by the cohesive strength alone.

4) The mechanism proposed for the upward migration of thin hydrocarbon pockets inside normal faults is practically the same as in the process of hydrocarbon-filled fractures migrating upward through intact shales of small tensile strength ('critical stress intensity'), which was postulated by Mandl and Harkness (1987b) in close analogy to work by Weertman (1970) on the migration of water-filled crevasses in glaciers, and by Secor and Pollard (1975) on the stability of hydraulic fractures in granite.

5) See, for example, H. Lamb: "Hydrodynamics", 6th edition, p. 581-584; Dover Publ., 1932.

6) According to the theory of partial differential equations, a partial differential equation of first order may be replaced by an equivalent system of ordinary first-order equations (characteristic equations), which, in the case of eq. I.3-12a reduces to eq. I.3-13.

One may also arrive at this equation more directly by first considering the solution $h = h(x,t)$ to eq. I.3-12a in the implicit form $F \equiv h - h(x,t) = 0$ and then changing the roles of the dependent variable h and the independent variable x, i.e. considering $x = x(h,t)$. Applying the rules for the differentiation of implicit functions of several variables yields:
$(\partial x/\partial t)_{h=\text{const.}} = -(\partial F/\partial t)/(\partial F/\partial x) = -(\partial h/\partial t)/(\partial h/\partial x)$, which becomes eq. I.3-13 after inserting the right-hand expression from eq. I.3-12a.

7) The difference between flow velocity and the travel velocities of disturbances of the interface is not a strange phenomenon, but is rather typical for processes that can be described by a first-order equation that is non-linear in the dependent variable (h). Such processes may be as different as, for instance, the deformations of a glacier surface, which travel at speeds quite different from the creep

rate of the glacier, and the generation of flood waves on large rivers, which move faster than the water of the river.

8) Stationary creep flow of completely dry salt would be described by a power law (see Note 1) in Ch. II.7). When the shear rate is proportional to the n−th power of the shear stress (n>1), h^2 in eq. I.3-13 has to be replaced by h^{n+1}.

9) With eq. I.3-15 and by replacing the independent variable x in eq. I.3-12 by the new independent variable y, and the dependent variable h by H, where
$y = e^{kx/3}$ and $H = h.e^{-kx/3}$
eq. I.3-12 attains the form
$H^2\, \partial H/\partial y + (12\eta/k\gamma' \tan\delta_0)\, \partial H/\partial t = 0$

10) To express the slip condition in terms of Σ_1 and Σ_2, insert into $\tan^2\varphi = |\vec{\tau}|^2/|\vec{\sigma}|^2$ from eqs. II.1-8 and II.1-11a of Ch. II.1 and the components $n_1 = \sin\delta\cos\omega$, $n_2 = \sin\delta\sin\omega$, $n_3 = \cos\delta$ of the unit normal in the coordinate system of Fig. II.1-17.

I.4. WRENCH FAULTING

In Chapter I.2 we have considered certain features of strike-slip faulting as particular expressions of mechanical processes which play a more general role in tectonic faulting. Here we resume this discussion to examine in a more systematic way the mechanical genesis of these faults, which are characterised by dominantly horizontal shear displacement along steeply dipping surfaces.

1. MECHANICAL CLASSIFICATION

From the mechanical point of view a first, and fundamental, distinction has to be made between strike-slip faults that form in a tectonic stress field with *stationary principal stress directions,* and those that form in a stress field whose *principal axes rotate* in response to imposed kinematic constraints. We have dealt with this distinction in Chapter I.2 (Section I.2.8) in connection with the primary stress changes that may accompany the formation of a fault, and we have referred to the first class of strike-slip faults as 'stress-controlled' faults, or Anderson-type faults (Fig. I.2-68), since they correspond with the strike-slip faults of Anderson's fault classification (Fig. I.2-9B). These faults occur mainly in external fold belts; a classical example are the conjugate strike-slip faults that intersect the fold belt of the Jura Mountains (Fig. I.4-1). The Anderson-type strike-slip faults are typical features of 'thin-skinned tectonics', which accommodate the shortening of a detached overburden.

Kinematically controlled strike-parallel shears form under the tectonic conditions that we have referred to, somewhat loosely, as 'simple shear tectonics', because of the similarity with the conditions of simple shearing in a shear apparatus. The main types are presented schematically in Figs. I.2-72 B,C. Type B is of the 'thin-skinned' type, and involves the horizontal shearing of a detached sheet between stronger boundary rocks, which may occur when the sedimentary fill of a graben is sheared by graben-parallel horizontal displacement of the graben boundaries. An example of this is shown in Fig. I.2-57. In the more important type C of Fig. I.2-72 basement is involved by shearing the overburden along a strike-slip zone in the basement. Such 'simple shearing' of the overburden is induced on any scale where strong plates in contact slide horizontally along their interface – on the largest scale along lithospheric plate boundaries. It is also induced by horizontal slip along a subvertical plane or zone of weakness inside a crustal plate or some rock layer that serves as mechanical basement for overlying weaker rocks. A well-known text-book example of a basement-induced zone of 'simple shearing', the Newport-Inglewood Trend in California, is shown in Fig. I.4-2.

Vertical or subvertical zones of horizontal shearing may also be generated by differential horizontal extensional or compressional straining of the overburden base as sketched in Fig. I.4-3A in plan view for the extensional case. This, too, is a kinematic condition enforcing a change in the principal stress directions. Note that here the difference in straining is not restricted to the terminations of the wrench zone, as, for instance, in transform faults, but persists throughout the wrench zone. Figure I.4-3B postulates a 'bookshelf'-type connection of normal overburden faults across the wrench zone by rotating cross faults with a strike-slip component.

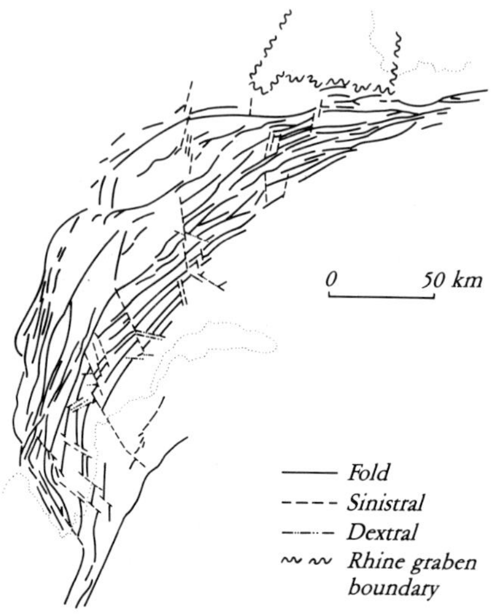

Fig. I.4-1
SIMPLIFIED TECTONIC MAP OF THE JURA FOLD BELT (After Laubscher, 1972; by permission of American Journal of Science) *Sedimentary sheet shortened by folding and conjugate wrench faulting under Alpine compression transmitted through thick Molasse sequence; gliding on Triassic evaporites.*

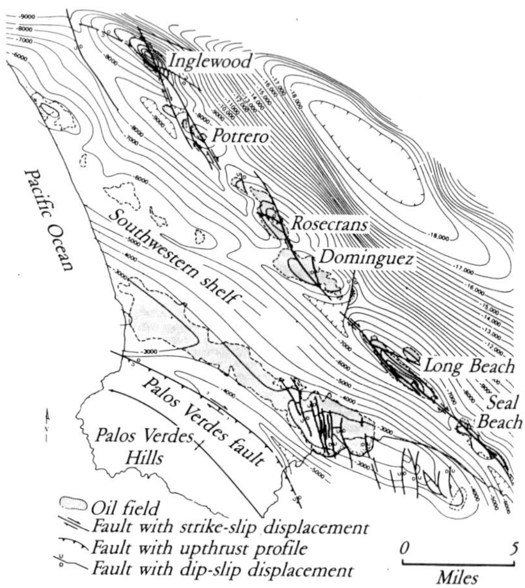

Fig. I.4-2
THE NEWPORT-INGLEWOOD WRENCH FAULT ZONE, CALIFORNIA (After Harding, 1973; reprinted by permission of "American Association of Petroleum Geologists")
Narrow belt of en échelon folds and strike-slip faults produced by relatively small right lateral displacement.

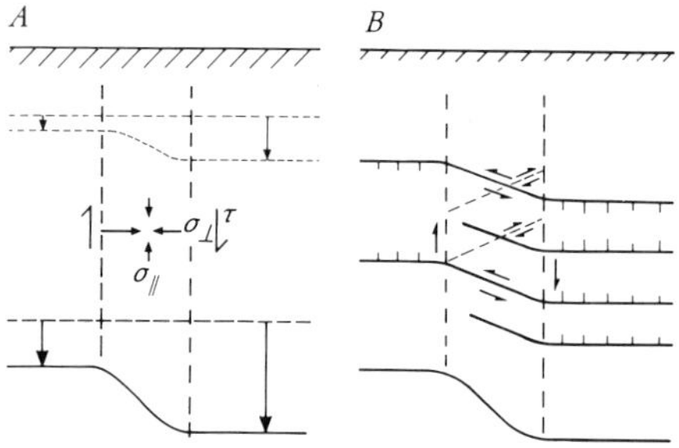

Fig. I.4-3
VERTICAL ZONE OF HORIZONTAL SHEARING GENERATED BY DIFFERENTIAL HORIZONTAL EXTENSION OF OVERBURDEN BASE (Plan view)
Reduction of shear-zone-parallel normal stress $\sigma_{\parallel}$ by extension (Fig. A) allows shear accommodation by 'domino-style' rotation of cross faults (Fig. B) and adds an antithetic strike-slip components to the dip-slip of these faults. The rotating cross faults would seem better suited to connecting normal faults across the shear zone than non-rotating cross faults (stippled in Fig. B) with synthetic strike-slip.

Because of the stress couples, which operate in simple shearing, the term 'wrench fault' would seem to be more apt for strike-slip faults of the 'simple shear'-type than for strike-slip faults of the Anderson type. But leaving aside such terminological subtleties, we shall follow the practice of many structural geologists in using the terms 'wrench faults', 'strike-slip' or 'transcurrent faults' as synonyms.

The formation of Anderson-type wrench faults is accounted for in Coulomb-Mohr's theory of shear failure simply enough by associating it with a vertical direction of the intermediate principal stress throughout the region in the limit state. Since this requires that shear stresses along horizontal planes are negligible, the faulted rock mass must be allowed to move on a base of very low frictional resistance. An interesting aspect of this type of wrench faulting, which was mentioned at the end of Section I.2.3, is the interplay of wrench faulting and thrusting or compressional folding, which commonly takes place in arcuate thrust belts. In thrusting and compressional folding the intermediate principal stress is directed along the strike of the thrust faults and along the fold axes. This stress will be reduced by the bending of thrust fronts or fold axes and may become the smallest principal stress, while the overburden stress assumes the role of the intermediate principal stress. It was illustrated in the Mohr diagram of Fig. I.2-5 that relatively small changes in the horizontal principal stresses may be required for the roles of principal stresses to switch and to allow conjugate wrench faults to operate (see the conjugate wrench faults of the Jura fold belt in Fig. I.4-1). The conjugate wrench faults contribute to shortening in the general thrust direction by wedging fault-bounded segments into the thrust belt. This wedging action restores the conditions for thrusting or folding by raising the principal stress along the strike of the thrust belt. After this, the switch between thrusting and wrenching may be repeated (Laubscher, 1972). Although in principle, from a mechanical point of view, this process is relatively simple, the palinspastic reconstruction of a wrench-faulted thrust belt may be complicated by the movements in the mosaic of fault-bounded plates and by the secondary structures induced by the concomitant local stress changes.

In contrast with the relative simplicity of Anderson-type wrench faults, kinematically controlled wrench faulting is much more complex. This is particularly the case for the type of wrench zones with which we shall be mainly concerned because of their practical importance; these wrench zones are specifically those generated in an overburden by strike-slip faulting in the (mechanical) basement (Type C of Fig. I.2-72). Of all styles of fault tectonics, these basement-induced wrench zones probably present the greatest structural complexities. Hence, in order to gain an understanding of the genesis of these structures and a rational framework for their identification and interpretation, we shall resort to simplifying models, which will enable us to study the role of controlling parameters one by one. In other words, we shall study the development of basement-induced wrench structures under ideal, well-controlled laboratory conditions.

2. *WRENCH ZONE ABOVE A SINGLE PLANAR BASEMENT FAULT WITH PURE STRIKE-SLIP*

Consider first a uniform isotropic overburden with a horizontal surface, which rests on a horizontal basement (i.e. a mechanically stronger rock) that is intersected by a vertical fault plane. The situation is sketched in Fig. I.2-72C of Section I.2.8 where it was pointed out that the overburden material above the basement fault and in the close vicinity of the free surface is subject to simple shearing parallel to the basement fault, when the basement is displaced along this fault in purely horizontal relative movement. The simple shear is then imposed by the undeformed outer parts of the overburden which are carried along passively by the moving basement. In the zone of simple shearing σ_I and σ_{III} are horizontally oriented and are rotated by the imposed shearing deformation towards a 45° orientation with respect to the general shear direction. This and the surface expressions of the associated fault features have already been discussed in Section I.2.8 and need only be summarised here.

The first faults to form are en échelon Riedel shears with a sense of motion synthetic with the general shear direction. These precursory wrench faults strike at an angle to the basement fault that depends on the orientation of σ_I at the onset of shearing. Initial horizontal *transpression* causes first faults to strike at a high angle with respect to the basement fault, while initial horizontal pre-compression of the overburden parallel to the basement fault results in first wrench faults that strike at a very small angle with the basement fault or are even aligned with it. In the literature this state of pre-stress is usually referred to as '*trans-tension*' – a term which is somewhat misleading since it may give the (wrong) impression of the overburden being under horizontal tensile stress in a direction normal to the basement fault, while normally this stress will still be compressive. In the absence of any horizontal pre-stress, which may occur in an 'undisturbed' sedimentary basin, the first set of synthetic Riedel faults will have an intermediate strike direction. For a summarising plan-view representation of the shear-induced changes in σ_I-orientation and associated orientations of the first precursory wrench faults, the reader is referred to Fig. I.2-73. The striking difference in orientations of the first set of Riedel shears, which develop under anisotropic horizontal stress conditions, is demonstrated in model experiments using a shear table of the type shown schematically in Fig. I.4-4A, together with equipment for pre-stressing the overburden by means of rubber base plates, described in Fig. I.4-4B. The photographs of Fig. I.4-5 show the surface expressions of synthetic Riedel shears, which have been produced in model experiments with dry sand under (A) initially isotropic horizontal stresses, (B) initial extra horizontal compression parallel to the 'basement' fault, and (C) under initial transpression. The pictures represent a first phase in the development of a wrench zone, which is reached in the experiment at a displacement along the basement fault of about 20 to 30% of the overburden thickness. Although in these experiments the pre-stresses were applied before wrenching began, other experiments confirmed that the results were essentially the same when pre-stressing and wrenching were simultaneous.

Transpression was also maintained in our ring-shear experiments (Mandl et al., 1977) in which an extra vertical load was applied to the horizontal shearing platens and caused the first Riedel shears to make a relatively steep angle with the horizontal shear direction (Fig. II.7-28). A variety of tectonic situations can cause anisotropic horizontal pre-stresses. Such conditions are, for instance, produced by subsidence and/or uplift of elongated basins (Dallmus, 1958; N.J. Price, 1974). Large-scale tectonic transpression might also be expected to act on major strike-slip zones on the margins of crustal plates, that are underridden by oblique subduction (Oxburgh, 1974).

High compressive pre-stresses parallel to the imposed shear direction may play an important role in the formation of the *'tear' faults.* These are wrench faults that intersect a thrust belt fairly straight and, unlike the conjugate wrench faults discussed in the preceding section, strike normal to the thrust front and fold axes and form the lateral boundaries of separate thrust units. Although the genesis of these faults is not yet well understood, it may be expected to be associated with drastic lateral changes in the advance of the thrust front or in the shortening of the thrust sheet, caused by lateral variations in push force, lithology or frictional resistance. The direction of the shearing resulting from these lateral changes in displacement is (sub)normal to the strike of the thrust belt. The shearing deformation is therefore combined with a very high shear-parallel compressive stress, which enables the first wrench faults to form parallel to the general thrust direction and to align into a through-going wrench fault.

Subsurface geometry. Whereas the principal stress axes near the free surface remain vertically and horizontally oriented in basement-induced wrench zones, at depth they will be diverted from these directions by the action of horizontal and vertical shear stresses, which become more effective with decreasing distance from the basement fault. Figure I.4-6 presents typical vertical sections, normal to the 'basement fault', through sandpacks that have been sheared on the shear table until a first set of synthetic Riedel faults had developed. The fault traces become visible after impregnation of the sandpacks with a dilute gelatin solution and slicing of the wetted pack, since the dilated shear bands are less saturated with fluid than the surrounding material. Typically, the Riedel shears that form in the absence of horizontal pre-stressing (Fig. I.4-6A) and under transpression (Fig. I.4-6C) are *concave-upwards* or straight. This cross-sectional 'tulip' shape was observed in model experiments by Emmons (1969). In all three cases the synthetic Riedels diverge from the basement fault. The degree of symmetry and the width of the pattern varies strongly with the position of

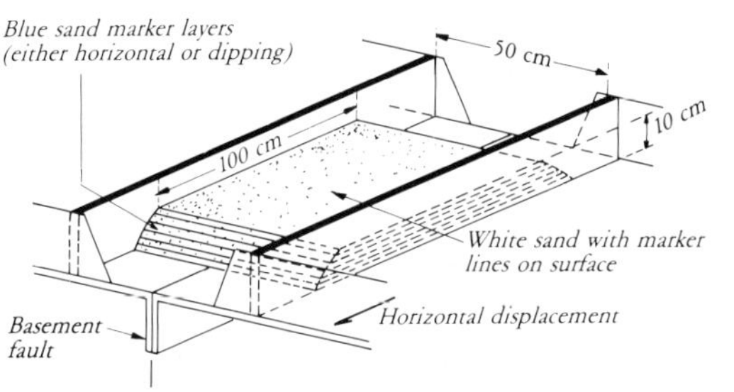

A Simplified sketch of basic equipment. Dip of basement fault can be varied and one table-half can be moved by two geared motor drives to simulate any combination of strike-slip and dip-slip along the basement fault.

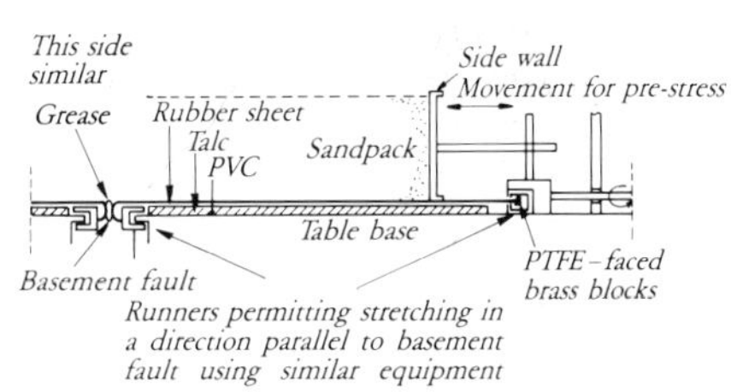

B Modification for pre-stressing of sandpack. Sectional view of pre-stressing device consisting of two rubber sheets, one fixed to each table-half by runners on all sides. The runners allow uniform stretching of the rubber sheets in the direction parallel or perpendicular to the basement fault before deposition of the sandpack and relaxation of the stretched sheets after deposition to pre-stress the sandpack (Naylor et al., 1986).

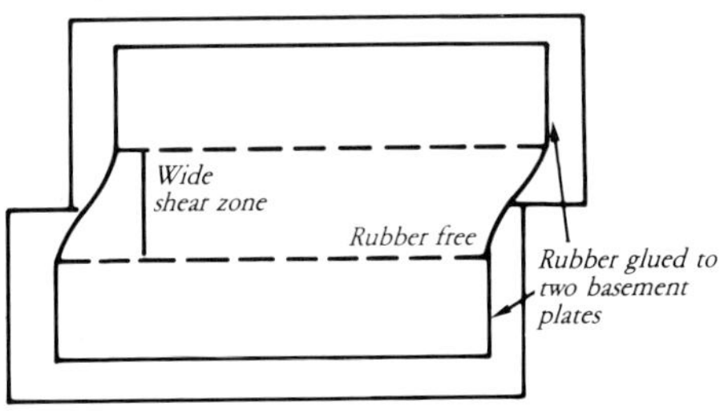

C Modification for wide basement shear zones. Horizontal shear displacement in basement is distributed over a broad zone by partially glueing a rubber or foam-plastic sheet to each half of the shear table.

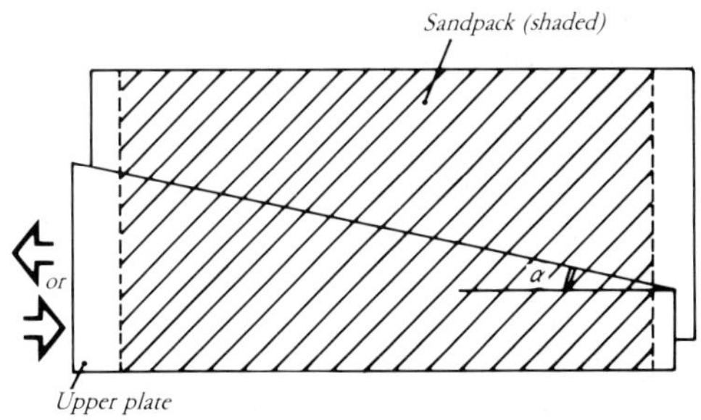

D Modification for modelling divergent or convergent wrenching. Two overlapping metal plates are connected to the shear table with the edge of the upper plate oblique to the direction of shear table movement

Fig. I.4-4 SHEAR TABLE USED FOR WRENCH FAULT MODELLING IN DRY SAND

the cross-section. This is illustrated in Fig. I.4-7 by a series of cross-sections through a sheared sandpack under initial transpression. While in this case the shape of the Riedel faults is drastically flattened and broadened in comparison with the fault shape in the absence of pre-stresses, pre-stressing parallel to the basement fault results in faults that are almost straight and vertical in cross-section. (Figure I.4-6B is somewhat special in that respect as it shows the traces of two such faults, while usually only one fault is cut by the section plane.)

The three-dimensional geometries of many individual Riedel shears have been reconstructed from vertical and horizontal serial sections of sheared sandpacks. Each synthetic Riedel shear has a helicoidal or 'plough-share' shape, as pictured in the block

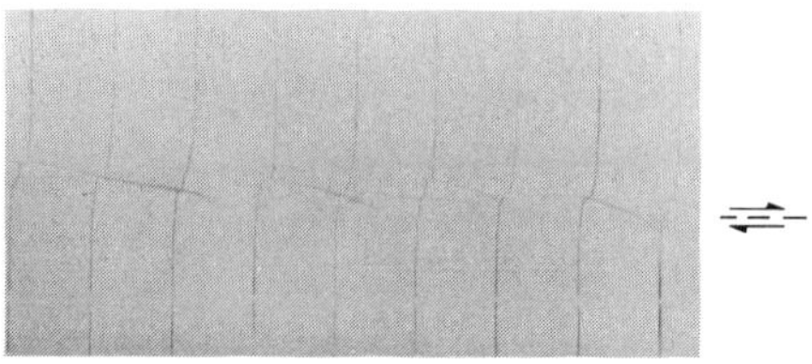

A No pre-stress

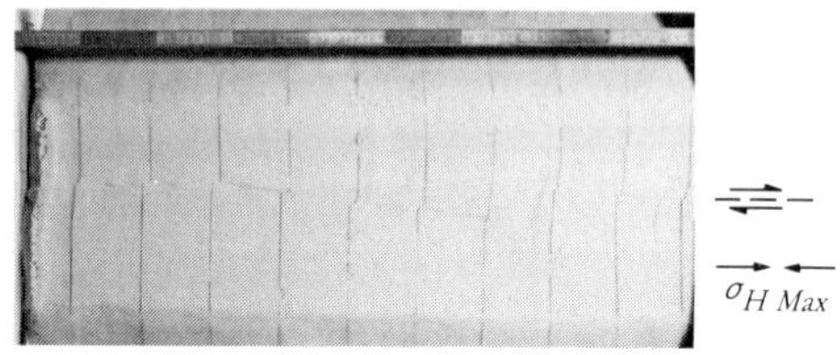

B Pre-stress parallel to basement fault

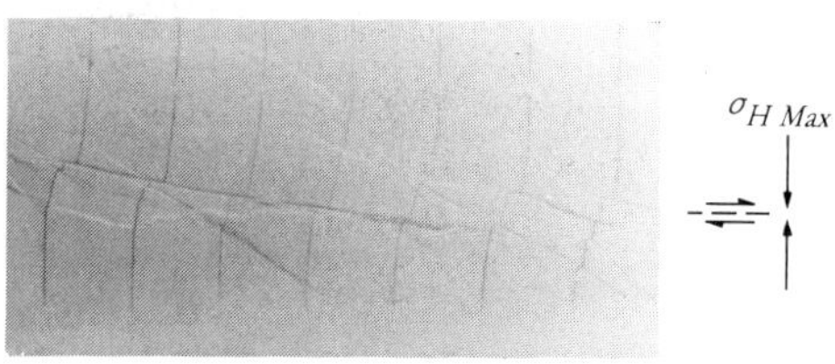

C Pre-stress normal to basement fault

Fig. I.4-5
TYPICAL SURFACE EXPRESSIONS OF RIEDEL SHEARS PRODUCED BY HORIZONTAL MOVEMENT ALONG STRAIGHT BASEMENT FAULT IN SANDPACKS UNDER DIFFERENT PRE-STRESSES.
In all cases, sandpack is 10 cm thick and illuminated from the northern side of the picture. (After Naylor et al., 1986).

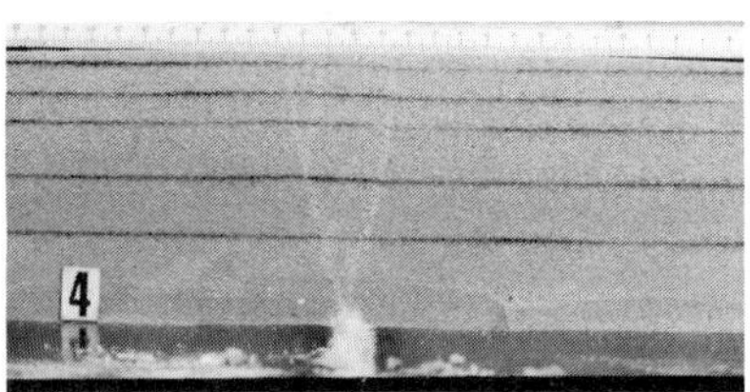

A | *Basement fault*

B | *Basement fault*

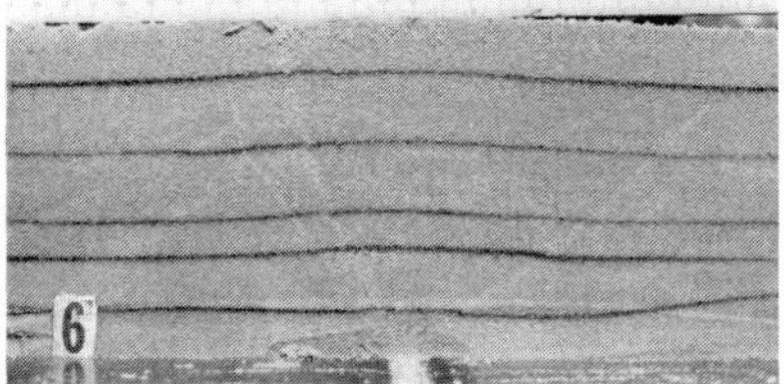

C | *Basement fault*

Figures A,B,C correspond to the three stress states of the previous figure. (After Naylor et al., 1986)

Fig. I.4-6
TYPICAL CROSS-SECTIONAL VIEWS (perpendicular to strike of basement fault) OF EXPERIMENTAL RIEDEL SHEARS, AFTER IMPREGNATION AND SLICING OF THE SANDPACK;

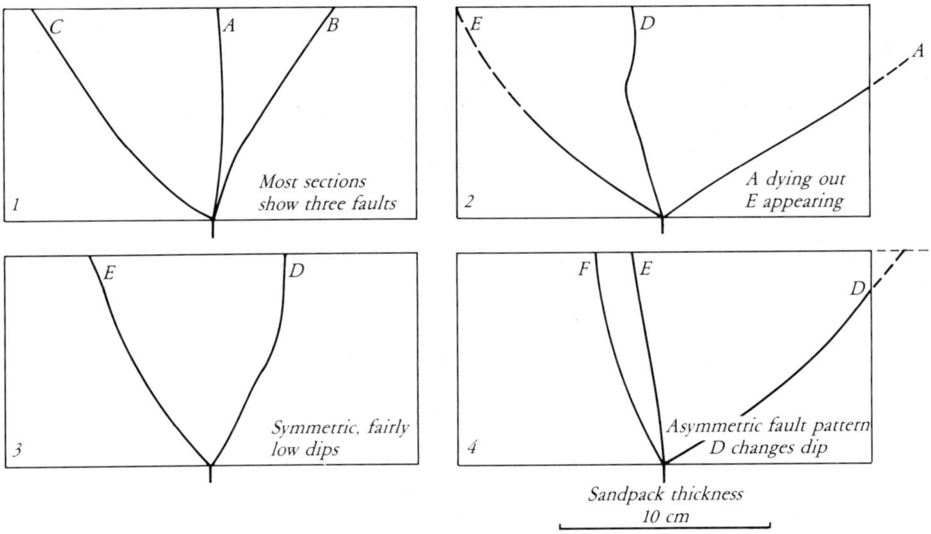

Fig. I.4-7 *SERIES OF CROSS-SECTIONAL VIEWS (perpendicular to strike of basement fault) OF SYNTHETIC RIEDEL SHEARS IN SANDPACK UNDER INITIAL TRANSPRESSION (after Naylor et al., 1986). The sections, which are from one experiment and 10 cm apart, illustrate the drastic flattening and broadening of the helicoidal shape of the Riedel shears which is caused by the pre-compression of the overburden in the direction normal to the strike of the basement fault.*

diagram of Fig. I.4-8A for the reference situation in which no pre-stresses were applied. Also indicated in the diagram is the tendency of the material to be squeezed upwards (U) in the region between two Riedels where the strike-slip motion is obstructed most. This small vertical displacement component reverses (D) where the Riedel fault crosses the basement fault, as is also shown by the 'fault escarpments' on the photographs of Fig. I.4-5. The helicoidal shape of the Riedel shears may be understood by considering the re-orientation of the principal stresses, which is caused by the shear stresses imposed upon the overburden by the movement along the basement fault. These shear stress components are shown schematically in Fig. I.4-9A. The resultant changes in the σ_I- and σ_{II}-directions are shown in a schematic and qualitative way in Fig. I.4-9B for the overburden half at the right-hand side of the basement fault. Note that the shear stress on horizontal sections must decrease with the distance from the basement fault both in the vertical and horizontal directions. This leaves the principal stress directions vertical and horizontal at the whole free surface and also in the subsurface at a lateral distance from the basement fault where the initial state of stress is no longer affected by the shearing along the fault. Considering the orientations of the potential slip elements, which are associated with the principal stress directions at different levels of the inner wrench zone (Fig. I.4-9B), one recognises that the synthetic wrench faults, which fit into the slip pattern and merge with the basement fault, have a helicoidal shape. It can also be seen that the conjugate slip elements, which are the potential constituents of antithetic Riedel faults, cannot combine into faults that merge with the single basement fault. This may explain the absence of antithetic Riedel shears in experiments with a single

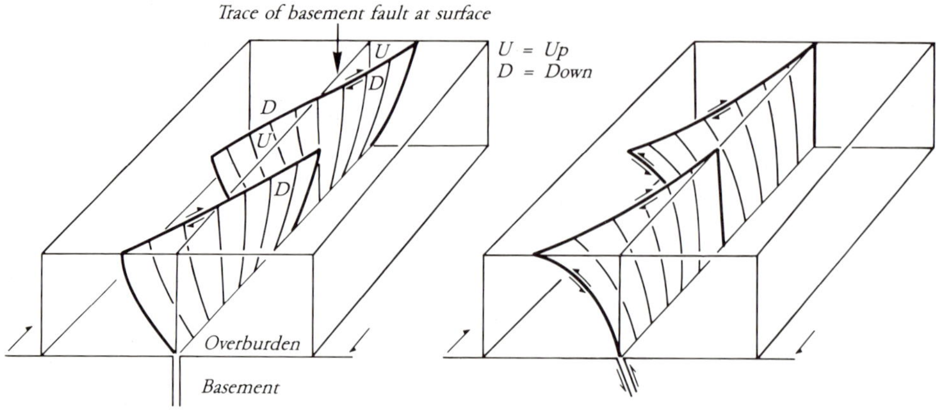

A *Pure strike-slip along basement fault. Curvature of Riedel shears is concave upwards ('tulip'-shape) in vertical sections perpendicular to the strike of the basement fault. Uplift of region between Riedel shears.*

B *Oblique slip along steeply dipping basement fault with predominant strike-slip and additional component of reverse dip-slip. Synthetic Riedels attain upward convex shape ('palm tree' or 'flower' structures) and are preferentially located above the downthrown side of the basement fault.*

Fig. I.4-8 *SCHEMATIC BLOCK DIAGRAM ILLUSTRATING THE THREE-DIMENSIONAL SHAPE OF SYNTHETIC RIEDEL SHEARS FORMED BY RIGHT-LATERAL STRIKE-SLIP DISPLACEMENT ALONG A SINGLE BASEMENT FAULT (from sandbox experiments by W.T. Horsfield, M.A. Naylor, N. van Ooyen & C.K. Sijpesteijn, 1979; unpublished work)*

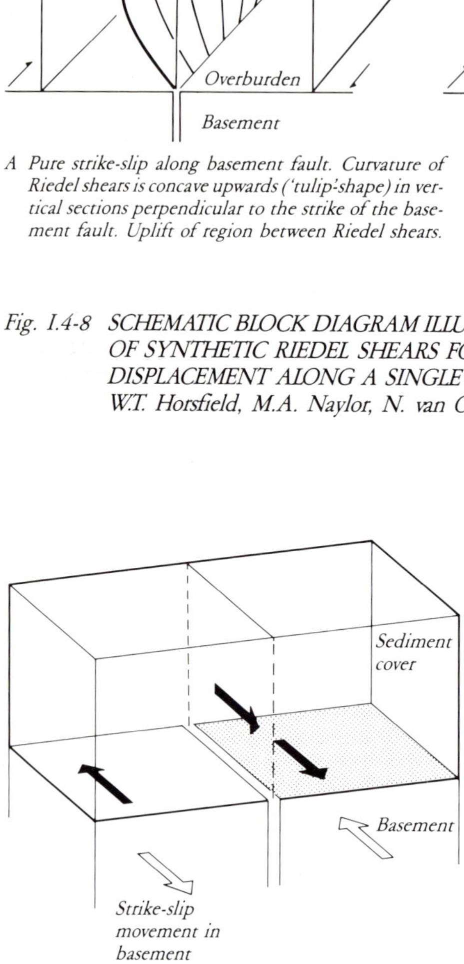

A *Direction of shear stresses (closed arrows) on overburden base and vertical extension of basement fault;*

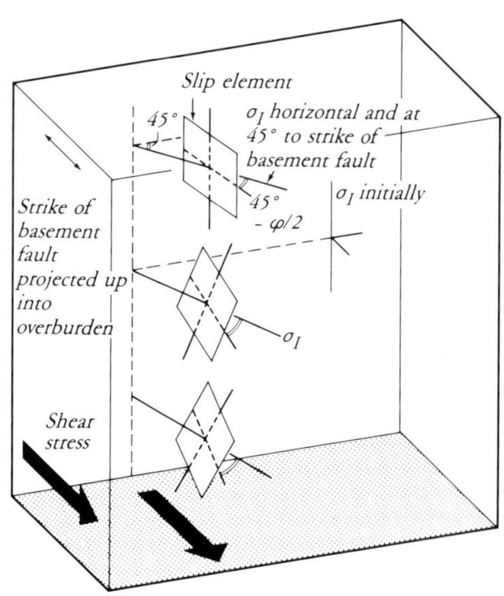

B *Overburden half with qualitative model of principal stress directions (heavy lines) and associated Coulomb slip elements with slip direction synthetic with movement on basement fault (elements of synthetic Riedel shears).*

Fig. I.4-9
SHEAR STRESSES IMPOSED UPON OVERBURDEN BY STRIKE-SLIP MOVEMENT OF BASEMENT

basement fault and a brittle overburden that remains attached to the moving basement. We shall see later, however, that these faults do form when the shear at the overburden base is distributed over a wider zone, either by a wider shear zone in the basement, or by ductile deformation and detachment of the overburden from the oppositely moving basement blocks.

Faulting of the overburden is, of course, restricted to a plastic zone, the lateral boundary of which is – for reasons of symmetry – parallel to the basement fault. In model experiments without pre-stress the surface width of the faulted zone was about equal to the thickness of the overburden (Naylor et al., 1986). Unfortunately, theoretical results on the cross-sectional geometry of the plastic wrench zone are not available yet because of the excessive computational demands of a three-dimensional elastic/frictional plastic analysis.

In many field situations the wrench zone does not develop beyond the formation of a first set of en-échelon wrench faults. It is obvious, however, that these first faults cannot accommodate continued movement along the basement fault, and that new wrench faults, better aligned with the imposed shear direction, have to form unless the first Riedel shears are already well aligned as a result of a pre-stress parallel to the basement fault. The various stages of the development of a fully developed wrench zone above a single basement fault are summarised in Fig. I.4-10, which presents the surface expressions observed in a typical sandpack experiment. It is of note that between the formation of the first set of synthetic Riedels in Fig. I.4-10A and the formation of better aligned shears in Fig. I.4-10C the first Riedels prolong their activity by extending and bending towards their receding sides (Fig. I.4-10B). Freund (1974) has drawn attention to the simple kinematics shown in Fig. I.4-11, which allows the displacement along a wrench fault to die out by bending of the fault ends towards the receding side of the fault. Later we shall return to the problem of wrench-fault termination. After this short episode of a last widening of the wrench zone, the first-formed Riedels become inactivated and new lower angle Riedel shears and the first synthetic P-shears (i.e. wrench faults that cross the basement fault in the direction opposite

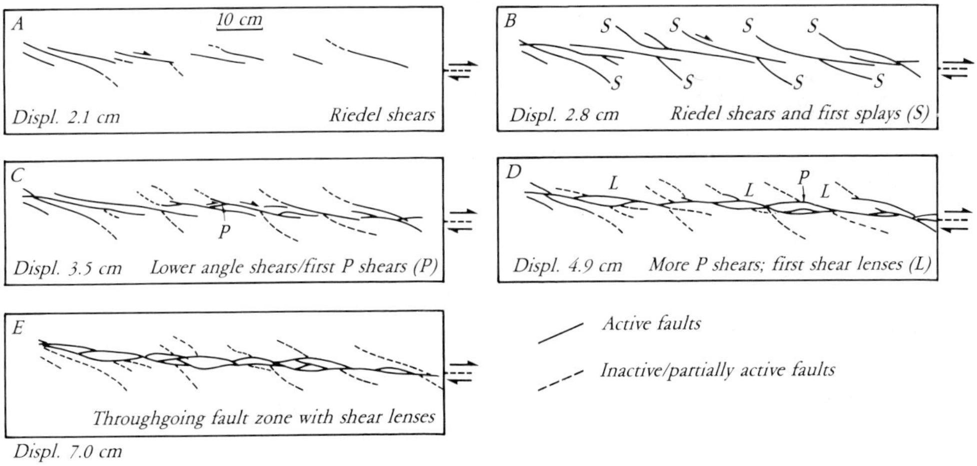

Fig. I.4-10 EVOLUTION OF EXPERIMENTAL SHEAR ZONE ABOVE SINGLE BASEMENT FAULT (after Naylor et al., 1986). Plan views drawn from photographs of sandpack surface at different stages of horizontal displacement along basement fault. Sandpack is 10 cm thick; isotropic initial state of horizontal stresses.

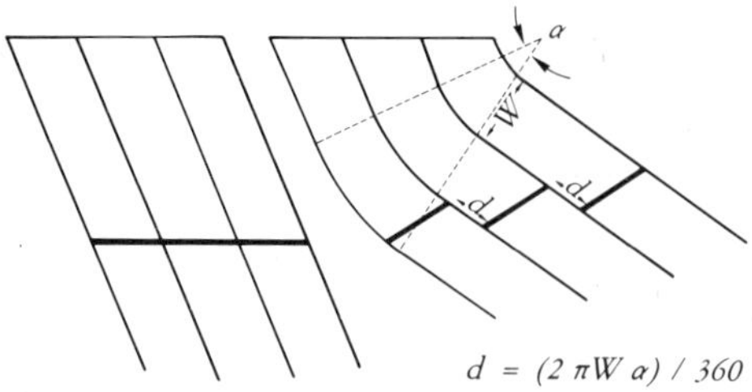

Fig. I.4-11
REDUCTION OR TERMINATION OF STRIKE-SLIP BY BENDING OF FAULTS TOWARDS RECEDING SIDE

to the Riedel faults) develop (Fig. I.4-10C) in the inner part of the wrench zone. The stress changes in the overlap region of the first synthetic Riedel faults that generate these new faults have been discussed and illustrated (Fig. I.2-75) in Section I.2.8. In Fig. I.4-10D the low-angle Riedels and P-shears have formed shear lenses (of the type already shown in greater detail in Fig. I.2-76), which eventually make up the typical anastomosing pattern of a throughgoing wrench fault with an average strike direction parallel to the basement fault.

In concluding this section it should be recalled that it was assumed that the overburden had uniform mechanical properties and, in particular, constant shear strength parameters. In reality, *lithologically contrasting layers* may also differ in elastic stiffness and shear strength parameters. Although little is known about the effects of such contrasts on the development of basement-induced wrench faults, it is clear that precursory shears, which have been generated in a comparatively strong layer, will act as 'basement faults' for a superincumbent weaker layer. At the interface between the two layers the wrench faults therefore tend to branch in the upward direction, thereby widening the zone that is affected by precursory wrench faults. Under such circumstances, extrapolations of subsurface configurations from surface observations of secondary wrench faults could be very misleading.

3. FAULTING ABOVE A WIDE ZONE OF BASEMENT WRENCHING

In nature, basement wrenching may be distributed over a number of sub-parallel basement faults or the shear displacement along a single basement fault may be distributed over a wider zone by an intervening ductile layer, such as shale or salt. To simulate in model experiments the effect of a wide zone of basement wrenching on the faulting of the overburden, the single 'basement fault' in the shear table was bridged by a rubber or foam plastic sheet as sketched in Fig. I.4-4C. Figure I.4-12A presents typical surface expressions of a sandpack in an advanced state of wrench zone development. Rather similar structures were produced in clay-cake experiments by Wilcox et al. (1973), in which the cohesive clay distributed the shear over a wider zone (Fig. I.4-12B). It is immediately clear that, compared with the surface expressions above a single basement fault (Figs. I.4-5A,10A), the synthetic Riedel shears (R) are much longer and overlap over considerable length. In addition, arrays of short antithetic shears (R′) have formed, most of which are confined between synthetic R-shears. In sandpack experiments M.A. Naylor observed that these R′-shears formed after the confining R-shears, and that they are subvertical and die out at relatively shallow depth. In contrast, the R′-shears, which are generated between the ends of synthetic R-shears, appeared simultaneously with the synthetic shears and extend down to the base. These R′-shears may be considered as the conjugates of the R-shears (Fig. I.4-12C). The R′-shears formed later inside the overlap region of first-generation R-shears, however, are not conjugates of these

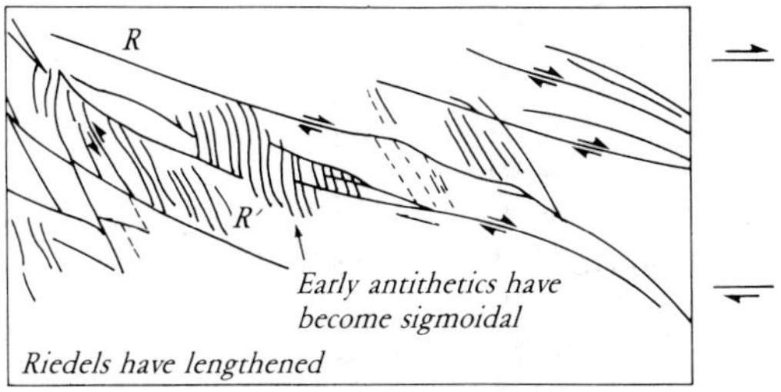

A Sandpack (10 cm high) after 15 cm horizontal basement displacement distributed over wide (30 cm) shear zone (after M.A. Naylor, private communication, 1982)

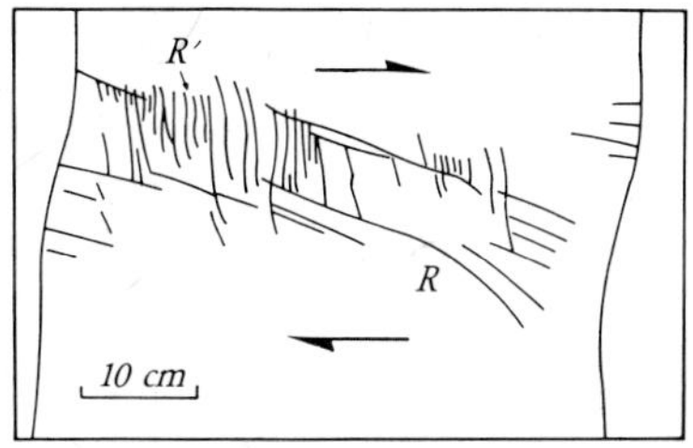

B Sheared clay cake (after Wilcox, Harding and Seely, 1973; reprinted by permission of "American Association of Petroleum Geologists")

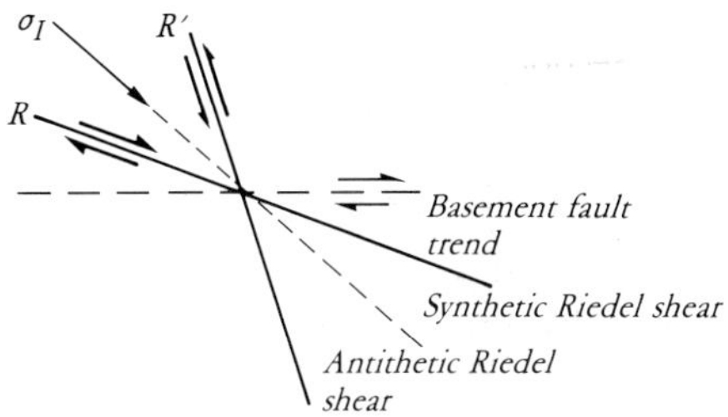

C Strike orientation of conjugate Riedel shears (plan view)

Fig. I.4-12
PLAN VIEW OF MODEL EXPERIMENTS WITH WIDE SHEAR ZONE CONTAINING BOTH SYNTHETIC (R) AND ANTITHETIC RIEDEL SHEARS (R′)

shears, but are associated with a σ_I that has been re-oriented more parallel with the strike of the R-shears by the increase in R-shear-parallel compression which accompanies continued basement wrenching (cf. Fig. I.2-75 and the related discussion in Section I.2.8).

A good field example of wrench faulting above a broad zone of basement wrenching is the Najd fault system in Saudi Arabia (Fig. I.4-13), which contains long synthetic Riedel faults and numerous en-échelon sets of antithetic shears (Figs. I.4-13B,C).

In both the experiments and the field example, the R′-shears show clear evidence of *rotation*, which accommodates basement-induced wrenching. From the experimental evidence it may be inferred that the R′-shears outside the overlap region of neighbouring R-shears start rotating and acquiring a sigmoidal shape after their formation as conjugates of the first-generation R-shears, while the R′-shears inside the overlap region form as rotating cross faults when the unfavourably oriented R-shears can no longer accommodate further basement wrenching. The rotation of these antithetics allows the accommodation of some further wrenching before new synthetic shears have formed that are better aligned with the direction of the basement shear. The sense of rotation is that of the 'dilational' bookshelf mode of shear accommodation (Fig. I.2-2A,B), which was discussed in Section I.2.6 on antithetic faults. It is the increase in normal stress parallel to the shearing between the R-shears which provides the mechanical conditions for the operation of the 'dilational' bookshelf mechanism. (These conditions were explained in Fig. I.2-61, where, for the present application, the 'main shear direction' has to be identified with the strike of the synthetic Riedel shears.)

The rotation of antithetic faults between overlapping synthetic Riedel faults poses a space problem because of the tendency of the set of rotating faults to thicken. It may be solved by upbulging of the free surface. The effect of this will be restricted to a relatively shallow depth range, which is in accord with the termination of these faults observed in model experiments.

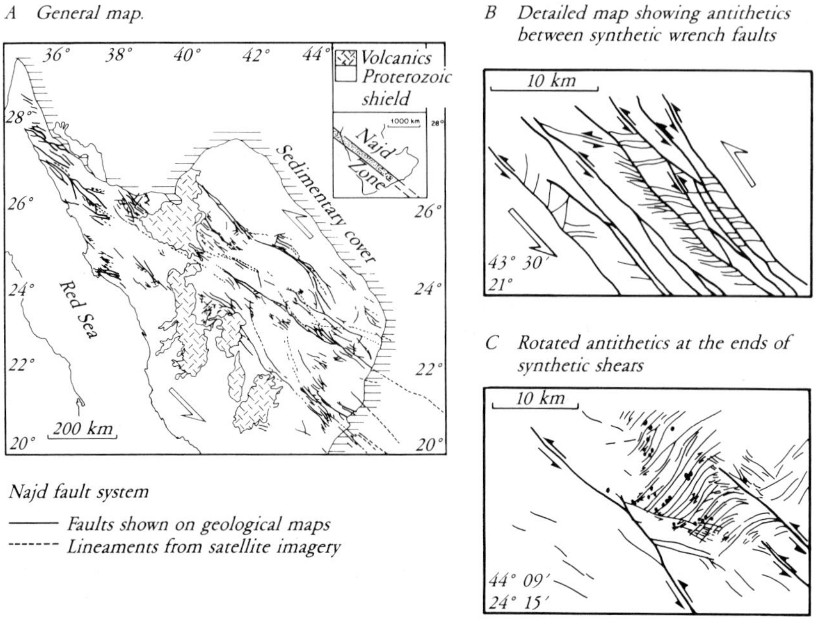

Fig. I.4–13 *THE NAJD FAULT SYSTEM, SAUDI ARABIA (after Moore, 1979; reproduced by permission of the Geological Society) Proterozoic left-lateral wrench system, 300 km wide and more than 2000 km long. Aeromagnetic lineaments indicate that a broad zone of more continuous, subparallel faults exist at depth. The exposed structures can thus be regarded as faults in the 'cover' above a broad zone of basement wrenching.*

When a set of rotating antithetics is situated at the margin of the wrench zone proper, as for example at the western margin of the segment of the left-lateral Icotea fault zone in Fig. I.2-56B, and the rotating rock slices are still 'anchored' at one side in undeformed rock, the rotating faults may open up as sketched schematically in Fig. I.2-2B. The normal stress that acts across these faults will drop accordingly and allow dip-slip to take place. Fault blocks may therefore drop down along the steeply dipping fault planes, as is indeed well documented for the marginal cross faults of the Icotea fault zone in Fig. I.2-56B. Such faults are expected to act as *migration paths for fluids*, since the low total rock pressure on the rotating fault planes may be overcome by fluid pressures that are only moderately in excess of the pressure of the hydrostatic fluid column.

The SW-heading faults in the narrow graben basin in Fig. I.2-57 play a similar role, although these cross faults are not rotated by the shearing action of synthetic Riedel faults, but by the sinistral shearing of the competent graben walls. Small strike-slip displacement along the cross faults has been proved (Letourneur, 1953).

In general, rotating cross faults will experience only small displacements along the fault planes and may therefore be difficult to identify on seismic records.

4. *OBLIQUE SLIP ON SINGLE BASEMENT FAULT – 'TULIP' vs. 'PALM-TREE' STRUCTURES*

We concluded from wrenching experiments with a single basement fault and from mechanical considerations that synthetic Riedel shears are concave-upwards in vertical sections. The same type of curvature was also observed in the wrenching experiments with wide basement shear zones. Hence, we might expect 'tulip'-type cross-sectional fault structures to be a common characteristic of basement-induced wrench zones. Surprisingly, however, the opposite curvature is also observed in the field (Sylvester and Smith, 1976; Harding and Lowell, 1979; Harding, 1985), on wrench faults that diverge from depth into upward convex branches with a reverse dip-slip component. Such 'palm-tree' structures seem to be as common in wrench zones as 'tulip' structures. How can this be explained?

When dealing with the mechanical origin of fault curvatures in Chapter I.2 (see Section I.2.4), we discussed the stress changes and precursory overburden faults that are induced by vertical movements of basement blocks. Potential faults appeared as upward-convex steep reverse faults bending over the subsiding basement block (see Fig. I.2-21A) – a result that is also corroborated by computer analyses and sandbox experiments, discussed in Chapter II.7 (see Section II.7.5, Example No. 5, Figs. II.7-31,32,33; and Section II.7-7, Example No. 7, Fig. II.7-46). The shape of these precursory reverse faults immediately suggests a dip-slip component on the basement wrench fault as the main cause of 'palm-tree' wrench structures in the overburden. This has indeed been confirmed by shear-table experiments in which one-half of the table was moved in an oblique direction along the partition, and the dip angle of the 'basement fault' and the ratio of strike-slip and dip-slip were varied systematically. A few cross-sectional views of typical experiments are presented in Fig. I.4-14. In the first experiment (A) the basement fault is vertical and the two upward-convex faults with reverse dip-slip are first-generation R-shears, while the steep fault with normal dip-slip is a later (throughgoing) wrench fault. A schematic picture of the three-dimensional fault configuration is presented in Fig. I.4-8B. This experimental 'palm-tree' structure is comparable with natural structures, such as the San Gabriel fault zone, California (Fig. I.4-15). In the experiments of Figs. I.4-14B,C the basement fault dips at 45° and the dip-slip component is normal. The associated extension across the basement fault causes a component of normal dip-slip on the wrench faults in the overburden. The pattern of the overburden faults is typically of the 'tulip'-type (B) when the strike-slip component predominates along the basement fault. As might be expected, a small strike-slip component (C) has less effect on the shape and pattern of the overburden faults, which resemble those of a single graben. The faults are relatively straight and only slightly en-échelon in plan. When the oblique slip on the basement fault has a reverse dip-slip component (D), in addition to a greater strike-slip component, the cross-sections show one or two, straight or slightly upward-convex reverse faults and a steep strike-slip fault close to the basement fault. The cross-sectional fault pattern is again of the 'palm-tree' type.

When we interpret the experimental patterns in terms of real wrench tectonics, we have to recall a limitation of the sand model that was discussed in Section I.2.8 in connection with the stress changes associated with differential uplift. The high volume dilatancy of sand under shear makes it possible for the reverse faults shown in Fig. I.4-14A to operate, which would be impossible without the volume increase of fault zone material, since the subsidence of the basement block would cause the reverse faults to open up (see also Fig. I.2-70). We have therefore concluded that the formation of these faults under conditions of suppressed or strongly reduced dilatancy requires supply of material by horizontal compression. Therefore, the formation of 'palm-tree' structures by oblique slip along a vertical basement fault will usually require some transpression, such as may be provided, for instance, by the convergence of wrench zones. If the basement fault is inclined, the transpression may be generated by a component of reverse dip-slip along the fault. If the dip-slip component is normal, the associated extension will prevent the formation of 'palm-tree' structures. Instead, the pattern will be of the 'tulip'-type (with a persistent dip-slip component) or of the graben-type, depending on the predominance of the strike-slip or dip-slip component in the basement.

Although mainly based on the results of laboratory studies, these conclusions are in good

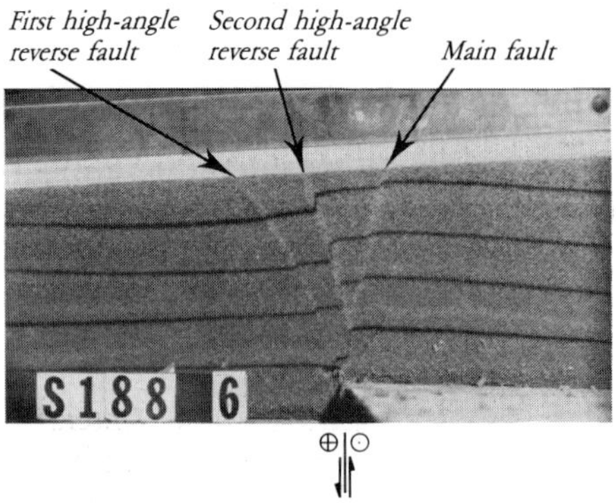

A Vertical basement fault (strike-slip/dip-slip = 2:1)

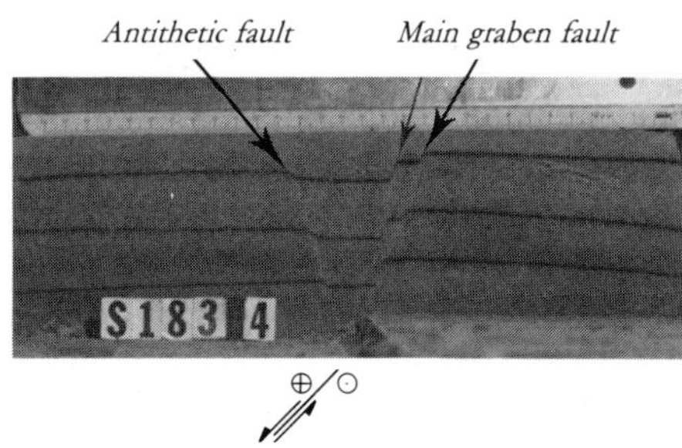

C 45° basement fault with normal dip-slip component (strike-slip/dip-slip = 1:1)

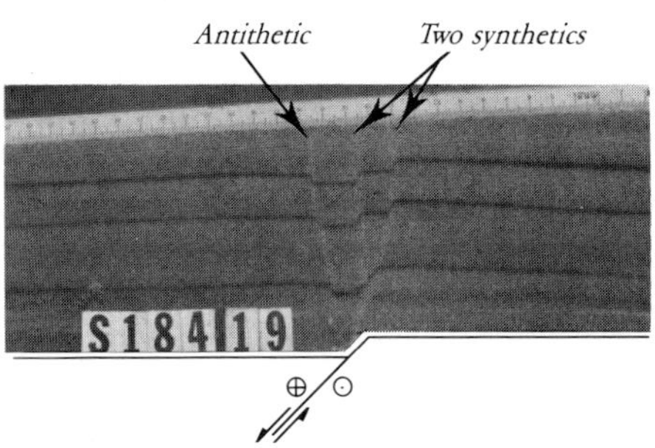

B 45° basement fault with normal dip-slip component (strike-slip/dip-slip = 2:1)

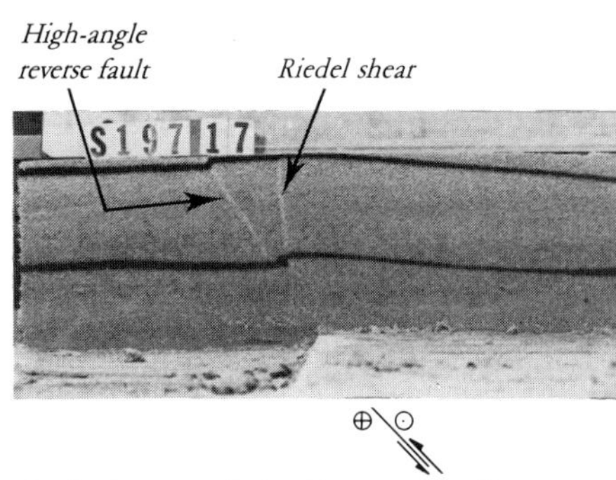

D 45° basement fault with reverse dip-slip component (strike-slip/dip-slip = 2:1).

Fig. I.4-14 CROSS-SECTIONAL APPEARANCE OF TYPICAL OBLIQUE SLIP FAULTS IN SANDPACK (After M.A. Naylor, 1981, unpubl. work)

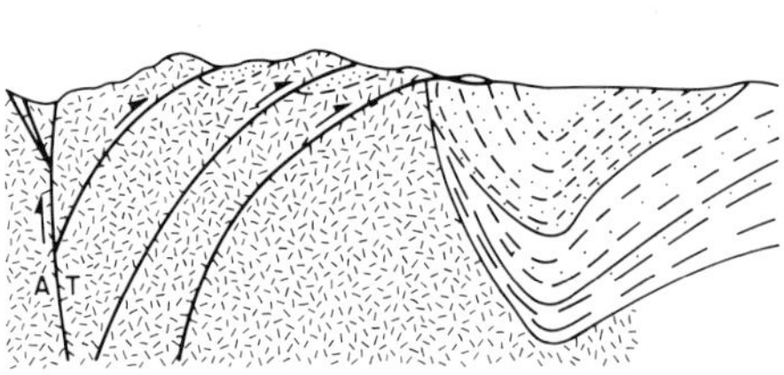

San Gabriel fault zone, California (after Wilcox, Harding and Seely, 1973; reprinted by permission of "American Association of Petroleum Geologists").

Fig. I.4-15 'PALM TREE' STRUCTURE OF RIEDEL SHEARS

agreement with analyses of field cases, in particular with those by Hardy (1985), in whose nomenclature 'tulip' and 'palm-tree' structures appear as, respectively, 'negative' and 'positive' flower structures.

5. *TERMINATION OF WRENCH FAULTS*

At the ends of a wrench fault the strike-slip displacement has to be accommodated by some other mode of tectonic deformation. The strike-slip displacement may be transformed into compressional folding or thrusting at the advancing side of the wrench fault and into normal faulting at the receding side; or the wrench fault may terminate by bending and distributing the displacement over a fan of secondary wrench faults. Large-scale transformation of strike-slip displacement into other modes of tectonic deformation is typical for the 'transform' faults of plate tectonics, which form boundaries of crustal plates and terminate at ridges, trenches, and thrust and fold belts, which absorb the strike-slip displacement along the wrench fault. Naturally, the analogous kinematical situation can be encountered on a wide range of scale. The situation is different for a wrench fault that lies completely inside a competent unit, the internal deformation of which only allows accommodation of a very limited amount of strike-slip. Again, strike-slip may be transformed into thrusting in the compression region at the advancing side of the fault edge, and into normal faulting in the extension region at the receding side, but often the termination of an 'internal' wrench fault is accomplished by bending and by 'splaying', i.e. by the displacement being distributed over a fan of numerous wrench faults (Freund, 1974). A large-scale example of splay-type termination of a wrench fault is shown in Fig. I.4-16, which the reader may compare with the splay termination in Fig. II.7-22, which is five orders of magnitude smaller.

The kinematics of wrench-fault termination by bending and splaying into a fan of wrench faults has been discussed by Freund (1971, 1974). In general, this mode of termination is asymmetric with respect to the strike of the main wrench fault. Bending is always to the receding fault side, as has been explained in Fig. I.4-11, and has also been accounted for in the structural reconstruction of the splay faults in Fig. I.4-16 (dotted fault traces). In contrast, the splays themselves in Figs. I.4-16 and II.7-22 extend towards the advancing side of the main fault. Other field examples indicate, however, that this is not a general rule and that the opposite asymmetry may occur too. In previous sections we have explored the conditions that determine whether the main fault will terminate by splaying towards the receding side or towards the advancing side.

We recall from Section I.2.8, in which we discussed secondary stress changes during faulting, the result of a numerical analysis (Fig. I.2-82A) which stated that, under plain strain conditions, a wrench fault approaching a barrier at a right angle, will terminate by splaying towards the *receding* side, when the lithological conditions and the fault-parallel constraints are the same on both sides of the fault. This asymmetry is caused by the preferential development of a plastic region at the receding side of the fault edge, where extension lowers the compressional stresses, which reduces the pressure-sensitive shear strength of the material. We have discussed this point in Section I.2.4 (Fig. I.2-28), where we also pointed out (Fig. I.2-29), that a reorientation of σ_I subparallel to the fault plane, as may be caused by an extreme reduction in frictional resistance of the fault plane, will shift the plastic splay region towards the advancing side. More commonly, however, splaying towards the *advancing* side will be associated with one-sided fault displacement, in which only the advancing side of the fault moves relative to the undisturbed foreland, and at the same time the material at this side of the edge region is allowed to move out sideways (see also Fig. I.2-82B).

In the Hope fault termination of Fig. I.4-16, one may envisage that, in addition to one-sided advancing displacement along the main fault, sideway-spreading of the splay fan is facilitated by a syndirectional movement of the Hope-fault side of the (right-lateral) Alpine fault.

Wrench-fault termination by bending and splaying implies that the deformations are accomplished essentially by strike-slip displacement. But why should the displacements in the termination region remain horizontal? In other words, when should we expect a wrench fault to terminate by splaying into secondary wrench faults, and when should we expect it to turn into faults with normal dip-slip at the receding (extensional) side of the termination region or with reverse dip-slip at the advancing (compressional) side? A complete answer would require three-dimensional stress analyses based on elastic/frictional plastic material behaviour. In view of the formidable numerical effort involved, we are content with a three-dimensional analysis of the termination region of a basement-induced wrench fault, which assumes a purely linearly elastic material behaviour. Although an elastic analysis cannot describe the formation of faults, as we have stated repeatedly, and the magnitudes of the stresses may change drastically after the onset of plastic yielding, the principal stresses' orientations produced by the elastic deformations may represent the orientation at the onset of faulting reasonably well. True enough, the principal stress axes may be changed considerably by shear softening of the fault zones, but these 'post-peak' changes cannot change the fault type that was determined by the principal stress directions at the very onset of faulting, when the state of stress was still close to the elastic limit. The elastically determined field of principal stress axes will therefore indicate whether the secondary faults will be of a predominantly wrench, thrust or normal fault type.

Such a three-dimensional elastic analysis of basement-induced wrench faulting was carried out by H.J. Neugebauer and B. Stappert (unpublished work). The problem is defined in Fig. I.4-17A, which shows in plan view a vertical basement fault extending through half of the basement plate. The dimensions and the finite-element net of the overburden are shown in Fig. I.4-17B. To simulate infinite extent of basement and overburden in the direction of the strike of the terminated basement fault (y-direction) a vertical barrier perpendicular to the basement fault is assumed, which may deform elastically by displacements in the

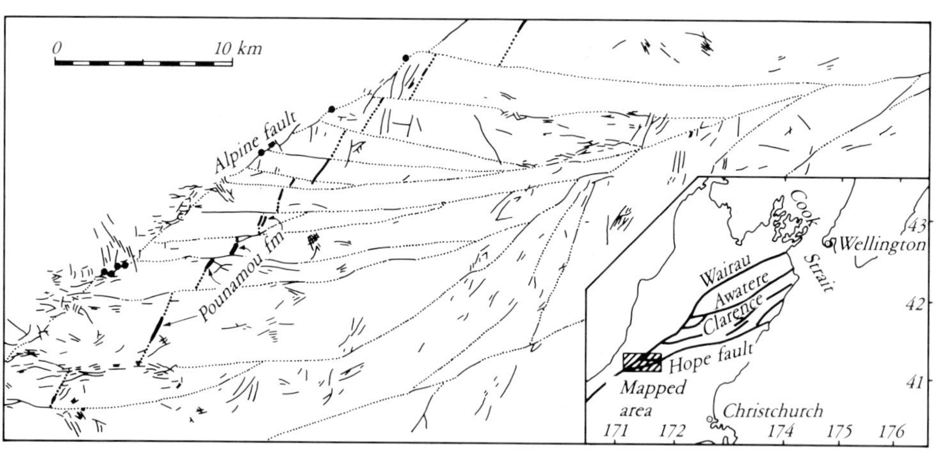

Fig. I.4-16 *THE SOUTHWEST TERMINATION OF THE HOPE FAULT TOWARDS THE ALPINE FAULT IN NEW ZEALAND BY SPLAYING AND BENDING*
Displacement along both faults is right-lateral (Freund, 1971)

y-direction. The lateral boundaries allow free movement parallel to the basement fault, in both the horizontal and vertical directions, but no movement in the direction perpendicular to the basement fault (x-direction). The basement is dislocated along the basement fault in the horizontal direction by 0.2% of the length of the fault segment considered. This

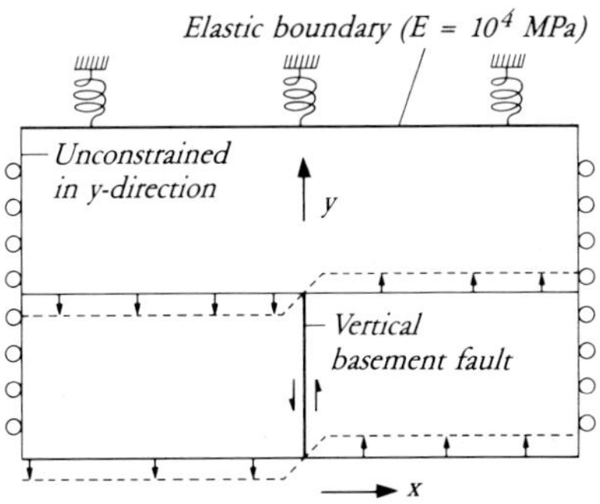

A Plan view of horizontal overburden/basement contact with terminating strike-slip fault in basement. Fault blocks are rigidly displaced along vertical fault plane, while the region ahead of stationary fault edge is elastically deformed; shear displacement along fault is 0.2% of the length of the fault segment considered.

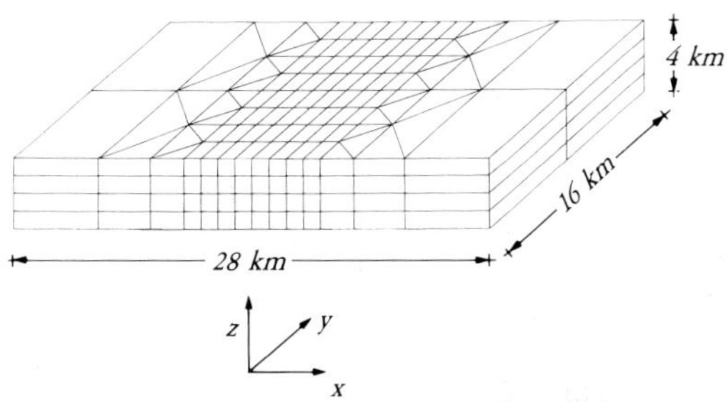

B Finite element mesh and dimensions of overburden sheet considered.

C

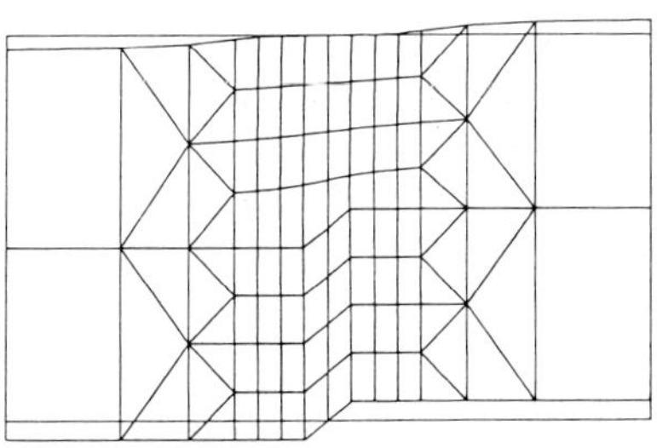

D

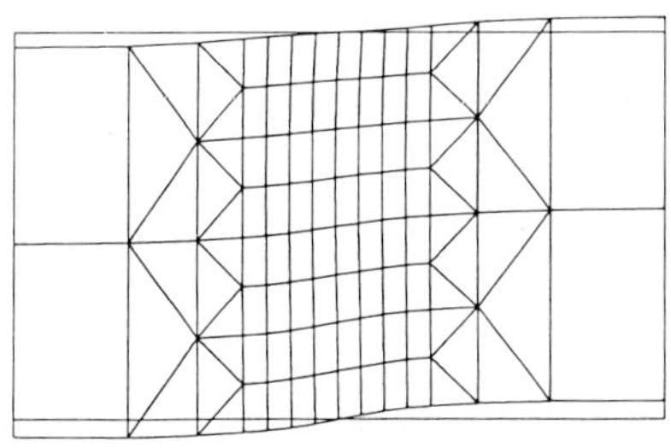

C, D Horizontal displacements near base (C) and near surface (D).

E

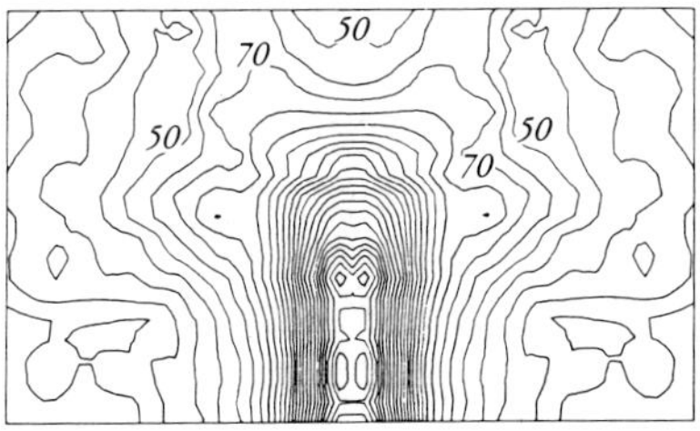

F

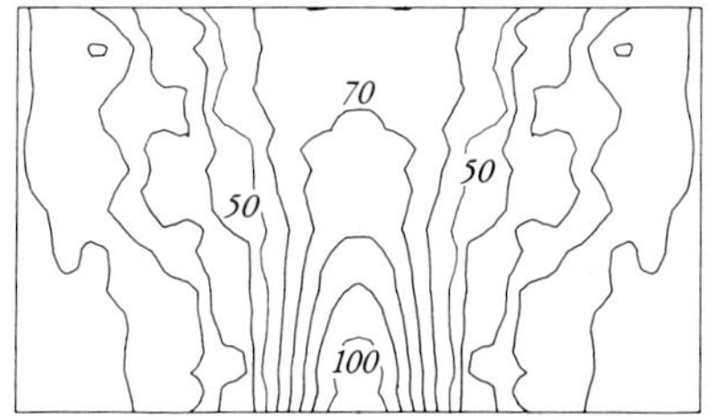

E, F Contours of maximum shear stress (in bar) in horizontal planes near base (E) and near surface (F). (Distortion of outer contours is due to increasing size of outer elements.) Complete symmetry with respect to strike of basement fault implies that in pressure-sensitive materials the limit state will first be attained at the receding (i.e. extensional) side of the fault.

Fig. I.4-17 THREE-DIMENSIONAL NUMERICAL ANALYSIS OF ELASTIC OVERBURDEN STRESSES ABOVE A TERMINATING BASEMENT WRENCH FAULT (H.J.Neugebauer & B. Stappert, 1984, unpubl. work.)

displacement is accommodated by elastic strain in the foreland of the basement, in the overburden and the elastic barrier. Uniform elastic moduli were assumed for the whole system. The horizontal elastic deformation of the finite-element mesh and its boundaries is shown – greatly exaggerated – in Figs. I.4-17C,D at two different depth levels. The values of the maximum shear stresses induced at these depth levels by the displacement of the basement are contoured in Figs. I.4-17E,F. The contours are completely symmetric with respect to the basement fault, as is to be expected from the symmetry of the configuration considered and from the prescribed displacement conditions. Since the maximum shear stresses are equal at points that are situated symmetrically with respect to the strike direction of the basement fault, again we see that the limit state will be reached first at the receding side, where the mean compressive stress is smaller than at the advancing side.

Of greater interest are the directions of the principal stresses, which, unfortunately, are difficult to present graphically. It was found that in the region ahead of the terminated basement fault, at any depth level of the overburden the maximum and minimum principal stresses were almost perfectly horizontal, which strongly suggests that secondary faults, which develop at the receding side, are wrench faults. Thus it would seem that dislocation along the basement fault can be accommodated by a splay of secondary wrench faults. A quite different result was obtained when the elastic foreland barrier was replaced by a rigid one. In the foreland of the basement fault the principal stress directions at any depth level became indicative for a component of thrusting at the advancing side and normal dip-slip at the receding side of the wrench zone. We therefore expect that in nature it will depend on the compliance of the barrier that forces a wrench fault to terminate, whether the termination will be accomplished by splaying wrench faults or by transforming horizontal displacement into vertical movements by secondary faults with a dip-slip component. Further, wrench zone termination that starts as splaying wrench faults may turn into thrusts or normal faults when the compliance of the obstructing foreland becomes exhausted and further strike-slip along the splay faults cannot be accommodated. Obviously, much more research is required to arrive at a sufficiently detailed mechanical model of wrench-fault termination.

6. LATERAL OFFSETS OF GRABENS AND WRENCH FAULTS

Graben offsets. Graben and rift systems often shift their positions along their lengths and develop transform-type wrench zones connecting the offset segments. An example is the Rhine-Bresse graben offset in the Rhine rift system, which is shown in Fig. I.4-18A and further schematised in Fig. I.4-18B. There are various reasons for such offsets. When a graben develops in the extensional zone above a mantle upbulge or, on a smaller scale, above the updoming of some other ductile material such as shale or salt, the location of the graben axis is very sensitive to the curvature of the top of the ductile uplift. This has been shown in finite element analyses (Neugebauer, 1978) and can be demonstrated by experiments of the type shown in Fig. I.2-23. Offsets may also reflect a jump in the location of basement fault zones.

Transform-type graben offsets have been modelled by M.A. Naylor and J.A.M. van Gool in dry sand with the shear table configuration shown in Fig. I.4-18C. The three-dimensional structures in the transform zone were reconstructed by sectioning the impregnated sandpacks. The essential features of a fully developed transform zone, which developed simultaneously with the opening of the offset graben segments, are summarised in Fig. I.4-18D. Some of the graben-bounding normal faults continue into Riedel-type wrench faults, which traverse the offset zone, similar to the synthetic Riedel faults in the offset zone between the

Rhine and Bresse graben in Fig. I.4-18A. The transform region remains a structurally high zone during rifting. As may be expected, the offset structures depend on offset length (L) and offset angle (α). When the offset length is smaller than the thickness of the overburden in the experiment and/or the offset angle is smaller than about 45°, the transform zone degenerates into a kink or bend in the graben. Surface expressions of graben-terminated transform zones were studied previously in kaolin by Tapponnier and Varet (1974).

It should be noted that a quite different fault pattern is formed in the offset zone of a graben, when offsetting takes place *after* the graben has formed. The graben is then cleanly cut by the wrench zone.

'Pull-apart' structures. In nature, wrench faults or wrench zones in the basement will rarely be completely straight as we have assumed so far. Side-stepping of the strike-slip motion in the basement causes extensional or compressional overburden structures, depending on the direction of the offset. A left-lateral basement fault, which is offset to the left (Fig. I.4-19A), or a right-lateral wrench fault offset to the right, will cause extensional structures in the overburden – termed 'pull-apart' or 'rhomb' grabens. The formation of these

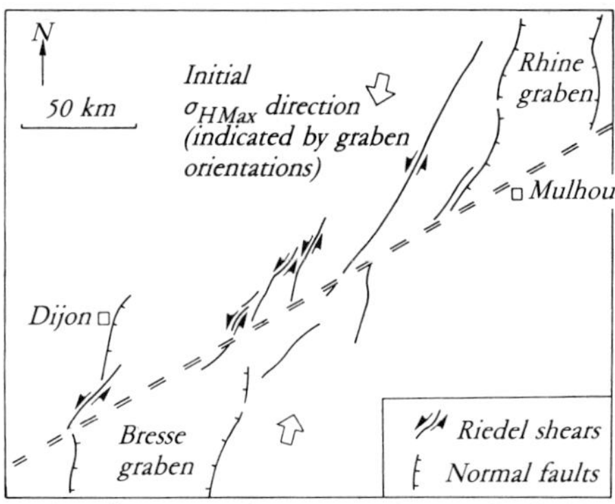

A 'Transform' zone between southern end of Rhine graben and northern end of Bresse graben with direction of maximum compressive stress in oligocene (after Bergerat, 1977; by permission of Masson, Ed.s.a., Paris)

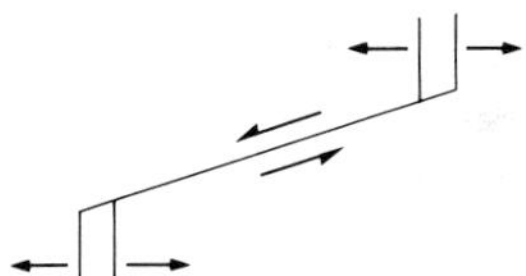

B Schematic sketch of graben-graben transform zone

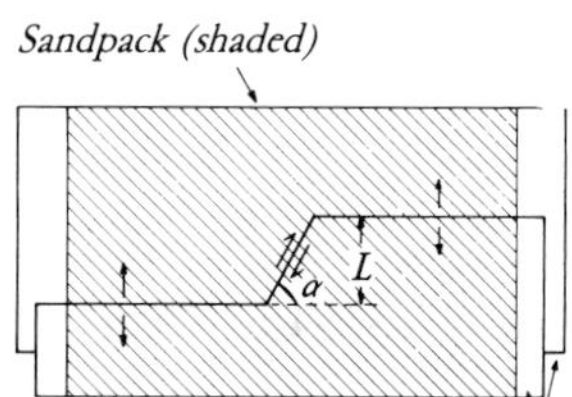

C Shear table equipment for modelling offset development of grabens in sandpack (schematic): Sandpack overlies two metal sheets whose relative movement is indicated by arrows.

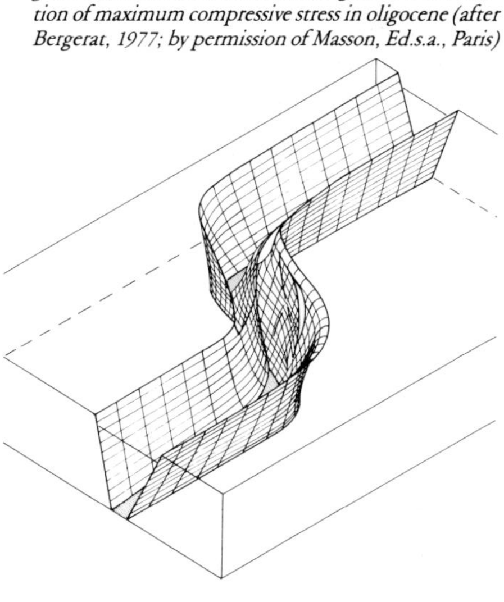

D Block diagram (simplified) showing fault pattern in fully developed transform zone (M.A. Naylor & J.A.M. van Gool, 1982, unpubl. work)

Fig. I.4-18
SYNTHETIC RIEDEL FAULTS IN 'TRANSFORM' SHEAR ZONE INDUCED BY OFFSET GRABEN DEVELOPMENT

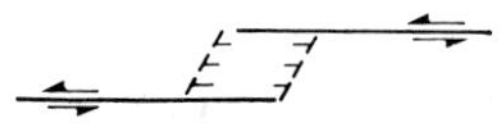

A Extensional fault offset (schematic)

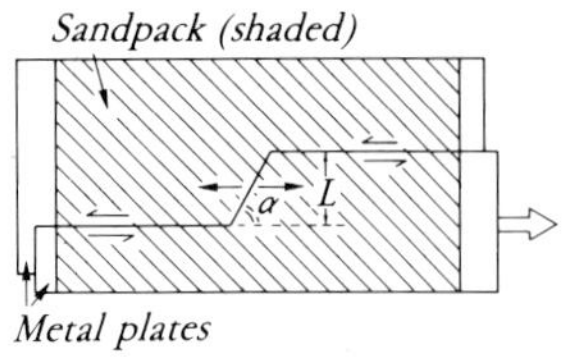

B Shear table equipment

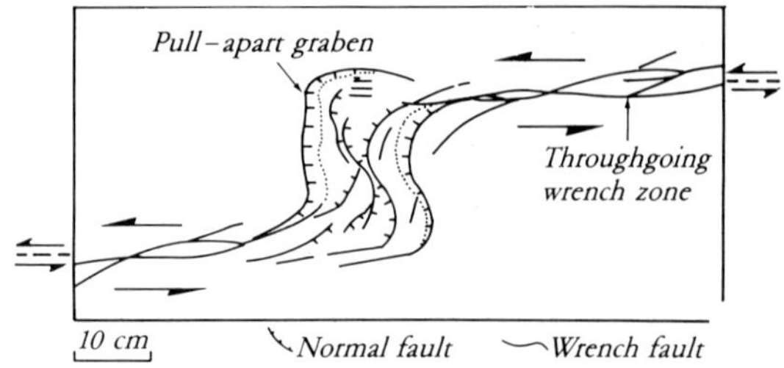

C Line drawing of surface expressions of fully developed offset graben (offset angle α = 135°, sandpack thickness 10 cm)

Fig. I.4-19 MODELLING OF 'PULL-APART' GRABEN AT THE EXTENSIONAL OFFSET BETWEEN TWO OVERLAPPING BASEMENT WRENCH FAULTS (from experiments by W.T. Horsfield and M.A. Naylor)

grabens is one of the principal mechanisms for the development of sedimentary basins in wrench zones (see, for example, Mann et al., 1983).

The orientation of the graben axis may vary between (sub)perpendicular and parallel to the wrench zone. The reason for this variation is made clear by model experiments using the shearing equipment sketched in Fig. I.4-19B. It is found that the graben axis reflects faithfully the orientation (α) of the offset in the basement, when the offset length (L) is large compared with the thickness of the overburden (L > 2× overburden thickness in the experiment). The graben axis is parallel to the wrench fault, regardless of the offset angle (α), when the offset length is smaller than the overburden thickness (< 0.5× in the experiment). Figure I.4-19C shows in plan view a fully developed 'pull-apart' graben between two slightly overlapping parallel wrench zones. The rhombic graben is bounded by normal faults, which swing round into synthetic Riedel shears, changing displacement from dip-slip to strike-slip. The graben is crossed by faults striking roughly parallel to the offset direction. When the two master wrench faults do not overlap, the diagonal faults connecting the tips of the two master faults acquire a strike-slip component, whose relative importance increases with decreasing offset angles.

It is of interest that the pull-apart grabens develop earlier than the first Riedel shears of the wrench zones themselves. This may be understood by comparing the limit stress circles for normal and wrench faulting (see, for instance, Fig. I.2-5). For normal faulting to take place, the constant overburden stress (σ_V) remains the maximum principal stress and a horizontal normal stress component must be reduced to become the minimum principal stress. In wrench faulting one of the horizontal normal stresses must be increased beyond the value of the overburden stress. This shifts the centre of the stress circle towards higher compressive stresses. This also implies a larger diameter of the limit circle. In other words, a higher differential stress than that required for normal faulting has to build up to form wrench faults.

Compressional offsets – 'pop-up' structures. A right-lateral basement fault offset to the left (Fig. I.4-20A), or a left-lateral fault offset to the right, will cause compressional overburden structures. In contrast with the development of overburden structures at extensional offsets, strike-slip displacements appear at the surface before thrust features, reflecting the higher level of differential stress required for thrusting (see again Fig. I.2-5). The line drawings of surface expressions in Figs. I.4-20C,D,E show characteristic states in the development of a typical 'pop-up' structure that was modelled in dry sand with the shearing equipment sketched in Fig. I.4-20B. First, the Riedel shears of the wrench zones develop, then boundary thrust faults which connect with wrench faults of the wrench zones, and finally the interior of the rhombic 'pop-up' is transected by conjugate wrench faults of the Anderson type.

The offset length (L) mainly affects the total size of the pop-up rhomb, while the offset angle (α) has only a minor influence on the pop-up shape. Only when the offset length is smaller than the overburden thickness does the pop-up region lose its characteristic shape and become elongated parallel to the main shear direction.

Inspection of vertical cross-sections revealed that the conjugate wrench faults in the interior of the pop-up rhomb are subvertical and terminate at a detachment surface. Strike-slip on these minor faults accommodates internal deformation and external rotation of the pop-up rhomb. It should be noted, however, that the formation of a detachment horizon is likely to be associated with a very localised shortening of the basement in the offset region, similar to that produced in the experiment by the overlapping base plates.

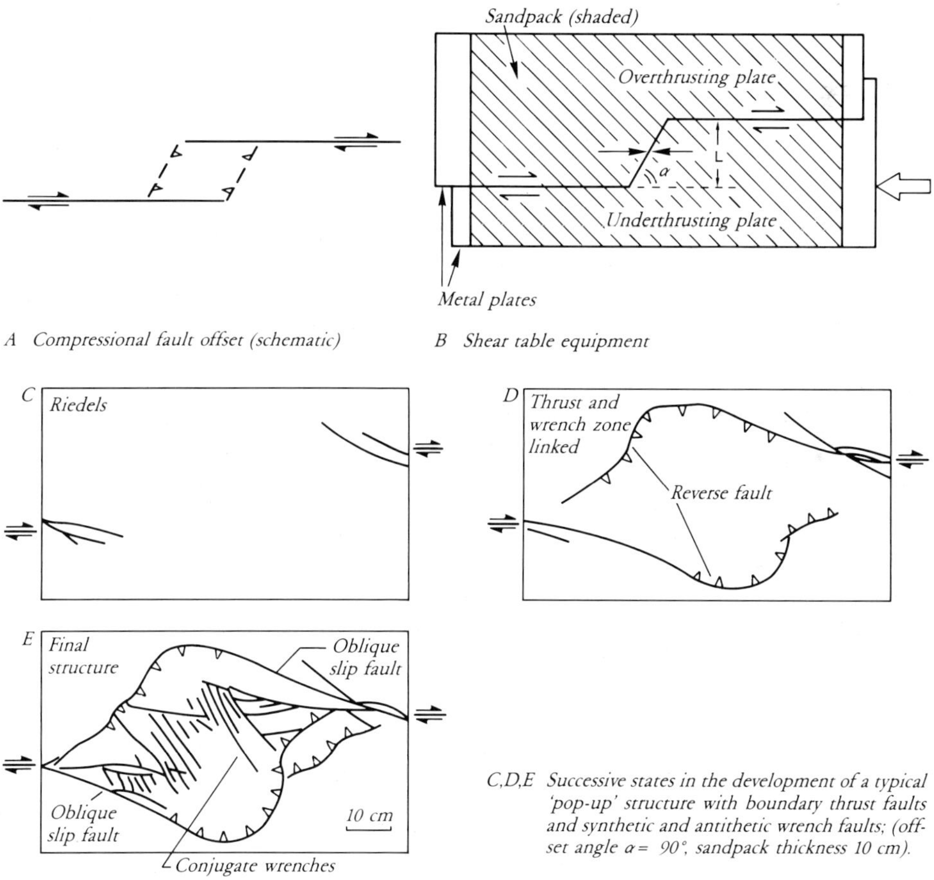

A Compressional fault offset (schematic)

B Shear table equipment

C,D,E Successive states in the development of a typical 'pop-up' structure with boundary thrust faults and synthetic and antithetic wrench faults; (offset angle α = 90°, sandpack thickness 10 cm).

Fig. I.4-20 MODELLING OF 'POP-UP' STRUCTURES AT THE COMPRESSIONAL OFFSET OF A BASEMENT WRENCH FAULT (from experiments by W.T. Horsfield and M.A. Naylor).

Numerical analyses. Two numerical studies of wrench fault offsets have been published recently, both considering two separate, offset planar strike-slip faults in a uniform elastic material. The analyses are two-dimensional, assuming that stresses and displacements do not vary with depth. In Rodger's (1980) analysis the faults are assumed to be sufficiently distant from each other so that mechanical interaction can be neglected and stress and elastic displacement fields can be obtained by superimposing Chinnery's (1966) elastic dislocation solutions for a single shear crack.

Segall and Pollard (1980) consider closely spaced shear fractures, which interact mechanically, so that the stresses on one fracture are not only determined by the far-field stresses, but also by the slip along the other fracture. While Rodgers considered extensional offsets only, Segall and Pollard also studied compressional offsets. Since the analyses are two-dimensional, they predict elastic stress and displacement fields between two *pre-existing* faults, which in the experiments would correspond with the stresses and displacements in the basement between the two basement faults. Therefore, in spite of certain similarities between the theoretical and experimental results, concerning for instance the orientation of the axis of a pull-apart graben, the theoretical models do not apply to an overburden subject to its own weight and sheared along its base.

I.5 THRUSTING

Horizontal shortening of crustal elements is mainly achieved by faulting and compressional folding. Although, of course, the two processes can operate simultaneously, often one process dominates the structural evolution, which one depending on lithology, strain rate, temperature and pressure. Here we consider tectonic shortening which is primarily accommodated by faulting. As pointed out before, horizontal shortening can be produced by conjugate wrench faulting (see Figs. I.2-68, I.4-1) and/or by thrusting. Wrench faults operate when a uniaxially shortening rock sheet, detached from its basement, is allowed to lengthen in the horizontal direction perpendicular to the direction of shortening. This, of course, requires that σ_{III} acts in the direction of horizontal extension. When horizontal extenstion is obstructed by lateral constraints (e.g. the frictional resistance of the detachment plane), the horizontal shortening is compensated by thickening in the vertical direction, which becomes the direction of σ_{III}, and faulting is of the thrust type, in which one rock body moves over another along a fault surface. As pointed out before (Section I.2.2), in a compressional belt, thrusting and conjugate wrench faulting may occur alternately, as the orientation of σ_{III} switches between vertical and horizontal.

1. THIN-SKINNED AND THICK-SKINNED THRUSTING

Considering thrusting from the mechanical point of view, one has to define the boundaries of the rock body whose structural development one wants to study, and to specify the interaction with its surroundings. As in extensional faulting, we therefore have first to distinguish between 'thin-skinned' and 'thick-skinned' processes. In thin-skinned thrust tectonics shortening is restricted to an overburden without involvement of the underlying rocks, and the compressive forces are applied more or less horizontally to the overburden. It should be noted that the term 'thin-skinned' is used here not in its literal sense, but to indicate the presence of a decoupling layer or surface, which may lie at considerable depth, even in the brittle/ductile transition zone of the crust. The character of the deformation will differ according to the shear resistance at the overburden base. When this shear resistance is low, a sheet of rock may be sheared off the base and may be moved as a *'thrust sheet'* ('nappe') over a flat detachment plane, occasionally 'stepping' up through the layered sequence to a higher gliding horizon. This process of *'overthrusting'*, which is the major type of thin-skinned thrust tectonics, is the main subject of this chapter. It is sketched (Fig. I.5-1A) in the very schematic survey of thrust styles which is presented in Fig. I.5-1 in close analogy with the survey of extensional fault styles in Chapter I.3. Needless to say, the various styles are independent of scale. If the basal shear resistance is too high for the detachment of a thrust sheet, or when further gliding of a detached sheet is obstructed, a suite of thrust faults will be formed that branch off the base, similar to the imbricate thrusts that can be observed in a sandbox (see Fig. II.7-24).

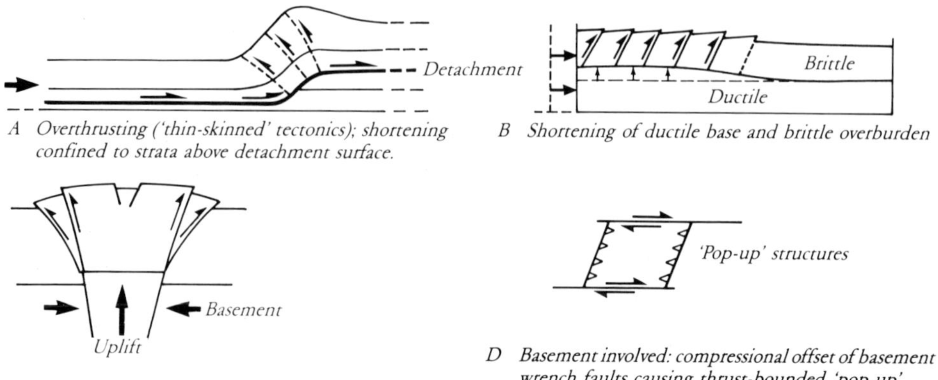

A Overthrusting ('thin-skinned' tectonics); shortening confined to strata above detachment surface.

B Shortening of ductile base and brittle overburden

C Basement involved: differential uplift

D Basement involved: compressional offset of basement wrench faults causing thrust-bounded 'pop-up' structures

Fig. I.5-1 STYLES OF THRUSTING (schematic)

In thick-skinned thrust tectonics the basement may be involved (1) as upward vertical movement of basement blocks caused by horizontal compression in the stronger basement (Fig. I.5-1C) or by vertical movements within ductile material at greater depth (e.g. mantle material), and (2) as horizontal movement along offset wrench faults (Fig. I.5D).

Overburden faulting by the *uplifting of a basement block* is discussed at several places in this book with reference to a vertical-sided basement block (see Sections I.2.4, I.2.8, Figs. I.2-21, 22, 70; and Sections II.7.5, II.7.7, Figs. II.7-31, 32F, 33, 46). The first overburden faults to form are steep, upward convex reverse faults as sketched in Fig. I.5-1C. It may be recalled that these faults are active longer, less steeply inclined and extend farther over the subsiding side, when extra horizontal compression is applied to the overburden (Fig. I.2-70B). This is of course the case when the uplifting of basement blocks is caused by horizontal shortening of the basement, which also requires that the sides of the uplifted blocks diverge in the upward direction (Fig. I.5-1C). From sandbox experiments we know that the final overburden faults, which are produced by the uplift of a wedge-shaped block with sides that dip rather steeply (say, at 60°), tend to continue the basement faults smoothly. The formation of vertical faults is therefore only associated with the uplift of vertical-sided basement blocks. It should be added here that at shallow depth the dip of the 'upthrusts' may be further reduced by a flattening of the σ_I-trajectories (Fig. I.2-21A, Fig. II.7-32E) as a result of the topography or previous uplifting of the whole region. This flattening of σ_I-directions is associated with an increase in a horizontal normal stress component, which, at the foot of slopes or underneath a valley floor, is caused by the gravitational stresses of the slopes (see, for example, Jaeger & Cook, 1968; McTigue & Mei, 1981), while in the upper few hundred metres of a regionally uplifted overburden, residual horizontal normal stresses that exceed the overburden stress may be 'locked in' (see Chapter II.6, Example No. 3).

Side-stepping of right-lateral (left lateral) basement wrench faults to the left (right) as sketched in Fig. I.5-1D causes subhorizontal compression in the overburden and 'pop-up' structures, which have been dealt with in Section I.4.7 (Fig. I.4-20) of the preceding chapter on wrench faulting.

Finally, we draw attention to a mode of thrusting (Fig. I.5-1B) that, similar to the 'domino-style' of extensional faulting, takes an intermediate position between thin-skinned and thick-skinned thrusting. In this mode, horizontal shortening is imposed upon an overburden by the continuous shortening of a ductile substratum and is accommodated by a suite of parallel thrust faults that operate by rotating into steeper positions. This is the *'dilational bookshelf'* mode (Fig. I.2-2A,B) which has been discussed in previous sections (Sections I.2.6 and I.4.3) as the accommodation mechanism for simple shearing. Although the literature abounds in well-documented cases of series of steep parallel thrust faults, it seems that these faults are usually interpreted as branching off a deep-seated decollement plane and hence being associated with thin-skinned thrusting. In many of these cases the decollement plane is purely hypothetical and the listricity of the faults (after time-depth conversion of the seismic records) is not sufficient by far to suggest smooth branching from a common gliding plane. One may be rather reminded of the toe region of a typical delta complex at a passive continental margin, where parallel toe thrusts extend down into thick mobile substrata that have been shortened and thickened by creep flow driven by the differential load of the whole delta wedge. Most likely, such suites of thrust faults have been initiated in the 'bookshelf' mode, sketched in Fig. I.5-1B, by the continuous shortening of the underlying ductile layers of shale or salt.

2. *MECHANICS OF SEDIMENTARY THRUST BELTS*

Thrust sheets may consist of non-metamorphosed sedimentary rocks or wholly or partially of metamorphic and/or igneous rocks. Both types, sedimentary and crystalline thrust sheets, are formed in orogenic belts and, according to plate tectonic concepts, are therefore somehow the result of the collision of lithospheric plates. But how are thrust belts and individual thrust sheets formed? Many ideas have been offered and various geological models have been developed and tested, to provide at least partial answers to this very complex problem. However, many uncertainties still exist, in view of which it will be no surpise that it is only fairly recently that a few mechanical models have been proposed that address the problem of thrust belt formation, these being restricted to the somewhat less complex problem of sedimentary thrust belts. These belts develop in the sedimentary basins in the foredeep of advancing mountain fronts, in the 'back-arc' basins on the continental side of orogenic belts at active plate margins with subduction of oceanic crust, and in the heavily imbricated accretionary prisms ('fore-arc' basins) on the oceanic side of these orogenic belts.

Typically, all these basins are wedge-shaped and rest on a base that slopes towards the thicker end of the wedge. The mechanical models, which we shall review, therefore deal with idealised sediment wedges that rest on a backwards-sloping basement. Compression is applied upon the back end of these wedges by horizontal forces.

Fluid-Push Model by A.G. Smith (1981). This model applies to the 'back-arc' basins of active continental margins and is based on the idea that the material of the thermally weakened zone in the volcanic arc of the central part of the orogenic belt behaves mechanically as a liquid rather than a solid and exerts a hydrostatic pressure upon the back end of the continental sediment wedge. Hence, in this model the push force is not directly imposed by the collision of the two plates. The sediment is considered to behave as a solid without further specification of its material properties. The configuration of the model is shown in Fig. I.5-2A. The surface of the sediment wedge slopes at a constant angle (α) towards the foreland, and the magmatic welt may be capped by a layer of solid rock. Then

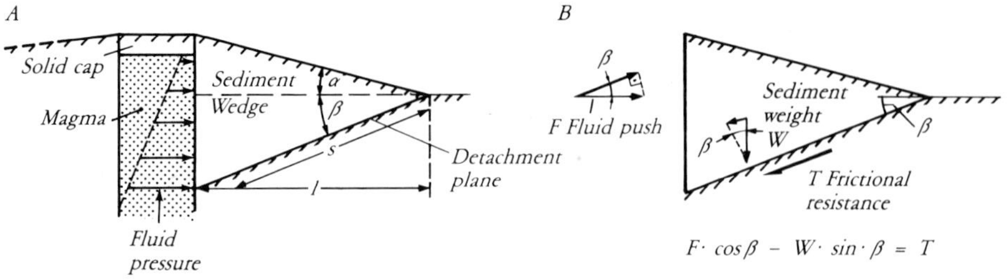

Sediment wedge pushed by the hydrostatic pressure of igneous fluid intrusion. (After A.G. Smith, 1981)

Fig. I.5-2 BALANCE OF FORCE COMPONENTS PARALLEL TO SLOPING BASE

a simple force balance is applied to the wedge as a whole (see Section II.2.3, 'Static equilibrium in the 'large'). In static equilibrium the component of fluid-push force parallel to the sloping base must be balanced by the downslope weight component of the wedge and the total frictional resistance along the base, which is expressed in the simple balance equation in Fig. I.5-2B (see also Example No. 7 in Ch. II.2). In Fig. I.5-3 the expressions for the fluid-push force F, the weight W of the wedge and the total frictional resistance T of the base are inserted in the force balance, which leads to a simple equation for the average shear stress ($\hat{\tau}$) that is induced on the sloping base in reaction to the fluid push. (For simplicity it is assumed that fluid and solid have the same specific weight γ, which actually underestimates somewhat the push force of the fluid welt. Notice also that the average $\hat{p}$ of the linearly varying fluid pressure in the fluid welt is half the sum of the pressure at the base of the welt and at the base of the solid cap.) The formula for $\hat{\tau}$ is further simplified by replacing the tangent and sine of the small slope angles by the angles (in radians) themselves. For practical purposes we express the angles in the final formula of Fig. I.5-3 in degrees, γ in gf/cm³, the heights in km, and the shear stresses in MPa, and obtain

$$\hat{\tau} = 0.09 \cdot \gamma \cdot H \cdot [\alpha^{o} - (h_3/H)^2(\alpha^{o} + \beta^{o})] \qquad (I.5\text{-}1)$$

Assuming, for example, the slope angles $\alpha = 2^{o}$, $\beta = 4^{o}$, a specific weight of 2.8 gf/cm³, a total height H = 10 km of the rear of the sediment wedge, and a thickness $h_3 = 0.5$ km of the solid cap of the fluid welt, the average shear stress in equilibrium with the fluid push becomes $\hat{\tau} = 5$ MPa. In other words, the fluid push will start moving the whole wedge upslope when the average sliding resistance along the base remains smaller than 5 MPa. The motion will continue as long as the supply of magmatic material maintains the pressure head of the widening fluid welt.

In considering this result and equation I.5-1, several points should be noted. First, the fluid pressure can only move the wedge if there is a surface slope. In fact, the upslope push is proportional to the slope angle when the cap thickness h_3 is negligible with respect to H. Further, the average shear resistance along the base has to be very small, as may be the case when the wedge can move on a ductile interlayer consisting, for instance, of salt or some limestones, the viscosity of which is drastically reduced by the high temperatures in the deeper parts of the wedge. The existence of a thin viscous interlayer was also postulated by Price and Johnson (1982) in a mechanical analysis of thrust sheet emplacement in southeast Nevada, which related the movement of the thrust sheet to the emplacement of the Sierra Nevada batholith. In the brittle domain, one might think of extreme overpressuring of the

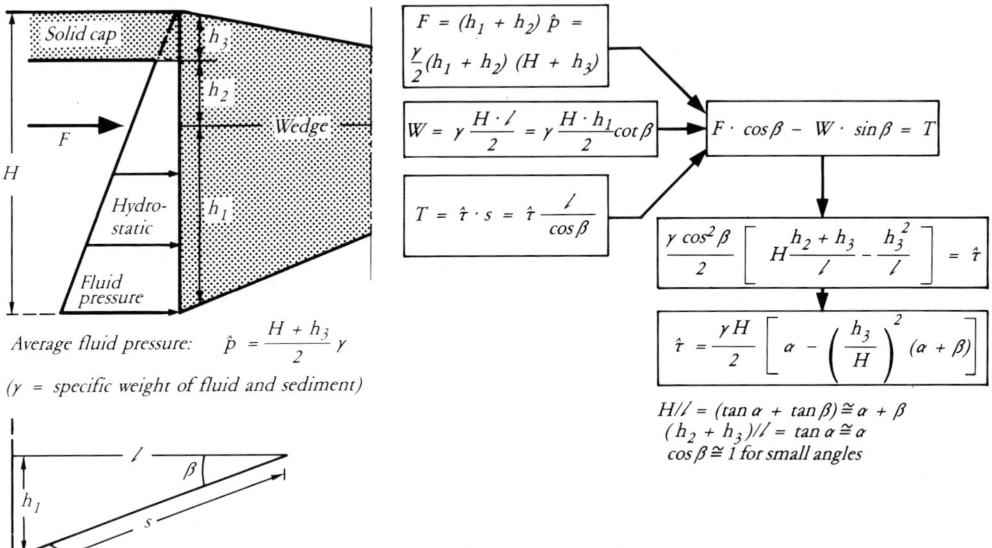

Fig. I.5-3 COMPONENTS OF FORCES

pore fluid as a mechanism for reducing the shear resistance to the low values required by eq. I.5-1. However, the brittle sediments will also have some cohesive strength, which is not affected by a reduction in effective stress and which is unlikely to provide an average shear resistance as low as required by eq. I.5-1. To overcome the cohesive resistance, at least in parts of a potential sliding plane, the fluid pressure would have to exceed temporarily the overburden load. In principle, this could be achieved by heating of the pore water in layers where drainage is impeded, and one might envisage the thermally raised fluid pressure to disrupt the cohesive bonds preferentially on some weak bedding planes, along which individual thrust sheets might be sheared off the sedimentary wedge. Moreover, since the magmas are heavier than the sediments, the magma pressure will exceed the (total) overburden stress and hot magma is likely to form sill-type intrusions along weak bedding planes; this might also contribute to the localisation of thrust planes. But these conjectures would have to be substantiated by a detailed analysis of the transient temperature distribution in the sedimentary wedge, taking into account the heat flow from the igneous intrusion body.

The Purely Cohesive Ideally Plastic Wedge (Chapple, 1978). Smith's fluid-push model does not consider deformation of the sediment wedge, which is pushed from the rear, and which in reality will undergo shortening and thickening as sketched in Fig. I.5-4. This deformational aspect is the objective of Chapple's theory. The basic assumptions of the theory are as follows. The rock deforms under the applied compression as a rigid/ideally plastic material (see the rheological model of Fig. II.7-1C), which means that deformation begins when the maximum shear stress τ_{Max} reaches a pressure-independent constant limit value K. The material considered is therefore purely cohesive (τ_0 = K, φ = 0) and the limit lines in Mohr's stress plane are parallel to the σ-axis and intersect the τ-axis at $\tau = \pm$ K (Fig. I.5-5A). Shortening and concomitant thickening starts at the rear and spreads until the whole wedge is deformed. At this stage the whole wedge is supposed to be in the plastic limit state. This is the basic assumption of the theory, which may or may not be adequate, but enables the adoption of one of the few analytical solutions of the plasticity theory – the Prandtl wedge (see also our remarks on this most elementary form of the plasticity approach to tectonic problems, at the end of Ch. II.7).

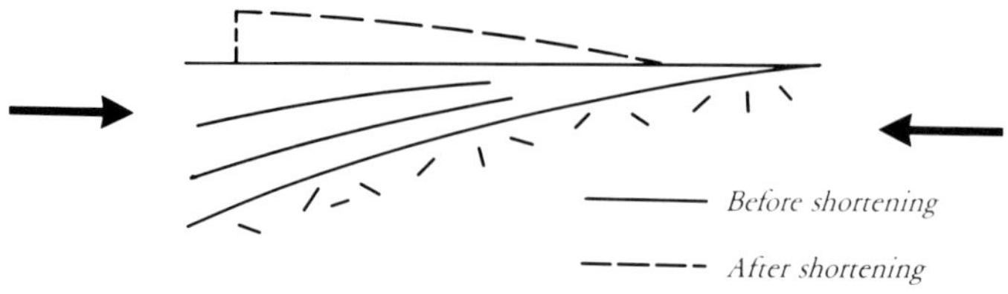

Fig. I.5-4 *SHORTENING OF SEDIMENT WEDGE UNDER COMPRESSION (schematic)*

Figure I.5-5B shows such a wedge, symmetric with respect to the horizontal and with an apex angle 2γ. The material confined between the rigid walls of the wedge is compressed in the radial direction towards the apex and is in the plastic limit state. The stress components are referred to a polar coordinate system which is centred at the apex of the wedge. The distributions of stresses in the plastic wedge are shown in Fig. I.5-5C. The tangential stress component $\tau_{r\vartheta}$ is constant on radial lines and attains the limit value K at the outer walls of the wedge to enable sliding of the compressed material along the rigid boundaries. At the midplane ($\vartheta = 0$) the value of this shear stress is zero, as it should for reasons of symmetry. The midplane is therefore only loaded by a distribution of normal stresses σ_ϑ. Note also the special distribution of the radial normal stress σ_r on the back end of the wedge, which is required to produce plastic yielding throughout the whole wedge. In Fig. I.5-6A a smaller wedge (shaded) is cut out of the Prandtl wedge, which is bounded by the horizontal midplane and a base sloping at the angle ϑ. It is further assumed that this base is a plane of weakness, the shear strength K being reduced to $\chi \cdot K$, with $\chi < 1$. For this condition to be compatible with the shear stresses in the plastic Prandtl wedge $\chi \cdot K$ must be equal to the value of the shear stress $K \cdot \vartheta/\gamma$ (Fig. I.5-6A). In other words, half the apex angle of the Prandtl wedge must be chosen as $\gamma = \vartheta/\chi$. Although the horizontal surface of the smaller wedge is free of shear stress, it is loaded by σ_ϑ, the transverse stress response of the plastic material under radial compression. In Chapple's model this surface load is applied by a sloping overburden (Fig. I.5-6B), which is supposed to exert only a vertical overburden stress upon the horizontal plane. This is a reasonable assumption, since tangential stresses may be neglected when the slope angles are small. The overburden slope tan δ must then equal the gradient of σ_ϑ along the horizontal plane. This derivative is inversely proportional to the distance r from the apex, since σ_ϑ itself depends logarithmically on r (Fig. I.5-5C). Hence, the surface slope required to maintain the plastic yield state in the wedge-shaped sediment between the horizontal plane and the inclined base would have to steepen in the downslope direction, which, certainly near the foot of the surface slope, becomes quite unrealistic. Chapple argued that for small values of ϑ and χ ($\chi < 0.3$) the surface load may be approximated by a slope with constant angle δ by the formula stated in Fig. I.5-6B for angles given in radians. To illustrate this result, let us consider again a basal slope angle of 4° (= 0.07 rad), $h_0 = 10$ km and let us assume, following Chapple, that the rock has the relatively high shear strength K = 100 MPa and a density $\varrho = 2.5$ g/cm³. From Chapple's formula then follows that for a coefficient of weakness $\chi = 0.23$ the surface slope angle $\delta = 2°$. For $\chi = 0.14$ no slope would be required, and for smaller values of χ the overburden surface would even have to slope in the same direction as the weak base ($\delta < 0°$) to maintain the plastic state throughout the whole wedge underneath the horizontal base of the surface slope.

We find it difficult to reconcile the last result with the leading idea of the model that the process of shortening itself creates a surface slope ('critical taper') that enables the whole wedge to be in the plastic yield state while gliding along a basal plane of weakness. The results suggest rather that only under very special conditions may the plastic yield state exist throughout the whole sediment wedge and that, in general, the deformation will remain confined to subregions of changing position and shape.

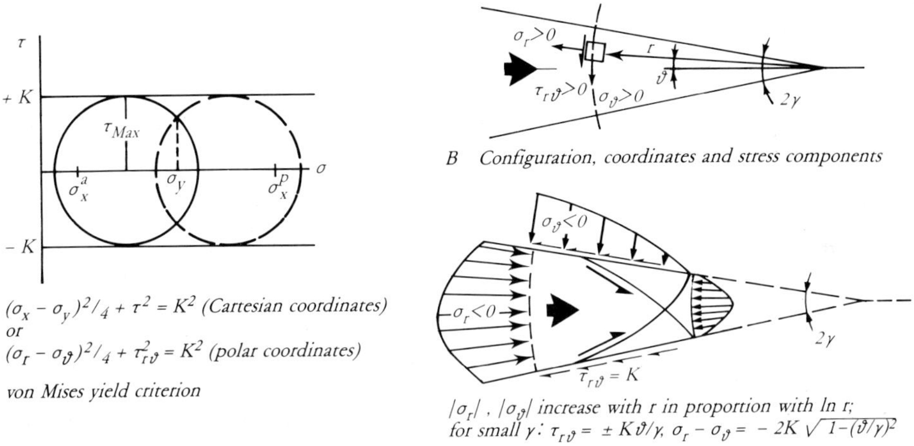

A von Mises yield criterion satisfied by the plastic material

C Distribution of stresses; (K pressure-independent shear strength)

Fig. I.5-5 PERFECTLY PLASTIC MATERIAL PRESSED BETWEEN INCLINED PLATES –THE 'PRANDTL WEDGE'– (cf. R. Hill, 1950; pp. 209–212)

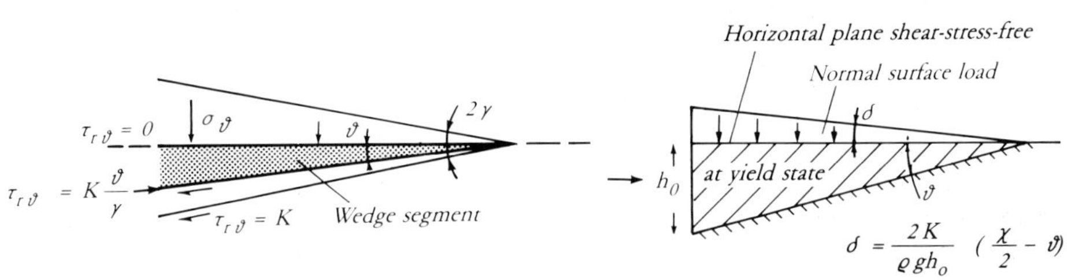

A Plastic wedge segment with horizontal, shear-stress-free top surface and back-sloping base. Given the slope angle ϑ of the base, the half-angle γ of the total wedge has to be chosen as $\gamma = \vartheta/\chi$ to ensure that the basal shear stress of the segmentary wedge equals the reduced shear strength $\chi \cdot K$.

B In Chapple's final approximation for small angles the non-linear distribution of the normal stress σ_ϑ on the horizontal top surface is replaced by a linear distribution, produced by a non-plastic overburden with constant slope angle δ.

Fig. I.5-6 CHAPPLE'S MODEL OF A PURELY COHESIVE PLASTIC SEDIMENT WEDGE ON A SLOPING BASE OF REDUCED SHEAR STRENGTH (Chapple, 1978)

Frictional Plastic Wedge (Dahlen, 1984; Lehner, 1986). Following work by Davis et al. (1983) on the mechanics of fold and thrust belts, Dahlen (1984) derived a rigorous solution for the 'critical taper' that a submarine sediment wedge of *noncohesive frictional* material on a basal plane of weakness will attain, when sufficiently compressed from its back end. Thus the problem differs from that treated by Chapple mainly in that the rock is treated as a frictional plastic (Coulomb-type) material without cohesion, instead of a cohesive plastic material without frictional shear strength. The material model is of the rigid/frictional plastic type (Fig. I.1-7A) which neglects elastic strains and strain hardening and softening. Although the constant shear strength of the (von Mises) limit condition used in Chapple's model is now replaced by the effective-stress-dependent internal friction of Coulomb-Mohr's limit condition, which enters an additional complication, the solution obtained is simple and not restricted to small slope angles. The solution can also be obtained graphically in Mohr's stress plane, as has been shown by Lehner (1986). We follow this approach in Fig. I.5-7.

Consider an infinite submarine slope of uniform material, with planar surface and dip angle δ. From the discussion of thin-skinned gravitational sliding in Section I.3.1 (and Example No. 5 in Section II.2.2) we know the effective normal stress $\sigma'_\perp$ and the shear stress τ that act on surface-parallel planes (eqs. I.3-1). To concentrate upon the essence of the problem we restrict the discussion to hydrostatic fluid pressure, which implies that the seepage term F_z in equations I.3-1 vanishes and the ratio of the two stress components becomes the constant $\tau/\sigma'_\perp = \tan\delta$. The states of stress on surface-parallel planes at any given depth then map in the stress plane of Fig. I.5-7A on a straight line L inclined at an angle δ with respect to the σ'-axis, when the latter is considered to represent the horizontal reference direction in physical space. The compressional (passive) limit state is represented for any given depth by the passive limit circle (see, for example, Fig. II.7-11) that intersects L at the corresponding stress point (Q) defined by the coordinates $\sigma_\perp$, τ. Then the σ_I-direction is determined in the Mohr diagram of Fig. I.5-7A by means of the pole method (Section II.4.2), as was repeatedly done in previous sections (for example, in Fig. I.2-16). The angle ψ_0 between the σ_I-direction and the slope surface is expressed in terms of the slope angle δ and the friction angle φ by applying the sine rule to the triangle OQM. Next we consider the basal plane of weakness (Fig. I.5-7B). For any point on this plane the normal and tangential stresses must define a point that lies on both the limiting stress circle and one of the pair of limit lines which represent the reduced shear strength (φ_b) of the plane of weakness. Since each of the two limit lines intersects the stress circle in two points, there are four pairs of normal and shear stresses defined, each of which is associated with a possible plane of weakness of well-defined orientation. Consider in Fig. I.5-7B the intersection point R of the limit line in the upper half of the stress plane. The orientation of the associated plane of weakness in real space is readily found by means of the stress pole P (Fig. I.5-7A). Hence, the orientation, referred to the horizontal direction in real space, is determined by the line S passing through P and R in Fig. I.5-7B. Similarly, we can determine the orientation of the plane of weakness that is associated with the second intersection point N. It has the orientation of the chord PN and is therefore much steeper than the plane of weakness associated with R. Slip along both planes of weakness has the direction indicated for S, i.e. synthetic with the surface slope. Slip along the potential planes of weakness associated with the intersection points on the lower half of the stress plane (not shown in the figure) would be in the opposite direction, i.e. antithetic to the surface slope. These planes of weakness therefore cannot form the base of the sediment wedge considered. The only candidate for a gently back-sloping base of the sedimentary wedge is the plane of weakness associated with intersection point R. Its slope angle must be related to the surface slope δ, since for given friction angles φ and φ_b the pole in Fig. I.5-7A was determined by the line L, which made the angle δ with the σ'-axis. Conversely, and more relevant to our problem, the existing basal plane of weakness determines the surface slope that is required to maintain the whole wedge in the limit state.

In determining the relation between the two slope angles δ and ϑ, we should bear in mind that in Fig. I.5-7B $\delta + \vartheta$ equals the difference between the angles ψ_1 and ψ_0, which the σ_I-direction makes with respectively the basal plane of weakness S and the surface. The angle ψ_1 is expressed in terms of φ and φ_b by means of the auxiliary angle γ, which is determined in Fig. I.5-7C. With the expression for ψ_1 and the expression for ψ_0 derived in Fig. I.5-7A, $\delta + \vartheta$ is finally expressed in terms of δ, φ and φ_b as stated in the figure.

For an extensive discussion and comparison with experiments and field examples of this interesting relation between the dip of the decollement base, the strength parameters and

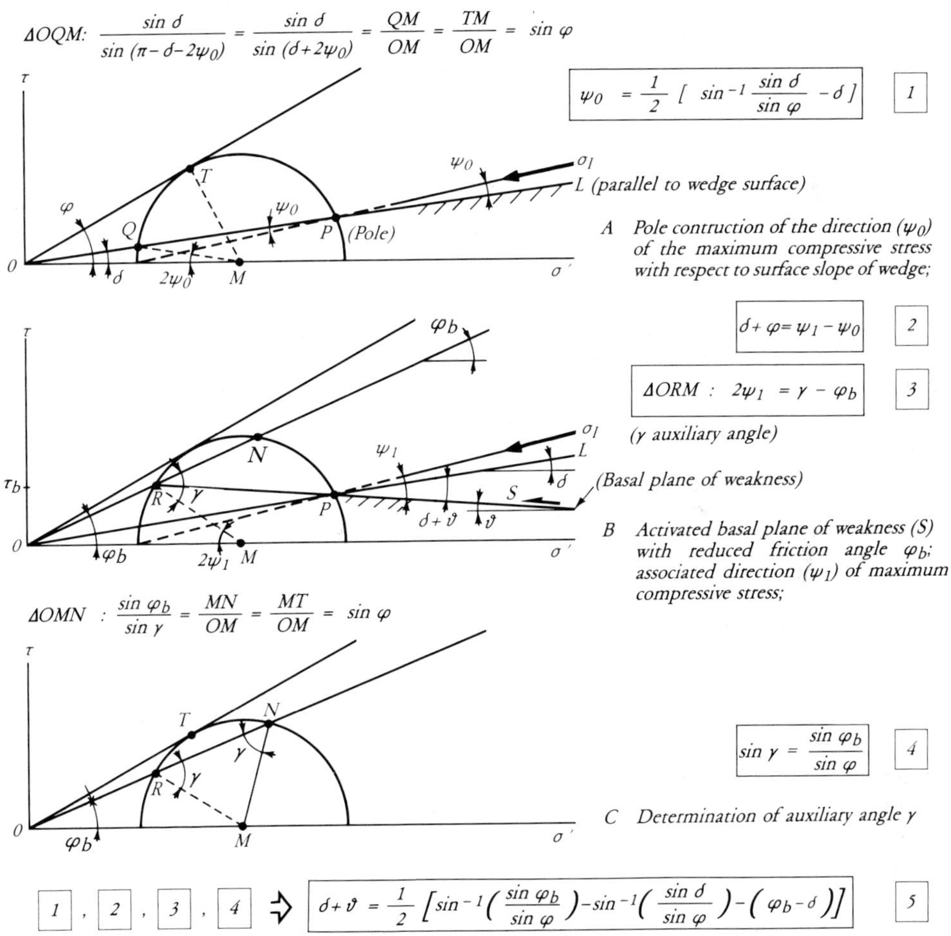

Fig. I.5-7 DAHLEN–LEHNER MODEL OF NON-COHESIVE FRICTIONAL PLASTIC WEDGE ON SLOPING PLANE OF WEAKNESS (Dahlen, 1984; Lehner, 1986) Mohr's stress diagrams to determine the surface slope (δ) which is required for the whole wedge to be in the passive limit state.

the surface slope that is required for the whole sediment wedge to be in the plastic limit state while moving upslope of its base, the reader is referred to the originasl papers by Davis et al. (1983), Dahlen (1984) and Dahlen and Suppe (1984). Here it may suffice to illustrate the result for the narrow plastic taper that is associated with the intersection point R in Fig. I.5-7B by assuming the friction angles $\varphi = 35°$, $\varphi_b = 15°$. For the dip angles $\vartheta = 6°$, $3°15'$, $30'$ and $0°$ the associated surface slope angle δ attains the respective values $0°$, $2°$, $4°$ and $4°22'$. This is the angular range of the narrow taper. But fully plastic wedges are also possible for higher slope angles of the basal plane of weakness, in which case we have to turn to the second point of intersection N in the Figs. I.5-7B,C. It may be readily seen that the only change now required in the formalism of the figure concerns the relation between ψ_1 and γ, which becomes $\psi_1 = \pi - \gamma - \varphi_b$. We then find that for ϑ varying from $69°$ to $6°30'$ the associated surface slope angle δ varies from $0°$ to $35°$, which for the chosen value of φ is the maximum dip the cohesionless slope can attain ('angle of repose').

From Figs. I.5-7A,B one can also read off the ratio between the shear stress τ_b that acts at a given depth on the basal plane of weakness and the shear stress τ that acts at the same depth on a plane parallel to the surface slope. Since τ_b and τ are the coordinates of respectively the points R and Q, one has

$$\tau_b = \tau.\ \sin 2\psi_1 / \sin 2\psi_0 \tag{I.5-2}$$

Since in the literature one often finds the two shear stresses considered to be approximately equal for small slope angles, it is worth noting that the coefficient of τ in this equation attains, for instance, the value 8 for the narrow plastic taper considered above with $\delta = 2°$ and $\vartheta = 3°15'$. While the shear stress τ that acts on surface-parallel planes is independent of the basal slope (see, for example, eq. II.7-10b), the basal shear stress τ_b depends on both slopes.

Dahlen and Suppe (1984) have extended the model by including an approximation for the influence of cohesion on the surface shape of a narrow wedge ($\delta + \vartheta \ll 1$) in the frictional plastic limit state. As a result of the cohesive strength component the critical surface slope is found to decrease towards the toe, while at the back end of the wedge it approaches the slope angle determined for noncohesive material. Cohesion will therefore cause some upward-concave curvature of the wedge surface.

In concluding our brief review of this interesting model, we recall its basic idea of a sediment wedge (in the brittle domain) which, when compressed from the rear, adjusts its shape to enable the whole wedge to be in the passive limit state. In general, this adjustment will require an internal deformation by imbricate thrust faults, the degree of which is likely to depend on the contrast between the initial shape and the critical shape. Once the wedge has attained the critically tapered shape, and is not affected by erosive changes in shape or by changes in shear strength, one might expect the prefaulted wedge to move as a stable mass. Although this expectation appears reasonable, it is not based on the limit state analysis of the wedge, which does not allow one to analyse the response of the wedge to externally imposed disturbances of its shape or to changes in the internal shear strength or basal shear resistance. Finally, the self-consistency and elegancy of the theory should not make one forget that it is still a hypothesis that a sediment wedge that is initially in a subcritical state will indeed attain a passive limit state throughout its whole interior. The observation of a whole sedimentary wedge transected by thrust faults does not exclude the alternative interpretation that active thrusting was confined to a narrower plastic zone, which gradually progressed through the wedge from back to front. In the next section we shall discuss mechanical models that emphasize this aspect.

The reader will have noticed that, in the models discussed above, the process was driven by horizontal forces acting upon the thicker end of a sediment wedge. At the higher crustal levels considered, these forces are usually associated with the collision of plates, which indeed appears quite plausible for accretionary prisms at that side of collision zones where an oceanic or continental plate is subducted. Of much more complexity is the situation for foreland basins (back-arc basins) on overriding plates, where the transmission of the collisional compression seems to be interrupted by the large-scale emplacement of magma in the volcanic arc. Smith's fluid-push model attempts to resolve this complication by decoupling the foreland thrust belt from the collisional forces, by replacing the collisional compression by the much lower hydrostatic pressure of the intervening igneous intrusions. While this procedure certainly elucidates the role of fluid pressures as driving agents of thrusts, it disregards the *shear interaction* between the subsiding plate and the overriding plate. This interaction has been studied by Neugebauer and Kracht (1984), who concluded that this is the mechanism responsible for the side-by-side coexistence of extensional and compressive regimes in the overriding plate, and that variations in the induced dynamic pattern are caused by variations in the dip of the descending plate.

Much older than the concept of thrust belts being formed by the 'push' of colliding plates is the idea of *gravity* being the driving force of thrust belt formation and thrust sheet transport. Today, gravity is considered to play only a minor role in shortening segments of the brittle crust (Bally, 1981). Nevertheless, it is the sole cause of the thrust belts at the toe of major deltas (Section I.3.1), but its role in thrust faulting of stronger rocks would seem very limited.

This may be illustrated in an approximate way by considering the maximum differential stress at the foot of a (water-saturated) rock sheet of uniform thickness H, length L and 'submerged' specific weight γ', on a slope of slope angle δ. We assume that at the foot of the slope any displacement of the sheet is blocked by an abutment, and that the frictional resistance on the slope is zero to allow for the greatest possible effect of the downslope-directed weight component of the sheet. With the exception of near-end regions, the principal stresses in the sheet are then parallel and perpendicular to the slope, since both slope and surface are free of shear stresses. Near the foot of the slope $\sigma_I' = \gamma' L \cdot \sin\delta$, since the downslope weight component of the sheet is uniformly distributed over the whole cross-section, and σ'_{III} increases linearly from surface to base, attaining at the base the value $\sigma'_{III} = \gamma' \cdot H \cdot \cos\delta$. Let us now take the average value of the maximum shear stress $q = (\sigma_I' - \sigma_{III}')/2$ taken over a cross-section near the foot of the slope $\hat{q} = \gamma' \cdot (L \cdot \sin\delta - \frac{1}{2} H \cdot \cos\delta)/2$ as a crude measure for the maximum shear stress experienced by the rock at the foot of the slope, and compare it with the average of the shear strength (eq. II.7-3a) of a noncohesive sheet $p \cdot \sin\varphi = \frac{1}{2}(\sigma_I' + \sigma_{III}') \sin\varphi = \gamma' \sin\varphi\, (L \cdot \sin\delta + \frac{1}{2} H \cdot \cos\delta)/2$. We obtain a very crude estimate of the critical length L_{cr} that would be required for a rock sheet with negligible cohesive strength to fail at the foot of the slope and thus to enable thrust faulting and overriding of the foreland:

$$L_{cr} = (H/2) \cot\delta\, (1 + \sin\varphi)/(1 - \sin\varphi) \qquad \text{(I.5-3)}$$

If, for example, a slope angle $\delta = 3°$ and $\varphi = 45°$ are chosen, the critical length of the cohesionless sheet would attain the high value of $55 \cdot H$, which, in view of the neglect of any basal friction and cohesive strength, would appear excessive. Note, however, that abnormally high pore pressures in the foot region would reduce L_{cr} by lowering the effective mean stress. Nevertheless, this consideration may illustrate sufficiently that very special slope conditions are required to enable the weight of a brittle rock mass to produce thrust faults.

Gravity Spreading. Naturally, gravity could greatly deform a rock mass and could make part of it override its foreland, if the rocks would behave as fluids without any or with only very low shear strength. This is the basic idea of the gravity-spreading model sketched in Fig. I.5-8. An initially created topography will be levelled out by the flow of a highly viscous fluid, whereby the lowering of the system's centre of gravity provides the mechanical energy that is dissipated by the deformation. This is essentially the content of the gravity-spreading model presented by Elliott (1976 a,b) with reference to earlier work by Goguel (1948). The model has an intuitive appeal because of its analogy with the flow of glaciers. But, although containing many interesting details, the model does not provide a clue to the basic question of the rate and rate history of the deformation process. Some criticism is also in order regarding a central point in the theory – the balance of the forces that act upon the ductile thrust wedge. The balance equation contains an expression for the shear stress on the base of the wedge, stating that the shear stress is equal to the overburden stress times the (small) slope angle of the surface, and hence is independent of the slope of the base. Although often quoted, this expression for the shear stress along a base whose slope differs from the surface slope, holds for a fluid at rest, but not for the flow of a viscous fluid – the essential feature of the model. This has been pointed out by Ramberg (1977, 1981), who has analysed theoretically and experimentally the internal flow and surface deformations of rectangular viscous bodies that are allowed to spread laterally on a rigid horizontal base under their own weight. Since in viscous flow the basal shear stress is proportional to the local velocity gradient perpendicular to the base, it should be clear that the basal shear stress depends on the flow pattern, and this was not considered in Elliott's theory. Interestingly, Ramberg (1981) finds the greatest basal shear stress at the ends of a spreading sheet that adheres to the base.

From Ramberg's analysis we infer that realistic spreading rates for nappes could only be expected for highly ductile rocks with viscosities not higher than those usually assumed for the mantle. Moreover, the 'fundamental shear strength' of the rocks, i.e. the shear strength that must be exceeded to enable creep to proceed beyond a phase of decelerating primary creep (see the end of Section II.7.7), must be very low. Even if one concedes that the value of the long-term strength is only two or three tenths of the values measured in the laboratory, the rocks of the brittle crust, with the possible exception of materials such as salt or overpressured shales, will not comply with the model of gravitational spreading. Even the observation of local flow structures in nappes can hardly be used as an argument for the application of the model. Such features may be the result of primary creep, that took place at locations of high stress concentrations, but died out under the prevailing stress, temperature and pressure conditions without developing into a secondary creep phase; or the 'flow' features turn out, at closer inspection, to be the result of diffuse brittle fracturing.

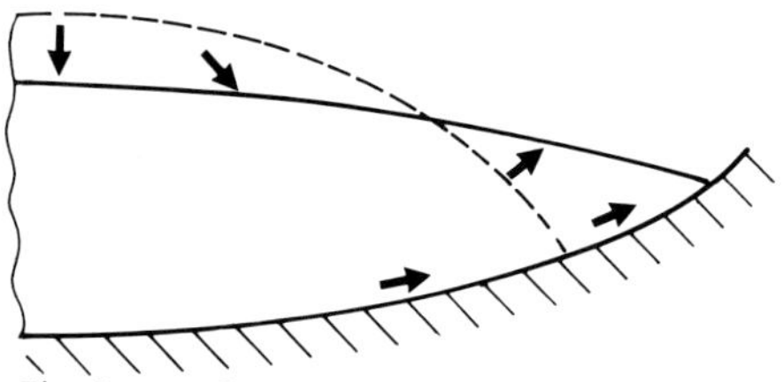

Fig. I.5-8. GRAVITATIONAL SPREADING

3. *MECHANICS OF A THRUST SHEET*

The models that we have reviewed in the preceding section have dealt with the compression of a sediment wedge as a whole and provided some insight into the conditions that would allow the whole wedge to be moved and plastically shortened. The models had in common that the mechanical interaction of the sediment wedge with the inner part of the orogenic belt was accounted for by horizontal stresses, rather than by displacements of the back end of the wedge. As a consequence of this, the ideally plastic wedge models cannot describe the internal and boundary *deformations* of the plastic wedge, since in ideal plasticity both stresses and displacements have to be prescribed at boundaries in order to determine plastic deformations (see Section II.7.1, "Plastic vs. Viscous"). (Note that the limit shape of the wedge surface was not derived as the result of plastic deformations of the fully plastic wedge, but was the special shape required for the whole wedge to be in the plastic limit state.) Therefore, the wedge models considered cannot tell us how individual thrust sheets are formed and moved over their foreland, and one may speculate on a variety of processes by which thrust sheets are thrust out, squeezed out, or 'peeled off' from the compressed sediment wedge. Probably the simplest mode of thrust sheet formation would be the forward thrusting of the slender tip region of the sediment wedge, by the thicker and stable main part of the thrust belt. This type of emplacement of nappes in the frontal zones of thrust belts may perhaps serve best as a rational frame for the mechanical analysis of single thrust sheets.

In studying the stresses and deformations in a single thrust sheet that is pushed from the rear, we distinguish three phases: the detachment phase (Fig. I.5-9A) in which the sheet is gradually detached from its base while still at rest in the frontal part, the overthrusting of the fully detached sheet on a basal layer of low frictional resistance (Fig. I.5-9B), and the imbrication phase when the resistance of frontal buttressing and/or the basal friction has increased to a point where the higher stresses cause faulting of the sheet.

The *detachment* of the sheet can only take place along a substratum of lower shear strength. Such substrata may consist, for instance, of weak limestones, whose shear strength may show little dependence on the effective stresses in the pertinent pressure range (see, for example, Riffault, 1969). Yet in many cases of overthrusting, the 'weak' rocks (e.g. shales), where detachment took place and which served as 'lubricants' for the moving sheet, have a shearing strength that increases with the effective mean stress. Therefore, the frictional shear strength can be drastically reduced in these rocks by high pore water pressures. This concept was employed in a classical paper on overthrusting by Hubbert & Rubey (1959), who were inspired by Terzaghi's (1950, 1956) concept of pore-pressure-induced landslides (see also Gretener, 1969, 1972). The essence of the detachment phase has already been shown in Section I.2.8, where we discussed the primary stress changes that accompany the formation of a tectonic fault. We considered in Fig. I.2-69 a rectangular sheet of strong

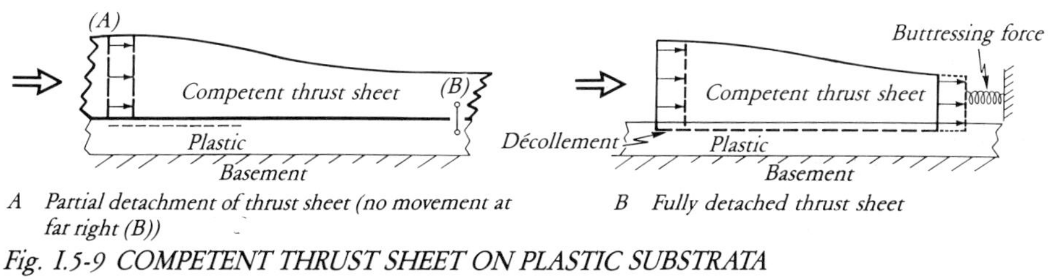

A *Partial detachment of thrust sheet (no movement at far right (B))* B *Fully detached thrust sheet*

Fig. I.5-9 COMPETENT THRUST SHEET ON PLASTIC SUBSTRATA

rock, separated from a rigid basement by a thin interlayer of weaker frictional-plastic material. In a finite element analysis the stress changes and deformations that result from a horizontal displacement of the back end of the competent sheet were studied. Both rocks were considered as elastic/frictional plastic materials with strain hardening and softening (Fig. I.2-69A). In the present context we refer to Fig. I.2-69B, which shows the extent of the plastic zones in the competent sheet and in the incompetent substratum after a total end displacement of 0.75% of the height of the sheet. It is clearly seen that plastic softening (+) has already penetrated far into the weaker substratum, while the competent rock is either still elastically deforming or in the plastic hardening range (o). From this and similar analyses we conclude that during the detachment phase the resistance of the weak substratum is not determined by the peak values of the shear strength parameters φ and τ_0 ('ultimate' shear strength), but by those of the residual strength parameters, which will be approached in the main part of the softened zone. Eventually, the completely detached thrust sheet will rest on a substratum, the whole of which is in a fully shear-softened plastic state.

In the absence of a weak interlayer or plane of weakness, a thrust sheet cannot be formed. This may be illustrated by a finite element analysis of the same thrust block as considered above, which now, however, is directly attached to the rigid base. The problem is outlined in Fig. I.5-10A. Fig. I.5-10B shows a slightly listric, softened shear zone, which extends from the lower corner of the sheet to the surface and which was fully developed after a total end displacement of 3.75% of the height of the sheet. At this stage, deformation in the other plastic regions had stopped. This can be inferred from Fig. I.5-10C, which shows the bundled contours of the magnitudes of the displacements that occurred during the last increment of the imposed back-end displacement. Note again the listricity of the shear zone, which confirms our statement in Section I.2.4 ('Curved Faults') that listricity of a fault may be caused by the basal shear stresses and the concomitant deflections of the σ_I-trajectories (Fig. I.2-20), which are induced by the deformations that precede the formation of the fault.

The Plastic Substratum of a Detached Thrust Sheet. Before considering the stresses and deformations in the competent thrust sheet, we shall be concerned with a mechanical analysis of the plastic substratum, which is required because thrust sheet and substratum interact mechanically, and the stresses and deformations in the thrust sheet therefore depend on those in the substratum. Secondly, the analysis of the plastic substratum will account for the formation and location of basal slip planes and for secondary faults underneath the competent thrust sheet – phenomena that can be abundantly observed on a macro and micro scale at outcrops of overthrusts.

The large lateral extent of thrust sheets and substrata permits us to make the simplifying assumption of plane strain, which means that the deformations are regarded as being the same for any thrust-parallel vertical cross-section. Furthermore, the plastic substrata will be treated as 'infinitely' long; a simplification justified by the high length/thickness ratio encountered in reality. To obtain an analytical solution, we assume that the substratal material behaves as an ideally plastic material with a stress-independent shear strength (Fig. I.5-5A); the same extreme idealization of plastic behaviour as used by Chapple for the plastic wedge. We shall, however, allow the shear strength to vary arbitrarily with depth; in other words, the thrust sheet may be underlain by a stack of plastic and even non-plastic layers of different shear strength. The problem is illustrated in Fig. I.5-11, in which the substrata are confined between two horizontal rigid plates at $y = \pm H$, which represent the base of the thrust sheet and the top of the supporting basement. Between these boundaries the material is compressed and sheared by the vertical normal stresses σ_y and horizontal stresses τ. Together with the horizontal normal stress σ_x, these stresses must satisfy the limit condition of perfect plasticity (Fig. I.5-5A)

$$(\sigma_x - \sigma_y)^2 + 4\,\tau^2 = 4\,K(y)^2 \qquad \text{(I.5-4)}$$

and the equilibrium equations (e.g. eqs. II.2-2a with subscripts 2≡x, 3≡y)

$$\partial\,\sigma_x/\partial\,x + \partial\,\tau/\partial\,y = 0,\ \partial\,\tau/\partial\,x + \partial\,\sigma_x/\partial\,y = \gamma\cdot y \qquad \text{(I.5-5)}$$

where K(y) is the shearing strength as function of depth and γ the specific weight of the substratal material.

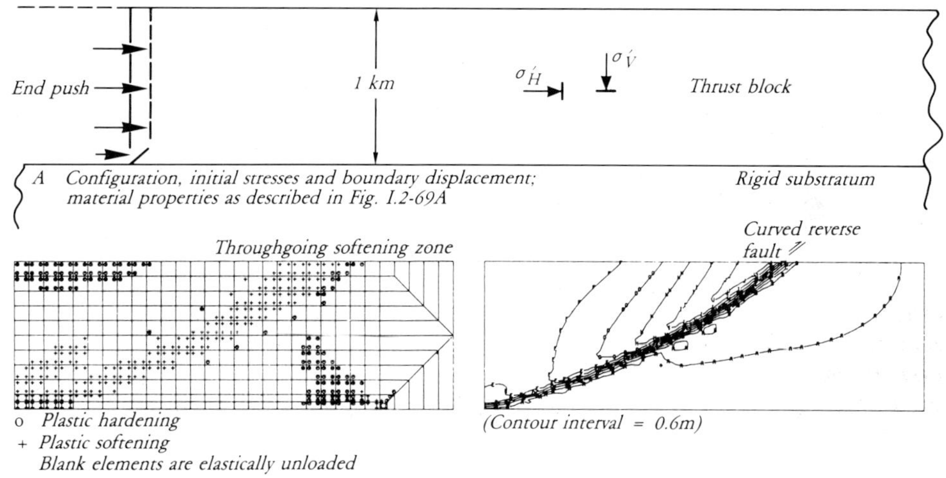

B Finite element mesh of rear part with throughgoing, slightly listric plastic softening zone after 37.5 m total end displacement. Notice second, steeper plastic zone, which stopped growing and unloaded elastically when the main shear zone reached the surface.

C Contours of the magnitudes of the displacement increments that took place during the last increment of boundary displacement; bundling of contour lines indicates concentration of simple shearing.

Fig. I.5-10 HORIZONTAL COMPRESSION OF RECTANGULAR ROCK BLOCK ATTACHED TO RIGID BASE. ELASTIC/FRICTIONAL-PLASTIC FINITE ELEMENT SIMULATION OF THRUSTING (after G.H. Mäkel & J.V. Walters, unpubl. work)

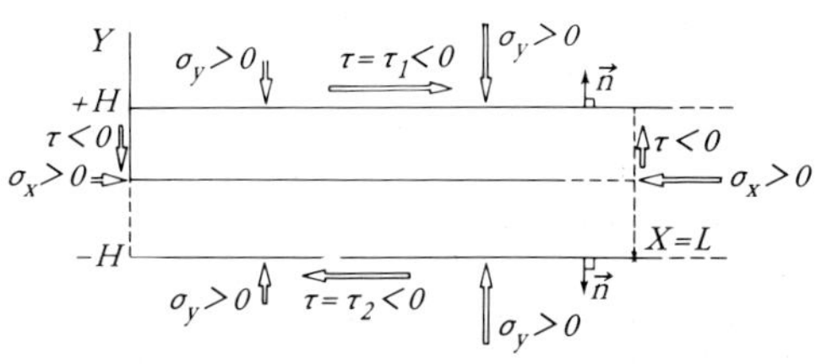

Fig. I.5-11 SHEARING OF PLASTIC MATERIAL BETWEEN RIGID PARALLEL PLATENS

The solution is as follows (Mandl and Shippam, 1981), as may be easily verified by substituting into eqs. I.5-4,5 (and paying attention to the signs of the stress components in Fig. I.5-11):

$$\tau = (\tau_1 + \tau_2)/2 - a \cdot y$$

$$\sigma_y = \sigma_y^\circ + \gamma(H-y) + a \cdot x \qquad \text{(I.5-6)}$$

$$\sigma_x = \sigma_y \pm 2\,[K(y)^2 - \tau^2]^{1/2}$$

with the positive constant

$$a = -(\tau_1 - \tau_2)/2H > 0 \qquad \text{(I.5-7)}$$

where τ_1, τ_2 are the (negative) values that the shear stress τ uniformly attains along the upper and lower boundaries, respectively. The constant $\sigma^\circ_y > 0$ denotes the compressive vertical stress exerted upon the strip at the reference point $x = 0$, $y = \pm H$ (Fig. I.5-11). [1)]

For $a = 0$ the solution would reduce to the trivial case of a constant shear stress τ throughout the layer and normal stresses that would not vary in x-direction. Such a special solution would not suit our problem, because an increase in the vertical stress in the direction of overthrusting along the base of the thrust sheet is necessary to balance the moment of the thrust force that acts upon the thrust sheet. If out of balance, the moment of the thrust force and the opposing basal friction force would exert a clockwise moment upon the thrust sheet which would tend to make it topple over. This point is extensively discussed in Ch. II.2 (Example No. 6: Overthrusting; Fig. II.2-12).

It is important to note that the linear increase of the vertical boundary stress in the shear direction is accompanied by a linear decrease in the magnitude of the horizontal shear stress with depth, since $d|\tau|/dy = -d\tau/dy = a > 0$ in the upward direction. Hence, the shear stress τ_1 imposed by the thrust sheet determines the shear stress τ_2 at the base of the substrata (Fig. I.5-11), which has a *smaller* magnitude than the shear stress at the top. Since the two shear stresses are not in balance, static equilibrium is maintained by an increase in the lateral 'confining' stress σ_x in the shear direction. The double sign of the square root in eq. I.5-6 expresses the fact that two different plastic limit states are compatible with the boundary conditions at $y = \pm H$. The plus sign corresponds with the highest possible lateral compressive stress that is associated with the passive limit state. Although locally the active limit state may exist, characterised by the negative sign, one would usually expect the passive limit state to predominate, since the flow of the substratal material in the x-direction will be obstructed.

It may be added, that in reality, where we deal with a very long, but finite, plastic layer, the 'confining' stress σ_x will adjust itself on near-end cross-sections to the boundary loads by means of distortion of the end regions. Such distortions are always observed in shearing experiments at the free ends of a cohesive plastic strip that is sheared between two parallel plates.

Basal thrust planes and secondary faults. In the plane plastic deformation of a perfectly plastic material to which the limit condition I.5-4 applies, the directions of potential slip lines coincide with the directions of maximum shear stress and consequently bisect at any point the right angle between the principal stress directions. Thus, conjugate slip lines of a purely cohesive perfectly plastic material intersect at right angles instead of at the angle

$\pi/2 - \varphi$ in a frictional plastic material, as it is to be expected since this situation may be considered to be a special case ($\varphi \Rightarrow 0$) of the more general theory for cohesive-frictional plastic materials. A discussion of the slip line field for the present problem has been presented by Mandl and Shippam (1981). Two typical solutions are presented in Figs. I.5-12,13.

In Fig. I.5-12A,B the slip lines are plotted for a case in which the tangential stress τ_1 at the base of the thrust sheet, i.e. at the upper boundary $y = H$ of the plastic strip, equals the shearing strength K of an homogeneous substratum. For both the 'weak' and 'strong' solutions, corresponding with the active and passive states respectively, the slip lines are cycloids, their radius of curvature being proportional to the parameter K/aH. For a plastic substratum of given height and shearing strength the radius of slip line curvature will therefore, according to the second equation I.5-6, be inversely proportional to the gradient of the vertical pressure exerted by the thrust sheet. The figures clearly demonstrate the characteristic differences in slip line pattern for the active and the passive states. The two solutions, however, have in common that one set of slip lines is tangential to the upper boundary; in each case slip can therefore occur along that boundary (thrust base).

Normally, the shear strength of the substrata may vary with depth. The solution for such a case is shown in Fig. I.5-13A,B. It is assumed that the shear strength varies as indicated in Fig. I.5-14 and that $\tau_1 = 0.6\ K_\infty$, K_∞ being the value the shear strength attains near the boundaries of the plastic layer. The active and the passive solutions each give a set of slip lines that intersect a plane SS tangentially. Along this plane, whose position is the same for both limit states, the material enclosed between the 'thrust sheet' and the plane will glide as a solid block, since movement along the slip lines is obstructed by the competent boundaries.

The plane SS in Fig. I.5-13A,B may be seen as the theoretical representation of real 'thrust planes', such as the one shown in the photograph of the Glarus nappe (Switzerland) in Fig. I.1-1 cutting through the top of the soft Malm limestone underneath the Permian sandstone of the nappe. The position y^* of SS is determined in Fig. I.5-14 by the point where the τ-line touches the strength profile. For $a > 0$, i.e. for increasing vertical compres-

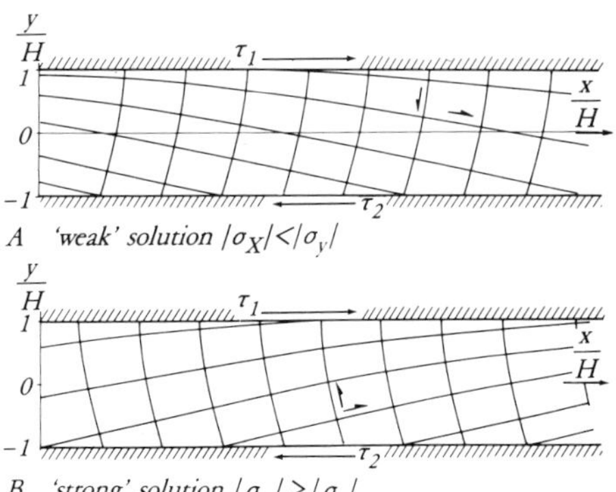

Fig. I.5-12
SLIP LINES IN HOMOGENEOUS PLASTIC LAYER UNDER SHEAR AND COMPRESSION (τ_1 = K, aH = K/20).

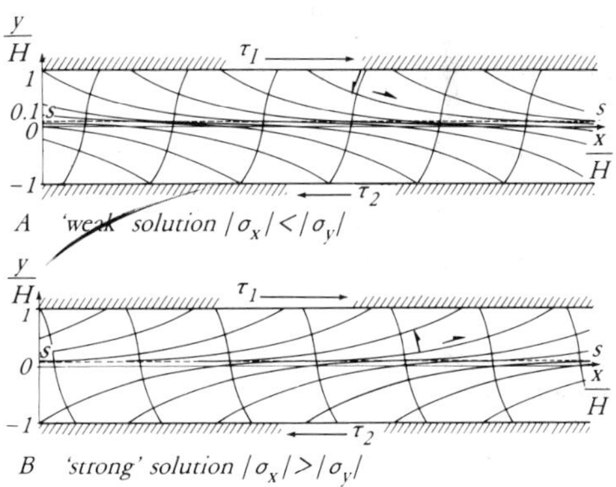

Fig. I.5-13
SLIP LINES AND 'THRUST PLANE' SS IN STRATIFIED LAYER UNDER SHEAR AND COMPRESSION ($K=[1-\frac{1}{2}\cos^2 y/H]K_\infty$ as shown in Figure I.5-14; $aH = 0.1K_\infty$, $\tau_1 = 0.6\ K_\infty$)

sive stress σ_y in the thrust direction, the thrust plane SS will lie above the plane of lowest shear strength and will shift towards the top of the plastic stratum as the shear strength decreases with depth in a more abrupt manner, as indicated by the dotted strength profile in Fig. I.5-14. The opposite would of course happen were σ_y to decrease in the thrust direction ($a < 0$).

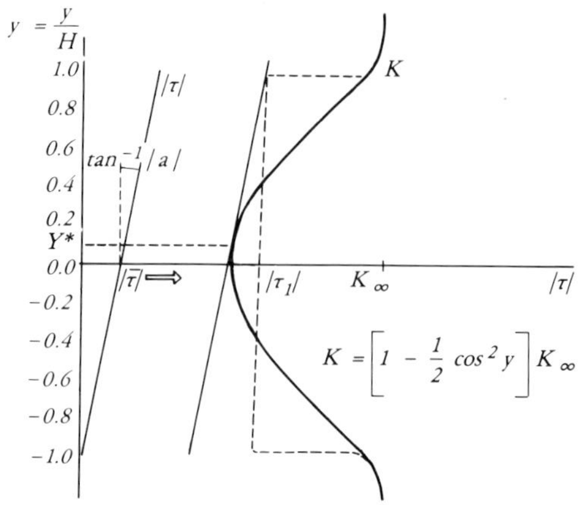

Fig. I.5-14
CRITICAL HORIZONTAL SHEAR STRESS IN STRATIFIED ZONE OF WEAKNESS

The strength profile K(y) will usually have a more irregular shape than that in Fig. I.5-14. In particular, the thrust sheet may rest on a stack of alternatingly incompetent (i.e. plastic) and competent (i.e. elastic) strata where, as shown in Fig. I.5-15, the horizontal shear stress varies only in the incompetent layers, the vertical normal stress remaining unchanged with depth throughout the stack of layers.

Since the variation of the horizontal shear stress with depth is determined by the horizontal gradient of the overburden stress σ_y and does not depend on the local shear strength, different overburden gradients may, in a given stack of substrata, mobilize different basal slip planes. As sketched in Fig. I.5-16, a stronger increase in σ_y in the thrust direction, as caused, for instance, by a local thickening or by some bending or buckling of the thrust sheet, will be associated with a slip plane at *shallower* depth and even in a layer of *higher* shear strength than in the case of a smaller gradient of σ_y. This would suggest the ***'stepping-up'*** of a sole fault underneath the location of a drastic increase in σ_y. This is illustrated by Fig. I.5-17.

The immobilization of the lower gliding horizon II in Figs. I.5-16,17 is accompanied by some shear stress reduction at that horizon, and this gives rise to a change in slip direction. This is demonstrated in Mohr's stress diagram in Fig. I.5-18, in which the new stress state at some point on the original slip plane II is represented by the dotted circle. Note that the stresses are still in the limit state, but that the stress point associated with slip plane II no longer lies on the limit line. For the sake of simplicity we have left the value of σ_y on plane II unchanged. Since the orientation of plane II and the associated stress point are known, we can use the stress pole to construct the new slip direction which is synthetic with the overthrusting. The chord drawn from the pole to the contact point on the limit line forms the same angle with the line $\tau = \tau^*_{II}$ as the new slip direction makes with the glide plane II. The former horizontal thrust direction is therefore replaced by a low-angle thrust element. Obviously, in reality the change in dip angle will be less abrupt than suggested in Fig. I.5-18, since, in passing the transition zone of increasing pressure gradient 'a' along plane II, the shear stress τ gradually decreases to the fully reduced value indicated in Fig. I.5-18. This enables a smooth branch-off of a 'step' fault connecting the two gliding horizons. Steepening of the step fault and final flattening can – at least qualitatively – also be accounted for in the construction of Fig. I.5-18 by allowing the K-value to change in the depth interval transected by the step fault.

We may now also understand why, usually, an overthrust plane does not step 'down', since this would be accompanied by an increase in the pressure gradient 'a' on the fault surface, which would tend to correct the 'misstep'.

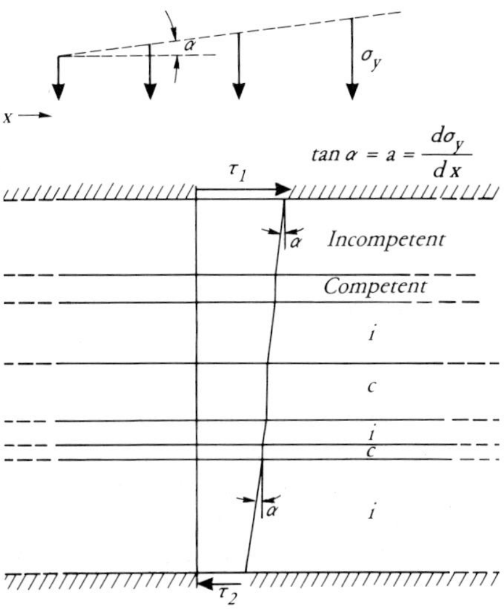

Fig. I.5-15
STACK OF INCOMPETENT AND COMPETENT STRATA UNDER COMPRESSION AND SHEAR

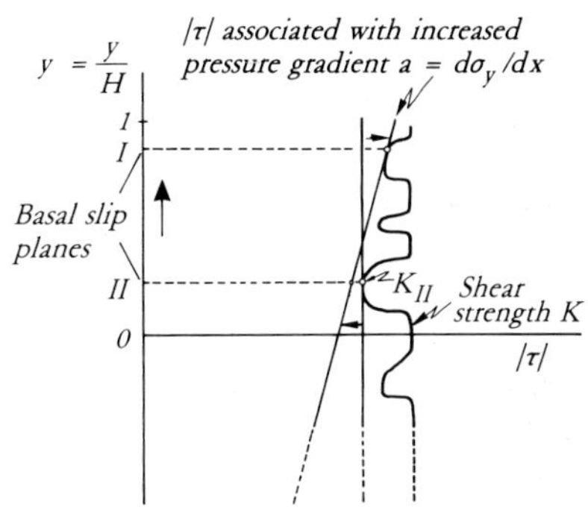

Fig. I.5-16
'STEPPING-UP' OF BASAL SLIP PLANES

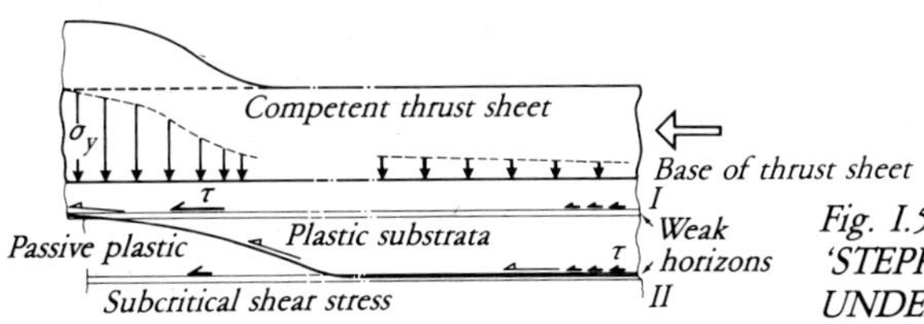

Fig. I.5-17
'STEPPING-UP' OF BASAL SLIP PLANE UNDERNEATH OVERTHRUST SHEET

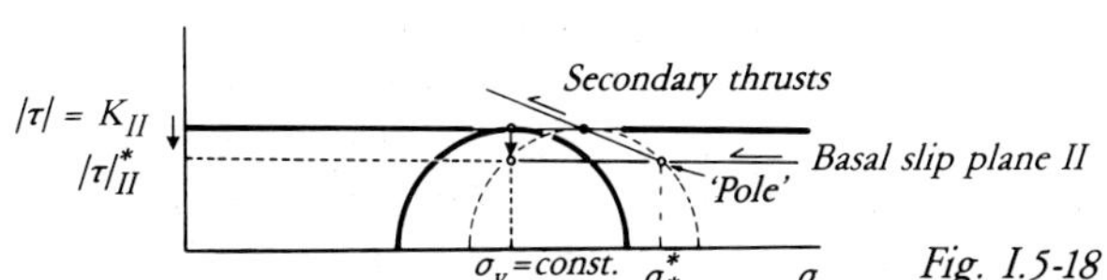

Chord drawn from pole to contact point on limit line forms same angle with line τ = const. as actual slip with basal slip plane.

Fig. I.5-18
IMMOBILISATION OF LOWER GLIDING HORIZON AND INITIATION OF SECONDARY THRUSTS

An increase in 'a' is not the only cause of a 'step-up' of the basal fault plane. Gliding on a weak horizon may meet with the 'buttressing' effect of a local increase in shearing strength. While the region of higher strength may not be in the plastic limit state, the weak substratum in front of the obstacle will still be in the passive limit state, but the horizontal compressive stress σ_x will be increased, while σ_y is less affected. Consequently, the centre of the Mohr circle in Fig. I.5-18, which lies at $(\sigma_y + \sigma_x)/2$ on the σ-axis, will shift to the right. For the shifted limit circle (dotted in Fig. I.5-18) this implies again a decrease in the value of the shear stress τ that is associated with σ_y on the new limit circle. The pole construction demonstrates again the formation of a step-up thrust.

Step-up thrusts always require a passive limit state and are therefore of the P-shear type, which we have considered before, for instance, as secondary shears contributing to the formation of shear lenses in wrench zones (Section I.2.8, 'Basement-induced wrenching, Fig. I.2-75). Also of the P-shear type are the set of slip lines that intersect the plastic strata in Figs. I.5-12B and 13B, and which have a slip direction synthetic with the direction of overthrusting. But these shears, which form in an undisturbed field of uniform stress gradients, should be clearly distinguished from the step faults above, which form at local disturbances of the stress field. Synthetic thrust shears of the type shown in Figs. I.5-12B and 13B may occasionally be mobilized, for instance, when the base of the thrust sheet deforms or when compaction in the incompetent strata enables minor slips. Fig. I.5-19 presents the synthetic thrust shears which in the case of Fig. I.5-13B could accompany overthrusting. As mentioned earlier (at the end of Section I.2.8, where we discussed the possible coexistence of active and passive limit states on the opposite sides of a fault plane), the lineaments in the weak limestone underneath the 'Glarus' overthrust in Fig. I.1-1 might be interpreted as secondary thrust shears corresponding with the lower parts of the P-shears in Fig. I.5-19. In contrast with a single step fault, these shears may show very little displacement because of the confinement by the thrust sheet, but may occur in whole families. In particular, such families of synthetic thrust shears may be observed between parallel glide planes. From Fig. I.5-16 it should be clear that situations may easily arise where two gliding planes are active at the same time. In this case a family of synthetic thrust shears may form and interconnect the two fault planes in a sigmoidal fashion, curving downward and upward into the bounding fault planes. This may be readily inferred from Figs. I.5-13B and 14, by replacing in Fig. I.5-14 the sinusoidal strength profile by a curve with two sinusoidal segments, both of which make contact with the straight shear stress line, and by superimposing upon the strip shown in Fig. I.5-13B a duplicate of that strip. In that way a new slip line solution is obtained with a set of synthetic slip lines that merge tangen-tially with two horizontal planes SS. This might account for a certain type of the *'duplex'* structures (Boyer and Elliott, 1982), a small-scale example of which is shown in Fig. I.5-20. [2)]

It should be recalled that our analysis was based on the assumption of ideally plastic behaviour with a stress-independent shear strength, and therefore cannot account for pore pressure effects. Although an analytical solution can be also obtained for a noncohesive frictional (i.e. pressure-sensitive) material[3)], we restrict ourselves to a semi-quantitative discussion of the additional effect that variations in pore pressure may have on cross faults between two gliding horizons.

Figure I.5-21A shows a frictional plastic layer in the passive limit state, enclosed between two parallel gliding horizons (bedding-plane faults). Concentrating on pore pressure effects, we neglect gradients of shear stresses and total normal stresses (a = 0) and assume that changes in the effective stresses σ_x', σ_y' are solely caused by pore pressure gradients. We also assume that the pore pressure along the main gliding horizons is higher than that inside the sheared interlayer (for example, because of liquefaction at the gliding planes).

Consequently, the effective stresses σ_y' decrease towards the bounding gliding planes, as is also indicated for the critical stress circles in the diagram of Fig. I.5-21B, which correspond with the positions 1, 2, 3 in Fig. I.5-21A. The poles of the stress circles lie on the line τ = const. and the chords drawn from the poles to the contact points on the Coulomb limit line indicate a decrease in dip angle with increasing pore pressure (i.e. decreasing σ_y'), which may account for the sigmoidal shape of the cross faults.

The reader may verify, by means of a similar stress diagram, that an increase in the friction angle φ from positions 1 to 3 in Fig. I.5-21A, i.e. a corresponding steepening of the limit line in Fig. I.5-21B, would have a similar effect on the shape of the cross faults.

In concluding the analysis of the plastic substrata of an overthrust we remind the reader that the plasticity approach does not account for strain rates. It therefore cannot provide any information on the *rate* at which the detachment front advances during the detachment phase of a thrust sheet, or on the rate at which a fully detached sheet would move when pushed by a certain force or on the rate at which the sheet is imbricated when stopped at its front. Considerations of rate would require the assumption of a viscous component in the response of the incompetent substrata (see Example No. 1 in Section II.7.2). Such an approach was, for instance, undertaken by Kehle (1970) to obtain an estimate of the rate at which a thrust sheet moves under its own weight or when being pushed by an 'orogenic force'.

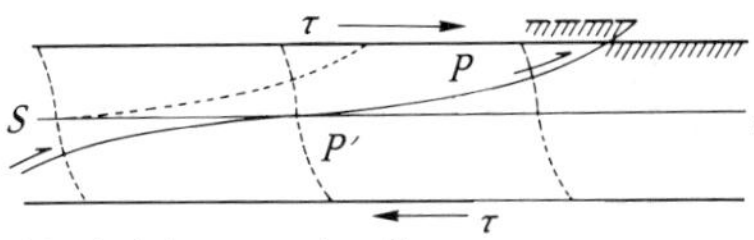

'Passive' (compressional) state

Fig. I.5-19
SECONDARY THRUST FAULTING IN PLASTIC SUBSTRATA

Fig. I.5-20 PARALLEL GLIDING HORIZONS WITH SECONDARY FAULTS IN TURBIDITES (Widemouth Bay, North Cornwall England); Slip surfaces quartz-coated and slickensided.

We wish to emphasize, however, that the model of a rigid/plastic material, which was applied above to the weak substrata of an overthrust, and in previous sections to a sediment wedge under horizontal compression and to thin-skinned gravitational sliding on a delta slope, is completely indifferent with respect to brittleness or ductility of the material, because in this simplest plasticity model the stresses are determined by equations and boundary conditions into which no strains or displacements have been entered. Therefore, as long as the material is in the limit state and subject to the same boundary stresses, the stresses will not be affected by any creep flow, and faults will be initiated along the predicted slip lines, irrespective of preceding creep deformations. Unfortunately, as pointed out at the end of Ch. II.7, there are only very few analytical solutions available and even fewer that can be applied to geological problems.

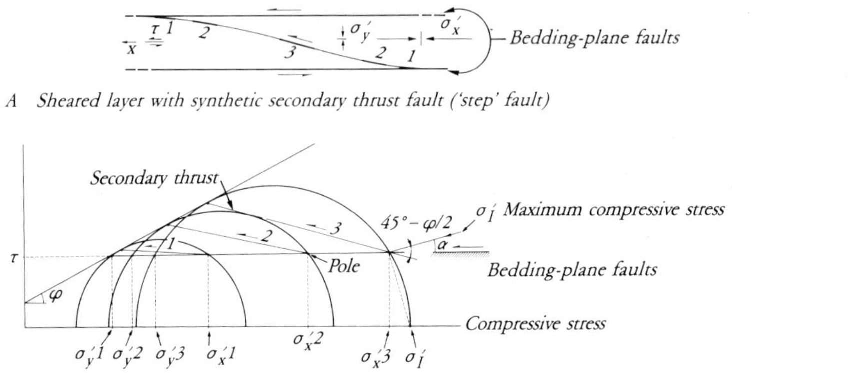

B Variation of effective normal stresses across sheared layer in passive limit state. (Pore pressure in interior of sheared layer lower than at boundary faults (τ = const.)

Fig. I.5-21 PORE-PRESSURE CONTROL OF STEP-BEDDING PLANE THRUSTS

Stresses and Faults in the Thrust Sheet. A thrust sheet that has been sheared from its substratum by a push from the rear and moves on a basal slip plane will eventually experience an increase in basal friction and buttressing forces at its front as the sheet 'ploughs' into its foreland or gets stranded on strata of higher frictional resistance. As the push from the rear continues, the increasing stresses will exceed the strength of the rock and induce faulting. When gravity is the main driving force, the stresses in a detached sheet whose motion is obstructed may be raised by an increase in slope angle; or, when the detachment is still proceeding, as in the thin-skinned gravity slide model of Section I.3.1 (Fig. I.3-6 A,B,C), the stresses may reach a critical value when the detached part of the sheet has reached a certain length.

To begin the discussion of stresses and deformations in a competent thrust sheet whose movement is obstructed at the front, we first note a basic mechanical difference between sheets propagated by a push from the rear and sheets that are moved by the downslope component of their own weight. In Fig. I.5-22 we consider a simple balance of the force components that act parallel to the thrust base on a rectangular segment of the thrust sheet. When the sheet is pushed from the rear (Fig. I.5-22A), the thrust-parallel normal stress decreases in the thrust direction since less force is required to overcome the basal friction ahead. If the sheet is moved downslope by the slope-parallel component $W_{\parallel}$ of its own weight W, the situation is different. Gliding then requires that the slope-parallel weight component of the section considered at least equals the shear force at the base of the element. Normally, however, it will exceed the basal shear resistance, because it has also to overcome an additional buttressing resistance at the front of the sheet. As shown in Fig. I.5-22B, the slope-parallel normal stress will then increase towards the front ('toe') of the sheet. In both cases of nappe transport the normal stresses, which act on base-parallel planes, change less in the thrust direction than the normal stresses that act parallel to the base. Along base-parallel planes the associated Mohr circles will therefore change in size as indicated in Fig. I.5-22C, where, for simplicity, $\sigma'_{\perp}$ is held constant. Obviously, the limit state will be first attained in that part of the sheet where $\sigma'_{\parallel}$ is highest. In a sheet pushed from the rear, faulting will therefore begin near the back end, whereas in the gravitationally gliding sheet faulting will begin at the front ('toe'). This tendency will hardly be affected by the slight increase of $\sigma'_{\perp}$ in the thrust direction along the base of a sheet moving over a plastic substratum, which we have found above (Fig. I.5-11). Naturally, this does not exclude the possibility of faults forming somewhere inside the sheet at places where the sheet is weaker or thinner, or where changes in the frictional resistance or the dip angle of the basal shear surface take place. In particular, a gravitationally gliding sheet is susceptible to localized normal faulting since, as was pointed out earlier (Section I.3.1, Fig. I.3-6D; Section I.2.8, Fig. I.2-85), only a small reduction in the base-parallel normal stress by a locally decrease in basal friction or by a local increase in the dip of the slip base is needed to induce locally the active limit state.

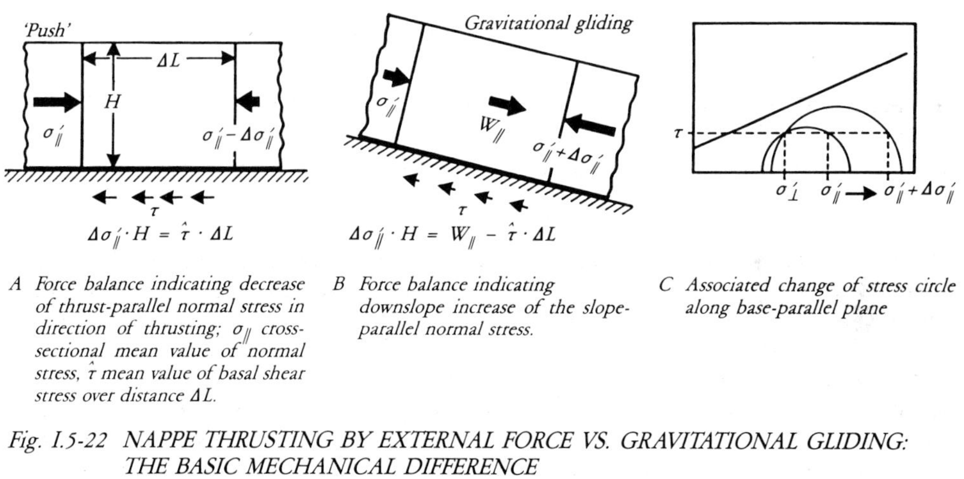

A *Force balance indicating decrease of thrust-parallel normal stress in direction of thrusting; $\sigma_{\parallel}$ cross-sectional mean value of normal stress, $\hat{\tau}$ mean value of basal shear stress over distance ΔL.*

B *Force balance indicating downslope increase of the slope-parallel normal stress.*

C *Associated change of stress circle along base-parallel plane*

Fig. I.5-22 NAPPE THRUSTING BY EXTERNAL FORCE VS. GRAVITATIONAL GLIDING: THE BASIC MECHANICAL DIFFERENCE

Let us now consider again an idealized thrust sheet of rectangular shape, which is propelled by a push from the rear and meets an increasing resistance at its front end. The vertical front and back walls of the sheet are to represent the boundary thrusts, which in reality usually confine a single sheet. The sheet has uniform material properties and sufficient 'brittleness' to deform predominantly by faulting, rather than by folding, when the elastic limit of the material is exceeded. Even for such a highly idealized thrust sheet various modes of the mechanical interaction between the sheet and its surroundings may be envisaged and, accordingly, different boundary conditions may be considered for our problem.

In Fig. I.5-23A different types of boundary conditions are summarized which have been applied to the same test sheet, which had a height of 1 km and a length of 10 km. In the first three cases tectonic compression is simulated by prescribing the 'push' in terms of horizontal normal stresses on the back face of the sheet [4)], while in the last three cases an incremental end displacement is applied parallel to the base. Figure a) in I.5-23A illustrates the boundary conditions of the earliest elastic stress analysis of a thrust sheet, which was presented by Hafner (1951) and which has found entrance into most textbooks of Structural Geology. In Hafner's formulation the pressure along the thrust base is constant and the anticlockwise moment of forces necessary to balance the clockwise moment of the net push force and to maintain the horizontal position of the sheet is produced by shear stresses at the end faces (see also the discussion in Section II.2.3, Example No. 6: 'Overthrusting', Fig. II.2-12A,B,C). These boundary conditions were dictated by the wish to obtain an analytical stress solution. In following the 'inverse' method, Hafner selected from a whole class of rigorous solutions to the stress equations those that define boundary stresses that looked acceptable from a geological point of view. Although the stress fields obtained are rigorous solutions for sheets of any rectangular geometry and are not marred by any shortcomings of numerical methods, the approach has the disadvantage that the boundary stresses at the front and rear ends cannot be varied in order to establish how typical the special stress solutions are for the tectonic problem. Such variation of the boundary conditions and comparison of the corresponding solutions are, however, necessary in view of the many uncertainties regarding the boundary conditions of the actual tectonic process. The means for such a flexible approach, which were not available at the time of Hafner's pioneering work, became available by the rapid developments in the computer technology and the accompanying progress made in the development or improvement of numerical methods. Therefore, the boundary conditions in Figs. I.5-23A,b-f were all tackled by a numerical approach using a finite-element program.

In Fig. I.5-23A,b the elastic sheet rests on the ideal-plastic substratum, which we have discussed above. Accordingly, the sheet is supported by a vertical pressure that increases slightly and linearly in the thrust direction, while the basal shear stress is constant (with the exception of the near end regions where it is gradually allowed to drop to zero to avoid singularities). Push and buttressing stresses are strictly horizontal and, although varying linearly with depth in the figure, were allowed in the calculations to vary with depth in various ways. The elastic moduli did not require changing, simply because Young's modulus does not enter the statically determined problem, and Poisson's ratio does not affect the in-plane stresses. The pore pressure was assumed to vary linearly with depth, but its gradient was varied in some calculations. A different situation is simulated by the conditions of the 'contact problem' in Fig. I.5-23A,c, in which the sheet rests in frictional, noncohesive contact ($\varphi = 26.6°$) directly on a rigid basement. The numbers at the base indicate the values of the basal shear stress that is induced by the normal stress in accordance with Coulomb's friction law. Under the applied load the elastically shortening sheet is not yet fully detached and has started sliding over only a part of the total base. In the models of Figs. I.5-23A,d,e,f the front of the sheet is not allowed to move in the horizontal direction, while the back end is uniformly displaced in the horizontal direction. In the first model the sheet rests on an ideally plastic substratum of the same purely cohesive type of material as that in the model of Fig. I.5-23A,b. The conditions sketched in Fig. I.5-23A,e are those of the sheet already considered in Fig. I.2-69, which rested on a substratum of elastic/frictional plastic material with strain hardening and softening. The last model (Fig. I.5-23A,f), that of a sheet still attached to its rigid base, has been added for completeness and is considered in Fig. I.5-10.

The major trends of the principal stress trajectories are the same in all models and are fairly predictable. In the rear part of the sheet σ_I is horizontal, its magnitude declining in the thrust direction and its direction being deflected towards the base by the basal shear stress (cf. Fig. I.2-69C). Refraining from a detailed discussion, we restrict ourselves here to a single aspect of the stress analyses, concerning the value of the shear strength the material must possess at any point in the thrust sheet to be capable of withstanding the local stresses without fracturing. The pressure-dependent part of this ***critical shear strength*** is determined by the stresses when – without much loss of generality – the angle of internal friction is assumed as a uniform constant throughout the sheet. Choosing for convenience a value of 30°, we may rewrite Mohr's limit condition for the principal stresses (eq. II.7-3 a,b) as

$$(\sigma_I' - 3\sigma'_{III})/2\sqrt{3} = \tau_o \qquad \text{(I.5-8)}$$

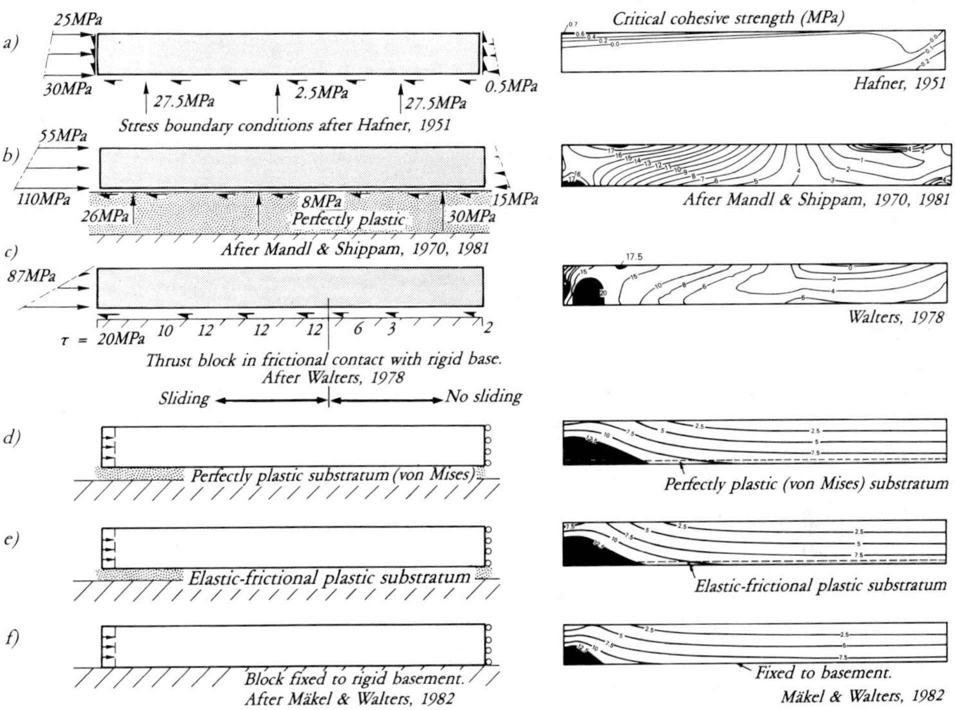

A Idealised thrust block under various boundary conditions.

B Contours of the minimal cohesive shear strength (in MPa) that would still allow the thrust block to deform elastically without fracturing under the conditions shown in the previous figure.

Fig. I.5-23 THRUST BLOCK ANALYSES

For a given stress field, equation I.5-8 then defines the minimum cohesive strength the thrust sheet material must possess to remain intact.

The contours of the calculated τ_0-values for the boundary problems of Fig. I.5-23A are presented in the corresponding sequence in Fig. I.5-23B, and locations of pronounced maxima are marked black. We note that Hafner's analytical solution differs strikingly from all the others: while all numerically tackled boundary problems show pronounced peak values of the critical strength at the lower rear corner and – to a somewhat varying degree – at a surface point at some distance from the rear, Hafner's solution indicates the upper corner at the rear as the spot where fracturing is most likely to start. Irrespective of the differences in boundary conditions, the results of all the cases b-f strongly suggest that a first, relatively steep, thrust fault will be formed that extends from the lower rear corner to the location of the surface peak. This contrasts with Hafner's case which suggests that only a very thin surface layer will be fractured or stripped off, while the remainder of the sheet remains intact. As has been pointed out by Mandl and Shippam (1981), this striking discrepancy is due to the very special and seemingly unrealistic distribution of shear stresses at the rear face of Hafner's thrust sheet (Fig. I.5-23A,a). In contrast, a distinct surface peak of τ_0 at some distance from the rear was persistently found, even when the dimensions of the sheet and the distribution of the stresses at the end faces in Fig. I.5-23A,b were greatly varied.

The formation of a ***first thrust fault*** extending from the lower rear corner of the sheet to the critical surface spot was simulated by Mäkel and Walters (1982, unpubl. work) by finite-element analysis of the problems in Figs. I.5-23A,d,e,f, assuming again elastic/frictional plastic behaviour of an homogeneous sheet with strain hardening and softening. A typical result is shown in Fig. I.5-24 for the case of the thrust sheet resting on a weak frictional plastic interlayer, which was considered earlier (Fig. I.2-69A). Figure I.5-24A shows a fully developed shear band. It is of interest to compare this fault zone with that generated in the same rock sheet directly attached to a rigid base. In the latter case (Fig. I.5-10B,C) the thrust fault is flatter and has a listric shape, because of the much higher shear stresses induced at the rear part of the base. The reader should be reminded, however, that the calculation assumed an isotropic and homogeneous shear strength. A subhorizontal layering and the associated anisotropy in elastic and strength properties would affect both the pre-faulting stresses in the rear part of the sheet and the orientation of the fault. Although we are not aware of numerical analyses addressing this problem, we would expect both effects to contribute to some flattening of the thrust faults (cf. Fig. I.2-10).

Comparing the field of σ_I-directions after the fault has been formed (Fig. I.5-24B) with the direction field before the formation (Fig. I.2-69C) we note a strong reorientation of σ_I: in a rear part, the direction of σ_I has been drastically changed, indicating a reversal in the direction of the basal shear stress, which implies a local reversal of displacement along the thrust base. This was caused by the elastic unloading during the formation of the thrust fault. However, the straightening of the σ_I-trajectories by the thrust fault is of more interest, since this could 'channel' the push from the rear further into the thrust sheet and may enable a new thrust to develop under quite similar stress conditions to those when the first one developed. This naturally requires that the first fault 'locks up', when the thrust block has risen sufficiently, and that the rear section has gained sufficient strength by thickening to withstand further deformation. The push would then be transmitted into the foreland of the first thrust fault, generating a second imbricate fault there and so on. Because of the large deformations involved, the process has not been modelled numerically for an elastic/frictional plastic material with strain softening.

Usually, a thrust sheet is bounded at its front by a thrust fault and the rocks in this zone are heavily fractured and their shear strength is therefore reduced. Consequently, the progress of *imbrication faulting* will come to a halt at a certain distance from the front thrust, since less push is required to mobilize the front thrust than to generate a new imbricate thrust in intact material. This has been quantified by Raleigh and Griggs (1963) in an elementary model, which is discussed in extenso in Ch. II.2 (Section 3, Example No. 7).

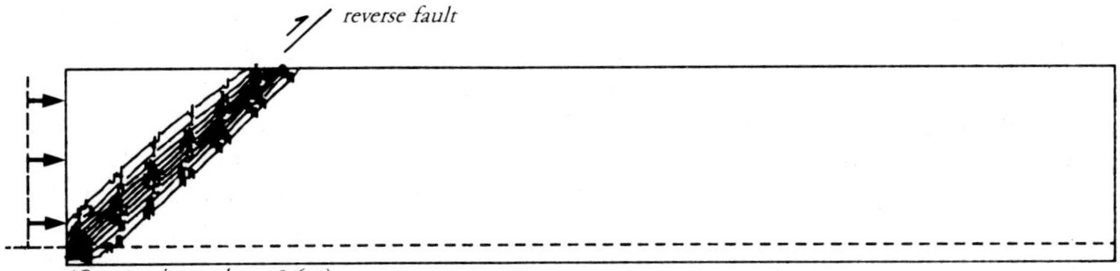

A Bundling of contours of magnitudes of incremental displacements at 22.5 m total end displacement indicates fully developed straight reverse fault.

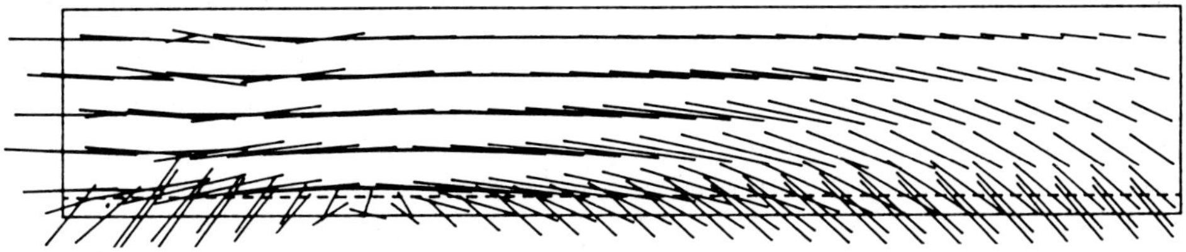

B Directions and relative magnitudes of σ_I after formation of the reverse fault. Compared with the direction field prior to faulting (Fig. I.2-69C) the σ_I-directions have been rotated into a horizontal position in the neighbourhood of the fault zone. Notice also the change in σ_I orientation at the base of the footwall of the fault, which indicates a reversal of basal shear stress caused by elastic rebound.

Fig. I.5-24 HORIZONTAL COMPRESSION OF RECTANGULAR ROCK BLOCK ON WEAK INTERLAYER (configuration and material properties as in Fig. I.2-69A). Elastic/frictional-plastic finite element analysis (after G.H. Mäkel & J.V. Walters, unpubl. work)

It should be clear that normally the stresses induced in a competent sheet and the process of imbrication will be the same whether the sheet is actively overriding its substrata or the sheet is at rest and *underthrusted* by its substrata. (See, for instance, the experiments by Malavieille, 1984.) A difference in faulting of the overlaying sheet may arise, however, when prior to underthrusting by ductile substrata and, hence, prior to the formation of a detachment plane, the ductile substrata are shortening under compression and imposing this pervasive shortening upon the overlying sheet. This may result in regularly spaced parallel thrust faults of the 'bookshelf' type as sketched in Fig. I.5-1B and discussed at the end of Section I.5.1. One may envisage that after its detachment further shortening of the underthrusted sheet will be accommodated by the activity of these faults.

The transmission of the push stresses through the thrust sheet may be greatly facilitated when the substratum behaves as a **viscous** material. Since, in contrast with plastic substrata, the shear resistance of a viscous layer depends on the rate of shearing (see Section II.7.2, Example No. 1), the stresses in the thrust sheet may change in time, even though the push force at the back end remains constant. In a simplified manner this dependence of the stress build-up on time is described by the *"stress diffusion" model,* which was applied by Bott and Dean (1973) to describe the transmission of collision stresses from plate boundaries into the interior of a plate. The model was adapted by Blay, Cosgrove and Summers (1977) to interpret model experiments of thrust sheet compression. Similar to the models in Fig. I.5-23A, the model considered consists of a rectangular elastic sheet. However, in their case it rests on a viscous interlayer. This is schematically shown in Fig. I.5-25A, where in a slightly more general representation, the interlayer is assumed to be a viscous fluid with some elastic properties (Maxwellian fluid; modelled by a viscous dashpot and an elastic spring in series). The addition of elastic properties not only adds an extra touch of reality but facilitates the verbal explanation of the model.

As explained in Figs. I.5-25B,C the application of a push force causes a rear segment of the sheet to shorten elastically, concomitantly with elastic extension and the onset of viscous flow in the adhering substratum. Initially, the resistance R of the deformed part of the substratum balances the push force P, but the creep relaxation of this resistance causes the transfer of shear load to neighbouring parts of the substratum, thereby allowing the overlying part of the thrust sheet to be put under higher stress. The movement of the transition zone (the 'stress front') between the stressed and unstressed part of the elastic sheet was demonstrated by Blay et al. (1977) by a photoelastic experiment in which a layer of thick "stiff" gelatine, representing the thrust sheet, lay upon a thin viscous layer and was loaded by a moving vertical wall. In other experiments the movement of the stress front was detected as the slowly propagating front of a slight surface relief. The observed travel times were interpreted in terms of the stress diffusion formula used by Bott and Dean (1973) (assuming a purely viscous substratum flowing parallel to the base of the overlying sheet at a velocity that decreased linearly with depth). The time T required for an elastic sheet of length L and thickness H to be stressed to within 90% of the constant end stress is [5)]

$$T = \frac{\eta}{E} \frac{L^2}{H \cdot h} \qquad (I.5\text{-}9)$$

where h is the thickness and η the viscosity of the substratum. For a low viscosity of 10^{20} poise and a Young's modulus E = 10^{11} dyne/cm^2 (= 10^4 MPa), a sheet length of 100 km and thickness H = 5 km, the 'passage time' of the stress front would be T (years) = 63500.h(km). This may illustrate the range of passage times that correspond with different thicknesses of the substratum. In spite of the extreme simplifications involved, the model of a travelling stress front may have important implications for the formation of a thrust belt or for the imbrication of a single thrust sheet. One may infer, for instance, that the moving stress front acts as a 'search' mechanism that locates weak 'spots', where it initiates thrust faults. When the push force at the rear of the sheet is applied in pulses, each pulse, when lasting at a constant level long enough in terms of eq. I.5-9, may be transmitted deep into the interior or to the front of a thrust sheet. When the duration of the push is too short, the effect will be restricted to a rear part of the sheet.

Tapered sheets. In nature, thrust sheets are commonly somewhat tapered. How would a slight taper affect the results obtained for sheets of rectangular vertical cross-sections? In a series of finite-element simulations Mäkel and Walters (1983-84; unpubl.) have shown that neither the detachment process nor the shape of thrust faults were affected by a slight taper, but that the first thrust did not necessarily occur at the rear of the pushed sheet. This was, of course, not unexpected, since the decreasing cross-sectional area tends to magnify σ_I in the thrust direction, while the subvertical σ_{III} decreases slightly. Hence, in the absence of basal shear resistance the differential stress would increase in the thrust direction and peak stress would be first attained at the front of the tapered sheet. The stress magnification is, however, opposed by the thrust-induced basal shear stress, and the first fault will be localized by the interplay between basal shear stress and the magnification effect of the taper.[6)] Since the effect of the shear stress decreases towards the surface, we may expect the faults to initiate at the surface.

When in the finite element simulation of Fig. I.2-69 the rectangular sheet was replaced by a sheet tapered at an angle of 2.5°, a pair of conjugate thrust faults developed from the surface and met the base at a distance of about 10% of the total sheet length from the back end. At this position both conjugates had a chance to develop, a situation quite similar to that at the end of a frictionless detachment zone. The result of a computer simulation of such a case is schematically shown in Fig. I.5-26. Both synthetic thrust and backthrust have been symmetrically developed from the base. Naturally, when the deformation continues and the synthetic thrust fault forms a ramp up which the thrust sheet moves, the existing backthrust will start acting as an accommodation fault, and new antithetic thrusts will form in sequence at the kink in the gliding path. The kinematics of this process was illustrated by the transparent model in Fig. I.2-41 ('toe' region). The successive formation of the antithetics at the foot of a ramp can easily be observed in sandbox experiments, and has been impressively demonstrated in high-pressure experiments on real rocks by Morse (1977). Small-scale field observations have been reported by Serra (1977).

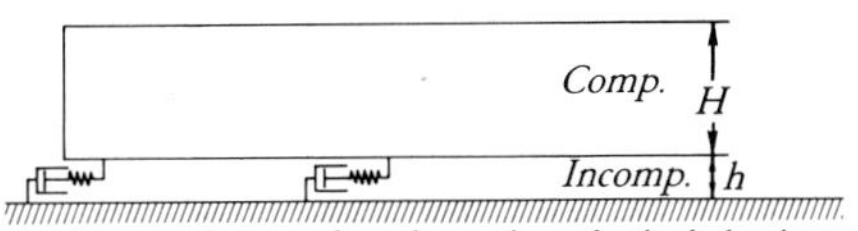

A State prior to thrusting; visco-elastic behaviour of substratum is modelled by elastic spring and viscous dashpot in series (Maxwell fluid).

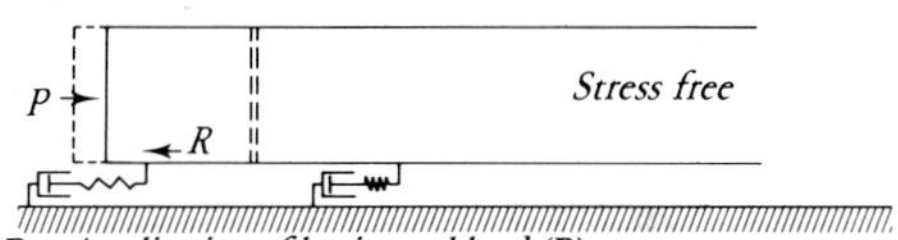

B Application of horizontal load (P) causes rear segment of rock sheet to shorten concomitantly with elastic extension and onset of viscous flow in the adhering substratum. Initial resistance R of the deformed part of the substratum balances the driving force.

C The relaxation of shear resistance, which accompanies the development of creep flow in the substratal segment, causes transfer of shear load to a neighbouring substratal segment and extension of the stressed region in the competet rock sheet

Fig. I.5-25
MOVING STRESS FRONT IN A COMPETENT ROCK SHEET ON A VISCO-ELASTIC SUBSTRATUM

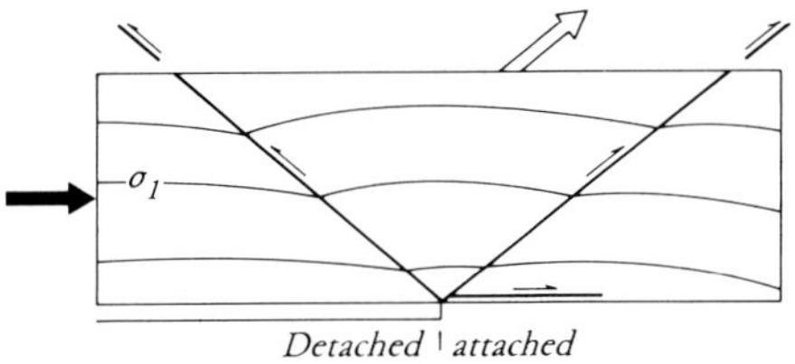

Rectangular block partially detached from a rigid substratum

Fig. I.5-26
CONJUGATE THRUST FAULTS DEVELOP FROM THE END OF THE DETACHMENT ZONE (schematised result of finite element simulation by Mäkel and Walters, unpubl. work)

NOTES

1) The solution I.5-6,7 also contains Prandtl's well-known solution to the *'squeezing'* problem (see, for example, Hill, 1960), in which rough parallel walls are either forced together or driven apart, either squeezing plastic material out or allowing it to enter. Prandtl's solution is obtained from eqs.

I.5-6 by imposing $\tau_2 = -\tau_1 = K(\pm H)$, which implies, with reference to the sign convention used in Fig. I.5-11, that *both* externally applied shear stresses act in the *same* direction upon the plastic strip – opposing the squeeze flow of plastic material. According to eq. I.5-7, the two plates must then be at a well-defined distance $2H = 2K/a$ to allow plastic squeezing.

This raises the question of whether plastic substratal material could be made to 'slip out' from underneath a thrust sheet. The answer depends on the magnitude of the pressure gradient $a = d\sigma_y/dx > 0$ which, as we have stated, is necessary to balance the moment of the push force that acts upon the thrust sheet. From a simple balance of the moments of the push force, basal friction and frontal buttressing that act on a detached thrust sheet, it follows that the value of 'a' is in the order of a few bar/km (Mandl and Shippam, 1981). According to the above formula for 2H the thickness required for the substratum to be squeezed out by slippage underneath a thrust sheet would, even for an extremely weak material, amount to tens of kilometres. Therefore, we exclude the possibility of *large-scale* thinning of the plastic substratum by its extrusion from underneath an overthrust. Such thinning by slippage may occur, however, on a *local* scale at places where the pressure gradient 'a' attains very high values, for instance as a result of local thickening or 'stepping-up' of the thrust sheet.

In Prandtl's solution the shear stress τ, which varies linearly with depth, decreases to zero halfway between the two boundary plates and then changes its sign and increases (see Fig. 4 in Mandl and Shippam, 1981). The plastic slip solution to our problem is therefore obtained directly from Prandtl's solution by placing the lower boundary inside the upper half of the Prandtl strip, i.e. at a distance from the upper plate that is smaller than K/a.

Thus, in a sense, our solution is obtained by 'tailoring' one of the few known analytical solutions of plasticity theory to suit a geological problem – the same procedure as applied to the 'wedge' problem.

2) It should be noted that these "duplex" structures with 'hinterland' dipping (Boyer and Elliott, 1982) cross faults are not necessarily associated with the passive state in the plastic substratum. Consider the steep antithetic slip lines in the slip line field of the active plastic state in Fig. I.5-13A. Once, slips of this type are induced they could accommodate the shearing between two horizontal gliding horizons (the sole and roof thrust) in 'bookshelf' fashion 'domino'-style. As discussed previously (Section I.2.6), the rotation of antithetic slip planes in 'domino' style requires that σ_x is considerably smaller than σ_y, which is, of course, incompatible with a passive limit state, and requires that the lengthening of the "duplex" is somehow accommodated. In contrast with the P-shears, the displacement along these faults would be antithetic to the direction of overthrusting. Unfortunately, because of the small displacement associated with such cross faults (or shear joints), it is often very difficult to define the slip direction on outcrops.

3) An analytical solution to Prandtl's squeeze problem (see note 1)) for a Coulomb-plastic material has so far only be obtained (Marshall, 1967) for a noncohesive frictional material with constant friction angle. The stress solution can be 'tailored' to suit our simple-shearing problem by prescribing along the upper boundary of the frictional plastic strip a constant ratio τ/σ_y equal to the ratio of the net push force to the total weight of the thrust sheet. (The constancy of this stress ratio along the boundary follows from the requirement that in the 'infinite' plastic strip the directions of the principal stresses do not vary in the x-direction.)

It may suffice here to mention that the slip lines are again isogonal cycloids, similar to those in Fig. I.5-12, but now intersecting each other at the acute angle $90° - \varphi$. Furthermore, the vertical stress σ_y no longer increases linearly in the thrust direction but *exponentially*, and so do the other stresses. Accordingly, the decrease of $|\tau|$ with depth inside the frictional plastic layer is also stronger than that in the purely cohesive plastic material.

4) Note that the stress solutions to an elastic problem for which the boundary conditions are exclusively formulated in terms of stresses (no displacements prescribed) remain valid even if the elastic response of the material is delayed by a component of Newtonian viscosity. The analogue model of such a Voigt-Kelvin material is represented by an elastic spring and a dashpot in parallel.

5) This follows from a classic 'diffusion equation' for the horizontal displacement movement u(x,t) of the sheet, and is derived by balancing the horizontal gradient of the push force with the basal shear

stress. According to Hooke's law the push stress is proportional to the horizontal displacement gradient, while the basal shear stress is proportional to the momentary vertical gradient of the horizontal flow velocity. Although the flow rates in the substratum have to build up from zero, the transient phase may be considered to be negligible (see Note 4) in Ch. II.7), and the basal shear stress may be associated with stationary flow profiles. The shear stress is then proportional to the local rate of sheet movement divided by the thickness (h) of the substratum.

The equation describing the displacement history of the sheet under a constant push stress then becomes:

$$K \cdot \partial^2 u / \partial x^2 = \partial u / \partial t$$

where the 'diffusivity' $K = E \cdot H \cdot h / \eta$

6) The competition between the stress magnification by tapering and the basal shear stress in localizing thrust faults may be quantified by considering the balance of horizontal force components that act on a thrust sheet with a horizontal base and a planar surface sloping at the angle α. Static equilibrium of any segment bounded by vertical cross-sections of the sheet requires that the horizontal net force on the vertical boundaries is balanced by the total shear force at the base. By expressing these forces in terms of the average value $\bar{\sigma}_{\|}$ of the horizontal normal stress on a vertical cross-section and the average $\bar{\tau}_b$ of the basal shear stress, one finds that an *increase* of $\bar{\sigma}_{\|}$ towards the thinner frontal part of the sheet is only possible when

$$\bar{\sigma}_{\|} \tan\alpha > \bar{\tau}_b$$

This condition must be satisfied, in particular, on vertical sections at the front and rear of the sheet. This implies that in a tapered sheet with a *stress-free front* stress magnification is completely suppressed by basal friction. In other words, magnification of the horizontal normal stress requires a horizontal *buttressing stress* at the front of the tapered sheet satisfying the above inequality.

When movement of the thrust sheet front is blocked, a horizontal buttressing stress may build up to the required magnitude, provided the concomitantly increasing differential stress has not first reached its limit value at the rear of the sheet, since otherwise thrust faulting would again start at the rear and not in the interior of the tapered sheet.

A thrust sheet may also be only partly tapered, the tapered part pushing ahead a part of uniform thickness. When the frictional resistance at the base of the frontal part increases, the above inequality may become satisfied.

Part II

BASIC CONCEPTS

II.1. STRESS

When a body is deformed the arrangement of its molecules is changed. This gives rise to extra internal forces. Since these forces are the result of changes in molecular interaction, they have an extremely short range of action. Hence, the range of the deformation-induced internal (intermolecular) forces is taken as zero in the mechanical theory of deformable bodies in which the molecular structure is disregarded and the body is considered to be a material continuum – or an assemblage of continua (as in 'jointed rocks'). (A 'very small' distance in the continuum is to be considered large in comparison with molecular distances.) It follows that: (1) the mechanical forces exerted on any part of a material continuum by surrounding parts act only on the surface of that part (Fig. II.1-1), and (2) internal surface forces and deformations are interdependent.

1. DEFINITION AND UNITS

The total force acting across a real or imaginary surface element of unit area is called 'stress'. Being a mixture of a physical ('force') and a geometrical ('surface element') quantity, the stress concept takes a key position in the mechanics of deformable bodies. It is indispensable in the determination of the deformation of a body under load – the primary objective of continuum mechanics.

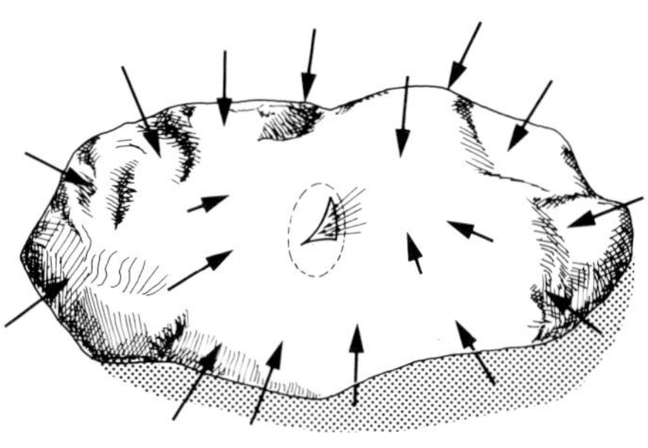

Fig. II.1-1
STRESSES INSIDE DEFORMABLE BODY UNDER LOAD

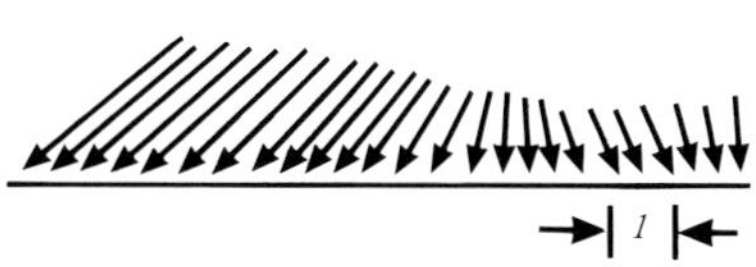

Fig. II.1-2
SIZE OF UNIT REFERENCE AREA

In defining 'stress' we must ensure that the actual size of a unit reference area is chosen correctly. If the forces transmitted through an outer or inner surface are everywhere the same, the size of the reference element does not matter, since the stress is not affected by the size chosen for the reference element. When, however, as is usually the case, the forces vary along a surface (Fig. II.1-2), the size of the unit element has to be chosen small enough to ensure an approximately uniform force distribution on the element.

According to its definition stress has the physical dimensions of force per area. Hence, the magnitude of stress is expressed in a variety of units, e.g. 1 bar = 10^6 dynes/cm^2 = 10^5 N(ewton)/m^2 = 10^2 k(ilo)Pa = 10^{-1} M(ega)Pa = 14.5 psi. (1 Pa(scal) = 1 N/m^2.) The unit 1 MPa is useful, for example, in calculating overburden pressures. (The average overburden pressure (lithostatic stress) at the base of a rock column of 1 km depth and 2.5 gm/cm^3 average density is 25 MPa.)

2. *STRESS VECTORS AND TENSORS*

Force is a vectorial quantity, i.e. a quantity characterised by direction and magnitude (intensity) – symbolised by an arrow in space (for more details, see Fig. II.1-3). Likewise, stresses that act on a given surface element are vectorial quantities and can be added as vectors. This is demonstrated in Fig. II.1-4 by superposition of a normal and tangential stress on *one and the same* surface element. The figure also illustrates how an arbitrary stress vector acting on a surface element can be resolved in a tangential ('shear') and a normal component vector. Similarly, different stresses applied to the surface of a volume element (Fig. II.1-5) can be combined vectorially into a single state of stress.

Opposite stresses. The stress exerted upon a surface element by the material on one side is counteracted by the stress exerted upon the same element by the material on the other side (Fig. II.1-6). According to Newton's law of action and reaction, the two stresses are opposite and equal in magnitude. To distinguish between the two stresses one has obviously first to define the front and back of the surface element. This is done by stating that the vector of the unit normal $\vec{n}$ of the surface element (Fig. II.1-6) is pointing from back to front. When the surface element is part of a closed surface its front is selected to coincide with the outer side, i.e. the unit normal is pointing outside ('outer normal'). The stress

A vector is an invariant object and as such independent of any frame of reference (coordinate system). To specify the vector numerically, however, a frame of reference is required; for instance, the orthogonal rectilinear (i.e. Cartesian) frame shown in this figure. With respect to such a frame of reference a vector $\vec{t}$ may be characterised

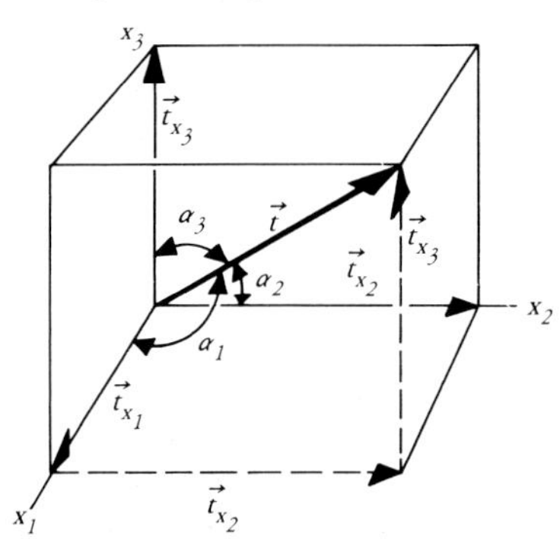

a) by its magnitude (length) $|\vec{t}|$ and two directional angles (the third angle being determined by the relation $\cos^2\alpha_1 + \cos^2\alpha_2 + \cos^2\alpha_3 = 1$)

or

b) by the magnitudes of the orthogonal projections $|\vec{t}_{x_i}|$ on the positive axes of the reference frame ('components' of the vector). (If $\alpha_i > \pi/2$ the corresponding projection has to be counted as negative.)

The two representations are related by $\cos\alpha_i = |\vec{t}_{x_i}| / |\vec{t}|$ (i = 1, 2, 3).

The vector $\vec{t}$ is also the sum of its three component vectors:

$$\vec{t} = \vec{t}_{x_1} + \vec{t}_{x_2} + \vec{t}_{x_3}$$

Fig. II.1-3 REPRESENTATION OF A VECTOR IN CARTESIAN REFERENCE FRAME

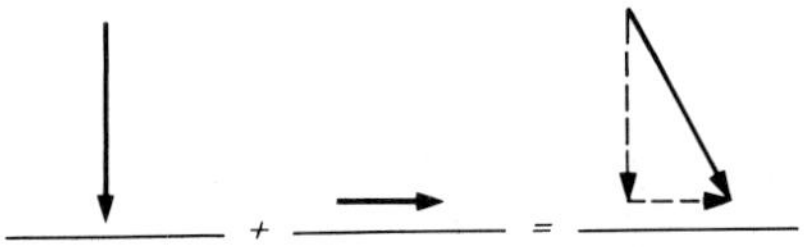

Fig. II.1-4
ADDITION OF STRESS VECTORS ACTING ON THE SAME ELEMENT

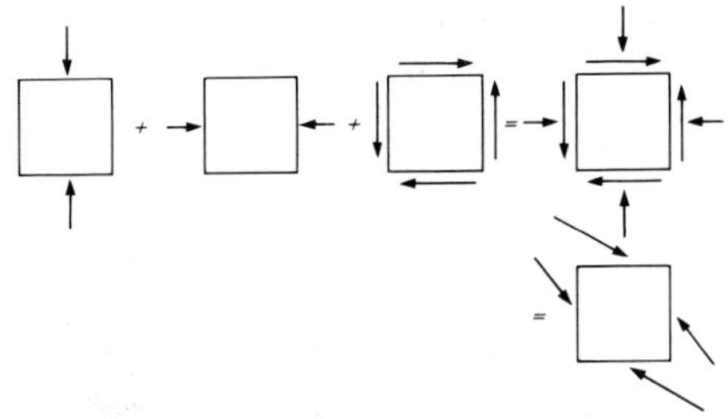

Fig. II.1-5
ADDITION OF STRESSES ACTING ON THE SAME VOLUME ELEMENT

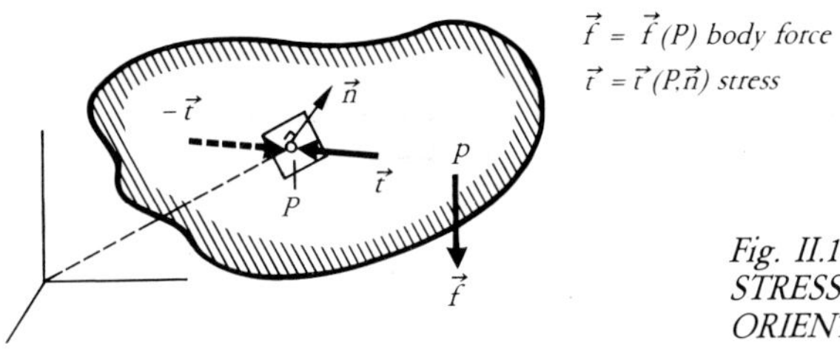

Fig. II.1-6
STRESS DEPENDS ON POSITION ***AND*** *ORIENTATION OF SURFACE ELEMENT*

exerted upon the surface element by the front material will be denoted by $\vec{t}$ and accordingly the stress exerted by the back material by $-\vec{t}$.

Surface and body forces. Before we continue with the discussion of the stress concept, it should be recalled that any part of a material body experiences not only surface forces but also body (or volume) forces. Forces of this type are the gravitational, centrifugal, electric and magnetic forces. The drag force a fluid flowing through a porous rock exerts upon the rock may also be considered a body force. In contrast with surface forces, the body forces act directly upon the particles within a body and are proportional to the volume of the element considered. The body forces are therefore measured in units of force/unit volume.

Although body forces and stresses are both vectorial quantities, there is an important difference between them, which makes stress the more complex quantity to work with. A body force $\vec{f}$ acts upon any 'point' of a material continuum (Fig. II.1-6). It may vary with position P and may therefore be considered a function of position ('force field'): $\vec{f} = \vec{f}(P)$. The *stress vector* will, in general, also vary with position, but in addition it *will vary with the orientation of the surface element* (specified by unit normal $\vec{n}$) it is acting upon. Hence, $\vec{t} = \vec{t}(P, \vec{n})$. A simple example may illustrate this essential point.

Example No. 1: Variation of stress vector

An homogeneous beam of cross-section A is loaded by an axial force $\vec{F}$ as shown in Fig. II.1-7A. The force is uniformly distributed over the ends of the beam and, obviously, over any internal cross-section parallel to the ends. The stress on any of these cross-sections is therefore $\vec{F}/A$. The force $\vec{F}$ also acts, uniformly distributed, on any inclined cross-section and since the area of an inclined section is larger than that of a vertical one, the stress on the inclined section will have a smaller magnitude than the stress on a vertical cross-section.

In this simple case of uniform uniaxial loading the direction of the stress vector remains the same for any surface element and, having prescribed the stress vector that acts upon the vertical elements, we could easily calculate the stress on any other element. In more general cases not only the magnitude but also the direction of the stress vector changes with the orientation of the surface element and knowing the stress vector on one surface element does not enable one to calculate the stress that acts on a differently oriented element. This can be immediately seen by imposing on the uniaxially loaded beam of Fig. II.1-7A a second uniformly distributed load system $\vec{F}'$ (Fig. II.1-7B). While the stress vectors acting on vertical sections are still horizontal and the same as in Fig. II.1-7A, the stress vectors on horizontal sections are vertical and the two stress vectors are completely unrelated. In other words, prescribing *the stress vector that acts at a point P upon a small surface element of given orientation* ($\vec{n}$) *is not sufficient to define the local state of stress in the material.* This has sometimes been overlooked in the mechanical approach to tectonic problems. To emphasise this point, we present a further example.

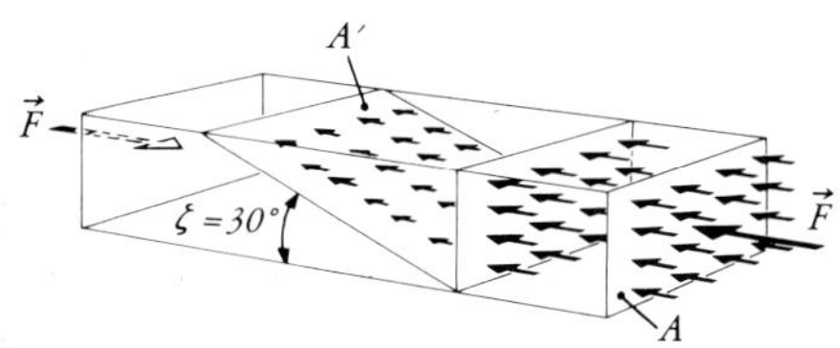

A. Uniform uni-axial loading

$A' = A/\sin\xi = 2A$ for $\xi = 30°$

As the same total force $\vec{F}$ acts on vertical and inclined cross-sections, the stress vector acting on the inclined section shown ($\xi = 30°$) has only half the magnitude of the stress acting on vertical sections.

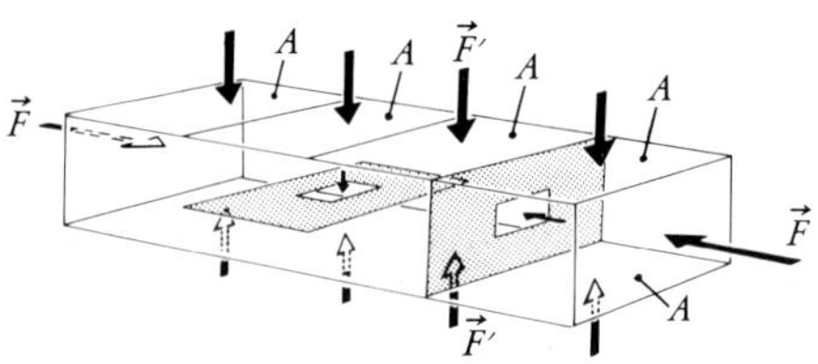

B. Uniform bi-axial loading

The stress vectors acting on vertical and horizontal sections are unrelated and differ in direction and magnitude.

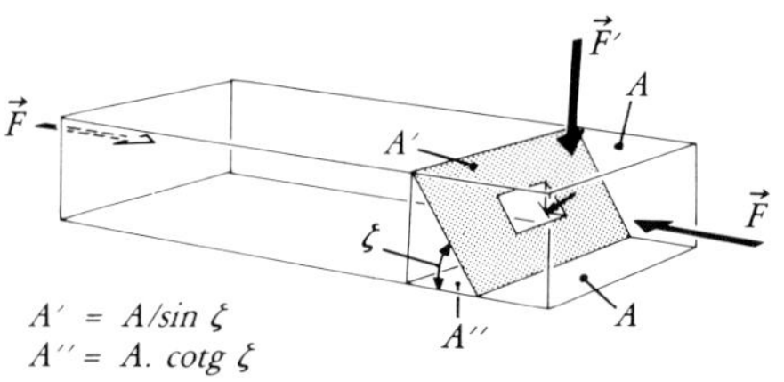

C. The stress vector acting on an oblique element is determined by the stress vectors that act on mutually perpendicular elements.

The horizontal load $\vec{F}$ distributed over area A contributes the horizontal stress $\vec{F}/A'$ to the total stress acting on the oblique section, while the vertical load $\vec{F}'$ distributed over the area A contributes the vertical stress $(\vec{F}'/A)(A''/A')$.

The resultant stress vector that acts upon the inclined element is therefore the vector sum

$$\vec{F}/A' + \vec{F}'A''/A\,A' = (\vec{F}/A)\sin\xi + (\vec{F}'/A)\cos\xi$$

where $\vec{F}/A$ and $\vec{F}'/A$ are the stress vectors that act on the vertical and horizontal elements, respectively.

FIG. II.1-7 STRESS VECTORS IN LOADED BEAM

Example No. 2: State of stress at a fault plane

Given is a fault plane, the dip-parallel trace of which is shown in Fig. II.1-8, together with the stress vector $\vec{t}$ that acts upon the fault plane and lies in the plane including dip direction and fault normal. With this stress vector, the magnitude $\sigma_\perp$ of the normal stress that presses the two fault blocks together and the magnitude τ and direction of the shear stress that tends to make the fault blocks slide on top of each other are also known. In general this is not true of the stress $\vec{\sigma}_\parallel$ that acts parallel to the fault plane upon surface elements perpendicular to the fault plane. This stress is often disregarded in tectonic fault analysis, although it causes fault-parallel compaction and buckling, and plays a decisive role in the generation of subsidiary faults that branch off the main fault. From the preceding discussion, it should be clear that only in the exceptional case of strictly uniaxial loading (Fig. II.1-7A) will the fault parallel stress $\vec{\sigma}_\parallel$ be determined by the 'fault stress' $\vec{t}$. One should keep this in mind, for instance, when analysing the changes in stress that are caused by a drop in shear strength along an active fault plane ('plane of weakness').

As the foregoing examples demonstrate, the definition of the local stress requires more than prescribing the stress vector that acts upon a single surface element. It is therefore the first problem in stress theory to establish quantities that completely describe the local state of stress.

Stress vectors on mutually perpendicular planar elements. To see how this may be accomplished, we return to the example of a beam (Fig. II.1-7B) that is loaded by an axial force and by a second force system acting perpendicular to the axis of the beam. Next, we consider an arbitrarily inclined cross-section (Fig. II.1-7C) of the beam and we see that each of the two load systems imposes a uniformly distributed stress on this cross-section. Simple trigonometry shows that the resultant stress vector that acts upon an element of the inclined cross-section is uniquely determined as a linear combination of the stress vectors that act on vertical and horizontal cross-sections, respectively (cf. Fig. II.1-7C).

Fundamental stress theorem. It is not difficult to abandon the restrictions on the stress directions in this example and to arrive at the following fundamental theorem: 'Any stress vector $\vec{t}$ that acts upon a surface element at a point P is determined by the stress vectors $\vec{t}_1$, $\vec{t}_2$, $\vec{t}_3$ that act upon mutually perpendicular planes at P.' Let the mutually perpendicular planes intersect along the axes of a local Cartesian coordinate system at P to which we may refer the direction angles α_1, α_2, α_3 of the normal to the arbitrarily oriented element at P (Fig. II.1-9). The fundamental theorem then takes the simple form

$$\vec{t} = \vec{t}_1 \cos\alpha_1 + \vec{t}_2 \cos\alpha_2 + \vec{t}_3 \cos\alpha_3 \qquad \text{(II.1-1)}$$

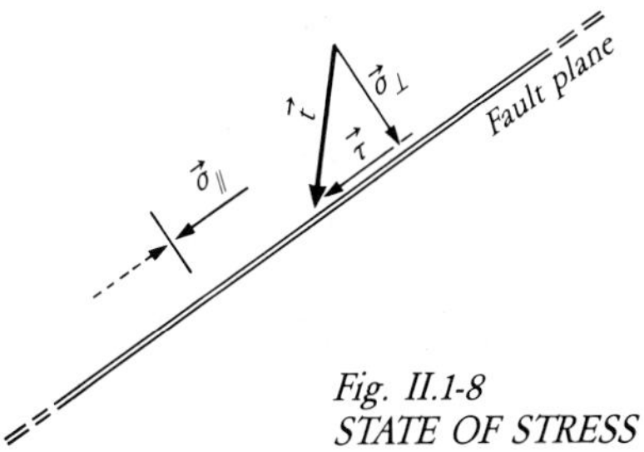

Fig. II.1-8
STATE OF STRESS IN THE VICINITY OF FAULT PLANE

In general, the perpendicular planes at any point P will be taken as coinciding with the coordinate planes of a single reference frame which is used for the description of the whole deformed body. The 'auxiliary' stresses $\vec{t}_1$, $\vec{t}_2$, $\vec{t}_3$, are then associated with planar elements of known orientation and, therefore, represent three vectorial stress 'fields' $\vec{t}_i = \vec{t}_i\,(P)$.

Plane Problems. In many tectonic problems one of the three 'auxiliary' stresses $\vec{t}_1$, $\vec{t}_2$, $\vec{t}_3$ vanishes or may be neglected. This is obviously the case when one of the three coordinate planes is stress-free (e.g. the x_2, x_3-plane in Fig. II.1-9). We shall see later that this implies that the remaining vectors lie in one plane. We encounter this state of 'plane stress' near any free surface of a rock body. More important in tectonic problems is the case of plane deformation ('plane strain') in which, by definition, the displacement of material 'points' takes place only within parallel planes (e.g. planes parallel to the x_2, x_3-plane in Fig. II.1-9) and, moreover, does not vary in the direction perpendicular to these planes. The latter implies that between planes of the set considered no relative displacements take place and, consequently, no tangential stresses will act between these planes (this holds at least for isotropic materials). Therefore, the stress vector acting upon the plane of deformation (for example, $\vec{t}_1$ in Fig. II.1-9) will be oriented perpendicular to that plane and will not contribute to the deformation. The stress vectors that do contribute are those acting upon planar elements with normals lying in the plane of deformation. Since one of the directional cosines of these normals vanishes (e.g. $\cos\alpha_1 = 0$), the stress vectors that are responsible for the plane deformation are determined – according to equation II.1-1 – by only two 'auxiliary' stresses ($\vec{t}_2$, $\vec{t}_3$). Again, these stresses lie in the plane of deformation, as will be shown in Section II.2.1 ('Plane problems'), but this may be inferred on intuitive grounds.

In tectonics, plane deformation may be approximately realised in areas where fault and fold structures show little variation over relatively large distances along strike (see Section II.5.3, 'Plane strain').

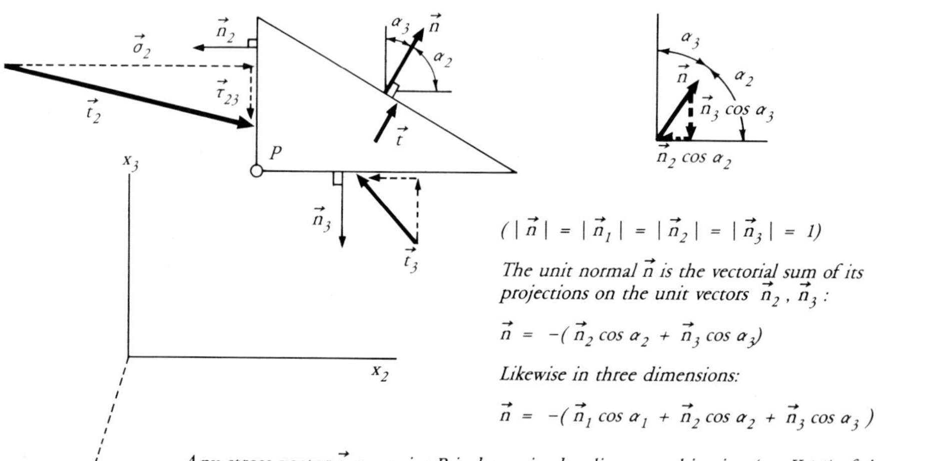

Any stress vector $\vec{t}$ at a point P is determined as linear combination (eq. II.1-1) of the stress vectors $\vec{t}_1$, $\vec{t}_2$, $\vec{t}_3$, that act across mutually perpendicular planes at P.

Fig. II.1-9 THE FUNDAMENTAL THEOREM OF STRESS

Description of State of Stress by Stress Components. Naturally, for actual calculations the stress vectors in the vector relationship II.1-1 have to be specified numerically. For a vector in three-dimensional space this requires three parameters. If all vectors considered lie in one plane, only two parameters are needed for each vector. We specify the vectors by their projections upon the axes of one and the same rectangular Cartesian frame of reference (see Fig. II.1-3). As shown in Fig. II.1-9, the frame is chosen parallel to the intersections of the three mutually perpendicular elements in P. Hence, each of the vectors $\vec{t}_i$ that act upon the intersecting elements at P is characterised by the magnitude of its projection upon the normal to the corresponding surface element and by the magnitude of two tangential projections or one, as the case may be, parallel to coordinate axes.

Notations. To distinguish between normal and tangential components, the letter σ is used to denote the normal stress components and the letter τ to denote the tangential (shear) components. A subscript designates the coordinate axis to which the normal of the particular plane element is parallel. For tangential stresses, a second subscript indicates the coordinate axis to which the stress component is parallel. Thus, the stress vector $\vec{t}_2$, for instance, which acts on a plane element with normal parallel to the x_2-axis (see Fig. II.1-9) will be characterised by the triplet of real measure numbers σ_2, τ_{21}, τ_{23}. Accordingly, the fundamental stress theorem (eq. II.1-1) implies that the numerical characterisation of the state of stress at a point P, i.e. the description of any stress vector acting on a surface element at P, requires *nine* stress components (four in the plane case). It will be shown in Section II.2.1 ('Symmetry of shear stresses') that the number of components reduces to *six* (3 in the plane case) because each shear stress has an equal counterpart

$$\tau_{21} = \tau_{12},\ \tau_{31} = \tau_{13},\ \tau_{32} = \tau_{23}\ . \qquad \text{(II.1-2)}$$

By projecting the vectors in equation (II.1-1) successively onto the three axes of the Cartesian frame, the vector equation II.1-1 splits into three 'componental' equations:

$$t_{x_1} = \sigma_1 \cos\alpha_1 + \tau_{21}\cos\alpha_2 + \tau_{31}\cos\alpha_3$$

$$t_{x_2} = \tau_{12}\cos\alpha_1 + \sigma_2\cos\alpha_2 + \tau_{32}\cos\alpha_3$$

$$t_{x_3} = \tau_{13}\cos\alpha_1 + \tau_{23}\cos\alpha_2 + \sigma_3\cos\alpha_3 \qquad \text{(II.1-1a)}$$

Stress Tensor. At this point it should be emphasised that the numerical representation of a vector always depends on the choice of the frame of reference (see. Fig. II.1-10), while

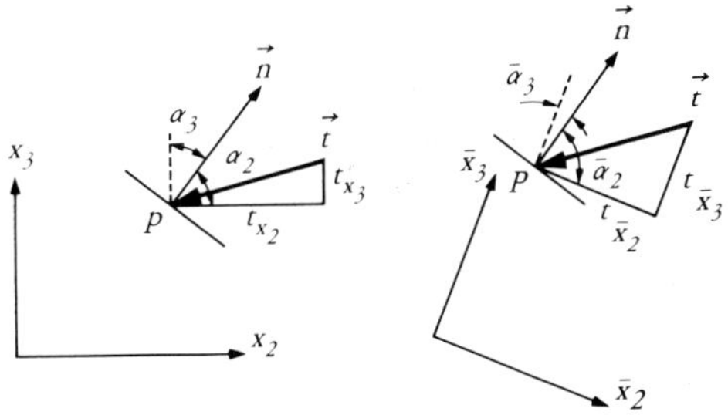

Fig. II.1-10
COMPONENTS OF STRESS VECTOR ON FIXED SURFACE ELEMENT (PLANE CASE)

the vector itself – like its geometrical image (the arrow) – is an invariant object and thus completely independent of any frame of reference. Analogously, the nine quantities σ_1, τ_{21}, ... σ_3, which are functions of position and characterise the state of stress at P, depend on the choice of the system of coordinate axes along which three mutually perpendicular surface elements intersect at P. In a new frame of reference the nine quantities attain new values at the old position P. The changes induced by a change of the reference frame are governed by well-defined 'transformation' rules. Actually, these rules, which will not be discussed here, identify the nine quantities σ_1, τ_{21}, ... as a numerical representation of an invariant object that is of higher order than a vector and is called a 'tensor' of second order. Hence, the nine quantities that characterise the local state of stress are usually referred to as the components of the 'stress tensor'. (The reader who wants to familiarise himself with the transformation formalism for vectors and tensors may find basic presentations in the literature cited at the end of this Chapter.)

3. INVARIANT PROPERTIES

Stress invariants. The study of the transformation of the stress components under a change of the reference frame reveals many interesting properties of these components. Two of these properties are of particular importance and will be stated here without proof. Although the stress components change by alteration of the reference frame, there are three combinations of stress components that remain unchanged ('stress invariants'). The first is the *mean normal stress*

$$p = (\sigma_1 + \sigma_2 + \sigma_3)/3 \qquad \text{(II.1-3a)}$$

which becomes equal to $p = (\sigma_{1'} + \sigma_{2'} + \sigma_{3'})/3$ when referred to an x_1', x_2', x_3'-frame. Further invariants are the expressions

$$\sigma_2\,\sigma_3 + \sigma_3\,\sigma_1 + \sigma_1\,\sigma_2 - \tau_{23}^2 - \tau_{31}^2 - \tau_{12}^2 \qquad \text{(II.1-3b)}$$

and

$$\sigma_1\,\sigma_2\,\sigma_3 + 2\tau_{23}\,\tau_{31}\,\tau_{12} - \sigma_1\,\tau_{23}^2 - \sigma_2\,\tau_{31}^2 - \sigma_3\,\tau_{12}^2 \qquad \text{(II.1-3c)}$$

In the case of plane stress (coordinates x_1, x_3) the invariants reduce to:

$$(\sigma_1 + \sigma_3)/2 \text{ and } \sigma_1\,\sigma_3 - \tau_{13}^2 \qquad \text{(II.1-3d)}$$

Principal Stresses. The second, important peculiarity of the stress components that shows up when the Cartesian frame of reference is rotated into different positions is described by the concept of 'principal stresses'. If we return to the simple example of uniaxial loading in Fig. II.1-7A, we notice immediately the existence of two perpendicular planar elements on which the stress vector has no tangential (shear) components. Moreover, on these elements the normal stress is maximal (or minimal), when compared with the normal components of the stress vectors on any other element. The tangential stress component, however, attains maximum magnitude on the two planes that bisect the angles between these two elements. This can be verified in Fig. II.1-7A by simple trigonometry. A similar result may be derived for a body subject to an arbitrary load system: there are always – at any point P of a deformable body under load – three mutually perpendicular planar elements upon which the stress vector is purely normal and, when differing in magnitude, the normal stress attains extreme values on two of these elements, and an intermediate value on the third element. These normal stresses are called 'principal' stresses and their values will be denoted by σ_I, σ_{II}, σ_{III} (where $\sigma_I \geqslant \sigma_{II} \geqslant \sigma_{III}$), or occassionally by σ_{Max}, σ_{Int}, σ_{Min}.

The directions of the principal stress vectors are called 'principal stress axes' or 'principal directions'.

Maximum Shear Stress. On the planar elements that bisect the right angles between the principal axes the shear stress attains the principal values

$$\tau_I = (\sigma_{II} - \sigma_{III})/2, \ \tau_{II} = (\sigma_I - \sigma_{III})/2, \ \tau_{III} = (\sigma_I - \sigma_{II})/2 \qquad \text{(II.1-4)}$$

The greatest shear stress is τ_{II}, which acts across a planar element the normal of which bisects the angle between the direction of highest (σ_I) and lowest (σ_{III}) principal stress (see also Section II.4.2, Fig. II.4-4A). The normal stresses that act on these shear planes are, respectively,

$$(\sigma_{II} + \sigma_{III})/2, \ (\sigma_{III} + \sigma_I)/2, \ (\sigma_I + \sigma_{II})/2 \qquad \text{(II.1-5)}$$

Principal axes as reference frame. When the principal directions are known, the discussion of the state of stress at a point P of a deformable body is greatly simplified by choosing the axes x_1, x_2, x_3 of the reference frame parallel to the principal stress axes. Since elements perpendicular to these axes are free of shear stresses, all off-diagonal (shear stress) terms in equation II.1-1a vanish and the stress vector $\vec{t}$ that acts on an oblique element at P is characterised by the components $t_{x_1} = \sigma_I \cos\alpha_1$, $t_{x_2} = \sigma_{II} \cos\alpha_2$, $t_{x_3} = \sigma_{III} \cos\alpha_3$, where α_1, α_2, α_3, are the direction angles of the normal of the element with respect to the principal axes. Likewise, the stress invariants (eqs. II.1-3) are simplified by the vanishing of all shear components.

Hydrostatic state. In the special case when all principal stress values are equal, $\sigma_I = \sigma_{II} = \sigma_{III} = \bar{\sigma}$, the state of stress is called 'hydrostatic' in accordance with the fact that a fluid at rest can only sustain normal stresses ('pressures'). Indeed, the equality of principal stress values implies that the stress acting on an element of arbitrary orientation is a purely normal stress of magnitude $\bar{\sigma}$.

The proof is simple and follows directly from the fundamental stress theorem (II.1-1). With reference to Fig. II.1-9 we assume that the coordinate axes are principal axes and the stresses $\vec{t}_1$, $\vec{t}_1$, $\vec{t}_3$ are therefore purely normal stresses of magnitude $\bar{\sigma}$ ('hydrostatic' state), i.e. $\vec{t}_1 = -\bar{\sigma}\vec{n}_1$, $\vec{t}_1 = -\bar{\sigma}\vec{n}_2$, $\vec{t}_3 = -\bar{\sigma}\vec{n}_3$. Substituting these expressions in equation II.1-1, and making use of the expression derived in Fig. II.1-9 for the unit normal $\vec{n}$ of an oblique element, we obtain $\vec{t} = \bar{\sigma}\vec{n}$. This means that the stress $\vec{t}$ acting on any element is directed parallel to the normal of that element and has the magnitude $\bar{\sigma}$.

Stress deviator. It is often useful to consider the deviation of the actual state of stress from the hydrostatic state. This stress deviation is obtained by subtracting the mean normal stress $p = (\sigma_1 + \sigma_2 + \sigma_3)/3$ ('mean pressure') from the normal stress components. Denoting the stress components of the deviatory state ('stress deviator') by the letter s, we obtain for the normal components

$$s_1 = \sigma_1 - p, \ s_2 = \sigma_2 - p, \ s_3 = \sigma_3 - p \qquad \text{(II.1-6)}$$

while the deviatory shear stresses are identical with the total shear stresses ($s_{ij} = \tau_{ij}$).

The principal axes of the deviatory stresses are the same as those of the total stresses. Furthermore, one may immediately notice that the mean normal deviatory stress vanishes ($s_1 + s_2 + s_3 = 0$). The mechanical significance of the deviatory state of stress is that it represents a state of pure shearing stress. We mention without proof that one can always find at any point P three mutually perpendicular elements on which the normal deviator stresses s_1, s_2, s_3 vanish. This means that at any point P of a deformable body we can define a small cubical volume element such that its sides are free of normal deviatory stresses and only subject to shearing stresses. This will be illustrated by the following example.

Example No. 3: Decomposition of plane stress

We consider the case of plane stress. As already discussed ('plane problems'), plane stress exists at any free planar surface of a rock mass. The absence of shear stresses on a free surface implies that one principal direction is perpendicular to the surface, while the remaining two principal directions lie in the planar surface. A slightly more general situation is depicted in Fig. II.1-11. A heavy rock plate, supported by a horizontal viscous substratum, is compressed by horizontal tectonic stresses σ_1 and σ_3. The top of the plate is free of stress and the base is acted upon by vertical normal stresses only, since a fluid at rest cannot sustain any shear stresses. If we now separate the overburden-induced vertical stress σ_2 from the total stress state (Fig. II.1-11), we are left with a plane state of stress on any horizontal section of the plate.

In Fig. II.1-12 the x_1, x_3-plane represents the stress plane. Consequently, no stresses act across that plane and the in-plane stresses act upon elements that intersect the plane. The quadrilateral in Fig. II.1-12 is the cross-section of a prism extending parallel to the x_2-axis. The coordinate axes have been chosen parallel to the principal stress axes and therefore the maximum shear stress $\tau = (\sigma_I - \sigma_{III})/2$ acts upon the two elements that bisect the angles between the principal directions. The normal stress on those elements is $(\sigma_I + \sigma_{III})/2$ (cf. 'maximum shear stress') and therefore also equal to the mean normal stress $p = (\sigma_I + \sigma_{III})/2$ in the plane. Therefore, when this mean pressure is separated from the general state of plane stress (Fig. II.1-12C), one is left with a plane deviatory state (Fig. II.1-12B) which is the result of the combined action of tensile and compressive principal stresses s_I, s_{III} of equal magnitude $|\sigma_I - \sigma_{III}|/2$. We also note that the elements of maximum shear stress in the deviatoric state are free of normal stresses. A quadrilateral, rotated through 45° with respect to the old one, will therefore experience only shear stresses (of the same magnitude as the principal deviatory stresses). This implies – as is further illustrated in Fig. II.1-13 – that the state of stress produced at a point P by tension and compression, with tension and compression mutually perpendicular and of equal magnitude σ, is also produced by the sole action of shear stresses of equal magnitude σ along elements inclined at 45° to the former stress directions.

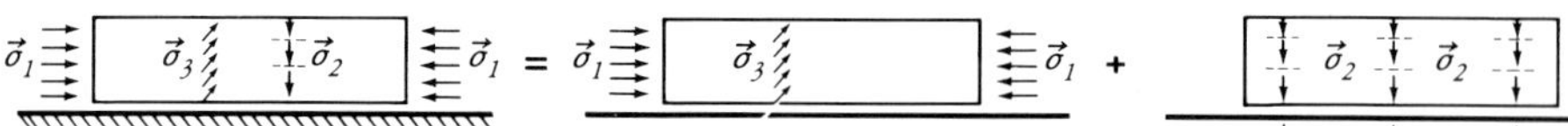

Horizontal compression by tectonic stresses $\vec{\sigma}_1$, $\vec{\sigma}_3$; vertical stress $\vec{\sigma}_2$ due to overburden — *plane stress in horizontal sections* — *stress field due to gravity*

Fig. II.1-11 COMPRESSED HEAVY PLATE ON VISCOUS SUBSTRATUM AT REST

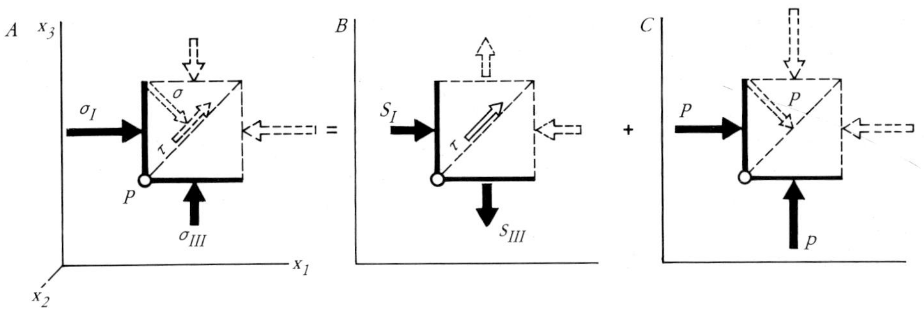

Coordinate axes parallel to principal directions
$$(\sigma = \frac{\sigma_I + \sigma_{III}}{2}\ ;\ \tau = \frac{\sigma_I - \sigma_{III}}{2})$$

Deviatory stress S_I and S_{III} are compression and tension of equal magnitude
$(S_I = \sigma_I - p,\ S_{III} = \sigma_{III} - p;$
$|\tau| = |S_I| = |S_{III}|)$

Hydrostatic stress
$$(p = \frac{\sigma_I + \sigma_{III}}{2})$$

Fig. II.1-12 DECOMPOSITION OF PLANE STRESS IN DEVIATORIC AND HYDROSTATIC STRESS

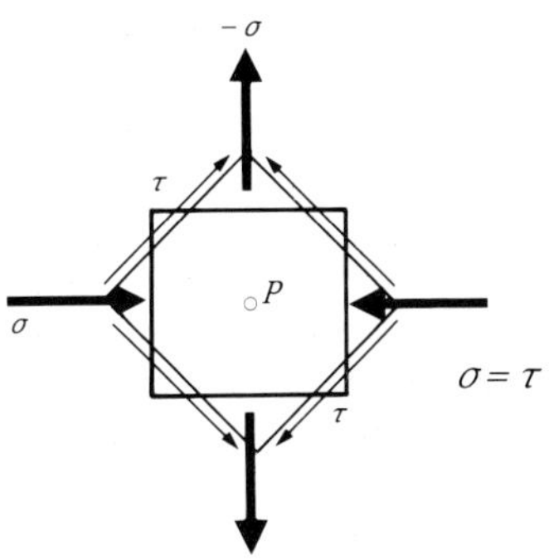

Fig. II.1-13
TWO EQUIVALENT REPRESENTATIONS OF THE SAME STATE OF STRESS AT POINT P
(Tension and compression, with tension and compression mutually perpendicular and of equal magnitude, is equivalent to a case of pure shear stress)

Example No.4: Stresses near vertical rock walls

As a further illustration of the stress concept we consider the state of stress at and near a vertical rock wall (Fig. II.1-14). We assume that the rock mass has not been subject to severe tectonism and that its top is approximately horizontal. Two principal tectonic stresses will then be horizontal – certainly near the top, but most likely also to some depth inside the rock mass, since abrupt stress changes are not to be expected. The orientation of the horizontal principal stresses is unknown, except in the vicinity of a vertical rock wall, where one horizontal principal stress must be perpendicular to the wall and vanish at the wall ($\sigma_W \approx 0$). In general, this principal stress will be the smallest, when we exclude the top and foot regions of the wall where topographic irregularities and 'notch' stresses may greatly disturb the stress field. This leaves us with the tectonically important question whether the vertical (σ_V) or the second horizontal principal stress (σ_H) is the greater compressive stress. In the absence of tectonic features, the type of weathering of the vertical rock wall may sometimes help us to decide this question (Gerber, & Scheidegger, 1965). A saw-shaped appearance of the wall (see Fig. II.1-14), for example, may be attributed to conjugate shear failures (see Section II.7.4, Figs. II.7-15-18) and since the planes of conjugate shear failure are known to intersect along the axis of the intermediate principal stress, the maximum compressive stress will be horizontal. Naturally, there is no reason why inside the rock mass – at some distance from the wall – the maximum compressive stress should remain parallel to the wall. Under the assumptions made above, it should, however, remain horizontal.

Geometrical representation of stresses. The state of stress can be described geometrically in different ways. The most important in the mechanics of rocks and soils is the stress circle diagram by O. Mohr. Because of its great convenience for the theoretical treatment of rock failure (faulting), it will be discussed separately in Chapter II.4.

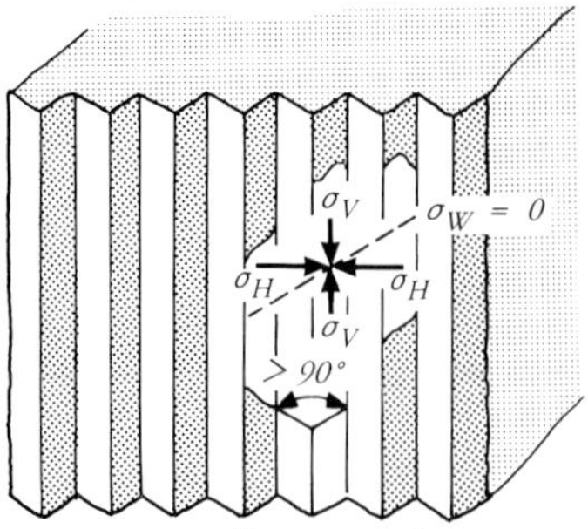

Fig. II.1-14
STRESSES NEAR VERTICAL ROCK WALL

4. SIGN CONVENTION

When working with stresses one has to agree on a sign convention, i.e. one has to define when a normal or shear stress component is counted positive or negative. Contrary to common usage in elasticity theory and following the example of outstanding rock mechanics specialists (Ref. 3), we chose to reckon ***compressive stresses positive.*** This choice is motivated by the fact that almost all normal stresses that are to be dealt with in Structural Geology are compressive, despite the frequent and misleading use of terms such as 'tensile forces', 'tension tectonics', etc., made in geological literature to characterise situations in which rock strata have been *extended* under the exclusive action of *compressive* stresses (see the illustration in Fig. II.1-15).

In addition to compressive stress components being counted as positive, the signs of the components of the stress tensor in equation II.1-1a, which have been introduced as the components of three vectors, should agree with the sign convention for vectors. Accordingly, stress components pointing in the direction of a positive coordinate axis should be given a positive sign. The sign convention, which is summarised in Fig. II.1-16, satisfies these requirements and will be used in most cases. Occasionally, a different sign convention may be more convenient and the reader is advised to take good note of differences in the sign conventions preferred by other authors.

5. STRESS FIELDS

Stresses inside a loaded body normally vary with position. We therefore speak of a stress 'field'. The internal stress field has to be determined if one wishes to know the deformations and, possibly, failure a body will suffer under a given load. How do the stresses vary with position and what determines the stress field? There are two laws that control the stress field. The first law is of universal validity and applies to any type of material. It ensures that at any point of the material continuum all forces acting on a small volume element are in equilibrium. The formulation of this law and its application will be discussed in the

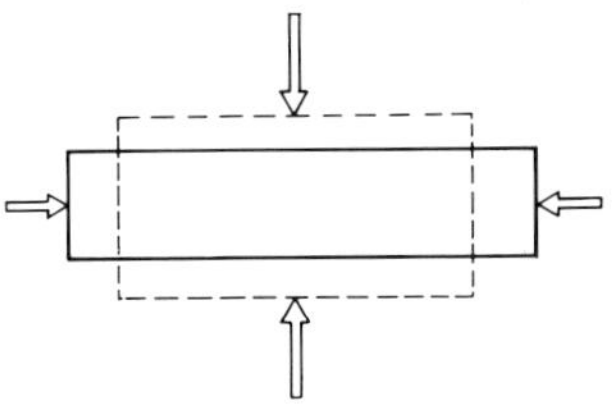

Fig. II.1-15
EXTENSION UNDER COMPRESSIVE LOAD

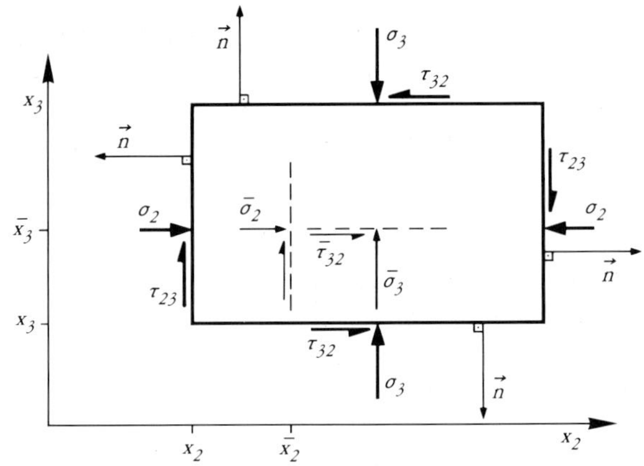

Fig. II.1-16 SIGN CONVENTION
Normal and shear stresses are counted ***positive*** *when oriented as shown.*

following Chapter (II.2).

The second law, which is generally required to determine the stress field, relates stresses and local deformations. It was mentioned in the introduction to this chapter that the internal stresses and deformations are interdependent. The character of this interdependence will depend on the type of material ('elastic', 'viscous', 'plastic', etc.). In other words, different types of materials exhibit different deformational response upon stressing. Material laws that relate stresses and strains will be dealt with in Chapters II.6 and II.7.

To get some idea of how this deformation response affects the stress field, we may think of replacing in an existing stress field the state of stress in the neighbourhood of a point P by another equilibrium state of stress. Without disturbing the equilibrium state of the stress field, this local operation would cause deformation of the material at P and, necessarily, of the surrounding material. This, in turn, would lead to a change in stress spreading throughout the material and reaching the boundary of the body. However, since conditions are fixed there, the whole process of stress changing would be suppressed. In other words, the material near P would react to any stress change at P by producing reactive stresses that restore the original state of stress which was in accordance with the equilibrium conditions, the material stress/strain law and the conditions at the boundary.

Example No. 5: Shear stress and oblique slip on plane of weakness

We shall conclude this chapter with the somewhat less elementary application of the stress concept to oblique slip faulting, i.e. fault movement composed of strike-slip and dip-slip components.

According to Coulomb's concept of shear failure (see Section II.7.4), a newly formed fault plane is parallel to the direction of the intermediate principal stress σ_{II}, while movement along the fault plane is in a direction perpendicular to this principal axis and coinciding with the direction of the maximum shear stress in the fault plane (see, for example, Fig. II.7-8G). Hence, when the σ_{II}-axis is horizontal or vertical, movement along the fault plane will respectively be of a pure dip-slip or strike-slip type. This will be different in general, when slip is induced on a pre-existing plane of weakness – be it a bedding plane, a tension fracture or an inactivated fault plane. Obviously, the direction of the induced maximum shear stress in the plane of weakness will deviate from the dip or strike directions when the principal axes of the slip-inducing stress system are not vertical or horizontal; the resulting slip will then be oblique. But the induced slip will even be oblique in a stress system with vertical and horizontal principal axes when the σ_{II}-axis is not

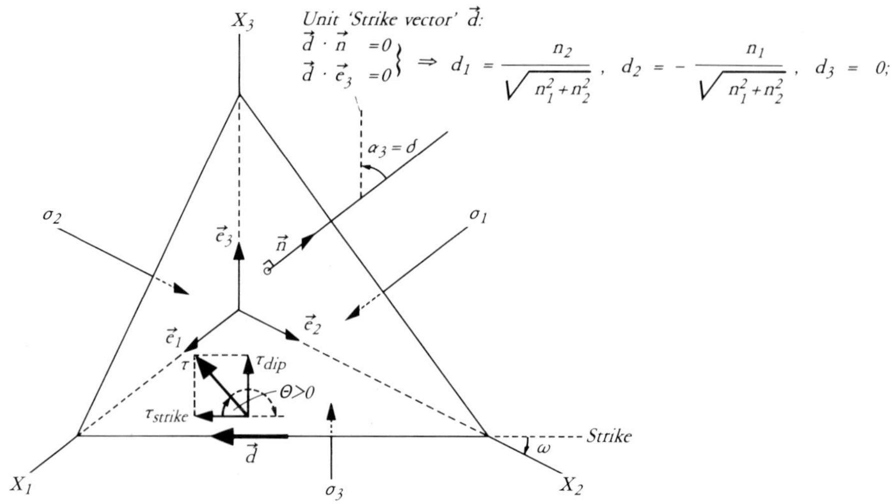

Fig. II.1-17 PLANE OF WEAKNESS IN STRESS SYSTEM WITH VERTICAL AND HORIZONTAL PRINCIPAL STRESS AXES
Principal stresses σ_1, σ_2, σ_3 are not yet ranked. Direction of maximum shear stress ($\vec{\tau}$) in plane of weakness is defined by angle Θ.

parallel to the plane of weakness. We shall consider this case and determine the direction of the maximum shear stress in the plane of weakness.

Consider an arbitrary plane of weakness in a Cartesian coordinate system with vertical and horizontal axes (Fig. II.1-17) that coincide with the principal axes of the stress system that induces slip on the plane of weakness. The components of the unit vector $\vec{n}$ of the plane of weakness are the directional cosines $n_i = \cos\alpha_i$ (i=1,2,3), the unit vectors $\vec{e}_1$, $\vec{e}_2$, $\vec{e}_3$ are directed along the positive coordinate axes, and the unit vector $\vec{d}$ lies in the intersection of the plane of weakness with the x_1,x_2-plane and is therefore parallel to the strike direction. Since in this reference frame the shear stress components vanish, the stress vector $\vec{t}$ that acts upon the plane of weakness has the components (eq. II.1-1a):

$$t_1 = -\sigma_1 n_1,\ t_2 = -\sigma_2 n_2,\ t_3 = -\sigma_3 n_3 \qquad \text{(II.1-7)}$$

where n_i (i=1,2,3) are the components (directional cosines) of the unit normal of the plane of weakness, and where the normal stresses σ_1, etc. are principal stresses, but not yet ranked according to magnitude. (We therefore deviate here from using the Roman numerals as subscripts which indicate ranked principal stresses.) The negative signs in eq. II.1-7 are in accord with $\vec{t}$ acting from *outside* upon the prism in Fig. II.1-17 (i.e. opposite to the direction of $\vec{t}$ in Fig. II.1-9).

The stress vector $\vec{t}$ is the sum of a vector $\vec{\sigma}$ normal to the plane of weakness and a vector $\vec{\tau}$ (Fig. II.1-17) parallel to that plane. It is easily seen that $\vec{\tau}$, being the orthogonal projection of $\vec{t}$ on the plane of weakness, represents the maximum shear stress in that plane. The magnitude of the vector $\vec{\sigma}$, which is directed oppositely to $\vec{n}$, is obtained by orthogonal projection of $\vec{t}$ on $-\vec{n}$, i.e. by performing the inner product of $\vec{t}$ and $-\vec{n}$, which because of eq. II.1-7 is

$$|\vec{\sigma}| = \vec{t}\,.\,(-\vec{n}) = \sigma_1 n^2{}_1 + \sigma_2 n^2{}_2 + \sigma_3 n^2{}_3 \qquad \text{(II.1-8)}$$

The vector $\vec{\sigma}$ is then obtained by multiplying the scalar $|\vec{\sigma}|$ by the unit vector $-\vec{n}$

$$\vec{\sigma} = -\ |\vec{\sigma}|\vec{n} \qquad \text{(II.1-9)}$$

Subtracting this vector from the stress vector $\vec{t}$ gives the tangential vector

$$\vec{\tau} = \vec{t} + |\vec{\sigma}|\vec{n} \tag{II.1-10}$$

Since the three vectors in this equation form a right-angled triangle, the squared magnitude of $\vec{\tau}$ becomes

$$|\vec{\tau}|^2 = |\vec{t}|^2 - |\vec{\sigma}|^2 = (\sigma_1^2 n_1^2 + \sigma_2^2 n_2^2 + \sigma_3^2 n_3^2) - (\sigma_1 n_1^2 + \sigma_2 n_2^2 + \sigma_3 n_3^2)^2 \tag{II.1-11}$$

By inserting $n_1^2 = 1 - n_2^2 - n_3^2$, this expression can be put in the more convenient form:

$$|\vec{\tau}|^2 = n_2^2 (\sigma_2 - \sigma_1)^2 + n_3^2 (\sigma_3 - \sigma_1)^2 - [n_2^2 (\sigma_2 - \sigma_1) + n_3^2 (\sigma_3 - \sigma_1)]^2 \tag{II.1-11a}$$

as may be verified by carrying out the squaring operations and comparing terms.

Next we determine the direction of the maximum shear stress vector $\vec{\tau}$ with respect to the strike of the plane of weakness, which is represented by the unit vector $\vec{d}$, whose components are stated in Fig. II.7.1-17. The direction of the vector $\vec{\tau}$ is defined by the 'pitch' angle θ (Fig. II.1-17) that is included by the vectors $\vec{\tau}$ and $\vec{d}$ and determined by the ratio of the components of $\vec{\tau}$ in the strike and dip directions:

$$\tan\theta = \tau_{dip} / \tau_{strike} \tag{II.1-12}$$

The component τ_{strike} is the inner product of the vectors $\vec{\tau}$ and $\vec{d}$. Since $\vec{d}$ is orthogonal to $\vec{n}$ ($\vec{d} \cdot \vec{n} = 0$), it follows from eqs. II.1-7, 10 that

$$\tau_{strike} = \vec{\tau} \cdot \vec{d} = -\sigma_1 n_1 d_1 - \sigma_2 n_2 d_2 - \sigma_3 n_3 d_3 \tag{II.1-13}$$

Inserting the components of the vector $\vec{d}$ from Fig. II.1-17 one eventually obtains for the strike component of the shear stress vector $\vec{\tau}$:

$$\tau_{strike} = (\sigma_2 - \sigma_1) n_1 n_2 / \sqrt{n_1^2 + n_2^2} \tag{II.1-14}$$

The expression for the dip component τ_{dip} is derived from

$$|\vec{\tau}|^2 = \tau^2_{dip} + \tau^2_{strike} \tag{II.1-15}$$

by inserting $|\vec{\tau}|^2$ from eq. II.1-11a and τ_{strike} from eq. II.1-14:

$$\tau_{dip} = n_3 [n_2^2 (\sigma_2 - \sigma_1) - (1 - n_3^2)(\sigma_3 - \sigma_1)] / \sqrt{(1 - n_3^2)} \tag{II.1-16}$$

Combining eqs. II.1-12, 14, 16 finally yields the formula for the 'pitch' angle of the maximum shear stress vector in the plane of weakness:

$$\tan\theta = [n_2^2 - (1 - n_3^2)(\sigma_3 - \sigma_1) \cdot (\sigma_2 - \sigma_1)^{-1}] \cdot n_3 (n_1 n_2)^{-1} \tag{II.1-17}$$

It should be recalled that in deriving this formula for the orientation of the maximum shear stress vector $\vec{\tau}$ in a plane of weakness, which was first presented by Bott (1956), it was assumed that the principal axes of the slip-inducing stress system were vertical and horizontal. Only for such a system does the angle θ, which is referred to the intersection line of the plane of weakness with the plane of the σ_1- and σ_2-axes, represent the inclination of the maximum shear stress vector with respect to the *strike* of the plane of weakness. In a differently oriented system the intersection line is not horizontal and therefore no longer parallel to the strike of the plane of weakness. Although the unit vector along this intersection line may be expressed in terms of the unit vector in the strike direction by a straightforward algebraic operation, the resulting expression for the inclination of $\vec{\tau}$ with respect to the strike direction in such a general system of principal stress axes becomes very complex and will not be discussed here.

For a plane of weakness, the orientation of which is defined by its unit normal, the angle θ depends on the ranking of the vertically and horizontally directed principal stresses σ_1, σ_2, σ_3. Assume, for instance, the following ranking: the vertical principal stress $\sigma_3 = \sigma_I$ and the horizontal principal stresses $\sigma_1 = \sigma_{III}$, $\sigma_2 = \sigma_{II}$. Let δ be the dip angle of the plane of weakness and, therefore, $n_3 = \cos\alpha_3 = \cos\delta$. Formula II.1-17 then turns into:

$$\tan\theta = \cos\delta\,[n_2^2 - (\sigma_I - \sigma_{III}) \cdot (\sigma_{II} - \sigma_{III})^{-1} \sin^2\delta]/n_1 n_2 \qquad \text{(II.1-17a)}$$

Considering the special case (Coulomb-type slip) of σ_{II} acting parallel to the plane of weakness, we let n_2 tend to zero ($\cos\alpha_2 \Rightarrow \cos 90°$) in eq. II.1-17a. Since $\tan\theta \Rightarrow -\infty$, θ becomes $-\pi/2$. Because θ is counted positive in the clockwise direction from the strike vector $\vec{d}$ (Fig. II.1-17), the maximum shear stress in the plane of weakness acts in the downdip direction producing a pure dip-slip. Now rotate the σ_{II}-axis in the horizontal plane towards the original x_1-axis through the angle ω. The components of $\vec{n}$ with respect to the rotated horizontal axes are obtained by first projecting $\vec{n}$ upon the horizontal plane, which gives a vector of length $\sin\delta$, and then projecting this horizontal vector on the new axes, giving the new components $n_1 = \sin\delta\cos\omega$ and $n_2 = \sin\delta\sin\omega$. Inserting these expressions in eq. II.1-17a gives:

$$\tan\theta = [\sin^2\omega - (\sigma_I - \sigma_{III}) \cdot (\sigma_{II} - \sigma_{III})^{-1}]\; 2 \cdot \cos\delta/\sin 2\omega \qquad \text{(II.1-17b)}$$

where δ is the dip angle of the plane of weakness and ω the angle by which the σ_{II}-axis deviates from the strike direction, as indicated in Fig. II.1-17.

COLLATERAL READING:

1) Jaeger, J.C., 'Elasticity, fracture and flow (with engineering and geological applications)', Methuen, London 1964.

2) Ramsay, J.G., 'Folding and fracturing of rocks', McGraw-Hill, 1967.

3) Jaeger, J.C. & Cook, N.G.W., 'Fundamentals of rock mechanics', Methuen, London 1968.

4) Turcotte, D.L. & Schubert, G., 'Geodynamics', John Wiley & Sons, 1982.

5) Fung, Y.C., 'Foundations of solid mechanics', Prentice-Hall, 1965.

6) Spencer, A.J.M., 'Continuum mechanics', Longman, London 1980.

7) Means, W.D., 'Stress and strain; basic concepts of continuum mechanics for geologists', Springer, 1976.

II.2. STRESS EQUILIBRIUM

A body which, under the action of body and surface forces, does not experience accelerations – neither as a whole nor in any of its parts – is said to be in static equilibrium. For this state to exist, it is necessary for the resultant (vectorial sum) of all forces acting on an arbitrary element of the body to vanish (Fig. II.2-1). If the forces should add up to a non-vanishing resultant, the body would be accelerated according to the fundamental laws of mechanics (Newton, Euler). Since the resultant force must vanish for *any* element as well as for the body as a whole, the equilibrium conditions can be formulated in the 'small' and in the 'large'.

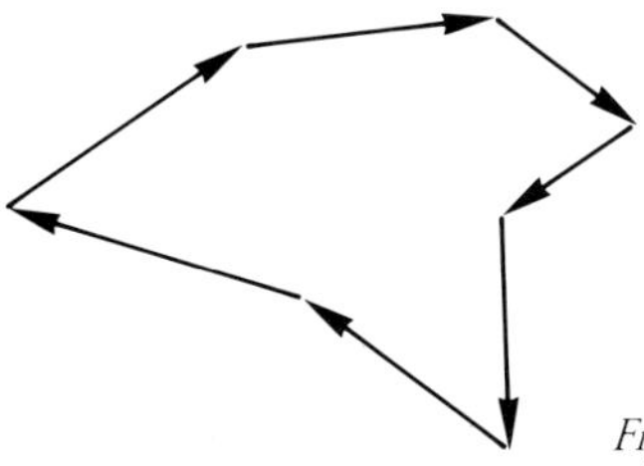

Fig. II.2-1 VANISHING VECTOR SUM

1. STATIC EQUILIBRIUM IN THE 'SMALL'

To formulate the conditions of static equilibrium in the 'small' we consider a small volume element (Fig. II.2-2) with faces perpendicular to the Cartesian coordinate axes and acted upon by body and surface forces. We assume that the surface forces are continuously distributed over the faces of the element and may therefore be expressed as stresses (see Chapter II.1) times areas. Since the body force $\vec{f}$ is referred to a unit volume (e.g. specific weight), the resultant of the body forces acting on the small volume of supposedly uniform material density is $\vec{f}\,\delta x_1\,\delta x_2\,\delta x_3$. As indicated in Fig. II.2-2, addition of all forces leads to the simple vectorial relationship expressing the vanishing of the resultant force

$$\frac{\delta \vec{t}_1}{\delta x_1} + \frac{\delta \vec{t}_2}{\delta x_2} + \frac{\delta \vec{t}_3}{\delta x_3} + \vec{f} = 0 \tag{II.2-1}$$

The first three terms on the left represent the rates at which the stresses $\vec{t}_1$, $\vec{t}_2$, $\vec{t}_3$ change when proceeding in the corresponding coordinate direction.

In the limit of vanishing dimensions of the volume element i.e. $\delta x_1, \delta x_2, \delta x_3 \rightarrow 0$, equation (II.2-1) turns into the differential equation

$$\frac{\partial \vec{t}_1}{\partial x_1} + \frac{\partial \vec{t}_2}{\partial x_2} + \frac{\partial \vec{t}_3}{\partial x_3} + \vec{f} = 0 \qquad \text{(II.2-1a)}$$

The partial differentiation symbols $\partial \ldots / \partial x_i$ denote a rate of change in the x_i-direction at a *point*, while the symbols $\delta / \delta x_i$ in equation II.2-1 denote an average rate of change over a small interval along the x_i-directions.

In dealing with the vector relationship II.2-1 numerically we have to specify the vectors in terms of components. This is done by projecting every vector upon the axes of the rectangular Cartesian frame in Fig. II.2-2 (see also Fig. II.1-3). Now, since the sum of vectors is a vector itself and since the vanishing of a vector means the vanishing of all its components, the vector equation II.2-1 is equivalent to three componental equations, each of which expresses the vanishing of one component of the resultant force.

In formulating the componental equations we have to exercise some care regarding the proper sign of componental terms. Note, that a component of a 'differential stress' vector $\delta \vec{t}_i$ defined in Fig. II.2-2, has to be counted positive when the projection of the vector upon the corresponding coordinate axis points in the positive coordinate direction (see Fig. II.1-3). However, such a positive component is equivalent to a negative change of a corresponding component of the stress tensor (introduced in Section II.1.2) in the direction of the positive coordinate axis. This may be seen on Fig. II.1-16 of the preceding chapter where, for instance, a differential stress vector $\delta \vec{t}_3$ pointing in the direction of the positive x_3-axis would represent a decrease of σ_3 in x_3-direction and, hence, would be equal to $-(\delta \sigma_3 / \delta x_3)\delta x_3$.[1] Therefore, when the notations σ for normal and τ for tangential stress components are used, as introduced in the preceding chapter (II.1.2), the vectorial representations for the differential stress vectors in equation II.2-1 become:

$$\begin{aligned} \delta \vec{t}_1 &: \; -\delta\sigma_1, \; -\delta\tau_{12}, \; -\delta\tau_{13} \\ \delta \vec{t}_2 &: \; -\delta\tau_{21}, \; -\delta\sigma_2, \; -\delta\tau_{23} \\ \delta \vec{t}_3 &: \; -\delta\tau_{31}, \; -\delta\tau_{32}, \; -\delta\sigma_3 \end{aligned}$$

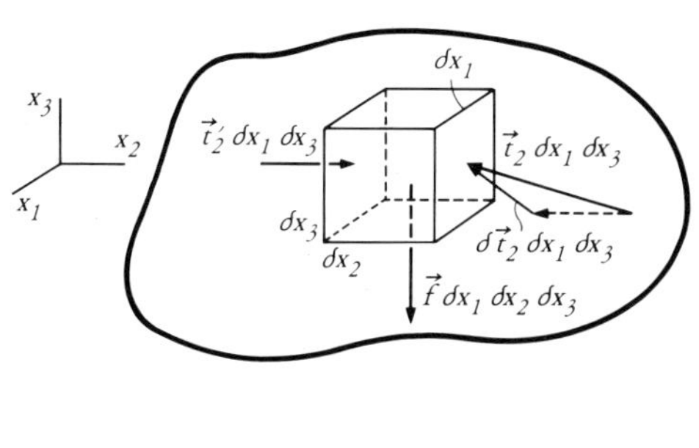

Fig. II.2-2
VECTORIAL EQUILIBRIUM EQUATION

Pairs of surface forces (only one pair shown in figure) add up to small differences, e.g.

$$\delta \vec{t}_2\, dx_1\, dx_3 = (\vec{t}_2 + \vec{t}'_2)\, dx_1\, dx_3$$

Vanishing of the sum of all forces acting on the volume element:

$$\delta \vec{t}_1\, dx_2\, dx_3 + \delta \vec{t}_2\, dx_1\, dx_3 + \delta \vec{t}_3\, dx_1\, dx_2 + \vec{f} dx_1\, dx_2\, dx_3 = 0$$

or after dividing by the volume $dx_1\, dx_2\, dx_3$:

$$\frac{\delta \vec{t}_1}{dx_1} + \frac{\delta \vec{t}_2}{dx_2} + \frac{\delta \vec{t}_3}{dx_3} + \vec{f} = 0 \qquad \text{(eq. II.2-1)}$$

The body force we deal with in geological problems is the specific weight $\vec{\gamma}$ of the (fluid-saturated) rock. Its components, i.e. its projections on the coordinate axes, are

$$\vec{\gamma} : \gamma \cos \alpha_1, \gamma \cos \alpha_2, \gamma \cos \alpha_3$$

where α_1, α_2, α_3, are the angles the vertical downward direction includes with the positive coordinate axes. (When, for instance, the coordinate axes are chosen as in Fig. II.2-2, the directional cosines become $\cos \alpha_1 = \cos \alpha_2 = 0$, $\cos \alpha_3 = -1$.)

Inserting, in turn, for the vectors in equation II.2-1 their first, second and third components, we obtain the *componental equilibrium equations*:

$$\begin{aligned} \frac{\delta \sigma_1}{\delta x_1} + \frac{\delta \tau_{21}}{\delta x_2} + \frac{\delta \tau_{31}}{\delta x_3} &= \gamma \cos \alpha_1 \\ \frac{\delta \tau_{12}}{\delta x_1} + \frac{\delta \sigma_2}{\delta x_2} + \frac{\delta \tau_{32}}{\delta x_3} &= \gamma \cos \alpha_2 \\ \frac{\delta \tau_{13}}{\delta x_1} + \frac{\delta \tau_{23}}{\delta x_2} + \frac{\delta \sigma_3}{\delta x_3} &= \gamma \cos \alpha_3 \end{aligned} \qquad \text{(II.2-2)}$$

where the equations represent, in turn, equilibrium in the x_1-, x_2- and x_3-directions.

In the limit of vanishing dimensions of the volume element (δx_1, δx_2, $\delta x_3 \rightarrow 0$) equations (II.2-2) turn into three differential equations where the 'average' rates of change $\frac{\delta ...}{\delta x_1}$, $\frac{\delta ...}{\delta x_2}$, $\frac{\delta ...}{\delta x_3}$ are replaced by the partial derivatives $\frac{\partial ...}{\partial x_1}$, $\frac{\partial ...}{\partial x_2}$, $\frac{\partial ...}{\partial x_3}$ in the respective coordinate directions.

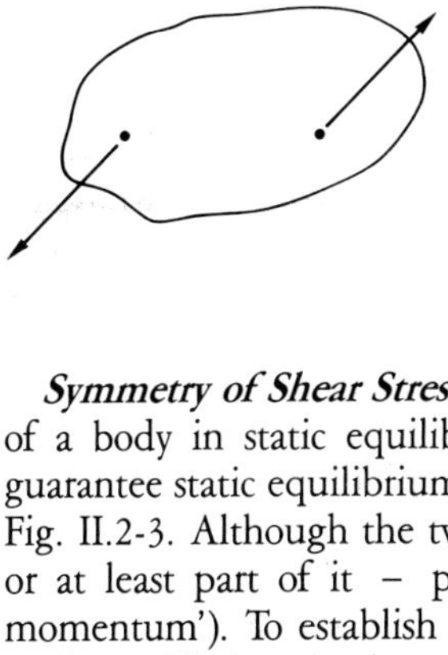

Fig. II.2-3
VANISHING OF RESULTANT FORCE NOT SUFFICIENT FOR STATIC EQUILIBRIUM ('COUPLE' WILL ROTATE BODY)

Symmetry of Shear Stresses. The equilibrium equations must be satisfied at any 'point' of a body in static equilibrium. Fulfilment of these conditions, however, does not yet guarantee static equilibrium. This is easily seen by considering, for instance, the 'couple' in Fig. II.2-3. Although the two forces are in local equilibrium, they tend to rotate the body or at least part of it – producing or changing an 'angular' momentum ('moment of momentum'). To establish the additional conditions that must also be satisfied to ensure static equilibrium in the small, we consider a cubical volume element (Fig. II.2-4A) with edges of unit length and parallel to the axes of a Cartesian frame, the origin of which coincides with the centre of the element. For the sake of simplicity only the intersection of the cubical element with the x_2, x_3-plane is shown. Choosing unit length for the edges of the cube makes surface forces and stresses numerically equal. If we now applied a shear stress couple of the type shown in Figs. II.2-4A, B, the condition of vanishing resultant force would still be satisfied, but the element would start spinning around the x_1-axis. Obviously, by combining two shear stress couples (Fig. II.2-4C) we could restore static equilibrium. More generally, we may say that a state of *pure* shearing stress is in static equilibrium when the shear stresses can be combined into pairs of equal magnitude and directed at right angles either to each other or away from each other (see the sign convention in Section II.1.4, Fig. II.1-16). This may be simply formulated as:

$$\tau_{23} = \tau_{32}, \tau_{12} = \tau_{21}, \tau_{13} = \tau_{31} \qquad \text{(II.2-3)}$$

Next, it can be demonstrated that this condition must be satisfied not only in case of pure shear stress but also for a general state of stress. Because of symmetry, it is already obvious that the body force will not tend to rotate the volume element considered. However, the effect of any normal stresses acting on the faces of the small cubical element has to be taken into consideration. We may distinguish two cases: first, the case in which the shear stresses satisfy equation II.2-3 and are, consequently, self-equilibrating. Normal stresses and body force must then balance each other irrespective of the shear stresses. It is obvious from Fig. II.2-4D that these stresses cannot disturb equilibrium by exerting a 'torque' upon the element. The second case to be considered is that the shear stresses do *not* satisfy equation II.2-3. Then at least one of the shear stresses is not balanced by the shear stress on the opposite face. As sketched in Fig. II.2-4E, the 'excess' shear stress $\delta\tau$ has to be balanced by the extra normal stress $\delta\sigma$. Because of its position with respect to the x_1-axis, the normal stress $\delta\sigma$ cannot compensate the tendency of the shear stress $\delta\tau$ to rotate the volume element. This demonstrates that violation of the symmetry condition II.2-3 would imply a state of non-equilibrium. Condition II.2-3 is therefore a necessary supplement of equation II.2-2 to ensure static equilibrium.[2)]

Because of condition II.2-3 the number of independent stress components in the equilibrium equations II.2-2 is reduced to six (three normal stresses and three shear stresses) and the order of stress indices in a shear stress component becomes immaterial. Clearly the three equilibrium equations will, in general, not suffice to determine the fields of the six stress components from the conditions at the boundaries. The additional relationships required are derived from the specific laws that govern the interdependence of stresses and deformations (see Chapters II.6 and II.7). Unfortunately, these laws do not possess the general applicability of the equilibrium equations, since they depend very much on the kind of idealised material behaviour considered. Nevertheless, valuable information on the stress field can sometimes be deduced from the equilibrium equations alone, when plausible assumptions permit sufficient simplification of the equations. Such results – independent of a detailed specification of the material behaviour – are of particular importance in tectono-mechanics where the choice of the stress/strain relationships often depends heavily on speculation.

Plane Problems. In many tectonic problems a considerable simplification is achieved by *approximating* the problems as one of plane deformation. By definition, *in 'plane deformation'* (see Section II.5.3) the displacement of material 'points' takes place within parallel planes and without relative displacements between these 'planes of deformation'. It is clear that in an isotropic material whose properties do not change in the direction perpendicular to the plane of deformation, plane deformation cannot induce shear stresses between neighbouring planes of deformation. When a Cartesian reference frame is chosen so as to make a plane of deformation coincide with say the x_2, x_3-plane (Fig. II.2-5), the absence of shear stresses along planes of deformation implies $\tau_{12} = \tau_{21} = \tau_{13} = \tau_{31} = 0$. Therefore, as sketched in Fig. II.2-5, the stress vectors acting on planes oriented perpendicular to the plane of deformation are parallel to the plane of deformation, whereas the latter is acted upon by a normal stress only.

In *'plane stress'* one of the coordinate planes – e.g. the x_2, x_3-plane in Fig. II.2-5 – may be chosen as being free of any stress. Then in addition to the shear stresses τ_{12} and τ_{13} the normal stress σ_1 vanishes as well. (According to the first equation II.2-2 this is, of course, only possible if no body force acts in the x_1-direction.) The equilibrium equations reduce, accordingly, to

$$\begin{array}{l} \\ \text{plane stress } (x_2,\ x_3\text{-plane} \\ \text{stress-free}) \end{array} \left\{ \begin{array}{l} \left. \begin{array}{r} \dfrac{\delta\sigma_1}{\delta x_1} = \gamma \cos \alpha_1 \\ \dfrac{\delta\sigma_2}{\delta x_2} + \dfrac{\delta\tau}{\delta x_3} = \gamma \cos \alpha_2 \\ \dfrac{\delta\tau}{\delta x_2} + \dfrac{\delta\sigma_3}{\delta x_3} = \gamma \cos \alpha_3 \end{array} \right\} \begin{array}{l} \text{plane deformation} \\ \text{in } x_2,\ x_3 \text{ plane} \end{array} \qquad (\text{II.2-2a}) \\ \qquad\qquad\qquad\qquad\qquad\qquad (\tau \equiv \tau_{23} = \tau_{32}) \end{array} \right.$$

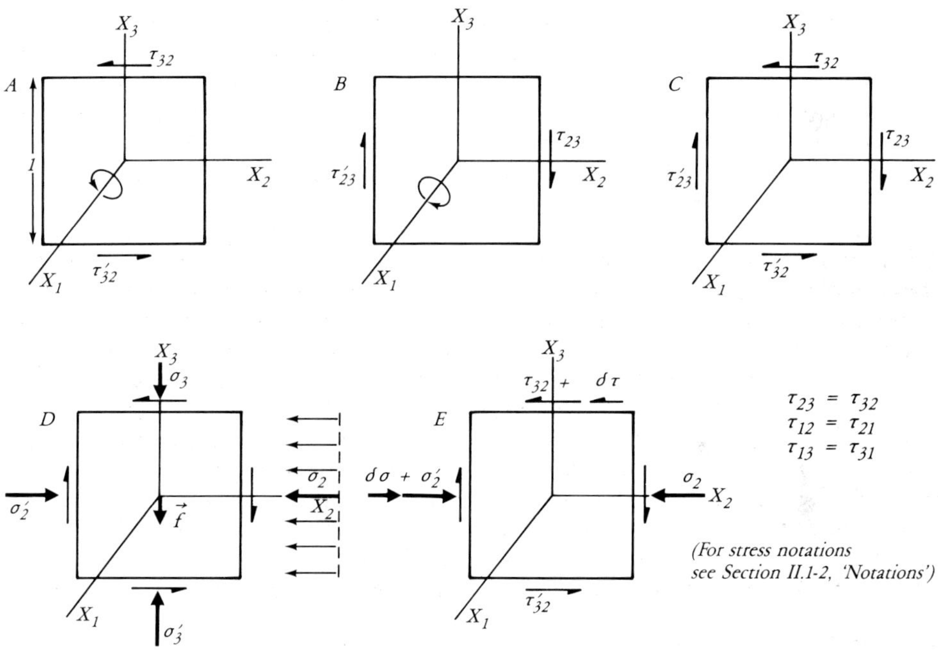

Fig. II.2-4 SMALL UNIT CUBE UNDER STRESS

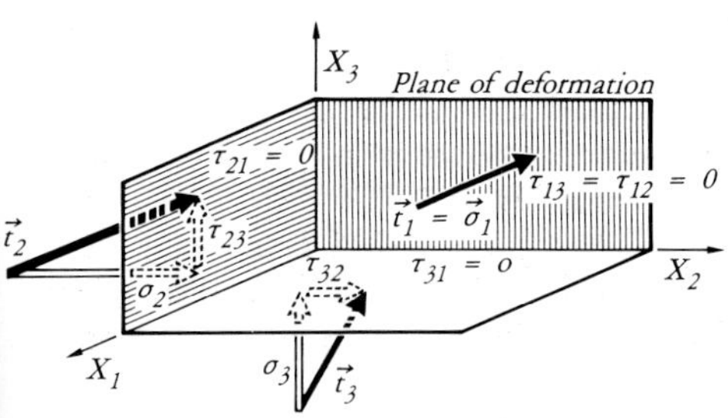

Fig. II.2-5
STRESSES IN PLANE DEFORMATION OF HOMOGENEOUS ISOTROPIC MATERIAL

We note that in plane problems the number of unknown stresses exceeds the number of available equations by only one.

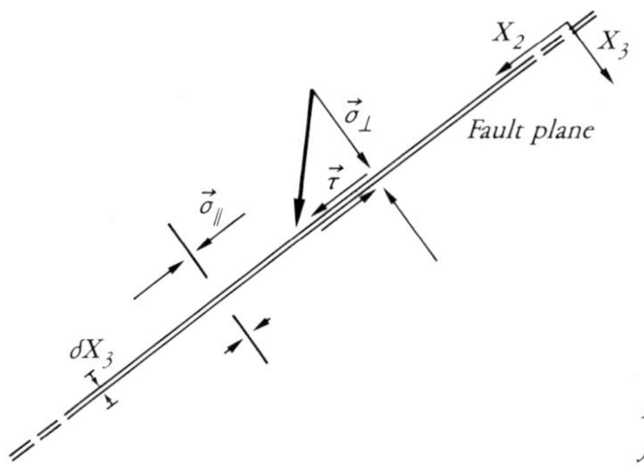

Fig. II.2-6
JUMP OF FAULT-PARALLEL NORMAL STRESS

Example No. 1: Stresses along fault planes

The equilibrium equations constitute certain relationships between corresponding stress components at both sides of a fault plane. If the coordinate axes are chosen as shown in Fig. II.2-6 for a dip section through a fault plane, the normal stress that acts upon the fault plane has the magnitude σ_3 and the normal stress that acts parallel to the dip has the value σ_2. Next, we define a small line element δx_3 that intersects the fault plane parallel to the x_3-axis, and multiply the second and third equations II.2-2a by δx_3. The equations become

$$\frac{\delta \sigma_2}{\delta x_2} \delta x_3 - \gamma \cos \alpha_2 \, \delta x_3 + \delta \tau = 0$$

$$\frac{\delta \tau}{\delta x_2} \delta x_3 - \gamma \cos \alpha_3 \, \delta x_3 + \delta \sigma_3 = 0$$

Letting δx_3 shrink to zero makes the first two terms in these equations vanish, since neither the weight components nor the rate of stress change in the x_2-direction attain infinite values. Consequently, we also obtain $\delta \tau = \delta \sigma_3 = 0$, which means that both the normal and tangential stresses that act *on* a fault plane have the same magnitudes at both sides. In other words, these stresses cannot experience a 'jump'-like change across a fault plane. More interesting than this almost trivial result is the fact that *a fault-parallel normal stress ($\vec{\sigma}_{\|}$) may suffer a 'jump' in magnitude across a fault plane*, since such a behaviour is perfectly compatible with the equilibrium equations which do not contain the rate of change of σ_3 in the x_3-direction [3)].

A jump in the magnitude of a fault-parallel normal stress may actually account for abrupt changes in tectonic rock deformation (e.g. folding) across a major fault plane, in cases where such changes cannot be attributed to a strong difference in the mechanical properties of the juxtaposed rocks. It should be noted, however, that the stress discontinuity – although statically admissible – can only exist if the associated difference in straining is accommodated by slip or creep along the fault plane.

Example No. 2: Sediment layer on infinite slope

We consider a parallel-sided slab of subaerial sediment of thickness h resting on a rigid plane inclined at an angle δ (see Fig. II.2-7A). The slab extends both in the dip and strike directions over a distance that is very large compared with h and which therefore may be considered infinite. It is assumed that the material properties of the sediment do not change in the dip or strike direction and that the sediment rests on the slope in uniform frictional or cohesive contact. The sediment slab is only subject to its own weight. Any deformation of the sediment under the action of this force will be of the plane type with dip sections being planes of deformation. We choose a coordinate system as shown in Fig. II.2-7A, with origin at the sediment surface and with the x_2-axis pointing down-dip and the x_3-axis perpendicular to the surface and oriented downwards.

Applying equations II.2-2a to the present case, we note at once from the first equation that the strike-parallel normal stress σ_1 cannot change along the strike, since there is no weight component in that direction (x_1-axis). The value of this stress is not of interest in the present context. We are left with two equations for the three stresses σ_2, σ_3, τ that act parallel to the dip section. Although it is impossible to determine all three of them without further specification of the type of material behaviour of the sediment, two stress components can be directly determined from the equilibrium equations by making use of additional information that follows from the assumed existence of an infinite (or very large) sediment slab. For the uniform, infinite slab considered, the dip-parallel normal stress σ_2 cannot change in the dip direction, since otherwise the rate of change has to be the same everywhere on the slope which would make σ_2 increase beyond any bound. Therefore, $\delta\sigma_2/\delta x_2 = 0$ for an infinite, uniform slab. Although σ_2 does not vary along dip, it will not be zero and may vary in the x_3-direction. It is obvious that the material tends to expand laterally under the overburden load but that lateral expansion of the sediment slab is impeded by the frictional or cohesive contact with the underlying 'rigid' substratum. Consequently, compressive stresses build up, which makes it understandable that σ_2 cannot be determined from equilibrium considerations alone but is essentially controlled by the deformational response of the material to loading. We shall return to the determination of σ_2 later.

With $\delta\sigma_2/\delta x_2 = 0$ and taking into account that the weight vector makes an angle $\alpha_2 = \pi/2 - \delta$ with the positive x_2-direction, for the second equation II.2-2a we obtain

$$\frac{\delta\tau}{\delta x_3} = \gamma \sin \delta \qquad \text{(II.2-4a)}$$

for the rate of change of the slope-parallel shear stress τ with 'depth'. For a constant specific bulk weight γ this rate of change does not vary with depth and since no shear stress is exerted upon the sediment surface, it follows that at 'depth' x_3

$$\tau(x_3) = \gamma x_3 \sin \delta \qquad \text{(II.2-4b)}$$

and, in particular, along the base of the sediment slab[4)]

$$\tau(h) = \gamma h \sin \delta \qquad \text{(II.2-4c)}$$

The bottom shear stress therefore completely balances the slope-parallel weight component of a cross-sectional element (see Fig. II.2-7A). It should also be noticed that, according to equation II.2-3, the shear stress τ also acts across elements perpendicular to the x_2-direction. Cross-sectional elements, as the one shown in Fig. II.2-7A, therefore interact with each other not only by compressional normal stresses but also by shear stresses.

Since τ does not change in the x_2-direction and $\alpha_3 = \delta$, the third equation II.2-2a yields

$$\frac{\delta \sigma_3}{\delta x_3} = \gamma \cos \delta \qquad \text{(II.2-5a)}$$

Since for a constant γ the stress σ_3 increases at constant rate with 'depth' x_3 and since $\sigma_3 (0) = 0$, the 'subvertical' normal stress becomes

$$\sigma_3 (x_3) = \gamma x_3 \cos \delta \qquad \text{(II.2-5b)}$$

and, in particular,

$$\sigma_3 (h) = \gamma \, h \cos \delta \qquad \text{(II.2-5c)}$$

along the base of the sediment slab [5].

Both stresses τ and σ_3 are known along the base of the sediment slab from the thickness h of the layer and its slope angle δ. Since these stresses have been found without reference to any specific material law, the present case of the uniform, infinite sediment slab may serve – in spite of the drastic idealisations involved – as a valuable reference case for more realistic, but theoretically less accessible modifications.

Example No. 3: Sediment layer with non-uniform bottom shear stress

Such a variant of the preceding example is sketched in Fig. II.2-7B. It is assumed that a local variation in lithology prevents, in a region of length L, the bottom shear stress τ (h) attaining the value required to equilibrate the dip-parallel weight component $\gamma \, h \sin \delta$. It is assumed that in this region the bottom shear stress τ' (h) lags behind the equilibrium value τ (h) by $\Delta\tau (h) \equiv \tau (h) - \tau' (h) = \gamma h \sin \delta - \tau' (h)$.

From the second equation II.2-2a then follows [6]

$$\frac{\delta \sigma_2}{\delta x_2} = \gamma \sin \delta - \frac{\delta \tau}{\delta x_3} = \frac{\delta \Delta \tau}{\delta x_3} = \frac{\Delta \tau (h)}{h} \qquad \text{(II.2-6a)}$$

Therefore, the total down-dip increase $\Delta \sigma_2$ of the 'slope pressure' that is required to compensate for the unbalance caused by the deficit in bottom shear stress becomes

$$\Delta \sigma_2 = \Delta \tau \, (h) \cdot \frac{L}{h} \qquad \text{(II.2-6b)}$$

Naturally, this increase in 'slope pressure' must in turn be balanced somewhere downslope. This may be accomplished, for instance, by the 'buttressing' effect of a regional decrease in slope angle or by a region of increased bottom shear stress (increased bottom friction or cohesion) as indicated in Fig. II.2-7B. The simple formula for $\Delta \sigma_2$ demonstrates that the increase in 'slope pressure' σ_2 can easily attain considerable values and may therefore be expected to cause local folding or other types of compressional features in the sediment slab.

At this point we would briefly mention the important case in which part of the substratum is not at rest but in a state of slow creep flow – possibly driven by a differential overburden load somewhere up-

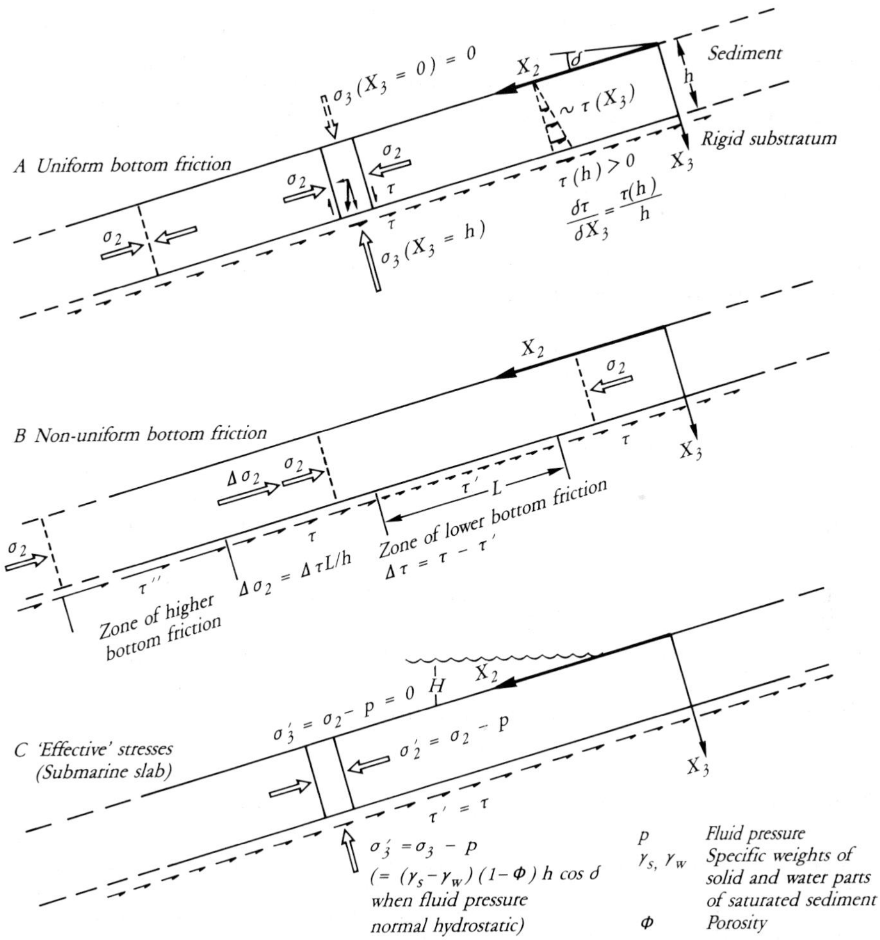

Fig. II.2-7 INFINITE SEDIMENT SLAB ON SLOPE

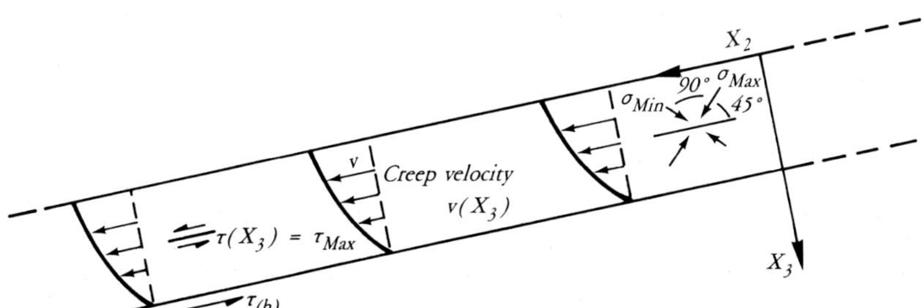

Fig. II.2-8 INFINITE SEDIMENT SLAB IN STATE OF STEADY CREEP (ORIENTATION OF PRINCIPAL STRESSES)

slope. When the sediment slab is prevented from being carried downslope – for instance by the buttressing effect of the slope foot – the creeping substratum will exert a shear stress τ' (h) upon the sediment slab, which is directed downslope, i.e. opposite to the shear stress in Fig. II.2-7. The quantity $\Delta\tau$ (h) then becomes greater than τ (h) = γ h sin δ and, consequently, the downslope increase of the 'slope pressure' σ_2 is much more drastic than in the previous case.

Example No. 4: Sediment layer in state of steady creep

Returning to the uniform, infinite sediment layer on a plane, rigid substratum, we now consider the case where the sediment slab is not strictly at rest but in a state of steady creep flow. This means that the material is creeping slowly down the slope at a speed that remains constant in time and does not vary in the x_2-direction (see Fig. II.2-8). Since there are no accelerations involved, the situation is still a static one and the equilibrium equations II.2-2a still apply. We assume that the flow is parallel to the slope, which means that we neglect a velocity component due to compaction in the x_3-direction. A likely velocity profile is shown in Fig. II.2-8. It is obvious for this type of slope-parallel creep flow that the highest shear stress τ_{max} is transmitted across plane elements parallel to the slope. In other words, slope-parallel planes are planes of maximum shear stress (see Section II.1.3). Consequently, they bisect the angle between the two principal stresses in the x_2, x_3-plane (σ_{Max}, σ_{Min} in Fig. II.2-8) and are acted upon by a normal stress equal to half the sum of the two principal stresses. It was further stated in Section II.1.3 (eq. II.1-3d) that the sum of the two principal stresses in the plane is equal to the sum of the two normal stresses that act on any pair of mutually orthogonal elements whose normals lie in the plane. Therefore, we have in particular

$$\sigma_3 = \frac{\sigma_{Max} + \sigma_{Min}}{2} = \frac{\sigma_2 + \sigma_3}{2} \qquad \text{(II.2-7a)}$$

from which the remarkable result follows that in steady slope parallel creep flow the 'slope-pressure' σ_2 is exactly equal to the 'subvertical' normal stress σ_3:

$$\sigma_2 = \sigma_3 = \gamma\, x_3 \cos \delta \qquad \text{(II.2-7b)}$$

As in example No. 2 the slope-parallel shear stress τ then attains the value II.2-4b and, therefore, varies linearly with depth independently of the rheological properties of the creeping material.

2. EQUILIBRIUM EQUATIONS IN EFFECTIVE STRESSES

When dealing with porous rock, one has to be aware that the equilibrium equations II.2-2 are formulated for 'total' stresses (i.e. the stresses transmitted through both the solid and fluid parts of a cross-sectional bulk element). In rock mechanics it is, however, often necessary to account for effects of the pore pressure and to separate therefore the pore pressure p from the total stress (see Chapter II.3). To this end, the concept of 'effective' stress was introduced, which implies for any normal component σ' of the effective stress

$$\sigma' = \sigma - p$$
and
$$\tau' = \tau \qquad \text{(II.2-8)}$$

for any shear stress.

With the expressions for the effective normal stresses substituted in equations II.2-2 the equilibrium equations read:

$$\begin{aligned} \frac{\delta\sigma_1'}{\delta x_1} + \frac{\delta\tau_{21}}{\delta x_2} + \frac{\delta\tau_{31}}{\delta x_3} &= \gamma \cos\alpha_1 - \frac{\delta p}{\delta x_1} \\ \frac{\delta\tau_{12}}{\delta x_1} + \frac{\delta\sigma_2'}{\delta x_2} + \frac{\delta\tau_{32}}{\delta x_3} &= \gamma \cos\alpha_2 - \frac{\delta p}{\delta x_2} \\ \frac{\delta\tau_{13}}{\delta x_1} + \frac{\delta\tau_{23}}{\delta x_2} + \frac{\delta\sigma_3'}{\delta x_3} &= \gamma \cos\alpha_3 - \frac{\delta p}{\delta x_3} \end{aligned} \qquad \text{(II.2-2b)}$$

In these equations the pore pressure gradient can be further expressed in terms of specific fluid weight γ_w and rate of volume flow through a unit bulk area. According to Darcy's flow law (see Chapter II.10; eq. II.10-1) the rate of flow V_i of a single fluid through a unit bulk section whose normal points along the positive x_i-axis (i=1, 2, 3) is related to the pore pressure gradient by

$$\frac{\eta}{k} V_i = -\left(\frac{\delta p}{\delta x_i} - \gamma_w \cos\alpha_i\right) \quad (i = 1, 2, 3) \qquad \text{(II.2-9)}$$

where η is the fluid viscosity, k the rock permeability, α_i the angle the specific weight vector includes with the positive x_i-axis. Substituting $\delta p/\delta x_i$ in equations II.2-2b and expressing the specific bulk weight in terms of the specific solid weight γ_s and fluid weight γ_w by $\gamma = (1-\Phi)\gamma_s + \Phi\gamma_w$ (Φ being the porosity) one obtains the equilibrium equations in the form

$$\begin{aligned} \frac{\delta\sigma_1'}{\delta x_1} + \frac{\delta\tau_{21}}{\delta x_2} + \frac{\delta\tau_{31}}{\delta x_3} &= (1-\Phi)\ (\gamma_s-\gamma_w) \cos\alpha_1 + \frac{\eta}{k} V_1 \\ \frac{\delta\tau_{12}}{\delta x_1} + \frac{\delta\sigma_2'}{\delta x_2} + \frac{\delta\tau_{32}}{\delta x_3} &= (1-\Phi)\ (\gamma_s-\gamma_w) \cos\alpha_2 + \frac{\eta}{k} V_2 \\ \frac{\delta\tau_{13}}{\delta x_1} + \frac{\delta\tau_{23}}{\delta x_2} + \frac{\delta\sigma_3'}{\delta x_3} &= (1-\Phi)\ (\gamma_s-\gamma_w) \cos\alpha_3 + \frac{\eta}{k} V_3 \end{aligned} \qquad \text{(II.2-2c)}$$

These equations allow a simple mechanical interpretation if – and only if – the rock is composed of particles whose contacts with other particles can be approximately considered as *'point' contacts* (see Chapter II.3; 'point' contacts). In that case we may think of a unit bulk element that is delineated by a closed 'wavy' surface that intersects the solid skeleton at interparticle point contacts only. The solid material of the unit bulk element may then be considered to be completely surrounded by fluid and to experience full weight reduction by buoyancy (see Chapter II.9). The first terms on the right of equations II.2-2c represent the components of this reduced solid weight ('submerged' specific weight)[7)]. The second terms on the right of the equations are the components of the drag force exerted by the moving fluid upon the solid material inside the unit bulk element. Hence, in the case of the interparticle point contacts the right-hand sides of equations II.2-2c may be interpreted as the sum of all body forces (including the fluid drag) that act upon the solid skeleton inside a unit bulk volume. This force is balanced by the net force exerted upon the solid material through its contacts with the surrounding skeleton. Since, in the case of point contacts, the stress transmitted across the contacts is identical with the 'effective' stress (see Chapter II.3; 'intergranular stress', 'point contacts') the left-hand sides of equations II.2-2b, c represent the components of just this net force.

Example No. 5: Submarine sediment on 'infinite' slope

In a plane case equations II.2-2b, c simplify in a similar way to the equations for the total stresses. To illustrate application of the effective stress equations we turn again to the uniform, inclined sediment slab of example No. 2, but we now assume (see Fig. II.2-7c) that the sediment is overlain by water (Terzaghi, K., 1950, Hubbert, H.K. & Rubey, W.W., 1959). Although in this case the sediment top is not free of stress but has to carry the load of the overlying water column, the effective stress σ_3' $(x_3=0)$ is zero and so is the shear stress $\tau(x_3=0)$, since fluid at rest does not exert shear stresses. It is furthermore assumed that a possible fluid flow will be perpendicular to the layer and independent of x_2 i.e. $V_2 = 0$, $V_3(x_3) \neq 0$. In nature, this would correspond with a sediment layer showing the same compaction behaviour all along the slope. Then compaction could not generate a driving pressure gradient (i.e. a gradient different from the hydrostatic one $\gamma_w \cos\alpha_2$) parallel to the slope (x_2-direction).

It is quite obvious that in such a uniform layer, with body and fluid forces independent of x_2 and its top surface free of effective stresses, the state of effective stress will not vary in the x_2-direction. Therefore, with $\alpha_2 = \pi/2 - \delta$, $\alpha_3 = \delta$ and $\tau_{12} = \tau_{13} = 0$ (plane conditions) equations II.2-2c reduce to

$$\frac{\delta\tau}{\delta x_3} = \gamma' \sin\delta \qquad \gamma' = (1-\Phi)(\gamma_s - \gamma_w)$$
$$\frac{\delta\sigma_3'}{\delta x_3} = \gamma' \cos\delta + \frac{\eta}{k} V_3 \qquad \text{(II.2-10a)}$$

From these equations, in complete analogy with Example No. 2, the effective stresses are found

$$\tau(x_3) = \gamma' x_3 \sin\delta$$
$$\sigma_3'(x_3) = \gamma' x_3 \cos\delta + F_3(x_3) \qquad \text{(II.2-10b)}$$

where $F_3(x_3)$ represents the resultant drag force exerted by the flowing water upon the material inside a column with unit cross-section and extending from the sediment surface down to the 'depth' x_3.[8)]

For water being squeezed out of the compacting sediment in an upwards direction the flow rate (V_3) is counted negative in accordance with the reduction in submerged sediment weight caused by the drag force of the percolating water. Clearly, the *pore pressure* in the flow region will be higher than hydrostatic, most noticeably in strata of very low permeability. It is a common observation that in rapidly sedimenting delta slopes below a certain depth and over a certain interval the pressure may increase at approximately the same rate as the total overburden stress σ_3. In this depth interval the effective overburden stress σ_3' therefore remains approximately constant, which means, according to the last formula, that in this depth interval the fluid drag compensates for the x_3-component of the submerged sediment weight. Hence, in that depth range the intergranular contacts only carry the submerged sediment weight of the upper strata in which the fluid drag is less (because of higher permeability of the material). In extreme cases, the drag force in a low permeability stratum may even become large enough to carry the submerged weight of the whole sediment column on top of that stratum. Since in this stratum σ_3' becomes zero, the intergranular stresses vanish – at least in the case of point-like contacts – and the material becomes *'liquefied'*.

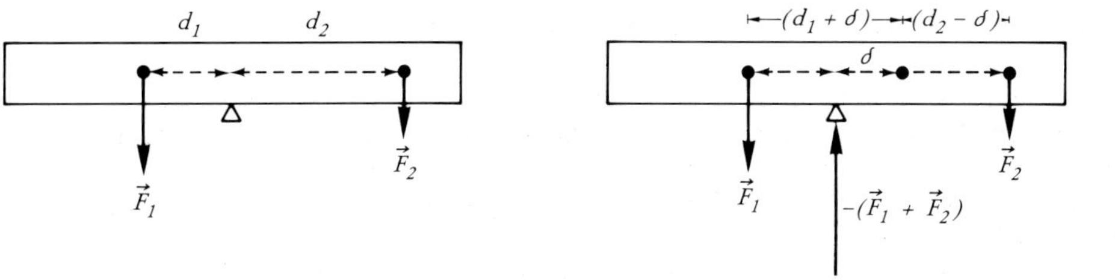

$|\vec{F}_1| d_1 = |\vec{F}_2| d_2$

A *Balance of moments*

$-|\vec{F}_1|(d_1 + \delta) + (|\vec{F}_1| + |\vec{F}_2|)\delta + |\vec{F}_2|(d_2 - \delta)$
$= -|\vec{F}_1| d_1 + |\vec{F}_2| d_2 = 0$

B *Shift of reference point does not affect sum of moments*

Fig. II.2-9 MOMENTS ON LEVER

3. STATIC EQUILIBRIUM IN THE 'LARGE'

When the equilibrium equations II.2-2 and II.2-3 are satisfied everywhere inside a material body, the body as a whole is in static equilibrium. For the external forces acting upon the body this implies that: (1) the vectorial sum of all these forces will vanish (Fig. II.2-1), (2) the forces do not make the body spin around.

The first condition, given at the very beginning of this chapter, means that the sums of the x_1-, x_2- and x_3-components of the force vectors will vanish separately. The second condition needs some comment and a more precise formulation. It obviously holds if none of the volume elements of the body is made to rotate by the forces acting upon it. This is guaranteed by the equilibrium condition II.2-3, as discussed earlier. But how can we establish, without knowledge of the internal stresses, whether or not a given system of external forces will make a body spin around? As we shall see later, this question is of great importance for a correct mechanical formulation of various tectonic problems.

Moment of Force ('Torque'). It has already been discussed (Fig. II.2-3) that the vanishing of the resultant external force does not exclude rotational accelerations. If we load a lever (Archimedes, 287-212 B.C.) that is in static equilibrium (Fig. II.2-9A) by two weights $\vec{F}_1$ and $\vec{F}_2$ in points at distances d_1 and d_2 from the fulcrum, it is not sufficient to support the lever by a vertical force of magnitude $|F_1| + |F_2|$ to maintain equilibrium. In addition, the weights must be inversely proportional to their distances from the fulcrum, i.e. $|\vec{F}_1| d_1 = |\vec{F}_2| d_2$, otherwise the lever will rotate out of its horizontal position.

Products of the magnitude of a force and its 'arm', i.e. the perpendicular distance of its line of action from an arbitrary reference point (here the fulcrum of the lever), are called *moments of force* ('torques'). Their values are counted positive or negative depending on whether the force tends to rotate the body clockwise or anticlockwise. In the case of the lever (Fig. II.2-9A) the two moments are $M_1 = |\vec{F}_1| d_1$ and $M_2 = -|\vec{F}_2| d_2$, and Archimedes' principle of the lever states that in equilibrium $M_1 + M_2 = 0$, i.e. the sum of the moments vanishes. It is important to note that, while the moments themselves depend on the choice of the reference point, the sum of all the moments does not. If it vanishes for one common reference point (e.g. the fulcrum of the lever) it will vanish for any other point as well, as is demonstrated in Fig. II.2-9B.

The definition of the moment of a force $\vec{F}$ with respect to a point P is described in detail in Fig. II.2-10. The plane defined by $\vec{F}$ and P may be assumed to be parallel to a coordinate plane, say the one with normal parallel to the x_2-axis of a Cartesian frame. Since the moment of $\vec{F}$ then operates about an axis parallel to the x_2-axis, we denote its magnitude by $M_{(2)}$. It is defined as

$$M_{(2)} = |\vec{F}| \cdot d = |\vec{r}| \cdot |\vec{F}| \cdot \sin \alpha \qquad \text{(II.2-11a)}$$

where $\vec{r}$ is the vector pointing to reference point P from the location where $\vec{F}$ acts upon the body. The moment can also be expressed in terms of components of $\vec{r}$ and $\vec{F}$ with respect to the x_1-, x_3-axes

$$M_{(2)} = r_3 F_1 - r_1 F_3 \qquad \text{(II.2-11b)}$$

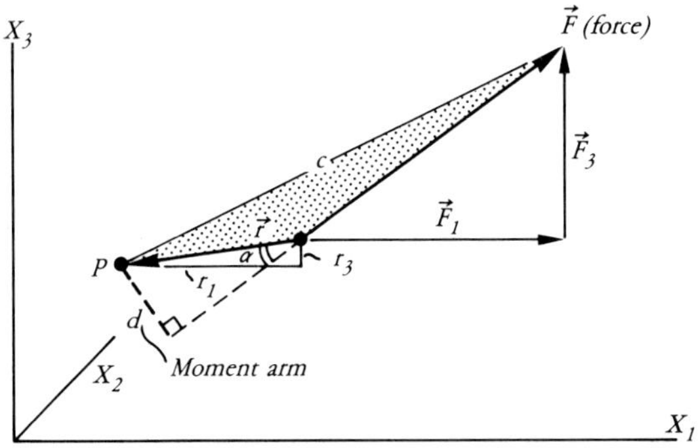

Moment of force $\vec{F}$ with respect to P (both $\vec{F}$ and P in X_1, X_3 – plane):

[1] $$M_{(2)} = |\vec{F}|\, d = |\vec{r}|\,|\vec{F}| \sin \alpha$$

or by adding moments of $\vec{F}_1$, and $\vec{F}_3$

[2] $$M_{(2)} = r_3 |\vec{F}_1| - r_1 |\vec{F}_3|$$

Proof that 1 and 2 are identical:

$$c^2 = (F_1 + r_1)^2 + (F_3 + r_3)^2$$

cosine rule applied to shaded triangle:

$$c^2 = |\vec{r}|^2 + |\vec{F}|^2 + 2|\vec{r}|\,|\vec{F}| \cos \alpha$$

[3] $$|\vec{r}|\,|\vec{F}| \cos \alpha = r_1 F_1 + r_3 F_3 \qquad \text{('scalar product' } \vec{r} \cdot \vec{F})$$

[4] $$|\vec{r}|\,|\vec{F}| \sin \alpha = \sqrt{(r_1^2 + r_3^2)(F_1^2 + F_3^2)(1 - \cos^2 \alpha)}$$

. . . .

[3], [4] $$\rightarrow |\vec{r}|\,|\vec{F}| \sin \alpha = \sqrt{(r_3 F_1 - r_1 F_3)^2}$$

Fig. II.2-10 MOMENT OF FORCE

where the terms on the right represent the moments of the force components. An elementary proof of the equivalence of the expressions II.2-11a and II.2-11b is given in Fig. II.2-10.[9]

The equilibrium condition established by Archimedes for the lever has been found to apply in straightforward generalisation to any body in static equilibrium. If all forces acting on a body are parallel to a single plane (Fig. II.2-11), the sum of the associated moments must vanish in static equilibrium:

$$\Sigma M_{(2)} = 0 \qquad \text{(II.2-12)}$$

If not all forces acting on a body are parallel to a single plane, one has to consider the orthogonal projections (components) of the forces and vectors $\vec{r}$ on three mutually perpendicular planes, and to calculate for each force the moments $M_{(1)}$, $M_{(2)}$, $M_{(3)}$ of its components in the respective planes. In equilibrium the three sums of corresponding moments must vanish.[10]

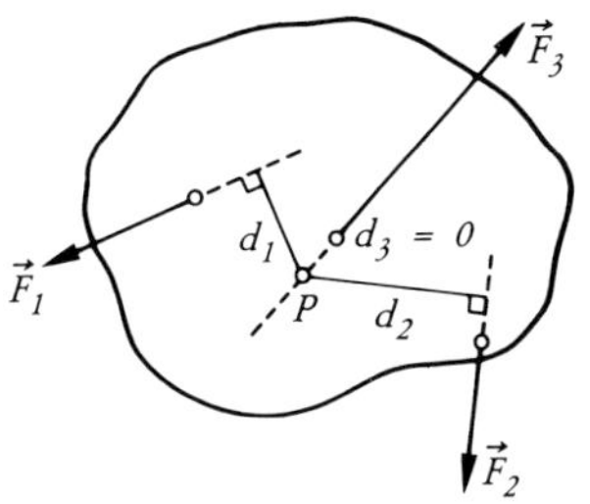

Equilibrium:

$\vec{F}_1 + \vec{F} + \vec{F}_3 = 0$

$- F_1 d_1 + F_2 d_2 + F_3 \cdot 0 = 0$

Fig. II.2-11
FORCES IN PLANE, EQUILIBRIUM OF MOMENTS ABOUT POINTS IN PLANE

Example No. 6: Overthrusting

The stress field, which in a rock body is induced by interaction with the surrounding material, is the key to the mechanical understanding of tectonic features in that body. Its calculation requires proper *specification of the boundary and volume forces* acting upon the body under consideration. In some tectono-mechanical problems one may assume the boundary forces to be known all around the boundary of the body[11], for instance in treating the case of a competent thrust sheet that has been detached from its substrata and subsequently been pushed over incompetent strata. Such an example is shown in Fig. II.2-12A, B, C together with the boundary forces (or stresses) that have been postulated in different mechanical treatments of overthrusting. The idealised thrust sheet is of rectangular cross-section and moves horizontally.

For a correct formulation of the problem the external forces acting upon the thrust sheet must be in equilibrium, i.e. resultant force *and* moment must vanish. In all three variants shown in Fig. II.2-12 the resultant force does indeed vanish; the weight of the sheet is supported by the vertical pressure at the base of the sheet, and the thrust force is balanced by the total basal friction and the buttressing force that may act upon the front face of the sheet. The thrust force must, therefore, equal or exceed in magnitude the basal frictional force in order to maintain motion. This simple equilibrium condition has an important consequence that was first recognised by the physicist M. Smoluchowski in 1909: If 'normal' rock friction were active along the base of a long, moving thrust sheet, the required thrust force would far exceed the limit the strongest rocks could withstand without destruction. Consequently, large overthrust sheets can only have moved when some 'lubricating' mechanism along the basis facilitated sliding.[12)]

Next, we inspect the moments of forces in Fig. II.2-12A,B,C. We may conveniently choose the midpoint at the base (Fig. II.2-12A) as our reference point P. We notice that the weight of the sheet and the basal frictional forces have no moment with respect to P, since their 'arms' vanish in all three cases of Fig. II.2-12. In Fig. II.2-12A the basal pressure is symmetrically distributed with respect to P and therefore has a vanishing net moment. This is not so with the thrust force, whose arm does not vanish and which therefore tends to rotate the thrust sheet in a clockwise direction. The boundary forces in the formulation of Fig. II.2-12A are therefore not quite correctly prescribed. The formulations in Fig. II.2-12B and C, however, are consistent with the requirements of equilibrium. In Fig. II.2-12B the moment of the thrust forces is mainly balanced by the moment of tangential forces that are assumed to act along the front face of the sheet and whose arms therefore equal half the length of the thrust sheet. In the formulation of Fig. II.2-12C, finally, no such tangential forces were assumed to act along the front of the sheet but, instead, an increase of basal pressure in the thrust direction provides the compensating moment. In the case of a high length/height ratio of the thrust sheet it takes only a small increase in basal pressure to produce the required moment, since associated moment arms are then much larger than those of the thrust forces.

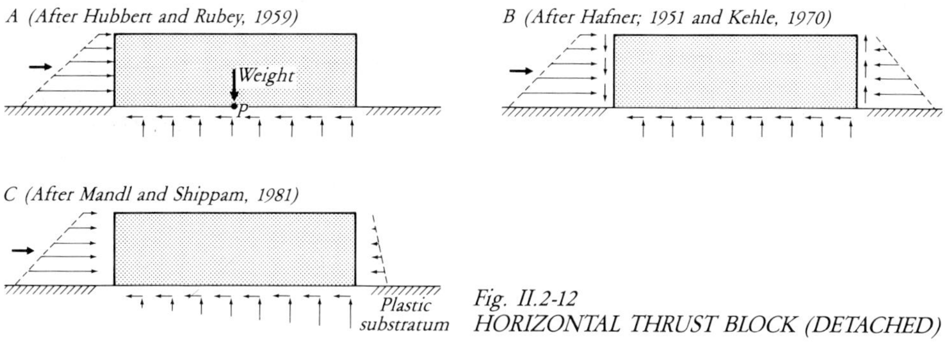

Fig. II.2-12
HORIZONTAL THRUST BLOCK (DETACHED)

Example No. 7: Overthrust 'toe'

When a thrust sheet is detached from its base, a frontal part will usually have to ride up an inclined fault plane – the 'toe' fault – as is shown schematically in Fig. II.2-13. The friction along the toe fault and the weight of the rock in the 'toe' region add an extra resistance to overthrusting. To account for this, Raleigh and Griggs (1963) have included the effect of an idealised triangular toe in the general force balance discussed in the previous example.

Restricting ourselves to the case of horizontal overthrusting, we notice first that the horizontal push force F, which acts on vertical cross-sections of the thrust sheet, will have to decrease from rear to front, since the total basal friction it has to overcome diminishes. If we consider, for instance, the part of the thrust sheet of length ΔL bordering the toe region (Fig. II.2-13), the balance for the horizontal push force becomes

$$F = F' + \tau_b \, \Delta L \qquad \text{(II.2-13a)}$$

where F and F′ act upon rear and front sections, respectively, and τ_b is the uniform basal shear stress. The force F′, in turn is balanced by the counteracting toe forces. With reference to Fig. II.2-13 the balance for the force components parallel to the toe fault becomes

$$F' \cos \beta = T' + W' \sin \beta \tag{II.2-13b}$$

As overthrusting proceeds, the toe will rise above the original surface. Consequently, both the frictional resistance of the toe and its weight will increase and the thrust force F has to be raised accordingly to maintain motion of the horizontal thrust sheet. Eventually, the increase in F may lead to large-scale rock failure and imbrication of the thrust sheet. The critical value of F at which imbrication of the thrust sheet would be initiated can be estimated very roughly by considering a potential thrust plane (dotted in Fig. II.2-13). In order to move the wedge-shaped block along the potential thrust plane, the component F cos β of the horizontal thrust force has to be in equilibrium with the cohesive and frictional resistance T of the unbroken rock and its weight component. Hence

$$F \cos \beta = T + W \sin \beta \tag{II.2-13c}$$

When the balances for the toe (II.2-13b) and for the potential imbricate thrust (equation II.2-13c) are compared, it should be noted that the resistance T along the potential thrust fault is higher than the resistance T′ along the toe fault. This has several causes. The cohesive strength of the unbroken rock is higher than that of the toe fault material. Likewise, the angle of internal friction at fault initiation is greater than that along a fully developed fault in the same material. Further, the effective normal stress that acts across the potential fault plane is higher than that across the toe fault since F > F′ and since, most likely, the pore pressure is higher along the toe fault. Therefore, for moderate differences in weights W and W′, conditions for the generation of new imbricate faults will not be satisfied within a certain distance of the toe. The critical distance from the toe is found as ΔL by inserting F and F′ from equations II.2-12b,c into equation II.2-13a:

$$\Delta L = \frac{T - T' + (W - W')\sin \beta}{\tau_b \cos \beta} \tag{II.2-14}$$

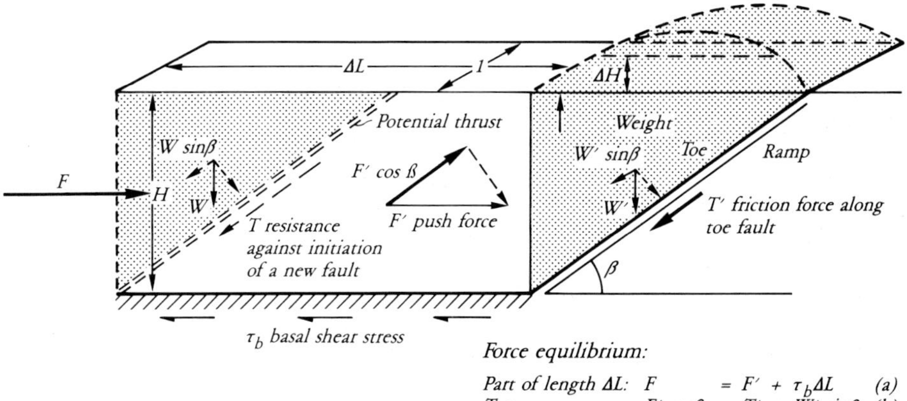

Force equilibrium:

Part of length ΔL:	F	$= F' + \tau_b \Delta L$	(a)
Toe:	$F' \cos\beta$	$= T' + W' \sin\beta$	(b)
Imbricate wedge:	$F \cos\beta$	$= T + W \sin\beta$	(c)

Fig. II.2-13 OVERTHRUST WITH 'TOE'

This formula permits an interesting interpretation. As overthrusting of an intact sheet starts, the toe wedge starts riding up the toe fault, at which instant $W = W'$ and ΔL has a positive value. Without specifying the basal stress τ_b and the shear resistances T and T′ along potential and fully developed faults, we know that in the initial stage of overthrusting ΔL must exceed the length of the thrust sheet, since otherwise the sheet would already have been imbricated. As the toe is gradually protruding above the original surface, W′ becomes larger than W and ΔL therefore decreases in value. As the 'buttressing' effect of the toe increases further ΔL may become smaller than the total length of the overthrust sheet and a first imbricate thrust may be formed *near the rear* of the sheet (i.e. outside the distance ΔL from the toe). Furthermore, we note that the length ΔL of the imbricate-free part of a thrust sheet is inversely proportional to the basal shear strength τ_b. According to Coulomb's law (Section II.7.3, eq. II.7-1c) this shear strength consists of a cohesive part and of a frictional part which is proportional to the effective pressure that acts upon the basal fault plane. A typical value for the shear strength at the base of a large thrust sheet would be $\tau_b = 20$ bar, while the average shear strengths of the unbroken rock and of the plane of the toe fault may easily differ by 200 bar. $T - T'$ would then become 200 bar $H/\sin\beta$, where H is the thickness of the thrust sheet. Inserting these values and $\beta = 30°$ into equation II.2-14 we obtain for ΔL at the beginning of the overthrust motion ($W = W'$) the result $\Delta L = 25$ H, which compares well with reality.

The simple formula II.2-14 that was derived from global force balances comprises much of the overthrust theories by Hubbert and Rubey (1959)[12)], Raleigh and Griggs (1963), and Hsü (1968, 1970)[13)]. Although the formula may demonstrate the usefulness of considering in tectonic problems the force equilibrium 'in the large', it should not be overlooked that the derivation of the formula implied several drastic simplifications, in particular concerning the shape of the faults and the complete neglect of any shear stress along the vertical cross-sections of the thrust sheet. One should therefore not expect formula II.2-14 to give more than a very rough estimate of the imbrication-free domain of a thrust sheet and to be of explanatory rather than of predictive value.

NOTES

1) In determining the gradient of σ_3 in x_3-direction, differentiation operates, of course, on $\sigma_3(x_3)$ as defined on the *same* side of parallel elements $x_3 = \text{const.}$

2) The proof of eqs. II.2-3 usually presented in textbooks makes use of the mathematical concept of moment of force and is based on the balance of moments. The proof given here avoids mathematics and may appeal to the geologist's preference for symmetry considerations.

3) The same result, also valid for σ_1 or any other fault-parallel normal stress, could easily be derived from the more general equations II.2-2.

4) If the sediment slab consists of n slope-parallel strata of different specific weights γ_i and individual thicknesses h_i, this formula becomes, as can be easily seen, $\tau(h) = (\gamma_1 h_1 + \gamma_2 h_2 + \ldots + \gamma_n h_n) \sin\delta$.

5) For a layered slab: $\sigma_3(h) = (\gamma_1 h_1 + \gamma_2 h_2 + \ldots + \gamma_n h_n) \cos\delta$.

6) This result has only an approximative character, since no allowance has been made for the effect of the finite length L of the zone with reduced shear stress on the stress field.

7) If the interparticle contacts cannot be approximated as point contacts, the equations II.2-2c, of course, remain correct, but their left-hand sides no longer represent a force that is the exclusive result of interparticle interaction with the surrounding skeleton material.

8) $$F_3\,(x_3) = \eta \int_0^{x_3} \frac{V_3}{k}\, dx_3$$

9) The reader familiar with vector algebra will recognise the vector cross product

$$\vec{M} = \vec{r} \wedge \vec{F}$$

which represents a vector of magnitude $|\vec{F}|\,|\vec{r}|\sin\Theta$ (Θ being the angle included by $\vec{r}$ and $\vec{F}$) with a direction normal to the vectors $\vec{r}$ and $\vec{F}$. The vector $\vec{M}$ is counted as positive when its direction coincides with the axial motion of a right-handed screw that turns $\vec{r}$ into $\vec{F}$. The Cartesian components of $\vec{M}$ are $r_2F_3 - r_3F_2$, $r_3F_1 - r_1F_3$, $r_1F_2 - r_2F_1$. As long as all forces and the common reference point lie in one plane, the associated moments have only one non-vanishing component, for instance, equation II.2-11.

10) This means, in terms of vector algebra, that the three components of the moment vectors 9) $\vec{M}$ have to be added separately and that their sums must vanish.

11) In other cases the boundary forces are only known on part of the total boundary, while on the remainder they are the a priori unknown reactions of the body to imposed constraints, such as fixed rigid walls. The determination of these reaction forces is then part of the problem.

12) Anomalously high pore pressures were introduced as 'lubricating' mechanisms in overthrusting by M.K. Hubbert and W.W. Rubey in their classic paper (1959) which adapted Terzaghi's concept of pore-pressure-induced land slides. The high pore pressure reduces the effective overburden stress which controls the frictional resistance to sliding (see Section II.7.2).

13) K. Jinghwa Hsü (1970) derived a slightly incorrect but otherwise similar formula for ΔL. Apparently, he made the tacit assumption that the frictional parts of resistances T and T′ were equal, which reduced T − T′ in equation II.2-14 to a difference in cohesive strengths. Continuous erosion of any protruding part of the toe will then leave the toe resistance of the thrust sheet unchanged, while the total frictional resistance along the basal thrust plane will be reduced by the concomitant shortening of the sheet. Under these conditions an imbrication of the thrust sheet would seem hardly feasible at all.

Rising of the toe above the original suface causes an increase in W′ and, therefore, a decrease in ΔL. This appears to be interpreted by Hsü as a decrease in the distance of the location of active frontal thrusting from the rear of the thrust sheet, and this creates the impression that, on mechanical grounds, imbrication of a horizontally moving thrust sheet will proceed from front to rear. Just the opposite is correct, as our discussion should show, since the reduced value of ΔL defines the distance of a new (imbricate) thrust from the original toe thrust.

COLLATERAL READING:

See the references given in Chapter II.1.

II.3. STRESSES IN POROUS ROCKS

Porous rocks and soils are highly heterogeneous systems of solid material and pore fluid. Stress, however, is defined for material continua only. The application of the stress concept to porous rocks and soils therefore requires that these materials are somehow idealised as continua (or as ensembles of continua) by disregarding 'small-scale' heterogeneities. It is important to have a clear understanding of the basic abstractions involved in the continuum approach to porous materials to avoid misconceptions – in particular concerning the mechanical effect of the pore fluid in porous rocks.

1. *MICRO- AND MACRO-SCALE*

Obviously, it would not make sense to talk about stresses that act between the atoms of a rock, but it does make sense to talk about the stresses that act inside the individual particles of which a rock is composed. We may, for instance, study the stresses that develop inside the individual grains of a sandy material under load, to learn which type of grain will be crushed first and how the process of cataclasis proceeds. In this case each grain is regarded as a separate material continuum and the size of the elements to which the stresses are referred is small compared with the grain size or other characteristic dimensions of the 'micro'-structure of the rock or soil material. In general, however, the mechanical theory of tectonic processes is not concerned with such 'micro'-stresses but with stresses that are transmitted across elements of a size much larger than the particulate structure of the rock or soil material.

Choosing the size of real or imaginary surface elements to be large with respect to material heterogeneities implies that the strong fluctuations in the micro-stresses that are usually associated with the material heterogeneities (e.g. variations of stresses from grain to grain) are smoothed out by forming a meaningful mean value of the micro-stresses that act across the 'large' element. It is a matter of judgement how large one might actually choose the scale of surface and volume elements to be. For rough estimates of a stress field it may, for instance, be tolerable to neglect heterogeneities due to stratification or jointing.

The fluid-filled porous rock as a single continuum. It should be immediately emphasised at this point that the choice of 'large' surface and volume elements does not mean that surface, cross-sections or bulk of a rock body are thought of as being divided into a finite number of elements. This would constitute a discontinuum rather than a continuum. The actual procedure is illustrated in Fig. II.3-1. Any point (P, P′, P″, ...) of a porous rock – irrespective of whether it is situated in the solid or fluid part – is made the centre of an imaginary surface or volume element. Thus the centres of 'neighbouring' elements are infinitely close. Furthermore, any 'macroscopic' quantity we define on such elements (e.g. porosity, the stress transmitted across the element, or the specific volume force acting upon a volume element) is associated with the element centres and, therefore, becomes a function of position throughout the region occupied by the bulk (solid plus fluid) of the material. This means that at any given point P not only quantities that are representative of the solid/fluid mixture are defined, as for instance the specific bulk weight or the total stress on a surface element, but also any macro-quantity associated with the solid *or* the fluid component of the rock.

2. *FLUID AND SOLID STRESSES*

Pore Pressure. As an example we may consider the pore pressure. It is defined at any point P – regardless of whether this point lies inside the pore space or the solid skeleton of the rock. The pore pressure is simply defined as the sum of all forces ($\vec{f}_f$; see Fig. II.3-1) that act perpendicularly across the fluid part of a 'large' surface element centred at P, divided by the *fluid-filled area* of the element (force per unit fluid area). This pressure would be measured by a standpipe (see Fig. II.3-2) with its 'large' orifice centred at P. Naturally, the pore pressure thus defined is the average of the 'micro' pressures that act in the many pores cut by the macroscopic surface element. (If the element is horizontal, the pore space interconnected, and the fluid at rest, then average pressure and micro-pressure become identical.) These remarks on the pore pressure concept should be sufficient to avoid the notorious confusion as to whether the fluid pressure is actually 'effective' inside the solid skeleton of a porous rock or not.

The example of the pore pressure demonstrates how one can account – within the context of a single material continuum – for the material properties and the state of stress pertaining on a macro-scale to either the solid or the fluid component of the fluid-filled porous material. This is of great importance for the mechanical theory of porous rocks, since it permits the incorporation of the pronounced effects of the fluid pressure on the deformation and failure behaviour of porous rocks and soils dealt with in the chapters on 'Plasticity of rocks' (II.7), Poro-elasticity (II.6) and Buoyancy in porous rocks (II.9). In addition to 'pore pressure' the theory utilises *different macroscopic stress concepts* which will be briefly discussed in the following.

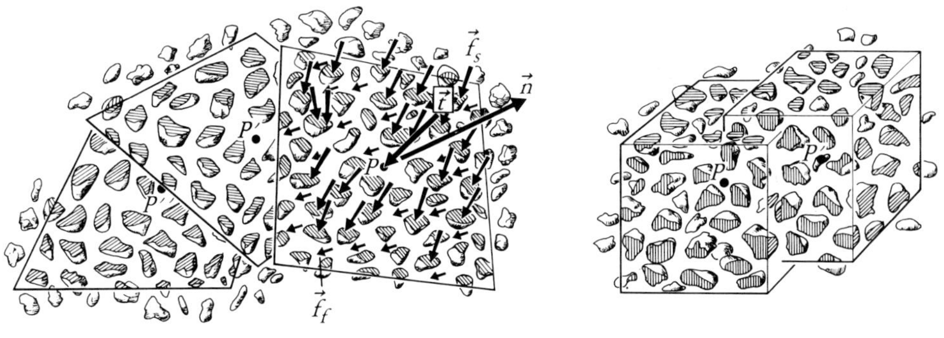

Fig. II.3-1 SURFACE AND VOLUME ELEMENTS IN POROUS MATERIAL
Elements are large as compared with the micro structure (pores, grains) of the rock. The elements overlap and the centres of 'neighbouring' elements are arbitrarily close.

Total Stress. The 'total stress' ('bulk' or 'gross' stress) $\vec{t}$ is the sum of all forces ($\vec{f_s}$ and $\vec{f_f}$; see Fig. II.3-1) that act across the fluid and solid parts of a macroscopic surface element, divided by the *total area* of the element.

Skeletal stress. The 'skeletal stress' $\vec{t_s}$ is the sum of all forces $\vec{f_s}$ that act across the solid parts of a macroscopic surface element, divided by the *total area* of the element.

Fluid stress. The 'fluid stress' $\vec{t_f}$ is the sum of all forces $\vec{f_f}$ that act across the fluid parts of a macroscopic surface element, divided by the *total area* of the element. When the fluid is at rest ('hydrostatic') the fluid forces act strictly perpendicular to the surface element and contribute the fluid stress $\vec{t_f} = \Phi \cdot p \cdot \vec{n}$, where Φ is the porosity (= pore area/bulk area of the element = pore volume/bulk volume), p the 'pore pressure' and $\vec{n}$ the vector of the unit normal of the element. The same relationship holds for 'seepage' flow, since tangential components of the microscopic fluid forces $\vec{f_f}$ on the macroscopic surface element cancel in the mean.[1)]

The three stresses are simply related by: $\vec{t} = \vec{t_s} + \vec{t_f}$, which further entails for any normal stress component (σ) and any tangential (shear-)stress component (τ):

$$\sigma = \sigma^s + \sigma^f = \sigma^s + \Phi \cdot p, \qquad \tau = \tau^s + \tau^f = \tau^s \qquad \text{(II.3-1)}$$

Effective Stress. The 'effective stress $\vec{t}'$ is defined as the total stress minus pore pressure:

$$\vec{t}' = \vec{t} - p \cdot \vec{n} \qquad \text{(II.3-2)}$$

which implies for any normal component

$$\sigma' = \sigma - p \qquad \text{(II.3-2a)}$$

and any tangential (shear-)component

$$\tau' = \tau \qquad \text{(II.3-2b)}$$

The 'effective stress' is very useful in accounting for the effects of pore pressure on the mechanical behaviour of porous rocks. The attribute 'effective' is justified by the fact that this stress controls the shearing strength of soils (Terzaghi 1923). To a large extent, it also controls the linearly elastic and visco-elastic deformation and the failure behaviour of fluid-filled porous rocks (see Chapters II.6 and II.7). 'Effective stress' also controls slip on existing fault planes and joints, the opening of cracks in porous rocks and the fluidisation of loose clastic material.

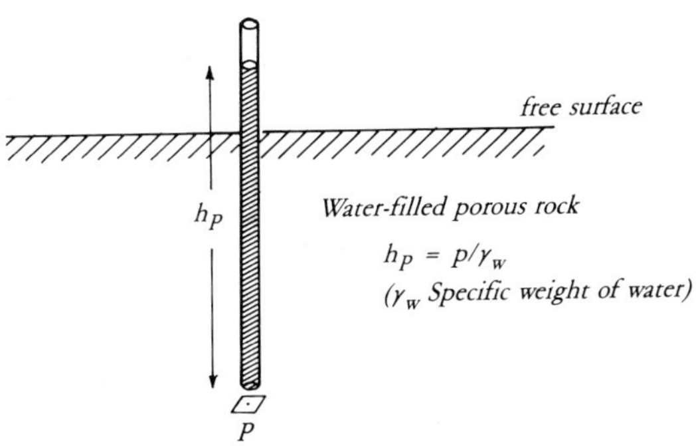

Fig. II.3-2
PORE PRESSURE p (P).
Height of water column h_P in standpipe above P measures pore pressure at P.

Although the definition of 'effective stress' by itself is perfectly clear, the concept has been a permanent source of controversy since its introduction by Terzaghi. This is understandable in view of the fact that the range of applicability of the concept has not yet been sufficiently established by theory. We shall deal with this problem in more detail in the chapters on 'Poro-elasticity' (Section II.6.3) and 'Plasticity of rocks' (Section II.7.2).

Intergranular stress. A stress whose mechanical significance is more evident is the 'intergranular stress' $\vec{t}_{gr}$. Recalling that the skeletal stress $\vec{t}_s$ is transmitted across a plane section that cuts at random through the solid skeleton of the rock (Fig. II.3-1), we may also define a macro-stress that results from the forces that act at the intergranular contacts of a porous rock of granular structure. To this end, we replace the plane surface element that is centred at a point P (Fig. II.3-3A) by a 'wavy' element (Fig. II.3-3B) that follows the plane element as closely as possible, but intersects the solid skeleton at intergranular contacts only. We next project the fluid part of the 'wavy' element on the associated plane element and define a *'porosity Ψ of the 'wavy' element*:

$$\Psi = \text{projected fluid area/total area of plane element} \tag{II.3-3}$$

The 'intergranular stress', defined as the sum of all forces $\vec{f}_c$ that act across the intergranular contacts of the 'wavy' element, is then related to the 'total stress' $\vec{t}$ and the 'skeletal stress' $\vec{t}_s$ by the simple relations [2)]

$$\vec{t}_{gr} = \vec{t} - \Psi \cdot p \cdot \vec{n} \text{ and } \vec{t}_{gr} = \vec{t}_s + (\Phi - \Psi)\, p \cdot \vec{n} \tag{II.3-4}$$

where $\vec{n}$ is the unit normal of the plane surface element.

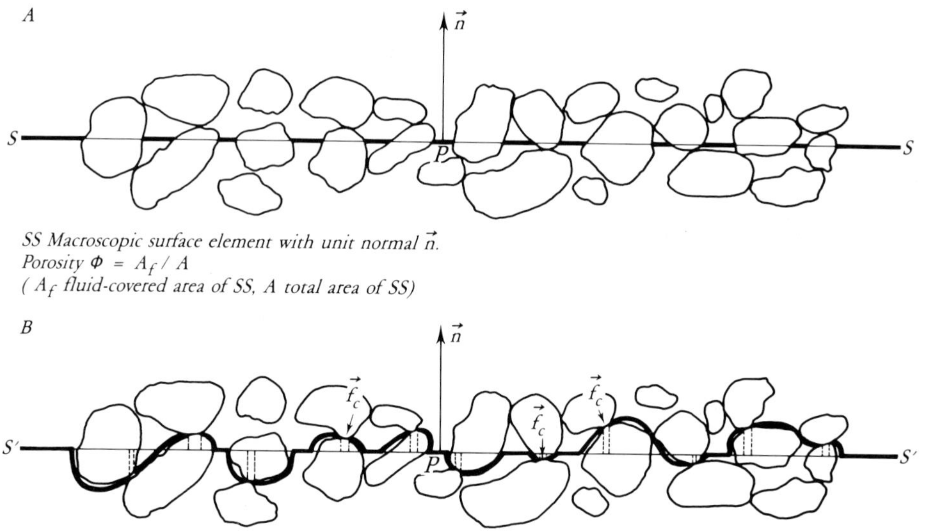

SS Macroscopic surface element with unit normal $\vec{n}$.
Porosity $\Phi = A_f / A$
(A_f fluid-covered area of SS, A total area of SS)

S'S' 'Wavy' element associated with SS intersects skeleton at intergranular contacts.
'Porosity' $\Psi = A_f' / A$
(A_f' area of fluid-part of S' S' projected on element SS)

Fig. II.3-3 SURFACE ELEMENTS IN POROUS MEDIA

'Point' contacts. Usually the two porosities Φ and Ψ differ, but become equal in the absence of a granular structure (or when diagenesis has obliterated that structure). Then 'intergranular' and 'skeletal' stresses are equal. Very often, however, the intergranular contacts are extremely small ('point' contacts) and Ψ is therefore very close to unity. Then the 'intergranular stress' coincides with the 'effective stress' – giving the latter a direct mechanical meaning.

Example: Porous rocks in contact

In concluding this section the application of different stress concepts may be illustrated by the simple, but in practice important, case of two porous rocks in contact. As sketched in Figs. II.3-4A, B, C the plane of contact may be 'open', sealed, or representing a low-permeability barrier to fluid flow. In any case the components of the total stresses that act upon the two sides of the plane of contact will be in balance, i.e.

$$\sigma_{(1)} = -\sigma_{(2)} = \sigma, \qquad \tau_{(1)} = -\tau_{(2)} = \tau \tag{II.3-5}$$

The pore pressure on both sides of the open contact will have identical values, but its values

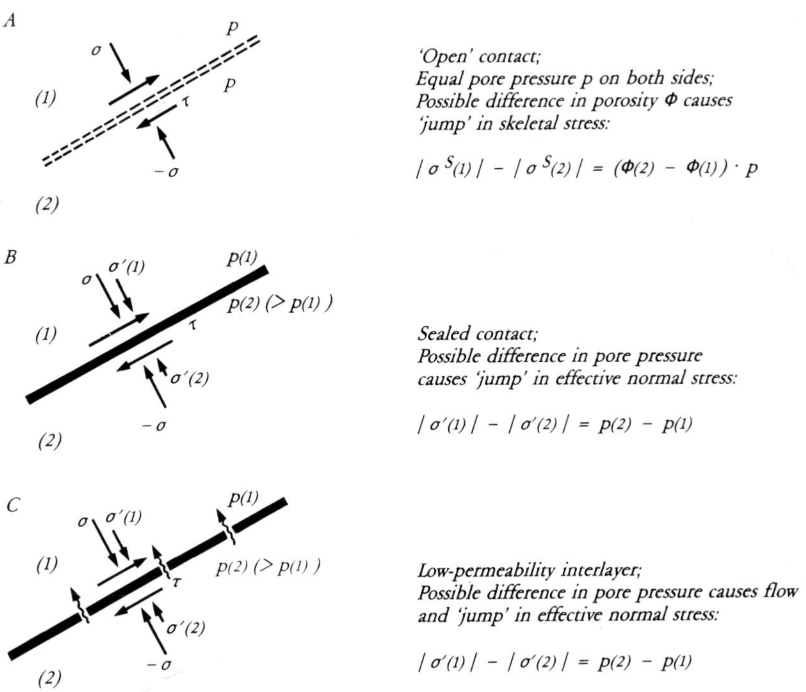

Fig. II.3-4 CONTACTS BETWEEN POROUS ROCKS
(σ and τ are components of total 'contact' stress.)

may differ when the contact is sealed or an interlayer of very low permeability impedes fluid flow (see Figs. II.3-4B, C). A 'jump' in pore pressure across the rock contact is accompanied by a *jump in the effective normal stress* σ'

$$|\sigma'_{(1)}| - |\sigma'_{(2)}| = p_{(2)} - p_{(1)} \qquad \text{(II.3-6)}$$

Since rock strength is controlled by effective stress, rather than total stress, the jump in effective stress may induce a considerable *strength contrast* between lithologically very similar rocks. While a *change in porosity* across the contact plane does not affect the effective stress, it does cause a jump in the magnitudes of the normal skeletal stresses σ^s

$$|\sigma^s_{(1)}| - |\sigma^s_{(2)}| = \Phi_{(2)}\, p_{(2)} - \Phi_{(1)}\, p_{(1)} \qquad \text{(II.3-7)}$$

pag.227

NOTES

1) A possible exception is formed by the very margin of the macroscopic flow region.

2) A proof of these relationships can be found in: Mandl (1964), 'Change in skeletal volume of a fluid-filled porous body under stress' (eq. 32). (Note that in this paper compressive stresses are counted negative but fluid pressure positive.)

A lucid discussion of 'intergranular' and 'effective' stresses, their history and applicability was presented by Bishop and Skinner (1977).

II.4. THE STRESS CIRCLE

The stress vector ($\vec{t}$) that acts on a planar element (Fig. II.4-1A) inside a stressed body normally varies with the orientation of the element (see Section II.1.2). Defining the local state of stress – say, at a point P – therefore implies that the stress vector is known for any orientation of planar elements at P. In the general case this requires the prescription of six 'stress components' at P, which complicates the analysis of stress fields. Geometrical representations of the stress state which can facilitate the analysis are therefore welcome. Such a representation is O. Mohr's stress circle diagram, which is particularly convenient in two-dimensional stress analyses.

1. DERIVATION

Forces on prismatic element. In Fig. II.4-1B a rectangular prism is shown, the oblique face of which is the planar element of Fig. II.4-1A and the other faces of which are chosen as to be acted upon by normal stresses only – the 'principal' stresses σ_I, σ_{II}, σ_{III} (see Section II.1.3). In many tectonic problems the orientation of the σ_I, σ_{III}-plane – the plane parallel to which the maximum and minimum normal stresses act – remains approximately constant over a wider region and is known. Figure II.4-1C shows the intersection of our prismatic element with this plane; the element itself is supposed to extend one length unit perpendicular to the plane. Next, we consider the equilibrium of the forces that act upon the prismatic element parallel to the plane of Fig. II.4-1C. Equilibrium requires that the total force and, hence, its components as well, must vanish. The total force is the vectorial sum of the body force (weight) and the boundary forces that act upon the prismatic element. The resulting components of the boundary forces are derived in Fig. II.4-1C with respect to directions perpendicular and parallel to the oblique face. With these expressions and denoting the corresponding body force components by γ' dV and γ'' dV, vanishing of the total force on our element implies

$$\sigma_\perp \, d\Sigma - \sigma_I \, d\Sigma \cos^2\alpha - \sigma_{III} \, d\Sigma \sin^2\alpha + \gamma' dV = 0 \qquad \text{(II.4-1)}$$

$$\tau d\Sigma - (\sigma_I - \sigma_{III}) \, d\Sigma \sin\alpha\cos\alpha + \gamma'' dV = 0 \qquad \text{(II.4-2)}$$

We finally divide these expressions by d Σ and let the prismatic element shrink towards a corner point P. Since the ratio dV/dΣ is proportional to a characteristic length of the prismatic element, it tends to zero as the element shrinks, and the body forces disappear

from the equations. By using the well-known identities
$\sin^2 \alpha = (1 - \cos 2\alpha)/2$, $\cos^2 \alpha = (1 + \cos 2\alpha)/2$, $\sin 2\alpha = 2 \sin \alpha \cos \alpha$,
equations II.4-1,2 turn into

$$\sigma_\perp = \frac{\sigma_I + \sigma_{III}}{2} + \frac{\sigma_I - \sigma_{III}}{2} \cdot \cos 2\alpha \qquad \text{(II.4-1a)}$$

$$\tau = \frac{\sigma_I - \sigma_{III}}{2} \cdot \sin 2\alpha \qquad \text{(II.4-2a)}$$

These extremely useful formulae relate the normal and tangential components of the stress vector that acts upon a surface element (whose normal is perpendicular to the σ_{II}-direction and includes the angle α with the σ_I-direction) with the principal stress values σ_I and σ_{III}. The relationships are independent of the intermediate principal stress σ_{II}.

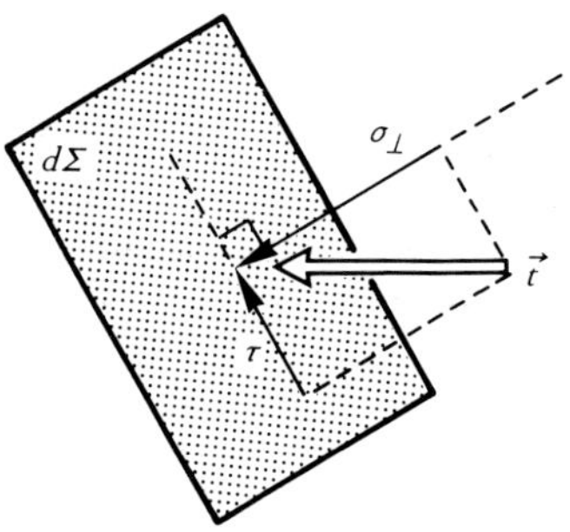

A Stress on infinitesimal element of area dΣ

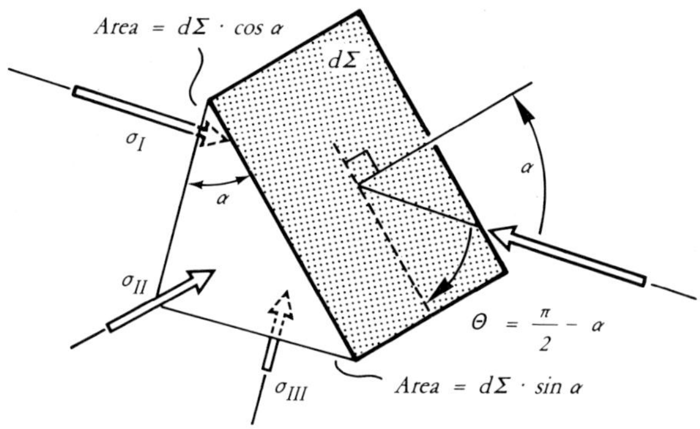

B Infinitesimal prism; three sides perpendicular to principal stress directions

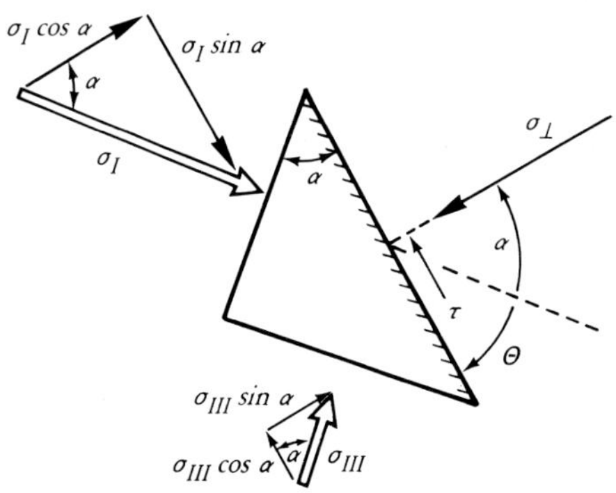

C Intersection of prismatic element with plane of maximum and minimum compressive stresses σ_I and σ_{III}

Sum of boundary forces acting parallel to $\sigma_\perp$ and τ, respectively:

1) $\sigma_\perp \, d\Sigma - \sigma_I \, d\Sigma \cos^2 \alpha - \sigma_{III} \, d\Sigma \sin^2 \alpha$

2) $\tau \, d\Sigma - (\sigma_I - \sigma_{III}) \sin \alpha \cos \alpha \, d\Sigma$

Fig. II.4-1 STRESSES IN PLANE OF MAXIMUM AND MINIMUM PRINCIPAL STRESS

To avoid any confusion, it should be mentioned that formulae II.4-1a, 2a are often used in the slightly different form obtained when introducing $\Theta = \frac{\pi}{2} - \alpha$ (Fig. II.4-1B):

$$\sigma_{\perp} = \frac{\sigma_{I} + \sigma_{III}}{2} - \frac{\sigma_{I} - \sigma_{III}}{2} \cos 2\Theta \qquad \text{(II.4-1b)}$$

$$\tau = \frac{\sigma_{I} - \sigma_{III}}{2} \sin 2\Theta \qquad \text{(II.4-2b)}$$

Mohr's stress circle. The angle α can be eliminated from the equations II.4-1a and II.4-2a. Since $\sin^2 2\alpha + \cos^2 2\alpha = 1$, rearranging terms in the equations, squaring and adding yields

$$\left[\sigma_{\perp} - \frac{\sigma_{I} + \sigma_{III}}{2}\right]^2 + \tau^2 = \left[\frac{\sigma_{I} - \sigma_{III}}{2}\right]^2 \qquad \text{(II.4-3)}$$

Since in a Cartesian x, y-diagram the equation $(x - a)^2 + y^2 = r^2$ (Fig. II.4-2A) represents a circle of radius r and with its centre on the x-axis at distance 'a' from the origin, equation II.4-3 suggests the following geometrical stress representation:
Along the orthogonal axes of a two-dimensional Cartesian frame (Fig. II.4-2B) are plotted the normal stress $\sigma_{\perp}$ and the shear stress τ that act on any material element at a point P inside the stressed material. The stresses acting on those elements the normals of which are perpendicular to the σ_{II}-direction satisfy equation II.4-3 and the corresponding stress points $(\sigma_{\perp},\tau)$ lie on a stress circle that is centred on the $\sigma_{\perp}$-axis at a distance $(\sigma_{I} + \sigma_{III})/2$ from the origin and which has the radius $(\sigma_{I} - \sigma_{III})/2$. Some typical stress circles are shown in Figs. II.4-3A-D for some simple, but important, stress states (in the σ_{I}, σ_{III}-plane). They hardly need further comment.

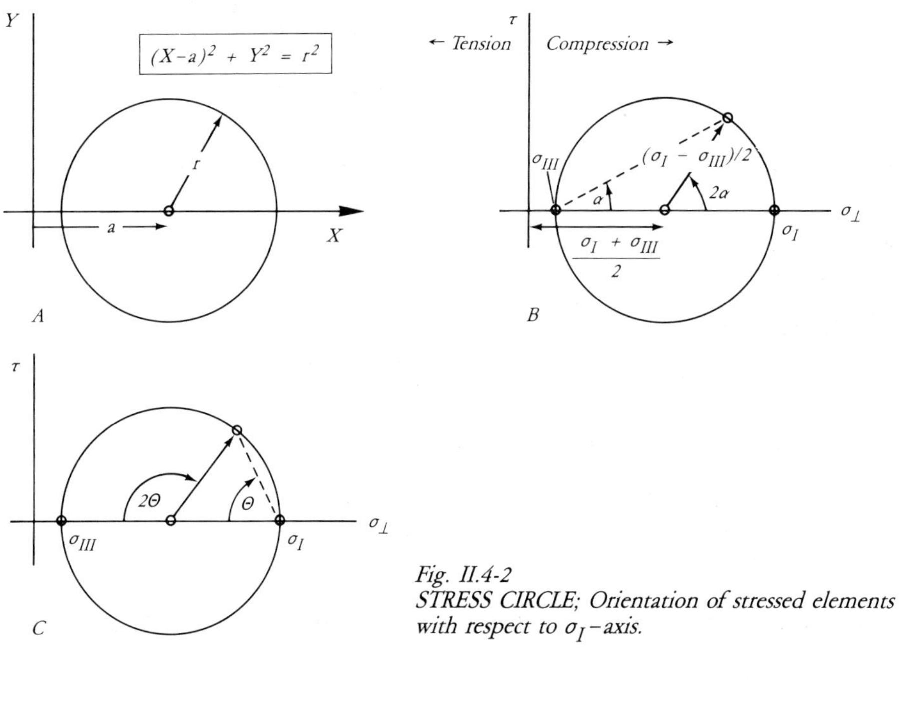

Fig. II.4-2
STRESS CIRCLE; Orientation of stressed elements with respect to σ_I-axis.

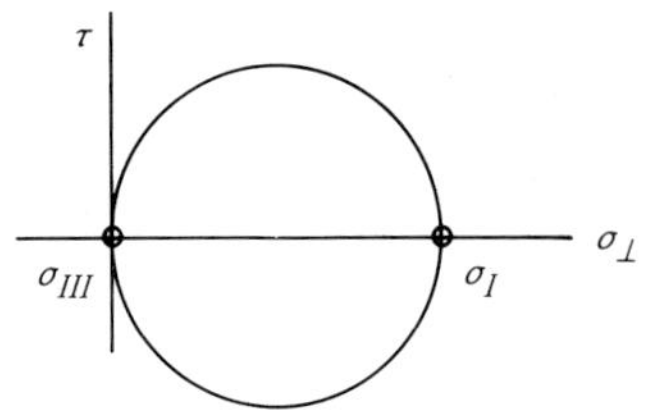

A Uniaxial compression
($\sigma_I > 0$, $\sigma_{III} = 0$)

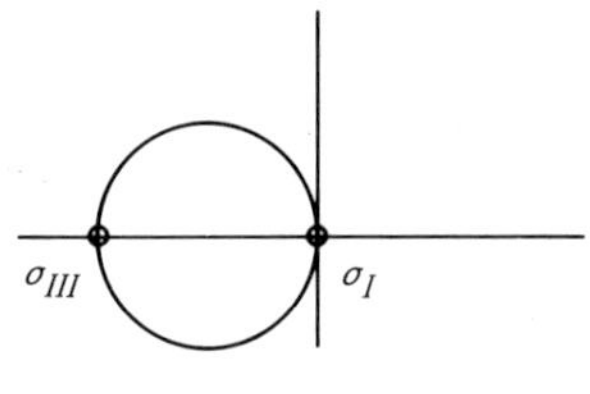

B Uniaxial tension
($\sigma_I = 0$, $\sigma_{III} < 0$)

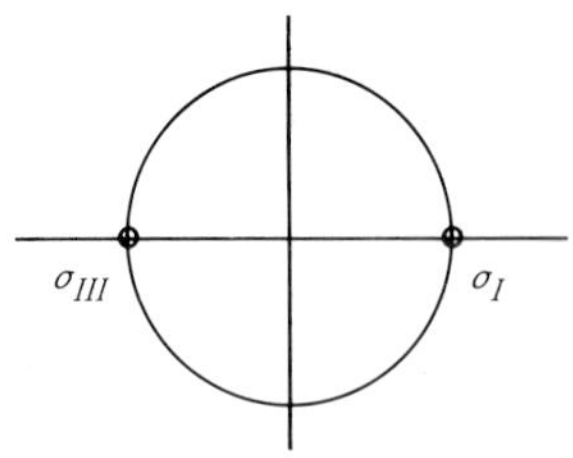

C Deviatoric stress ('pure shear')
(cf. Fig. II.1-12B)
($\sigma_I = -\sigma_{III} = \tau_{Max.}$)

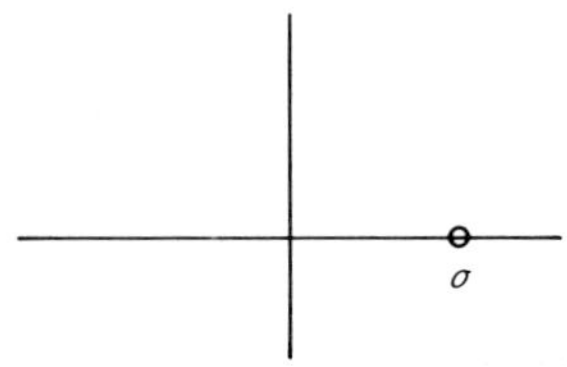

D Hydrostatic compression
(all normal stresses equal to σ)

Fig. II.4-3 STATES OF STRESS

2. PROPERTIES OF MOHR'S STRESS CIRCLE

Points to note are:

1) As the material element (Fig. II.4-1B and C) is rotated anti-clockwise around the σ_{II}-axis through an angle α, the radius vector to the corresponding stress point on the Mohr circle rotates through the angle 2α in the same sense (Fig. II.4-2B). Note also that the chord extending from the stress point (σ_{III}, 0) to the stress point in question ($\sigma_{\perp}$, τ) makes the angle α with the $\sigma_{\perp}$-axis, as follows from elementary geometry. To avoid ambiguity, it should also be mentioned that when the complementary angle $\Theta = \frac{\pi}{2} - \alpha$ (Fig. II.4-1B) is used to specify the orientation of the element, the associated radius vector of the stress circle is defined as shown in Fig. II.4-2C.
2) If the principal stress values σ_I and σ_{III} are given, the stress circle can be drawn (Fig. II.4-2B). To relate stress points with material planes, the orientation of the principal stress axes has to be known as well.
3) It is immediately apparent from the Mohr diagram (Fig. II.4-4A) that the maximum shear stresses occur on elements inclined at 45° to the σ_I-axis (i.e. $2\alpha = 90°$), and have a value of $(\sigma_I - \sigma_{III})/2$. The normal stress acting on these elements is $(\sigma_I + \sigma_{III})/2$.
4) Shear stresses that act on mutually perpendicular elements, as shown for instance in Fig. II.4-4A, appear in the Mohr diagram with opposite signs. It may be recalled, however, that stress equilibrium requires that these stresses are equal in magnitude *and* sign (see Section II.2.1, 'Symmetry of shear stresses'; and Fig. II.1-16). Whereas, for instance, the shear stress on BB′ in the insert of Fig. II.4-4A has the same sign as the shear stress on AA′, it attains the opposite sign in the Mohr diagram (point B). This may be easily understood by considering the set of all planar elements that are parallel to the σ_{II}-axis

and by marking on one element, say AA′ in Fig. II.4-4A, the side to be acted upon by the stress vector of the associated point on the Mohr circle. The stress on any other element (Fig. II.4-1C) will then act upon the side that coincides with the marked side of the first element when this is rotated, say, anticlockwise into the new position. Obviously, in Fig. II.4-4A this will be the hatched side of the element BB′, and the associated shear stress in the Mohr diagram will therefore be $-\tau_{max}$.

5) Instead of by prescribing the principal stresses, the Mohr circle is also fully determined by the normal and shear stresses that act on two arbitrary, mutually perpendicular planes (see insert of Fig. II.4-4B). The associated stress points lie symmetrically with respect to the centre of the stress circle (Fig. II.4-4B) since the associated radii include the angle $2\alpha = 180°$. The values of the principal stresses can be read from the diagram or determined from an analytical expression derived in Fig. II.4-4B by applying Pythagoras' theorem.
6) The inclinations of the principal stress axes are easily found in the Mohr diagram by making reference to a material element with known stress vector (e.g. the element in Fig. II.4-4B that is acted upon by the stresses σ_3, τ_{31}). The simple construction is shown in Fig. II.4-4C which at the same time provides the derivation of the important analytical expression for the orientation of the σ_I-axis:

$$\tan 2\Theta = -\tan 2\alpha = 2\tau / (\sigma_1 - \sigma_3) \qquad \text{(II.4-4)}$$

where σ_1 and σ_3 are the normal stresses that act upon the elements with normals parallel to the x_1- and x_3-axes of a Cartesian frame (Fig. II.4-4B), τ is the shear stress on these elements, and Θ and α are the angles which the σ_I-axis makes with the (positive) x_1- and the x_3-axis, respectively (see insert of Fig. II.4-4C).

The 'Pole' of the Stress Circle. A very useful concept in Mohr's stress representation is what is termed the 'pole'. It provides a link between Mohr's $\sigma_\perp$, τ-space and the physical space. Recall that the points of the Mohr circle that runs through the points σ_I, 0 and σ_{III}, 0 (Fig. II.4-5A), represent the normal and shear stresses that act on the planar elements that are located at a given point in the stressed material and are parallel to the direction of the intermediate principal stress σ_{II}, and that, moreover, the angle between the radii of two stress points in the Mohr diagram is twice the angle included by the corresponding planar elements in physical space. Note, however, that this correspondence between stress points in the $\sigma_\perp$,τ-plane and planar elements in the stressed material is not yet complete, since it does not indicate which stress point and which element are associated. To complete the correspondence, one has to know, first of all, the direction of the intermediate principal stress σ_{II} in the material, and secondly, one planar element (parallel to this direction) together with the normal and shear stresses that act on it.

This planar element may then serve as a reference element for correlating stress points on the Mohr circle with planar elements in the material. Let the normal and tangential stresses that act on our reference element be represented by point A in Fig. II.4-5A, and consider an arbitrary second point B on the stress circle. Since the planar elements associated with the two points include in physical space half of the angle between the radii of the two stress points in the Mohr diagram, the orientation of the planar element associated with B with respect to our reference element is also known. The angle ($\Delta\alpha$) between the two elements in physical space appears in the Mohr diagram as the angle extended from any point on the periphery of the stress cicle over the arc AB (Fig. II.4-5A), since according to an elementary theorem this angle is half the angle extended from the centre of the circle over the same arc.

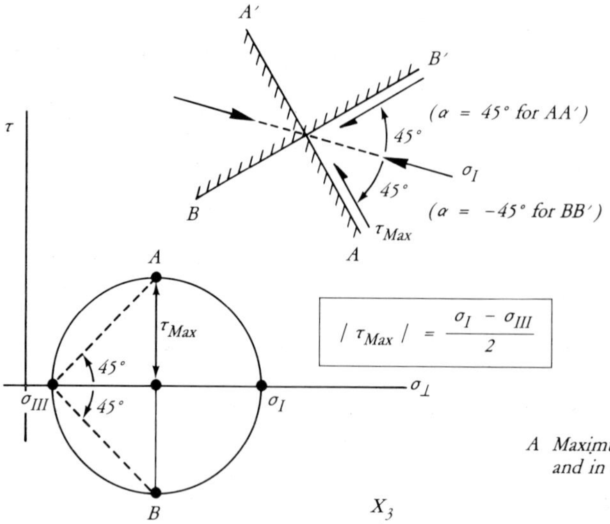

$$|\tau_{Max}| = \frac{\sigma_I - \sigma_{III}}{2}$$

A Maximum shear stress in Mohr diagram and in real space (insert)

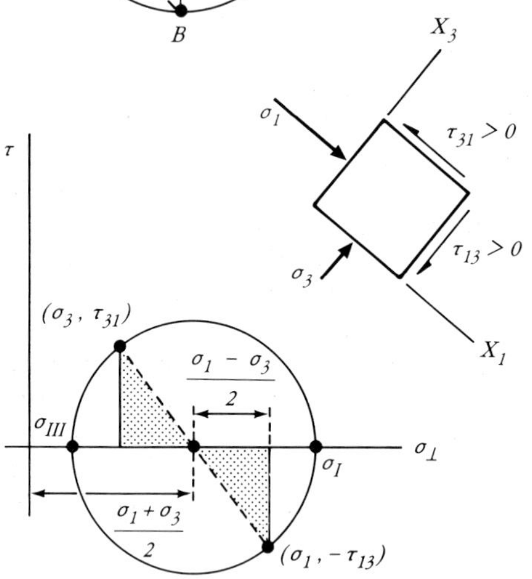

$$\left.\begin{matrix}\sigma_I \\ \sigma_{III}\end{matrix}\right\rangle = \frac{\sigma_1 + \sigma_3}{2} \pm \sqrt{\frac{(\sigma_1 - \sigma_3)^2}{4} + \tau_{13}^2}$$

$$|\tau_{Max}| = \frac{\sigma_I - \sigma_{III}}{2} = \sqrt{\frac{(\sigma_1 - \sigma_3)^2}{4} + \tau_{13}^2}$$

B Principal stresses and Cartesian stress components

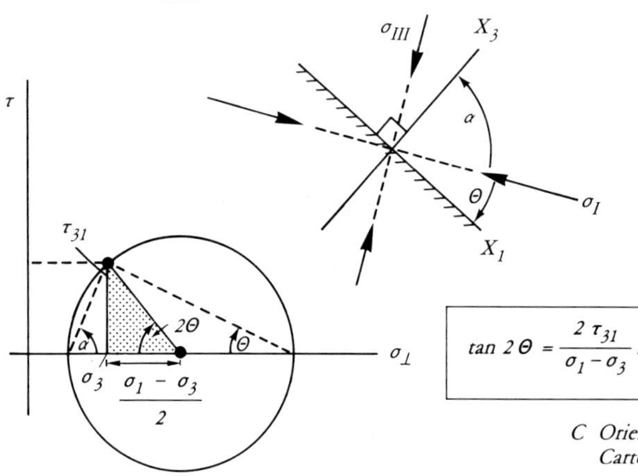

$$\tan 2\Theta = \frac{2\,\tau_{31}}{\sigma_1 - \sigma_3} = -\tan 2\alpha$$

C Orientation of principal stress axes in Cartesian reference frame

Fig. II.4-4 STRESSES IN MOHR DIAGRAM

To arrive at the pole construction we have to 'orient' the Mohr diagram by associating the $\sigma_\perp$-axis with a fixed direction in the physical σ_I, σ_{III}-plane, i.e. the plane perpendiculary intersected by the planar elements in physical space. As reference direction we may take, for instance, that of the x_1-axis of a two-dimensional Cartesian frame in the physical σ_I, σ_{III}-plane. Thus, we look now at the $\sigma_\perp$-axis in two ways: as axis in Mohr's stress plane and as x_1-axis in the σ_I, σ_{III}-plane of real space. We now draw in Fig. II.4-5A the chord from point A that makes the same angle with the $\sigma_\perp$-axis as our reference element with the x_1-axis in the σ_I, σ_{III}-plane. The point where this chord meets the circle is the 'pole'. Because of what has been said about the angle at the periphery, any chord drawn from the pole to some other point B of the stress circle includes the same angle with the reference chord AP as the angle included by the corresponding elements in real space. The chord PB therefore directly represents the orientation of the planar element associated with B with respect to the x_1-axis in real space.

To illustrate this remarkable property of the pole we construct in Fig. II.4-5B the direction of the principal stress σ_I. To find the orientation of the material element that is acted upon by this stress, we only have to connect the pole P with the stress point σ_I, 0. Since σ_I acts perpendiculary on the element represented by this chord, its direction in physical space is that of the line drawn from point σ_{III}, 0 to the pole P.

Effective stresses. The mechanical behaviour of fluid-saturated rocks and soils and, in particular, their shearing strength, is strongly affected by the pore fluid pressure (p). The pore pressure effects on the shear failure are sufficiently taken into account by introducing 'effective' instead of 'total', stresses. (For more details see Section II.7.2). The two stresses differ only in the normal components, the effective stress σ' being related to the total stresses σ by $\sigma' = \sigma - p$ (see Section II.3.2, 'Fluid and solid stresses'). Mohr's stress circle representation applies equally well to effective stresses when the effective normal stresses $\sigma'_\perp$ are plotted along the former $\sigma_\perp$-axis (Fig. II.4-6A). This is in accord with the fact that the equations II.4-1a-4 remain unchanged when the pore pressure p is subtracted from each normal stress term. In particular, the radius of the stress circle remains the same as in total stress space. The position of its centre $((\sigma'_I + \sigma'_{III})/2, 0)$ in the effective stress diagram will shift, however, as the pore-pressure changes at fixed total stress $(\sigma_I + \sigma_{III})/2$. It is obvious from Fig. II.4-6A that an increase of pore pressure, even though the total stresses may be kept constant, can render a 'safe' state into a failure state (stress circle touching the failure envelope).

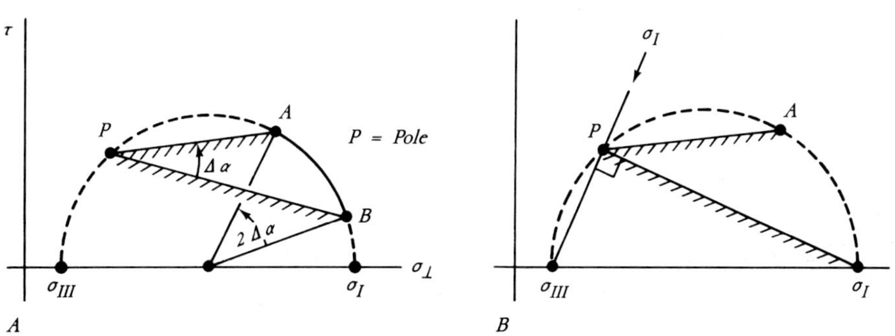

Fig. II.4-5 POLE OF STRESS CIRCLE
Chords are parallel to intersections of planar material elements with the σ_I, σ_{III}-plane in physical space.

3. APPLICATIONS

Example No. 1: Orientation of failure planes

Within the framework of Coulomb-Mohr's theory of shear failure the states of stresses that may lead to shear failure in geological materials are represented by stress circles that are tangential to pairs of experimentally determined 'limit' or 'failure' lines ('Mohr failure envelope'). According to the theory, which is discussed extensively in Chapter II.7, actual shear deformation or failure will, in general, be the result of slips along material elements that are associated with the stress points that Mohr circles and limit lines have in common. Usually the limit lines are approximately straight over a considerable range of compressive stresses (Fig. II.4-7A), but they tend to flatten at high compressive stresses (Fig. II.4-7B) and are certainly strongly curved where extended into the region of tensile effective stresses (Fig. II.4-7C).

When we use the notations for the angle between a material element and the σ_I-direction in Fig. II.4-4C, it becomes obvious that for fixed principal stress directions the orientation of failure planes is independent of the position of the stress circle for a straight failure envelope, but that it varies with circle position as the circles are enveloped by a curved failure line (Fig. II.4-7A, B, C). In the range of high normal stresses, failure envelopes may become parallel to the $\sigma'_\perp$-axis, in which case a failure plane will include a 45°-angle with the σ'_I-direction. In contrast, owing to the curved (parabolic) part of the failure envelope in the tension domain (Fig. II.4-7C), the failure planes should tend to align with the σ'_I-direction as the point of tangency moves farther into the tension region of the stress plane (Muehlberger, 1961; Brace, 1964).

Example No. 2: Increase of pore pressure in horizontal strata

In sediment accumulations the pore pressure may rise above hydrostatic values (e.g. because of compaction delay by poor drainage, or because of hydrocarbon accumulation). Let us consider the effect such a rise in pore pressure may have on the effective stress state in horizontal sediment strata and let us assume that the sediment responds to imposed load

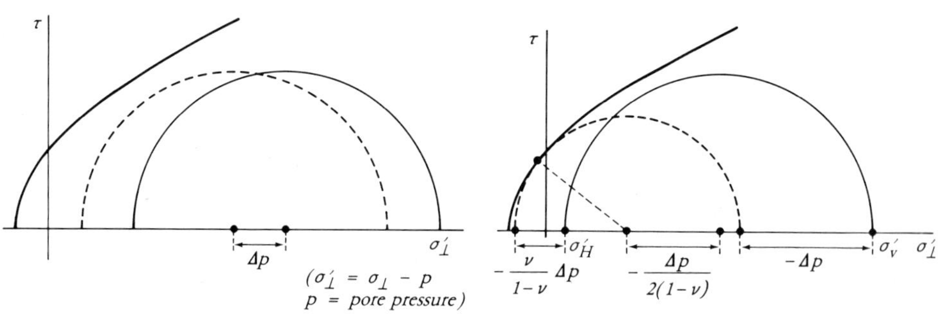

A Shift of circle under change Δp in pore pressure (total stress constant)

B Change of stress circle caused by increase in pore pressure in laterally constrained material

Fig. II.4-6 'EFFECTIVE STRESS' CIRCLE

and pore pressure *changes* by poro-elastic deformations (see Chapter II.6, Table). The vertical effective stress $\sigma_v' = \sigma_v - p$ is associated with the overburden weight σ_v per unit area, and the horizontal effective stresses $\sigma'_H = \sigma_H - p$ are assumed to be the same in all directions. The horizontal strains e_H are then related to changes of these stresses by

$$e_H = \frac{1}{E}[\Delta\sigma_H'(1-\nu) - \nu\Delta\sigma_v'] \qquad \text{(II.4-5)}$$

where E is Young's modulus (i.e. the ratio of effective normal stress and associated contraction on uniaxial compressive loading) and ν is Poisson's ratio (i.e. the ratio of the absolute strain values measured perpendicular and parallel to the direction of the uniaxial loading).

As the effective overburden is changed by $\Delta\sigma_v'$, the strata tend to change their horizontal dimensions. In general, this is prevented by interstratal friction and friction along the base of the sediment package. In the absence of horizontal straining ($e_H = 0$), the two effective stress changes are therefore related, according to equation II.4-5, by

$$\Delta\sigma_H' = \frac{\nu}{1-\nu}\Delta\sigma_v' \qquad \text{(II.4-5a)}$$

Thus, for a typical value $\nu = 0.25$ a change in vertical effective stress is accompanied by a change in horizontal effective stresses of one-third this magnitude.

In Fig. II.4-6B the case is considered in which under constant overburden (σ_v = const.) a sediment layer is 'invaded' by a higher pore pressure from an adjacent overpressured layer. For the layer in question, a pore pressure rise Δp then means a decrease of *effective* overburden stress by $-\Delta p$. According to equation II.4-5a, the Mohr circle then shrinks and shifts as illustrated in Fig. II.4-6B. If the sediment has acquired some tensile strength, a sufficient

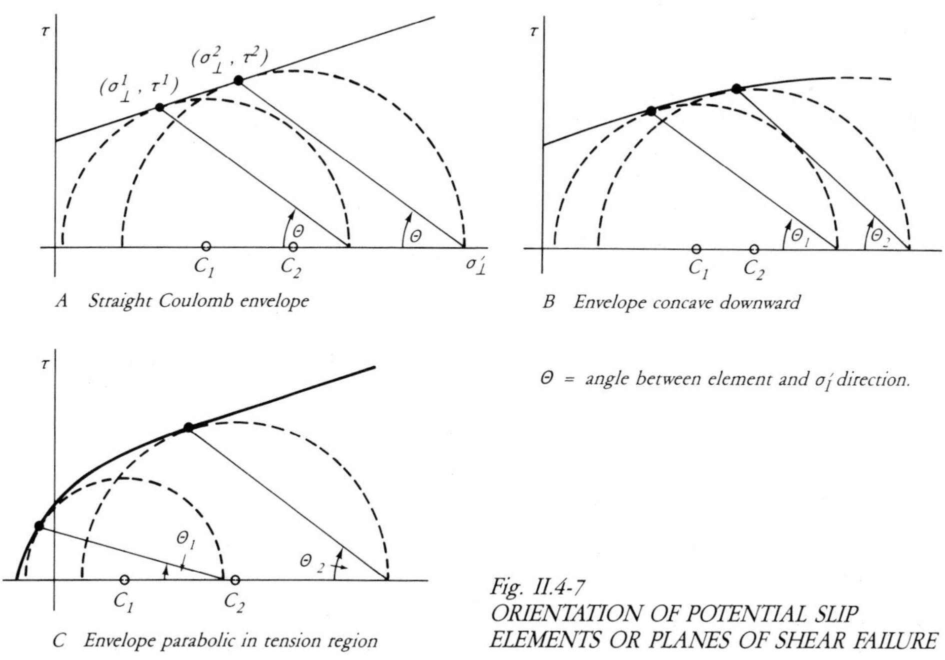

A Straight Coulomb envelope

B Envelope concave downward

C Envelope parabolic in tension region

Fig. II.4-7
ORIENTATION OF POTENTIAL SLIP ELEMENTS OR PLANES OF SHEAR FAILURE

rise in pore pressure may even lead to failure along planes that are acted upon by zero or tensile effective stresses ($\sigma'_{\perp} \leqslant 0$). This is of particular importance for the formation of open fractures.

Example No. 3: Slip along planes of weakness

A plane of weakness is a material plane or zone with shear and tensile strengths lower than those of the surrounding rock. We consider such a plane in the field of lateral tectonic compression as shown in Fig. II.4-8A. In Mohr's diagram (Fig. II.4-8B) the shear strengths of plane of weakness and surrounding rock are represented by the failure lines Γ_W and Γ, respectively. If we start with a stress state represented by the smaller circle in Fig. II.4-8B and maintain the overburden stress σ'_{III}, an increase of the horizontal tectonic stress σ'_I may eventually activate the plane of weakness, i.e. induce slip. For this to happen, it is not sufficient for the stress circle to intersect the Γ_W line, but in addition one intersection point W_c must represent the stresses acting upon the plane of weakness. As discussed earlier (Fig. II.4-2B), this point is defined as the end point of the stress circle's chord that extends from the point (σ'_{III}, 0) and includes the angle α with the $\sigma'_{\perp}$-axis. The critical stress circle that represents activation of the plane of weakness in our example is therefore easily found, as indicated in Fig. II.4-8B.

It should be noted, however, that not every orientation of the plane of weakness allows activation of that plane. Inspection of Fig. II.4-8C shows that orientation angles α which are smaller or greater than the angular interval indicated by radial shading are 'unfavourable' since the growing stress circle will touch the failure line Γ, and cause shear failure of the intact rock, before slip along the plane of weakness can be induced.

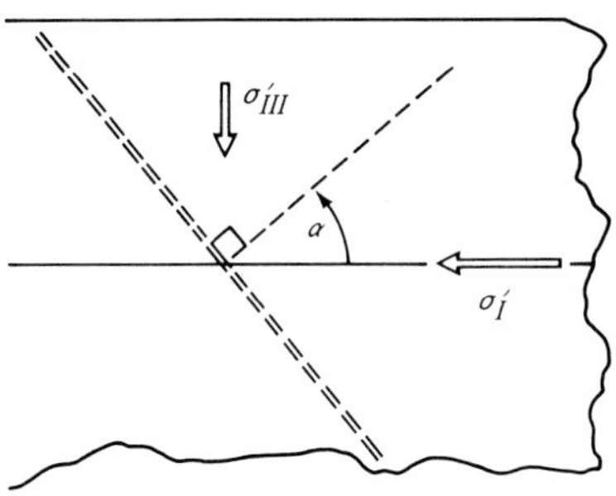

A Plane of weakness in field of lateral tectonic compression

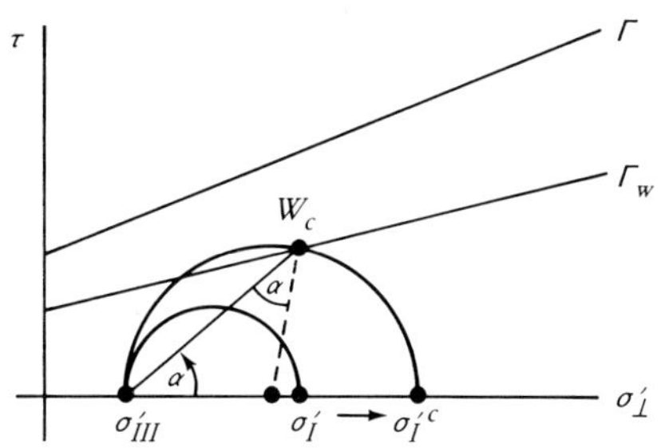

B State of stress that activates plane of weakness

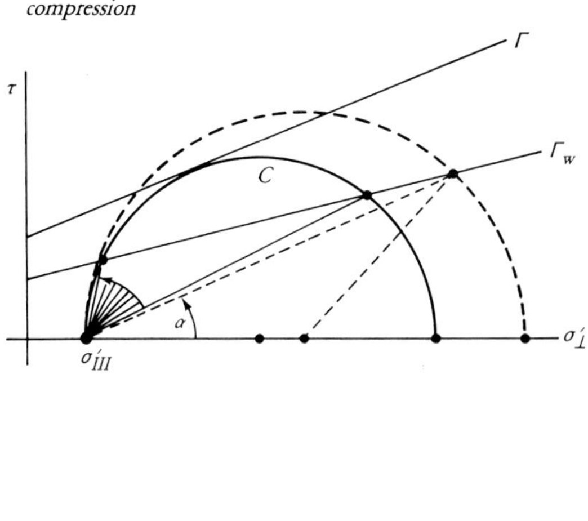

C 'Favourable' and 'unfavourable' orientations of the plane of weakness

Fig. II.4-8
SLIP ALONG PLANE OF WEAKNESS

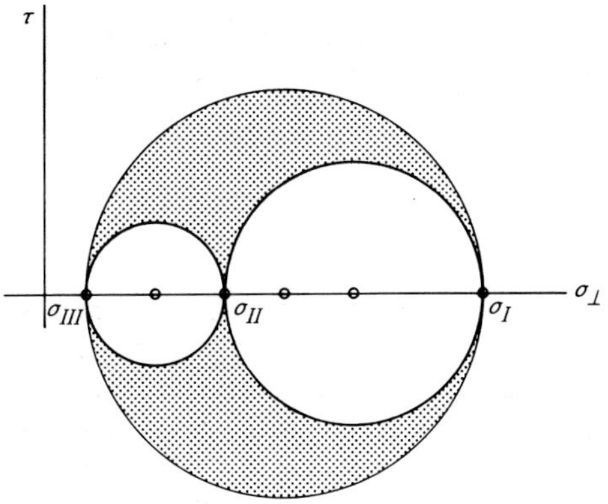

Fig. II.4-9
MOHR'S REPRESENTATION OF THREE-DIMENSIONAL STATE OF STRESS

4. THREE-DIMENSIONAL STATE OF STRESS

So far, we have dealt with stresses that act in the σ_I, σ_{III}-plane, i.e. upon material elements whose normals are perpendicular to the σ_{II}-axis (see Fig. II.4-1B). Quite analogously, one may consider all the elements with normals perpendicular to the σ_I-axis or to the σ_{III}-axis, and will find that their respective stresses map on the smaller circles shown in Fig. II.4-9. Without proof, we add that the stresses $\sigma_\perp$, τ which act on elements that are obliquely oriented with respect to all three principal stress directions map on the interior of the crescent-shaped region enclosed by the three principal stress circles (see Jaeger & Cook, 1968; Chapter 2.6).

II.5. STRAIN

Suppose that the beam in Fig. II.5-1A is deformed as in Fig. II.5-1B. The parts of the beam undergo displacements and relative changes in shape. Consider, for example, the very small volume element characterised by perpendicular straight-line edges A-B and A-C in the undeformed state. This volume element is subjected to translation, rotation, and *relative* changes in length, orientation and curvature of its edges during transition to the deformed state. (The relative change in orientation of A-B with respect to A-C is indicated by lack of perpendicularity in the deformed state.) Hence, we shall term the translation and rotation *rigid-body displacements,* and define the relative changes in shape as the *components of strain*.

*1. INFINITESIMAL STRAIN **

There is a large class of materials for which the components of strain are very small within the range of elastic deformation. If they are small, the relative change in curvature of very short line segments is negligible. In that case, we shall assume that short straight-line segments in the undeformed state remain straight-line segments in the deformed state. The deformation of an element bounded by initially perpendicular short-line segments (Fig. II.5-2A) may then be regarded as composed of rigid-body translation, rigid-body rotation, extensional strain and shear strain. These components are illustrated in exaggerated form in the diagrams (B), (C), (D) and (E) of Fig. II.5-2. In the following, attention is focused on the components of strain.

* This introductory paragraph closely follows the elementary presentation by Bombolakis, 1968.

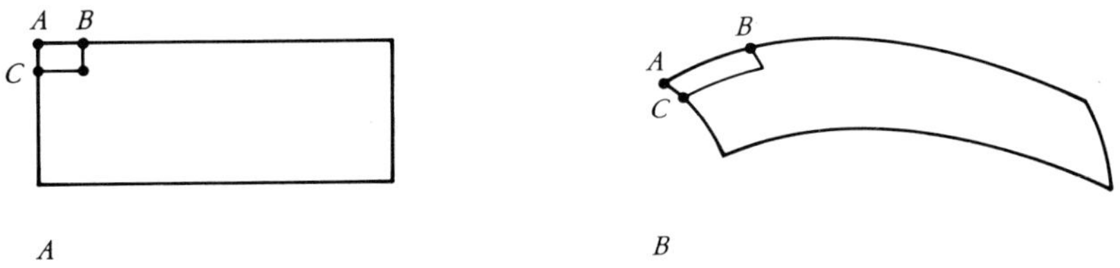

Fig. II.5-1 BEAM IN UNDEFORMED STATE (A), AND DEFORMED STATE (B)
Segments (AB) and (AC) represent two edges of a volume element of the beam.

As a quantity describing the change in length of a material line element, one may use the dimensionless ***elongation***

$$e = \frac{l - l^*}{l} \tag{II.5-1}$$

where l is the length of the unstrained element and l^* the length of the strained element. Note that, contrary to common usage in elasticity, we count a contraction as positive. This is in accordance with the sign convention adopted for stresses in Section II.1.4, in which compressive stresses are considered positive.

The dimensionless elongation is by no means the only measure that can be used for elongational strains. Other measures are the 'stretch' $l^* / l = l - e$, its square, and the 'natural' strain $(l - l^*)/l^*$. The extensional strains of line elements initially parallel to the x_1, x_2, x_3-axes of a rectangular Cartesian coordinate frame will be designated by e_1, e_2, e_3, respectively (see Fig. II.5-2C).

By means of the extensional strains, one can also describe the volumetric strain, i.e. the change in volume of a volume element referred to its volume V in the unstrained state. Considering a small parallelepiped with edges of length dx_1, dx_2, dx_3 parallel to the Cartesian coordinate axes, we see from Fig. II.5-2C that

$$\frac{\Delta V}{V} = \frac{V - V^*}{V} = \frac{1 - (1-e_1)(1-e_2)(1-e_3)}{dx_1 \cdot dx_2 \cdot dx_3} \, dx_1 \cdot dx_2 \cdot dx_3$$

Restricting ourselves to *small* strains, which implies that products of strain components can be neglected with respect to the strains themselves, the above expression for the volumetric strain attains the simple form

$$\frac{V - V^*}{V} = e_1 + e_2 + e_3 \tag{II.5-2}$$

Another type of strain – the ***shear strain*** – measures the distortion associated with angular changes. Consider, for instance, the two elements parallel to the x_1- and x_3-axes, respectively, in Fig. II.5-2D. Let the initial right angle between the two elements be changed by an angle γ (measured in radians). When any rigid body rotation is excluded, a line element bisecting the angle between the two elements must maintain its orientation and the angular change γ must consequently be produced by equal but opposite rotations through $\gamma/2$ of the initially perpendicular elements. This symmetric distortion by shear can therefore be characterised by either one of two equal shear strains

$$e_{13} = e_{31} = \gamma/2 \tag{II.5-3a}$$

Analogously, the changes in the angles between line elements parallel to the x_1, x_2- and x_3, x_2-axes, respectively, will be accounted for by the shear strain components

$$e_{12} = e_{21}, \; e_{23} = e_{32} \tag{II.5-3b}$$

In engineering literature shear strains are often used which, by definition, are twice as large as the ones just introduced. Hence, for these *'engineering' shear strains* (denoted by the letter γ) holds

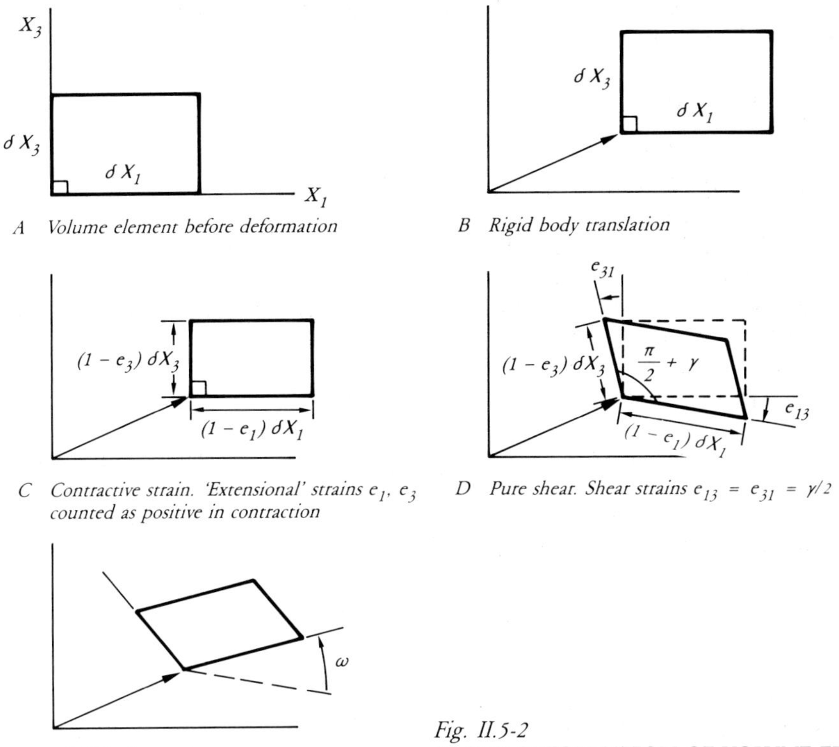

A Volume element before deformation

B Rigid body translation

C Contractive strain. 'Extensional' strains e_1, e_3 counted as positive in contraction

D Pure shear. Shear strains $e_{13} = e_{31} = \gamma/2$

E Rigid body rotation

Fig. II.5-2
PLANE DEFORMATION OF VOLUME ELEMENT

$$\gamma_{13} = \gamma_{31} = 2e_{13} = \gamma;\ \text{etc.} \tag{II.5-3c}$$

In accordance with the *sign convention* adopted for shear stresses (Section II.1.4; Fig. II.1-16) shear strains are here counted as positive when they enlarge the angle between line elements originally oriented parallel to the positive Cartesian axes. For convenience, the sign convention for shear strains and stresses is summarised in Fig. II.5-3.

Pure and Simple Shear. The shear deformation just discussed is pure shear, i.e. free of body rotations. We may specify the shear deformation a little further by the restriction that small shears should not affect the volume of the sheared element. Fig. II.5-3 shows such a shearing operation acting on a square in the x_1, x_3-plane. The diagonals of the square remain fixed, i.e. the deformation is *irrotational*, but while one diagonal is shortened, the other is lengthened by the same amount. Therefore, the infinitesimal volumetric strain (eq. II.5-2) vanishes, as can easily be seen by choosing the Cartesian axes so as to coincide with the diagonals of the square in Fig. II.5-3 ($e_1 = -e_3$, $e_2 = 0$)[1)].

Another simple deformation mode that also preserves the volume while changing angles between line elements is *'simple' shear.* It is characterised by the fact that it leaves a single family of parallel material planes undistorted. This is illustrated in Fig. II.5-4A which emphasises the analogy with a deck of thin cards that is homogeneously sheared. Figure II.5-4B presents a geological example of simple shearing – the bedding plane shearing in the limbs of a fold.

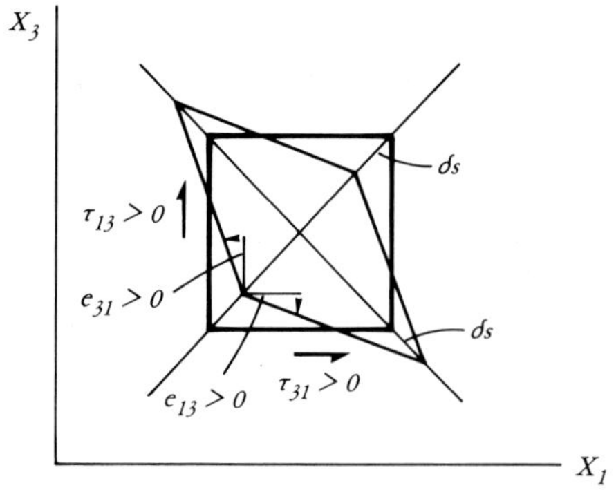

Fig. II.5-3
SIGNS OF SHEAR STRAINS AND STRESSES; 'PURE SHEAR'

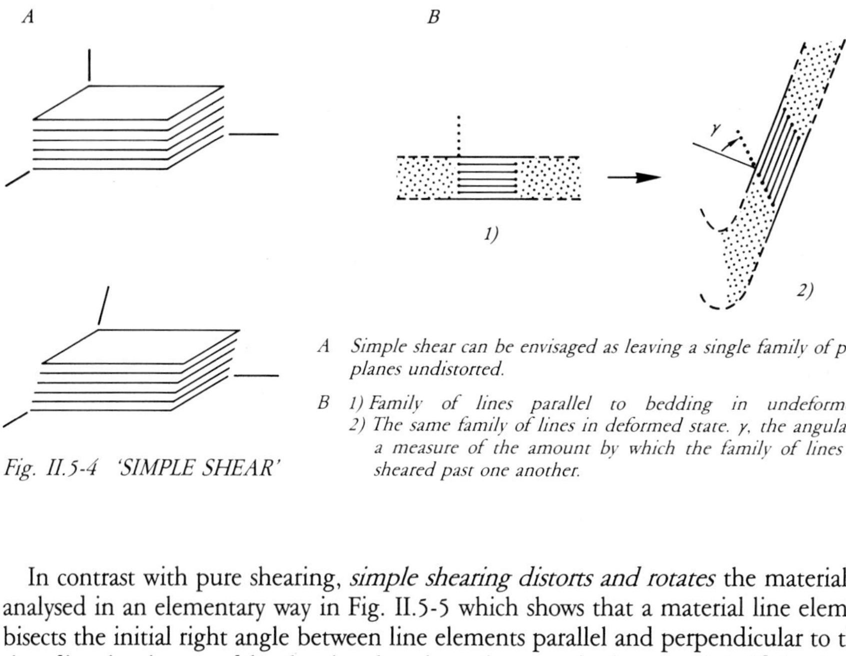

A Simple shear can be envisaged as leaving a single family of parallel planes undistorted.

B 1) Family of lines parallel to bedding in undeformed state. 2) The same family of lines in deformed state. γ, the angular shear, is a measure of the amount by which the family of lines has been sheared past one another.

Fig. II.5-4 'SIMPLE SHEAR'

In contrast with pure shearing, *simple shearing distorts and rotates* the material. This is analysed in an elementary way in Fig. II.5-5 which shows that a material line element that bisects the initial right angle between line elements parallel and perpendicular to the shearing direction is rotated by the shearing through an angle that equals half the shear angle γ. Subtracting this rigid-body rotation, i.e. rotating the deformed element (Fig. II.5-5C) in the opposite direction through $\gamma/2$, leaves the element in a state of pure shear. Thus *pure and simple shear deformation differ only by a rigid body rotation through half the (small) shear angle.*

More will be said about simple and pure shear in the discussion of the strain/displacement relationships and Mohr's strain circle.

2. DISPLACEMENTS AND STRAINS

The deformation of a body – including strains and translations and rotations of its elements – is completely described when the displacements that carry the material points of the undeformed body into their positions in the deformed state are defined. As illustrated in Fig. II.5-6, the displacements are vectorial quantities with components, say, u_1, u_2, u_3 in a Cartesian reference system. These quantities are defined at any point of the deforming body and thus constitute a displacement 'field'.

It is easy to express the components of infinitesimal strain in terms of displacement components. Consider first a uniaxial contraction of the cubical volume element in Fig. II.5-7A. The displacements take place parallel to the x_1-axis only, i.e. $u_1 \neq 0$, $u_2 = u_3 = 0$, and do not depend on the coordinates x_2 and x_3. Let the displacement of the left face at x_1^o be u_1^o and that of the opposite face at $x_1^o + \delta x_1$ be $u_1^o + \delta u_1$, where

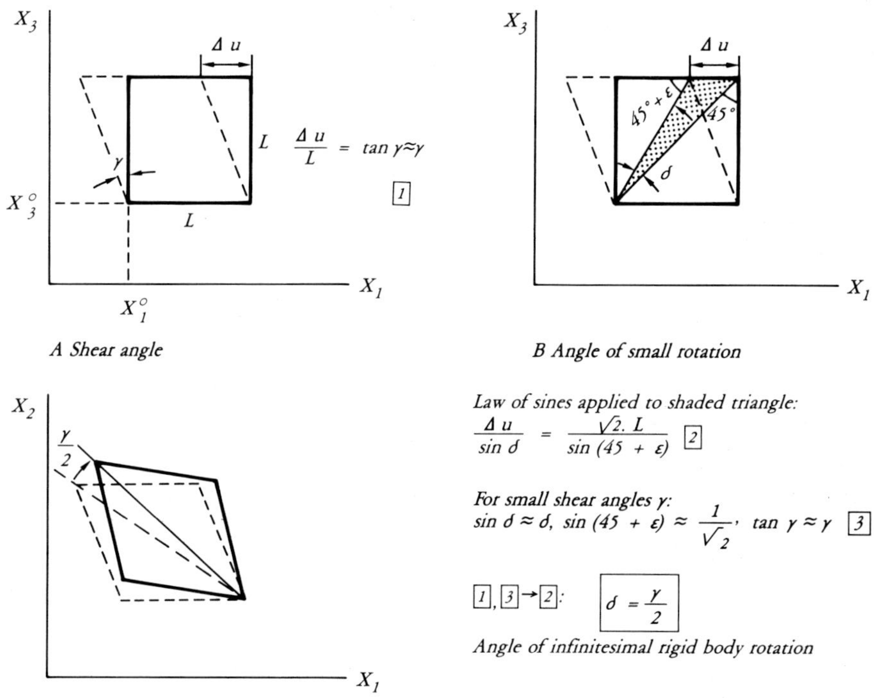

A Shear angle

B Angle of small rotation

C Simple and pure shear differ only by a rigid body rotation through half the shear angle

Fig. II.5-5
ROTATION IN 'SMALL' SIMPLE SHEAR

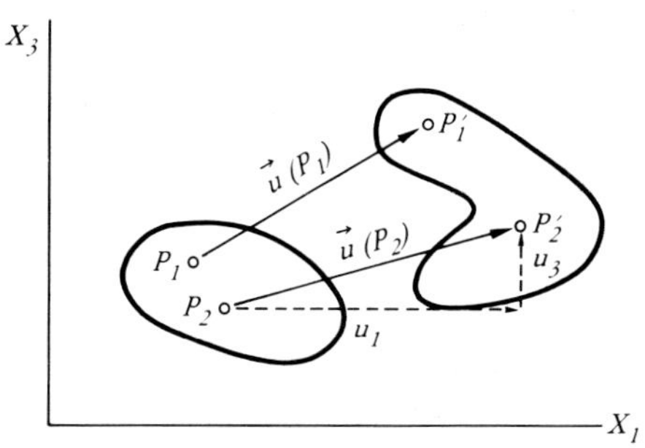

Fig. II.5-6
DISPLACEMENTS
The deformation of a body is completely described by the field of displacement vectors.

$\delta u_1 < 0$ in *contraction*. Associated with this displacement field is a uniaxial strain, which according to eq. II.5-1 is

$$e_1 = \frac{l - l^*}{l} = -\frac{\delta u_1}{\delta x_1} \qquad \text{(II.5-4)}$$

Analogous expressions hold for the uniaxial elongations in the x_2- and x_3-directions.

The ratio $\delta u_1 / \delta x_1$ is a *'displacement gradient'*, specifying the change in the displacement component u_1 with particle position varying in the x_1-direction. A constant value of this gradient, for instance, means that the distance between the two vertical faces in Fig. II.5-7A is changed under the applied deformation by the same fraction, irrespective of the original distance and position of the elements. Such a deformation is called *'homogeneous'*.

In inhomogeneous deformations a displacement gradient has to be assigned to each ma-

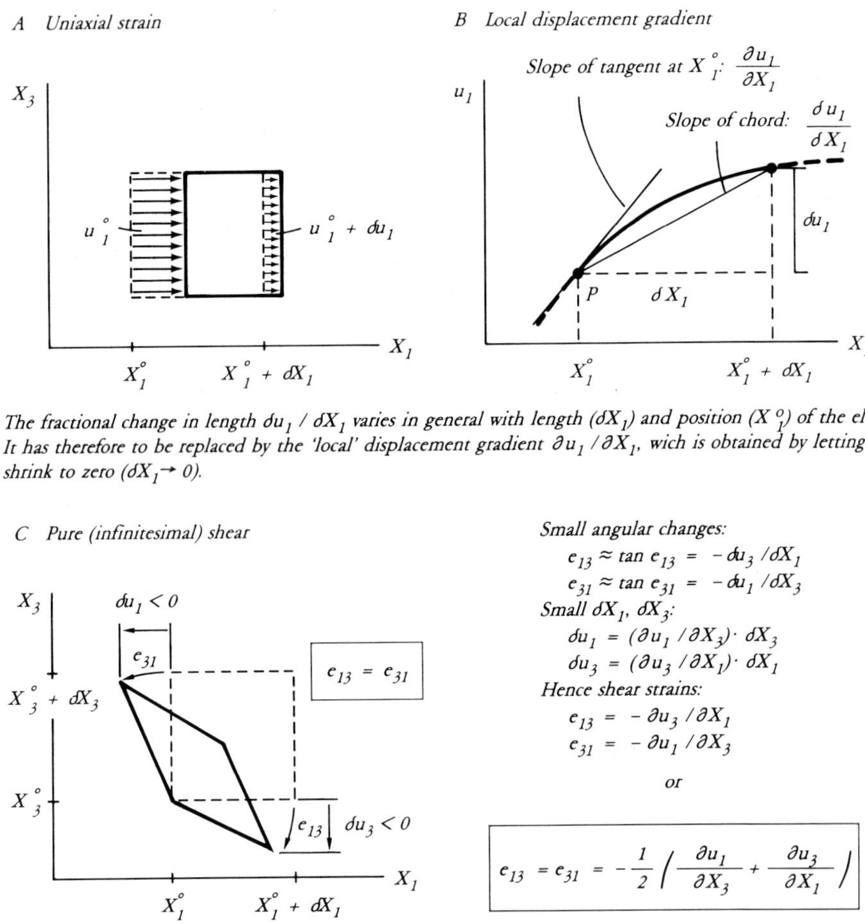

Fig. II.5-7 STRAIN COMPONENTS IN TERMS OF DISPLACEMENTS

terial point. To obtain this local displacement gradient one considers infinitely close parallel surface elements, i.e. with reference to Fig. II.5-7A, the distance δx_1 between the two elements at x_1^o and $x_1^o + \delta x_1$ is made to shrink to zero ($\delta x_1 \rightarrow 0$). The procedure is illustrated in Fig. II.5-7B which shows a graph of u_1 varying with x_1. As δx_1 shrinks towards zero, the chord connecting the points associated with x_1^o and $x_1^o + \delta x_1$ approaches the tangent at P (x_1^o) and, accordingly, the slope of the chord $\delta u_1 / \delta x_1$ approaches the slope of the tangent which is denoted by $\partial u_1 / \partial x_1$ (the 'partial derivative' of u_1 with respect to x_1). This limiting slope value is the *local displacement gradient.* Therefore, quite generally, the components of the local elongational strains become in terms of displacement:

$$e_1 = -\partial u_1/\partial x_1,\; e_2 = -\partial u_2/\partial x_2,\; e_3 = -\partial u_3/\partial x_3 \qquad \text{(II.5-5)}$$

and measure the elongational straining of an infinitesimal surrounding of a material point considered.

Expressions II.5-5 were derived for uniaxial straining. The expressions remain valid, however, for the general case of multi-axial inhomogeneous straining – provided the strains are *'small'.* This restriction, which was already made in deriving equation II.5-2, means that products of displacement gradients can be neglected with respect to the gradients themselves or, in other words, that the absolute values of the displacement gradients are much smaller than one. The reason for this severe restriction should be well understood and will be demonstrated next.

By way of generalising the uniaxial contraction of Fig. II.5-7A, we now have the material undergo additional displacements u_3 in the x_3-direction, which, moreover, vary linearly in the x_1-direction. As sketched in Fig. II.5-8, a line element, originally parallel to the x_1-axis and of length δx_1 will, in response to the displacements u_3, now change its direction. Its new length δl^* is found (Fig. II.5-8) by means of Pythagoras' theorem as

$$\delta l^* = \delta x_1 \sqrt{(1 + \partial u_1/\partial x_1)^2 + (\partial u_3/\partial x_1)^2} \qquad \text{(II.5-6)}$$

Consequently, the fractional change in length of the element originally parallel to the x_1-axis becomes

$$\frac{\delta x_1 - \delta l^*}{\delta x_1} = 1 - \sqrt{(1 + \partial u_1/\partial x_1)^2 + (\partial u_3/\partial x_1)^2} \qquad \text{(II.5-6a)}$$

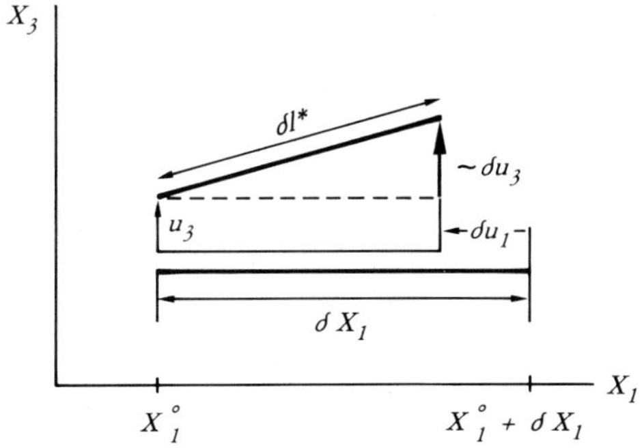

$$\delta l^{*2} = (\delta X_1 + \delta u_1)^2 + \delta u_3^2$$

$$\delta u_1 \approx \frac{\partial u_1}{\partial X_1}\,\delta X_1 \quad (<0)$$

$$\delta u_3 \approx \frac{\partial u_3}{\partial X_1}\,\delta X_1$$

$$\boxed{\delta l^* \approx \delta X_1 \sqrt{\left(1 + \frac{\partial u_1}{\partial X_1}\right)^2 + \left(\frac{\partial u_3}{\partial X_1}\right)^2}}$$

Fig. II.5-8 FINITE CHANGE IN LENGTH OF LINE ELEMENT

an expression which only when $|\partial u_3 / \partial x_1| \ll 1$ reduces to the corresponding expression II.5-5 for e_1.

With expressions II.5-5 inserted in equation II.5-2 we obtain the small volumetric strain in terms of displacement gradients:

$$dV/V = -(\partial u_1/\partial x_1 + \partial u_2/\partial x_2 + \partial u_3/\partial x_3) \qquad \text{(II.5-7)}$$

where a decrease in volume is counted as positive.

Next we express the shear strains (eq. II.5-3) in terms of displacement gradients. The simple procedure is outlined in Fig. II.5-7C for the shear component e_{13}, and shows that for small changes in the original right angle between two line elements that were parallel to the x_1-axis and x_3-axis, respectively, the shear strain becomes

$$e_{13} = e_{31} = -\frac{1}{2}\left[\frac{\partial u_1}{\partial x_3} + \frac{\partial u_3}{\partial x_1}\right]$$

and analogously for the other components (II.5-8)

$$e_{12} = e_{21} = -\frac{1}{2}\left[\frac{\partial u_1}{\partial x_2} + \frac{\partial u_2}{\partial x_1}\right]$$

$$e_{23} = e_{32} = -\frac{1}{2}\left[\frac{\partial u_2}{\partial x_3} + \frac{\partial u_3}{\partial x_2}\right]$$

To illustrate these relationships between strains and displacement field, let us again consider ***simple*** and ***pure shear*** deformation. In simple shearing, as sketched in Figs. II.5-4A and II.5-5A, material points move parallel to, say, the x_1-axis and the displacements u_1 vary linearly in the x_3-direction. With reference to Fig. II.5-5A (in which $\Delta u = u_1$) the simple-shear displacement field is therefore

$$u_1 = -\tan\gamma \cdot (x_3 - x_3^o), \; u_3 = 0 \qquad \text{(II.5-9)}$$

from which follows directly for the shear strains according to eq. II.5-8

$$e_{13} = e_{31} = \frac{1}{2}\tan\gamma \approx \frac{\gamma}{2}. \qquad \text{(II.5-8a)}$$

In pure shear (Fig. II.5-7C) both u_1 and u_3 differ from zero, u_1 varying linearly in the x_3-direction (from 0 to δu_1) and u_3 doing the same in the x_1-direction (from 0 to δu_3). Because of the symmetry of the deformation with respect to the chosen coordinate system, the corresponding gradients must have a common value, i.e. $\partial u_1 / \partial x_3 = \partial u_3 / \partial x_1 = -a$ with $a > 0$. For the displacement field in Fig. II.5-7C we therefore obtain

$$u_1 = -a(x_3 - x_3^o), \qquad u_3 = -a(x_1 - x_1^o) \qquad \text{(II.5-10)}$$

where, according to eq. II.5-8, $e_{13} = e_{31} = a$.

Comparing equations II.5-10 and II.5-9, one immediately notices that a state of small pure shear can always be interpreted as the net result of two consecutive simple shears along mutually perpendicular directions as indicated in Fig. II.5-9.

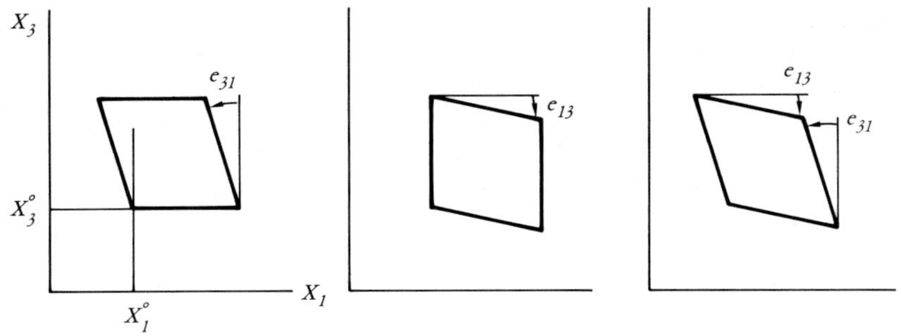

Fig. II.5-9 SUPERPOSITION OF TWO 'SMALL' SIMPLE SHEARS LEADING TO A PURE SHEAR

Strain Tensor. The strains that characterise the extensions (contractions) and shear distortions of a material element in three dimensions and with reference to a given Cartesian frame may be written in a systematic array (matrix)

$$\begin{bmatrix} e_1 & e_{12} & e_{13} \\ e_{21} & e_2 & e_{23} \\ e_{31} & e_{32} & e_3 \end{bmatrix} \qquad \text{(II.5-11)}$$

which is symmetric with respect to the dotted diagonal and therefore contains only six independent parameters. These 'components of strain' completely characterise the local state of strain, i.e. how the material in an infinitesimal surrounding of a point P has been distorted from its original state. (The strain components do not tell us, however, how this small material region may have been displaced as a whole by rigid-body motions.) In other words, the parameters II.5-11 completely describe how the shape of a small cubical element at P with edges parallel to the axes of a Cartesian frame is changed or, inversely, what the undeformed shape has been of a cubical element delineated in the deformed body.

It is clear that the actual values of the strain parameters in equation II.5-11 very much depend on the Cartesian reference frame chosen. For instance, in the simple case of Fig. II.5-7A the line elements parallel to the x_1-axis undergo a shortening, whereas the ones parallel to the other coordinate axes remain unchanged in length and, consequently, $e_1 \neq 0$, $e_2 = e_3 = 0$. If we were to describe the same state of strain with respect to axes that have been rotated with respect to the ones used in Fig. II.5-7A, the corresponding strains $\bar{e}_1$ and $\bar{e}_3$ would both attain non-vanishing values. These changes in the numerical representation of one and the same state of strain, which accompany a change in the Cartesian reference frame used, are of a purely geometrical nature and therefore governed by well-defined 'transformation rules' [2]). It should be mentioned that these rules identify equation II.5-11 as the componental representation of an 'invariant' object – termed a (symmetric) 'tensor of second order'. Hence, the parameters in equation II.5-11 are usually referred to as the components of the 'strain tensor'.

Principal Strains. At this point we draw attention to the complete formal analogy between (infinitesimal) strain and stress. Both are characterised by symmetric second-order tensors, and, as a consequence of this, what is said in Section II.1.3 on 'stress invariants', 'principal stresses', 'maximum shear stress', 'principal axes as reference frame', etc. applies almost verbatim to strains.

In particular, at any point of an infinitesimally strained body, there are three mutually perpendicular straight-line elements which during straining were exposed to only extensional (or contractive) strains, but show no angular changes. Thus the right angles between these material elements remain preserved under the deformation. The directions of these elements are known as 'principal' directions (or axes) of strain. If these principal axes are chosen as reference frame, all shear strain components vanish and equation II.5-11 retains only diagonal terms, because the right angles of a cubical element with edges parallel to the principal axes will not change. The extensional strains along the principal axes are called 'principal strains' and are denoted by e_I, e_{II}, e_{III}, where $e_I \geqslant e_{II} \geqslant e_{III}$.

In the case of pure plane shear (Fig. II.5-7C) the diagonals of the original square (dotted in the figure) represent the principal directions of highest contraction ($e_I > 0$) and highest extension ($e_{III} < 0$), the intermediate principal strain $e_{II} = 0$ being associated with the x_2-direction. In the case of simple shear (Fig. II.5-5A) the principal directions again coincide with the diagonals of the undeformed square and the x_2-axis. In contrast with pure shear, the axes of greatest and smallest strain are rotated during the deformation through half the shear angle, as demonstrated in Fig. II.5-5B. This indicates, of course, the rigid-body rotation that is associated with simple shear.

Maximum Shear Strain. Inspecting once more Figs. II.5-5A and II.5-7C we notice that in pure and simple shear the pairs of originally perpendicular line elements that suffer the greatest angular changes are those parallel to the x_1- and x_3-axes. These elements are also inclined at $\pm$ 45° towards the principal axes of greatest and smallest strain. Without proof, we mention here that this is quite generally true, i.e. the greatest shearing strain always occurs along directions at $\pm$ 45° to the axes of greatest and smallest principal strain. The value of the maximum shear strain (measured in radians) $\gamma_{max}/2$ is

$$\gamma_{max}/2 = (e_I - e_{III})/2 \qquad \text{(II.5-12)}$$

This result can be read off directly from Mohr's strain circle which will be discussed later.

3. PLANE STRAIN

Strain analysis is greatly simplified in the case of 'plane strain'. Plane strain in the strict sense is defined as a deformation where (1) all displacements are parallel to a single plane, and (2) these displacements do not change in the direction perpendicular to this plane. The plane itself is called 'plane of deformation' and may, for instance, be identified with the x_1, x_3-plane of a Cartesian frame (Fig. II.5-10A). The displacements then have the components:

$$u_1 = u_1(x_1, x_3),\ u_2 = 0,\ u_3 = u_3(x_1, x_3) \qquad \text{(II.5-13)}$$

In this type of deformation line elements parallel to the x_2-axis do not change in length or orientation. Hence, in accordance with eqs. II.5-5 and II.5-8 we have the vanishing strains

$$e_2 = e_{23} = e_{21} = 0 \qquad \text{(II.5-14)}$$

The number of non-vanishing strain components is thereby reduced from six to three (e_1, e_3, $e_{13} = e_{31}$).

Since no shearing takes place between planes parallel to the x_1, x_3-plane (i.e. the deformation plane), the x_2-axis is, by definition, a principal strain axis and the remaining two principal strains must lie in the x_1, x_3-plane. To illustrate this, Fig. II.5-10B shows the case of a plane strain where no volume changes take place. One principal strain (e_I) is therefore positive, while another (e_{III}) is negative. The vanishing strain e_2 becomes the intermediate principal strain e_{II}. Since there are no volume changes one has $e_I = - e_{III}$ as follows immediately from equation II.5-2.

What kinds of *real situations* lend themselves to the plane strain simplification? First of all, a state of plane strain is produced in a uniform material, if the material is confined between fixed frictionless, parallel plates and is loaded by forces that act parallel to these walls and do not vary in the direction perpendicular to them. 'Sand box' experiments used to model tectonic processes may approximate this plane strain situation fairly well. Fig. II.5-11 is a photograph of such a device. The sand is confined between two thick glass plates which prevent extension of the sand pack in one direction ($e_2 = 0$). A horizontal load is

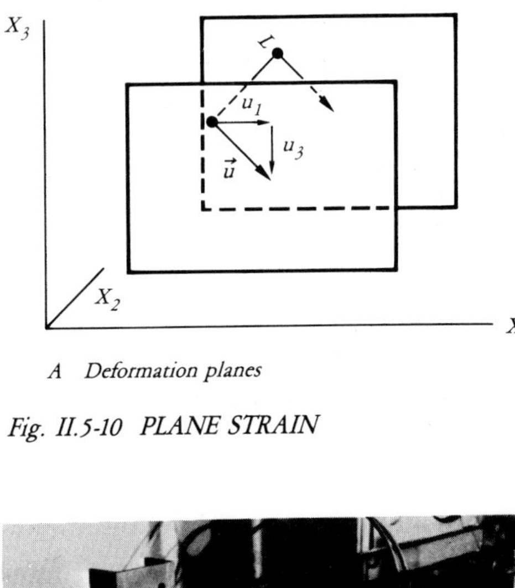

A Deformation planes

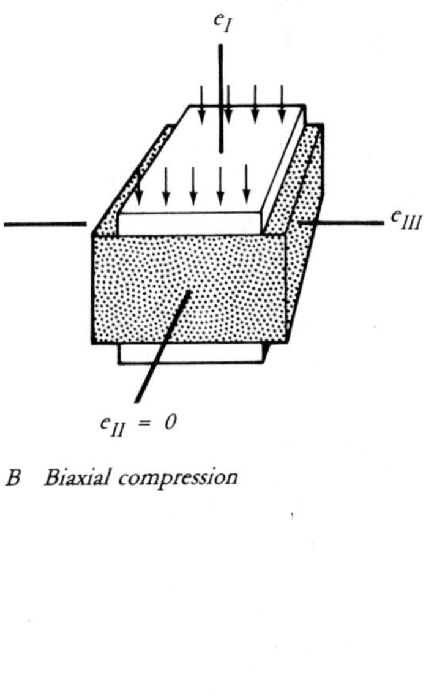

B Biaxial compression

Fig. II.5-10 PLANE STRAIN

Fig. II.5-11
PLANE STRAIN
IN 'SANDBOX' EXPERIMENTS
Faulting induced in sediment above subsiding basement block

applied parallel to the confining plates by vertical pistons (one is shown on the right of the photograph). Although the friction between the sand and the glass plates is low, it is not possible completely to avoid small shear strains (e_{12}, e_{32}) developing between planes parallel to the glass plates and causing some deviation from ideal plane strain during deformation of the pack. In the 'faulting' experiments of Fig. II.5-11 this deviation expresses itself as a slight curvature of the fault scarps. This indicates that the displacements, although still parallel to the glass walls, vary slightly across the pack. Obviously, this 'wall effect' becomes less as the glass plates are spaced more widely[3].

The situation is similar in the tectonic case schematised in Fig. II.5-12A, in which a thrust sheet, bounded by transcurrent faults, is pushed over a 'lubricating' substratum by forces that are approximately uniformly distributed along the rear boundary. In general, shearing along the parallel boundary faults will have little effect on the deformation of the inner part of the thrust sheet, where the deformation thus approximates plane strain.

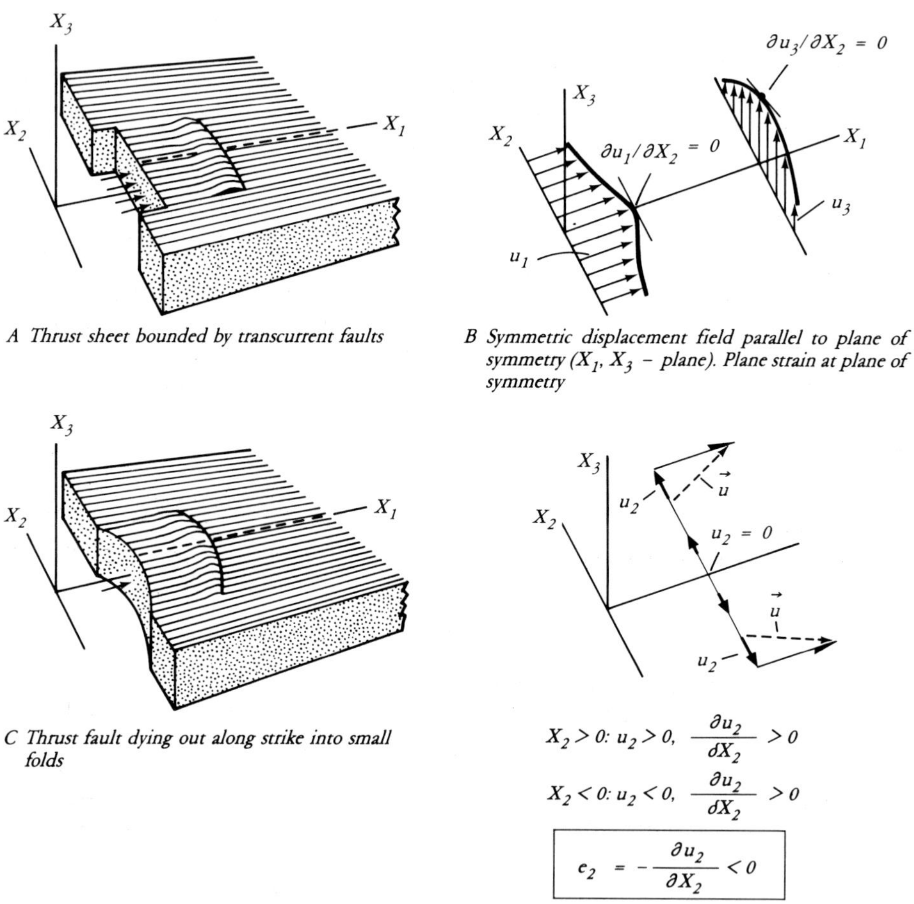

A Thrust sheet bounded by transcurrent faults

B Symmetric displacement field parallel to plane of symmetry (X_1, X_3 – plane). Plane strain at plane of symmetry

C Thrust fault dying out along strike into small folds

D Symmetric displacement with out-of-plane component $u_2 \neq 0$. Strain at plane of symmetry = plane strain + extensional strain $e_2 \neq 0$

Fig. II.5-12
SYMMETRIC DISPLACEMENT FIELDS

Even if in the last example the shears along the bounding faults were very intensive, the shear strains e_{23} and e_{21} would vanish exactly on the midplane ($x_2 = 0$) of the thrust sheet. This follows from the symmetry of the displacement field with regard to this plane. As sketched in Fig. II.5-12B, this symmetry makes the displacement gradients $\partial u_1 / \partial x_2$ and $\partial u_3 / \partial x_2$ – and hence the shear strains e_{21}, e_{23} – vanish at the midplane ($x_2 = 0$). The plane strain analysis therefore applies rigorously to the immediate vicinity of the plane of symmetry.

Next, we consider the thrusting process schematised in Fig. II.5-12C. Here again, forces and displacements are symmetric with respect to the plane $x_2 = 0$, but in the absence of lateral boundaries the displacement vectors may have a non-vanishing out-of-plane component u_2. It can easily be shown (Fig. II.5-12D) that even in this case the shear strains e_{21} and e_{23} vanish along the plane of symmetry ($x_2 = 0$), since the gradients of u_2 in the x_1- and x_3-directions have different signs for $x_2 > 0$ and $x_2 < 0$ and, therefore, vanish at the plane of symmetry $x_2 = 0$. The present case of symmetry differs, however, from the case discussed above as far as the out-of-plane 'normal' strain e_2 is concerned. As indicated in Fig. II.5-12D, u_2 must vanish along the plane of symmetry $x_2 = 0$. In the symmetric displacement field with non-vanishing out-of-plane component u_2 the gradient $\partial u_2 / \partial x_2$ will therefore have the same sign on both sides of the plane of symmetry. The strain e_2 will therefore not vanish on the midplane $x_2 = 0$ and the state of strain along this plane is the sum of a plane strain and an out-of-plane 'normal' strain. In many geological cases the latter strain can be regarded as approximately constant (extension or contraction) and then does not complicate the analysis of the problem.

Finally, tectonically deformed rock bodies frequently extend in one direction over a distance very much larger than the thickness of the body and, at the same time, show only little variation in cross-sectional geometry, as sketched in Fig. II.5-13. Long parallel folds of relatively uniform shape are well-known examples of this. One may further assume that the loads (and reactive forces) that produced the deformation were uniformly distributed and acted perpendicularly to the 'long axis' of the rock body (x_2-axis in Fig. II.5-13). These tectonic loads not only shorten the rock body in the direction of their action, but they also tend to produce lateral extension. Owing to the large lateral extent of the rock body these extensional strains (e_2) would result in large lateral displacements (u_2) if the base of the rock body could stretch freely. This, however, is commonly prevented ($e_2 = 0$) by friction along the base of the rock body, which thus produces a similar effect on the strain field as the parallel confining 'walls' discussed earlier[4)].

Mohr Circle for Small Strain. In the analysis of two-dimensional stress problems Mohr's stress circle diagram is a very convenient tool (see Chapter II.4). An analogous representation can be used for plane strain and most of what is said about the stress circle also applies to the 'strain circle', when 'stress' is replaced by 'strain'.

In analysing the state of stress at a given point, one associates two quantities with each planar element through that point – the normal stress $\sigma_\perp$ and the shear stress τ. Analogously, in analysing plane strain, we consider planar elements with normals in the deformation plane (Fig. II.5-14) and associated strains – the extensional or 'normal' strain $e_\perp$ and the angular change $\gamma / 2$ (i.e. the shear strain) which a line element (originally perpendicular to the planar element) undergoes. As the line element considered is rotated anticlockwise in the plane of deformation (Fig. II.5-14), the corresponding 'strain point' in the $e_\perp, \gamma / 2$-plane describes the strain circle[5)] (Fig. II.5-15A). The centre of the circle lies on the $e_\perp$-axis at a distance $(e_I + e_{III}) / 2$ from the origin and its radius equals the magnitude of the maximum shear strain $\gamma_{max} / 2 = (e_I - e_{III}) / 2$.

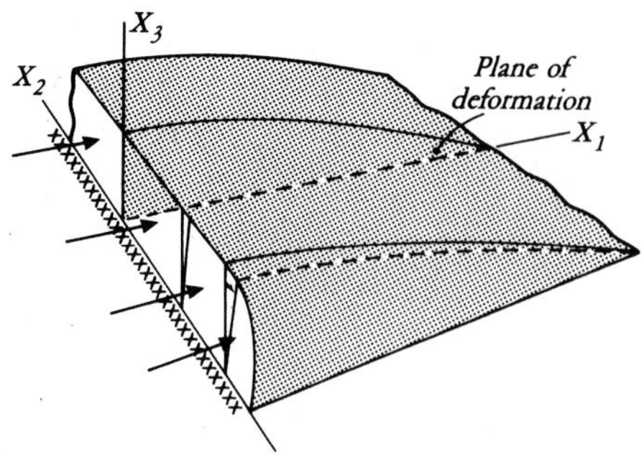

Fig. II.5-13
PLANE STRAIN IN GEOLOGICAL BODY OF LARGE LATERAL EXTENT
Lateral extension is suppressed by basal friction

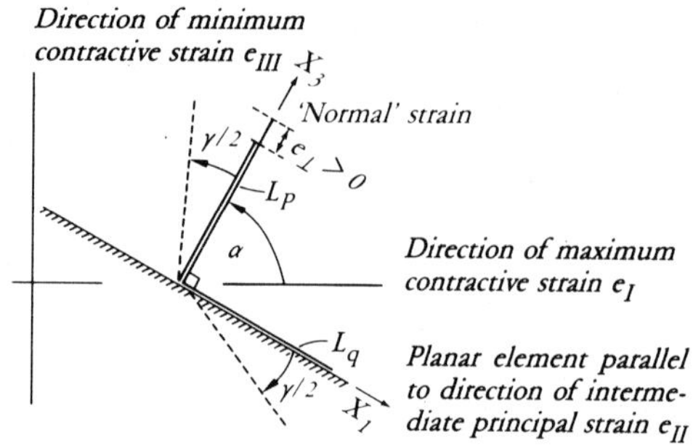

Fig. II.5-14
'NORMAL' AND SHEAR STRAIN ASSOCIATED WITH UNIT LINE ELEMENT (L_P) IN PLANE OF MAXIMUM AND MINIMUM PRINCIPAL STRAINS IN PHYSICAL SPACE

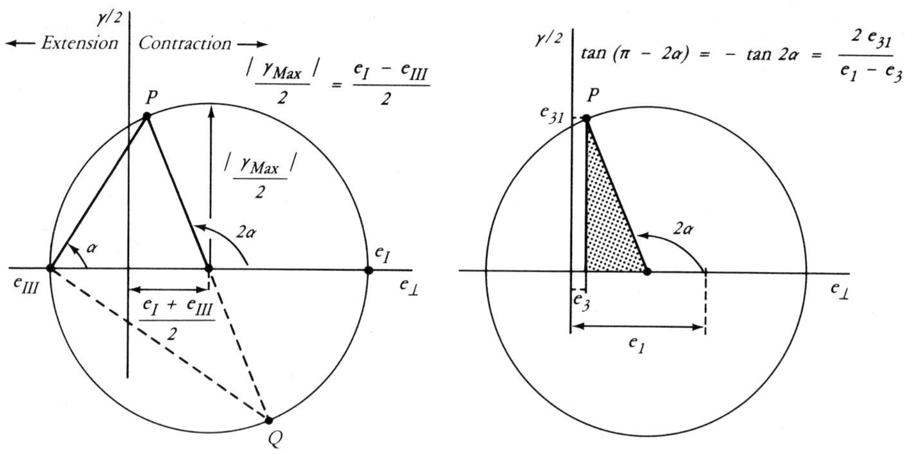

A Strain points associated with the line elements of Fig. II.5-14 *B Angle (α) between e_I-axis and positive X_3-axis*

Fig. II.5-15 MOHR'S CIRCLE FOR INFINITESIMAL STRAIN

When the principal strains e_I and e_{III} are known, the circle can be constructed immediately and the normal and shear strains associated with a line element, including the angle α with the e_I-axis (Fig. II.5-14), can be read off (point P on strain circle in Fig. II.5-15A). Conversely, when the strains $e_\perp$, $\gamma/2$ are given for two mutually perpendicular material line elements (L_P, L_Q in Fig. II.5-14), the Mohr circle can also be constructed (circle through P and Q in Fig. II.5-15A) and the values and directions of the principal strains can be determined.

Let us now consider two line elements L_P and L_Q (Fig. II.5-14) which are parallel to, respectively, the x_3- and x_1-axes of a Cartesian coordinate system. Element L_P, and hence the positive x_3-axis, includes the angle α with the positive e_I-axis. Associated with L_P are the strains $e_\perp = e_3$, $\gamma/2 = e_{31}$. They determine a point P on the strain circle (Fig. II.5-15A), while the strains associated with L_Q determine the opposite point Q. The normal strain associated with Q is e_1 and, therefore, the base of the shaded triangle in Fig. II.5-15B has the length $(e_1 - e_3)/2$. Applying elementary geometry to this triangle, one readily finds the important relationship

$$\tan 2\alpha = -\frac{2e_{31}}{e_1 - e_3} \tag{II.5-15}$$

which expresses the angle the e_I-axis includes with the positive x_3-axis (Fig. II.5-14) in terms of the plane strain components.

When in plane strain the volume remains unchanged, the strain circle is centred at the origin of the strain plane (Fig. II.5-16A) since, in accordance with equation II.5-2, $e_I + e_{III} = 0$. The circle then represents a state of pure shear, since two mutually perpendicular line elements, the strain points of which are A and B in Fig. II.5-16A and which bisect the angle between the principal strain axes, suffer maximum angular change, but remain unchanged in length ($e_\perp = 0$). This is further illustrated in Fig. II.5-16B, C by the deformation of a square (dotted), the sides of which bisect the angle between the principal strain axes. It will be noticed that the same state of pure shear is also produced by opposite straining of equal magnitude ($e_I = -e_{III}$) along the principal axes.

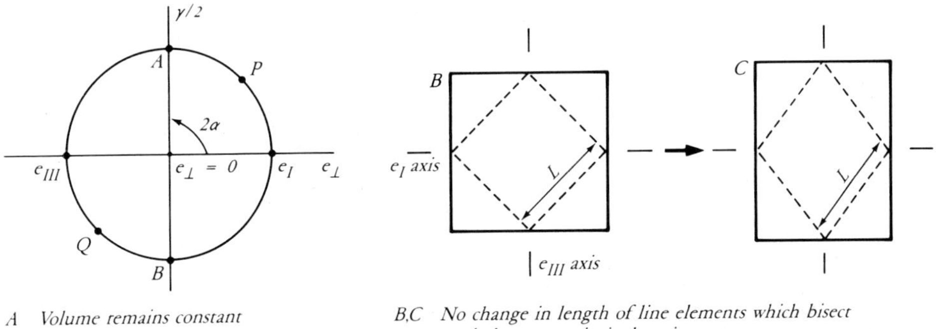

A Volume remains constant

B,C No change in length of line elements which bisect angle between principal strain axes

Fig. II.5-16 PURE SHEAR IN MOHR'S STRAIN CIRCLE

Strains in rocks. Rocks, whether porous or not, are highly heterogeneous materials when viewed on a 'micro' scale (e.g. grain size scale). When stressed under uniform load the displacements set up inside a rock sample will, therefore, vary strongly over micro distances, i.e. distances that are characteristic of the heterogeneities of the rock. Strains determined on such a scale will fluctuate from point to point. In general, however, one is not interested in these 'micro strains', which characterise the straining of the particulate constituents of a rock or rock mass, but rather in the straining of rock elements which are very large in comparison with the micro-structural dimensions. It is not difficult to see that by choosing elements of sufficient size the disturbing effect of the fluctuating micro displacements is 'smoothed out' and meaningful average displacements can be defined from which to determine the macro strains. For instance, in determining a small contraction of a porous rock in a certain direction, one may measure (Fig. II.5-17) the change in distance between two parallel macro elements, i.e. elements that cut through a great number of grains. When plane in the undeformed state, the material macro element will show a large number of microscopic 'warpings' in the deformed state, which are caused by the heterogeneities of the rock. In other words, the material points that constituted the original macro elements will in the deformed state define microscopically 'wavy' elements (Fig. II.5-17). The randomness of the deviations from an ideally plane cross-section allows, in principle, the wavy elements to be replaced by plane ones the distance l^* of which is then to be compared with the original distance l. In practice it will be sufficient to take the unstrained distance between two elements as being so large that even for very small straining the change in average distance δl is large compared with the 'micro warping' of the material elements.

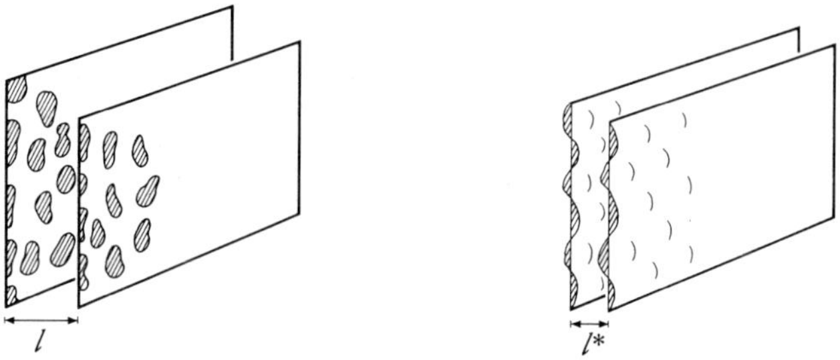

Fig. II.5-17 MACRO STRAINS IN ROCKS

Compatibility of Strain Components. We have seen that, in general, six strain components are related to the three components of the displacement field (eqs. II.5-5, 8). In plane strain three strain components are related to two displacement components. If one were to arbitrarily prescribe strain components that varied from point to point in the deformed body, one could hardly expect a unique and smooth displacement field to be produced. Let us think of the unstrained body as being cut up into a number of small square prisms and let us deform each prism in accordance with an arbitrarily prescribed strain distribution. Will it be possible to fit the deformed prisms back together – without additional straining – in such a way as to leave no holes or areas of overlap? Obviously, the strains cannot be completely arbitrarily prescribed, but have to satisfy certain 'compatibility' relationships in order to ensure that the strains of adjacent elements are compatible with a displacement field.

For plane strain (in the x_1, x_3-plane), in which three strain components (e_1, e_3, e_{13}) are related to two displacement components (u_1, u_3), the strain components will have to satisfy one compatibility relationship. Its derivation requires a little calculus. We differentiate the strains e_1 and e_3 in equations II.5-5 twice with respect to x_3 and x_1, respectively. The sum of the two derivatives becomes

$$\frac{\partial^2 e_1}{\partial x_3^2} + \frac{\partial^2 e_3}{\partial x_1^2} = -\frac{\partial^2 u_1}{\partial x_1 \partial x_3^2} - \frac{\partial^2 u_3}{\partial x_3 \partial x_1^2}$$

We notice that the right-hand side of this equals $2\partial^2 e_{13}/\partial x_1 \partial x_3$ (cf. eq. II.5-8) since the order of differentiation does not matter. In plane strain the strain components must, therefore, satisfy the (differential) compatibility equation

$$\frac{\partial^2 e_1}{\partial x_3^2} + \frac{\partial^2 e_3}{\partial x_1^2} = 2\frac{\partial^2 e_{13}}{\partial x_1 \partial x_3} \qquad \text{(II.5.16)}$$

Example: Compacting layer on slope (after Kröner, 1969)

Let us consider a uniform sediment layer on an inclined plane base of 'rigid' material (Fig. II.5-18). Owing to the compaction process a material which at a given instant occupied a cross-section perpendicular to the slope (x_1 = const.) will after some time occupy a curved surface as sketched in Fig. II.5-18. The displacement vector $\vec{u}$ with components u_1, u_3 parallel and perpendicular to the slope will decrease towards the base where it vanishes ($u_1 = u_3 = 0$), since the material is supposed not to slip along the rigid basement. Assuming that the sloping layer is very long, the displacements will vary only with 'depth' x_3: $u_1 = u_1(x_3)$, $u_3 = u_3(x_3)$ [6]. In addition, we postulate that the displacement vectors are parallel throughout the layer and inclined at the angle γ with respect to the slope. (A similar example, but with slope-parallel displacements, was treated in Section II.2, Example No. 4.). We wish to determine the direction of maximum contrative strain e_I associated with this compaction behaviour.

As indicated in Fig. II.5-18 we obtain from equations II.5-5 and II.5-8 for the small compaction strains produced during the time interval considered: $e_1 = 0$, $e_3 = -\tan\gamma \cdot \partial u_1 / \partial x_3$, $e_{31} = -\frac{1}{2}\partial u_1 / \partial x_3$. By inserting these expressions into equation II.5-15 we obtain for the angle that the e_I-axis includes with the positive x_3-axis (Fig. II.5-18):

$$\tan 2\alpha = 1/\tan\gamma = 1/\cot\left(\frac{\pi}{2} - \gamma\right)$$

The angle the direction of maximum compaction includes with the x_3-axis in Fig. II.5-18 is therefore

$$\alpha = 45° - \gamma/2 \qquad \text{(II.5-17)}$$

Taking the vertical direction as reference direction (Fig. II.5-18), we find for the angular deviation of the e_I-axis $(\beta - \delta)/2$, where β is the inclination of the displacement vectors with respect to the vertical and δ the slope angle.

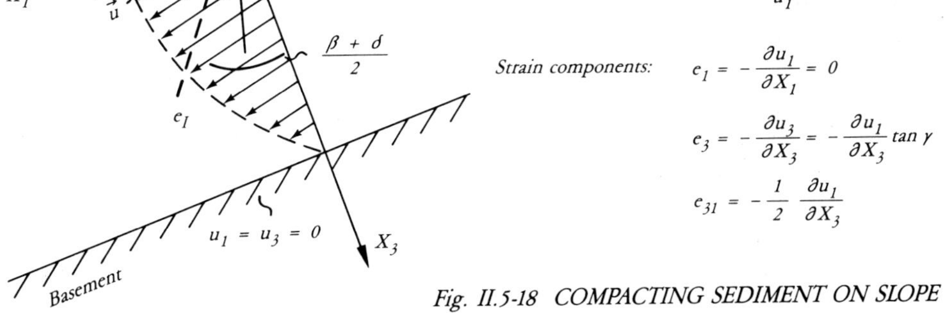

Fig. II.5-18 COMPACTING SEDIMENT ON SLOPE

4. STRAIN RATES

The infinitesimal displacements in a deforming body are related to the velocities at which the particles move during an infinitesimal time span dt. The elementary relationship between the displacement components u_1, u_2, u_3 and the velocity components v_1, v_2, v_3 reads:

$$u_i = v_i\,dt \qquad (i = 1, 2, 3) \tag{II.5-18}$$

Next, we consider the strain increments e_1, e_2, ... , e_{23} that are produced during the interval dt. Dividing these incremental strains by dt we obtain the 'strain rates', i.e. changes in strain per unit time, which we indicate by the symbols $\dot{e}_1$, $\dot{e}_2$, ..., $\dot{e}_{23}$. Replacing the displacements by the velocities in equations II.5-5,8, we have the strain rates expressed in terms of particle velocities:

$$\dot{e}_1 = -\frac{\partial v_1}{\partial x_1}, \quad \dot{e}_2 = -\frac{\partial v_2}{\partial x_2}, \quad \text{etc.}$$

$$\dot{e}_{12} = \dot{e}_{21} = -\frac{1}{2}\left(\frac{\partial v_1}{\partial x_2} + \frac{\partial v_2}{\partial x_1}\right), \quad \text{etc.} \tag{II.5-19}$$

Like the strains themselves, these expressions vanish if the material undergoes rigid-body motions [7)].

Since strain is dimensionless, strain rate has the physical dimension of a reciprocal time. The numerical specification 10^{-1}/s, for example, would mean a ten percent straining (contraction) per second. The often quoted geological strain rate of 3.10^{-14}/s states that it takes 3.10^{13} seconds, or approximately one million years, to bring about a 10 per cent strain. The range of geological strain rates has been discussed by Price (1975) and Ramsay & Pfiffner (1982)

5. SMALL vs. LARGE ROCK STRAIN

Our discussion of the kinematics of strain is restricted to 'small' strains, i.e. strains that do not exceed a few per cent. In view of the overwhelming evidence of much larger rock strains in nature, the concept of 'small' strains may appear highly inadequate – the more so as the study of 'finite' (large) strains has become a fashion among structural geologists and several textbooks offer the geological reader introductions to the rather complex theory [8)] of finite strains. A few comments, therefore, seem relevant.

In reconstructing the kinematics by which a rock mass has attained its present configuration and position, geologists often study strain indicators, i.e. small inclusions whose original, undeformed shape and structure are known and can be compared with the present, deformed state (e.g. ooids, mud pellets, spherical or cylindrical marine fossils, etc.). An illustration of this is given in Fig. II.5-19 for the two-dimensional deformation of an initially circular element (t_0). In its final stage (t_9) the element has attained elliptical shape, the lengths and orientations of the principal axes defining the local (large) straining the element has undergone. Setting aside the question as to how representative the straining of small inclusions may be of the straining of a mechanically different host material,

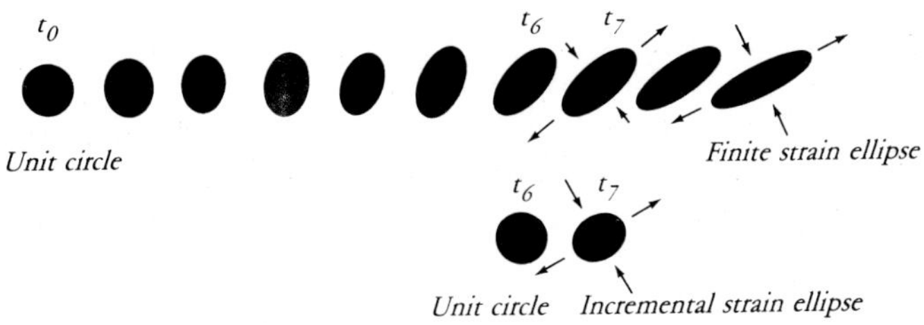

Fig. II.5-19 *FINITE TWO-DIMENSIONAL DEFORMATION OF AN INITIALLY CIRCULAR ELEMENT VIA ARBITRARY DEFORMATION SEQUENCE.*
Shape of 'strain ellipse' at different stages. To find the 'incremental' strain between two successive stages t_6 and t_7, one would have to consider the distortion of a new circle emplaced in the material at t_6

it should be clear that, in general, the final strain (t_9) does not tell us the actual deformation history of the inclusions. The same state (t_9) could be produced in an infinity of different sequences of straining and rigid-body rotations.

In contrast with this purely kinematic approach, mechanical theories attempt to trace step-by-step (and in a unique way) a tectonic process by focussing on the *stresses* in the deforming body and their interdependence with the strains. What is now the nature of the relationship between stress and strain in rock deformation? Nobody can seriously believe that the large rock strains observed in nature are the result of elastic deformations. Rather they are produced by viscous and/or plastic-type processes (see Section II.7.1) that may have been preceded or accompanied by small elastic deformations. It is typical of the viscous and plastic deformations that at any stage of the process the stresses are not related to the total ('finite') strain accumulated in the past, but depend on the *incremental changes in strain* between two successive stages of the process. In Fig. II.5-19 this incremental strain would correspond with the distortion a material circle, emplaced in the material at say stage t_6, would experience during the transition to a successive (infinitely close) stage t_7. Unfortunately, in nature, one will find little evidence of such incremental strains from which one could directly deduce the state of stress.

Since the concept of finite strains does not enter the stress/strain (rate) relationships for viscous and ideally plastic materials and since elastic rock straining will usually be small, the theory of finite strains is only marginally relevant in the context of this book [9)]. Readers interested in finite-strain theory are referred to references 1) and 2) for collateral reading.

COLLATERAL READING:

1) Ramsay, J.G., 'Folding and fracturing of rocks'. McGraw-Hill, 1967.
2) Means, W.D., 'Stress and strain; basic concepts of continuum mechanics for geologists'. Springer Verlag, 1976.
3) Jaeger, J.C. & Cook, N.G.W., 'Fundamentals of rock mechanics'. Methuen, 1968.
4) Hobbs, B.E., Means, W.D. & Williams, P.F., 'An outline of structural geology'. John Wiley & Sons, 1976.

NOTES

1) Simple geometry shows that the volume change is proportional to the square of the elongational strain involved.

2) Consider two Cartesian systems with x_1, x_2, x_3-axes, and $\bar{x}_1$, $\bar{x}_2$, $\bar{x}_3$-axes, respectively, and denote the angle between the x_i- and the $\bar{x}_k$-axes by α_{ik} (the subscripts i, k standing for any values 1, 2, 3). Then the strain components in the $\bar{x}$-frame, indicated by a bar, are obtained from those in the x-frame by means of the Cartesian transformation rules.

$$\bar{e}_{mn} = \sum_{i,\,k=1}^{3} \cos\alpha_{mi} \cos\alpha_{nk}\; e_{ik}$$

where the subscripts m and n may take any of the values 1, 2 or 3 and strain components with two identical subscripts are the extensional strains, which we agreed to indicate by one subscript only (e.g. $e_{11} = e_1$ etc.). Since $\bar{e}_{mn} = \bar{e}_{nm}$ and $e_{ik} = e_{ki}$, the transformation rules comprise six linear relationships between the strain components in the two different Cartesian systems. The transformation rules become somewhat more complex, although remaining linear in the strain components, when non-Cartesian reference systems are used (e.g. cylindrical or spherical coordinate systems).

3) In model experiments one should be aware that wall friction might affect the development of structures. If the displacement of the model material along side walls has a horizontal component, wall friction has such a component too, which, when distributed across the model material, is equivalent to a horizontal body force. Added vectorially to the vertical gravity force, it will produce some inclination of the resulting body force.

4) Cross-sections may be somewhat warped near the ends by tectonic loading as illustrated in Fig. II.5-13. With reference to this figure the reader may readily convince himself that the out-of-plane displacements u_2 decrease towards the interior of the rock body in common with the associated shear strains (e_{21}, e_{23}). In the main part of the extended rock body, deformation will thus still approximate plane strain.

5) The proof consists in deriving for the strains $e_\perp$ and $\gamma/2$ expressions that are the complete analogues of equations II.4-1a, 2a in Section II.4.1.

6) At first sight it may be surprising that u_1 does not increase upslope as it should if the slope-parallel weight component were also to produce compaction. Slope-parallel compaction, however, is prevented by the adherence of the slope material to the underlying base.

7) These expressions hold only for Cartesian vector components and become more complicated when – as often done in continuum mechanics – curvilinear (e.g. cylindrical or spherical) coordinate systems are used. In curvilinear systems the reference axes for the componental representation of a vector – the tangents to the curved coordinate lines – vary from point to point. Accordingly, a vector component, for instance v_1, would change from point to point even in the case of a completely uniform velocity field (i.e. a field of parallel vectors of equal magnitude) such as occurs in purely translatory motion. To compensate for such purely geometry-induced changes in the vector components and to maintain vanishing of the strain rates in rigid-body motions, the spatial derivatives of the velocity components in equation II.5-19 have to be replaced by more general expressions ('covariant derivatives').

8) In 'displacements and strains' we have touched on the complications that arise when the small-strain condition (magnitudes of displacement gradients $\ll 1$) no longer applies. The strain expression II.5-6a then remains non-linear in the displacement gradients and, as a consequence, the possibility of adding strains (without regard to the sequence of the individual straining operations and rigid body rotations) is lost.

9) It should be noted, however, that even in the context of this book, finite strains may enter the material laws when the mechanical properties are changed under large straining. This may occur in plastic strain hardening and softening of rocks, which is dealt with in Chapter II.7.7. Attention should also be drawn to the limitations in the application of linear elastic stress/strain relationships to rocks which, although subject to small strains only, are in a state of high prestresses. As discussed in Chapter II.6, small internal rotations may then impede the application of the linear elastic laws.

II.6. PORO-ELASTICITY

1. *ELASTIC STRAINING OF POROUS MATERIAL WITHOUT CHANGE IN PORE PRESSURE*

In the following we shall be concerned with the deformational response of a rock whose skeleton consists of homogeneous, isotropic and linearly elastic material. In addition, the internal pore geometry of the porous material is assumed to be, from a statistical point of view, homogeneous and isotropic. Small strains in the porous rock at zero (reference) pore pressure will then be linearly related to the total stresses by means of *Hooke's law for isotropic elastic materials*, which in unabridged notation reads:

$$\begin{aligned} e_1 &= \frac{1}{E}\left[\sigma_1 - \nu(\sigma_2 + \sigma_3)\right] \\ e_2 &= \frac{1}{E}\left[\sigma_2 - \nu(\sigma_3 + \sigma_1)\right] \\ e_3 &= \frac{1}{E}\left[\sigma_3 - \nu(\sigma_1 + \sigma_2)\right] \end{aligned} \qquad \text{(II.6-1a)}$$

for normal strains and stresses, and

$$\begin{aligned} e_{12} &= \frac{1}{2G}\tau_{12} \\ e_{23} &= \frac{1}{2G}\tau_{23} \\ e_{31} &= \frac{1}{2G}\tau_{31} \end{aligned} \qquad \text{(II.6-1b)}$$

for shear strains and shear stresses. Both compressive stresses and contractive strains are counted as positive.

Two of the three moduli in these equations have the dimension of stress: Young's modulus E (i.e. the ratio between normal stress and associated longitudinal strain in uniaxial loading) and the shear modulus G. The third modulus, Poisson's ratio ν, is defined as the ratio of the absolute strain values measured perpendicular and parallel to the direction of uniaxial loading. Since pure shear strain (Section II.5.1; Figs. II.5-3, 16B,C) can also be interpreted as a combination of an extension and a contraction, the shear modulus G can be expressed in terms of E and ν:

$$G = \frac{E}{2(1 + \nu)} \qquad \text{(II.6-2)}$$

Therefore, in the absence of pore pressure, the isotropic elastic solid is fully characterised by *two* moduli. It should be emphasised, however, that the moduli of the porous material refer to the response of the *bulk material* and will, in general, be quite different from the corresponding moduli of the skeleton material itself.

2. ELASTIC STRAINING WITH CHANGE IN PORE PRESSURE

When in a fluid-saturated porous body with continuous pore space the pore pressure is changed, the body will, usually, experience some straining. When, moreover, no external constraints are imposed on the body during the change of pore pressure, no reactive forces will be created at the boundary and the body will deform freely without any change in the total stresses, i.e. in the stresses transmitted through a bulk cross-section containing solid and fluid. To illustrate this, in Fig. II.6-1 we consider a porous body that has been cut out of a larger body, the solid skeleton being smoothly cut at random. The skeleton can then be loaded by means of perforated plates as sketched in the figure. The stresses applied to the boundary plates are transmitted as total stresses through the porous body. Now we envisage the whole system being immersed in fluid of uniform pressure. A change in fluid pressure Δp will then not only change the pressure on the pore walls by Δp but also the total normal stresses that act from outside upon the boundary plates. Now the loads on the boundary plates may be readjusted to compensate for the change Δp, and the total stresses inside the porous body will then not be affected by a rise or drop in pore pressure while, obviously, the bulk volume of the body will have changed. Therefore, the pore fluid pressure p has to be entered in the stress/strain relationships as a *separate state variable*.

In the isotropic porous material, the application of a pore pressure or pore pressure increment will affect all three normal strains in the same way, while the shear stresses are not affected at all [1]. Hence, a pressure term has to be added to the right-hand side of equations II.6-1a, the sign of this term being negative since a positive pore pressure causes a negative strain (i.e. extension):

$$e_1 = \frac{1}{E}\left[\sigma_1 - \nu(\sigma_2 + \sigma_3)\right] - \frac{p}{3H} \quad \text{etc.} \tag{II.6-3a}$$

$$e_{12} = \frac{1}{2G}\tau_{12} \quad \text{etc.} \tag{II.6-3b}$$

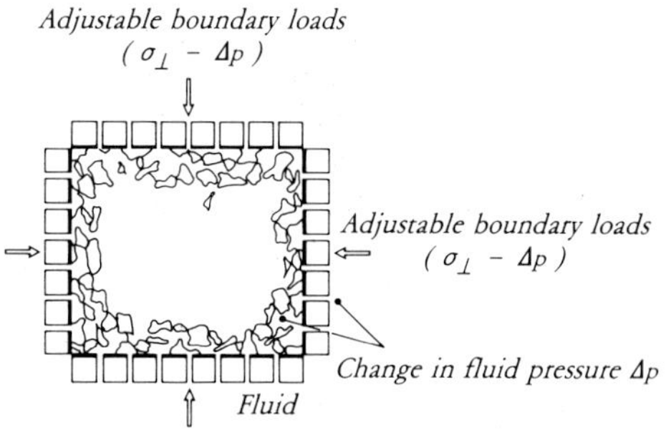

Fig. II.6-1
UNIFORM CHANGE OF PORE PRESSURE WITHOUT CHANGE OF TOTAL STRESSES

The change of total bulk volume V is obtained (see Chapter II.5.1; equation II.5-2) by adding the normal strains in equation II.6-3a:

$$\Delta V/V = e_1 + e_2 + e_3 = \frac{1}{K}\frac{\sigma_1 + \sigma_2 + \sigma_3}{3} - \frac{p}{H} \qquad \text{(II.6-4)}$$

where the bulk compression modulus

$$K = \frac{E}{3(1-2\nu)} \qquad \text{(II.6-5)}$$

represents the compressional effect of the mean total stress, while the new modulus H characterises the change in bulk volume that would be produced by applying a pore pressure or pore pressure increment without changing the total stresses.

Solved in terms of stresses, equation II.6-3a reads:

$$\sigma_1 = 2G[\ e_1 + \frac{\nu}{(1-2\nu)}\ (\ e_1 + e_2 + e_3\ )\ ] + \frac{K}{H}\ p \qquad \text{etc.} \qquad \text{(II.6-6)}$$

Compression Moduli. The two bulk moduli K and H are not independent, but related by means of the compression modulus K_s of the skeleton material. This can readily be seen by considering the whole porous body immersed in fluid of uniform pressure p. Then, both the pore walls and the outer surface of the skeleton are subject to the pressure p, which changes the skeleton volume V_s by

$$\Delta V_s/V_s = p/K_s \qquad \text{(II.6-7)}$$

Since loads other than the fluid pressure are absent, the total normal stresses will be equal to the fluid pressure p, and the change in bulk volume (II.6-4) becomes:

$$\Delta V/V = (1/K - 1/H)\ p \qquad \text{(II.6-8)}$$

In this very special case of isotropic loading the fluid pressure will reduce every linear dimension of the skeleton by the same factor. In other words, the geometry of the porous body is 'scaled down'. Consequently, the relative volume changes of skeleton and bulk must be equal, and the comparison of equations (II.6-7) and (II.6-8) gives the relationship

$$1/K - 1/H = 1/K_s \qquad \text{(II.6-9)}$$

It has to be noted here that, in contrast with the compression modulus of the skeleton material, the bulk moduli K and H may depend strongly on the mean effective stress and the magnitude and sign of stress changes (Bishop & Hight, 1977). This may be understood in terms of the internal changes of the pore space that may take place in a dry rock under external load. In particular, flaws and fractures will tend to close under increasing load and will make the rock less compressible (Lama & Vutukuri, 1978). One may expect that these internal changes will diminish at high effective mean stresses, to an extent depending on the rock type. At lower effective stress levels, K and H should be considered as constants only for small changes in effective stresses.

Change in pore volume. When use is made of relationship II.6-9, formula II.6-4 for the relative change in bulk volume can now be written

$$\frac{\Delta V}{V} = \frac{1}{K}\left[\frac{\sigma_1' + \sigma_2' + \sigma_3'}{3} + (1-a)\cdot p\right] \qquad \text{(II.6-10)}$$

where $\sigma_i' = \sigma_i - p$ (Terzaghi's effective stress; Section II.3.2)

and

$$a = \frac{K}{H} = 1 - \frac{K}{K_s} \qquad \text{(II.6-11)}$$

Associated with the change in bulk volume is, naturally, a change in pore volume $\Delta V_p/V$. When drainage is restricted or completely suppressed (undrained conditions), the change in pore volume will be accompanied by a change in fluid pressure p. When the compressibility of the skeleton material ($1/K_s$) is much smaller than the compressibility of the skeleton as a whole, as will often be the case, the volume change of the skeleton material under loading may be neglected and the change in bulk and pore volume considered equal. Because of $K_s \gg K$, the dimensionless parameter 'a' will approach 1 and the extra pressure term in equation II.6-10 will vanish. The change in bulk volume will then be determined solely by the mean effective stress.

The derivation of the appropriate expression for the relative pore volume change $\Delta V_p/V$ is not quite as simple when the compressibility of the skeleton material cannot be neglected. Various researchers have presented derivations; Biot (1941) was the first, followed by Gassman (1951), Geertsma (1957), Skempton (1960), Jaeger & Cook (1969), Bishop (1973) and Rudnicki & Rice (1975). A fairly elementary proof can be obtained from a theorem on skeleton stresses by Mandl (1964). It may suffice here to state the crucial element of the derivation only and to refer the interested reader to Note 2 at the end of this chapter for details.

Let us consider (Fig. II.6-2) an infinitesimal volume element δV_s *inside* the linearly elastic skeleton material and denote the local normal stresses that act upon the element by s_1, s_2, s_3. With respect to its former unstressed state, the element will have experienced a volume change

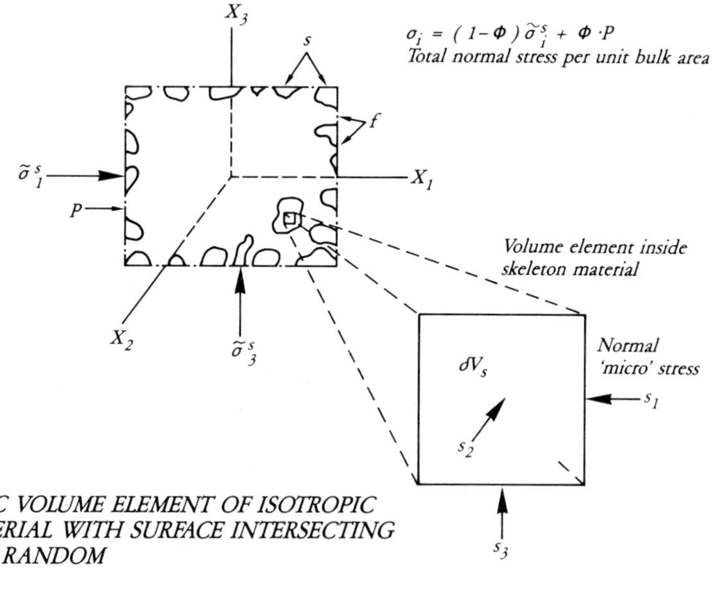

Fig. II.6-2
MACROSCOPIC VOLUME ELEMENT OF ISOTROPIC POROUS MATERIAL WITH SURFACE INTERSECTING SKELETON AT RANDOM

$$d\delta V_s / \delta V_s = \frac{1}{K_s} \frac{s_1 + s_2 + s_3}{3} \tag{II.6-12}$$

Integrating over the whole skeleton volume inside the porous body, one obtains the relative change in volume of the skeleton material

$$\frac{\Delta V_s}{V_s} = \frac{1}{V_s} \int_{V_s} d\delta V_s = \frac{1}{K_s V_s} \int_{V_s} \frac{s_1 + s_2 + s_3}{3} \delta V_s = \frac{\tilde{\sigma}^s}{K_s} \tag{II.6-13}$$

where

$$\tilde{\sigma}^s = \frac{1}{V_s} \int_{V_s} \frac{s_1 + s_2 + s_3}{3} dV_s \tag{II.6-14}$$

is the average 'skeletal pressure', i.e. the average of the mean normal stresses inside the skeleton material, referred to the skeleton volume.

Now consider a macroscopic volume element of the porous body. For simplicity, it may have the cubical shape shown in Fig. II.6-2. The faces of the cubical body cut the skeleton at random. The macroscopic normal stresses that act on the surface of the cube consist of a fluid part and a skeletal part. If we denote the porosity of the body by Φ, the fluid pressure contribution, referred to unit bulk (total) area, is $\Phi \cdot p$. The normal stress transmitted across the *solid* constituents of a face perpendicular to the x_i-axis is the avarage $\tilde{\sigma}_i^s$ of all normal stresses that act locally on the solid elements of the face. Since these macro stresses $\tilde{\sigma}_i^s$ are referred to the area covered by solid, their contributions to the total normal stresses, which are referred to unit *bulk* areas, become $(1-\Phi) \cdot \tilde{\sigma}_i^s$, and the total normal stresses are

$$\sigma_i = (1 - \Phi) \tilde{\sigma}_i^s + \Phi \cdot p \tag{II.6-15}$$

It may be added here that, for a completely disordered porous material, the porosity of a cross-sectional element that cuts the skeleton at random and the volume porosity are equal.

It is now the – by no means trivial – essence of our proof (Note 2) that the average skeletal pressure (eq. II.6-14) is also equal to the mean of the skeletal stresses $\tilde{\sigma}_i^s$ that act across three mutually perpendicular macroscopic cross-sections:

$$\tilde{\sigma} = \frac{\tilde{\sigma}_1^s + \tilde{\sigma}_2^s + \tilde{\sigma}_3^s}{3} . \tag{II.6-16}$$

Inserting the total stresses from eq. II.6-15 into the last expression, we now obtain from eq. II.6-13:

$$\frac{\Delta V_s}{V} = (1-\Phi) \frac{\Delta V_s}{V_s} = \frac{1}{K_s} \left[\frac{\sigma_1 + \sigma_2 + \sigma_3}{3} - \Phi \cdot p \right] \tag{II.6-17}$$

where the change in skeletal volume is referred to the original bulk volume V. In terms of Terzaghi's effective stresses this equation reads:

$$\frac{\Delta V_s}{V} = \frac{1}{K_s} \left[\frac{\sigma_1' + \sigma_2' + \sigma_3'}{3} + (1-\Phi) \cdot p \right] \tag{II.6-17a}$$

Because $\Delta V_p = \Delta V - \Delta V_s$, we finally obtain from equations II.6-10 and II.6-17a for the change in pore volume in terms of effective stresses:

$$\Delta V_p / V = \left(\frac{1}{K} - \frac{1}{K_s}\right) \frac{\sigma_1' + \sigma_2' + \sigma_3'}{3} + \frac{\Phi}{K_s} p = \tag{II.6-18}$$

$$= \frac{a}{K} \frac{\sigma_1' + \sigma_2' + \sigma_3'}{3} + \frac{(1-a)}{K} \Phi \cdot p$$

3. *THE PORO-ELASTIC EQUATIONS IN TERMS OF EFFECTIVE STRESSES*

Formally, we may simplify the stress/strain equations by introducing the normal stresses

$$\sigma_i^x = \sigma_i - a \cdot p = \sigma_i' + (1 - a) \cdot p \tag{II.6-19}$$

as summarised in the Table. The stress/strain relationships and the expression for the change of the bulk volume then become identical with the corresponding equations for a compact, linearly elastic, isotropic material.

Because of this formal identity, Nur and Byerlee (1971) termed the stresses σ_i^x, defined by equation II.6-19, '*true* effective stresses' – in contrast with Terzaghi's effective stresses σ_i'. This was unfortunate for two reasons. Firstly, the introduction of these stresses does not eliminate the pore pressure from the equations for the relative changes in the volume of the pore space (see Table) and of the skeleton material. Secondly, and more importantly, these 'true' effective stresses do *not*, in general, control the shear strength of a porous rock, which is usually controlled by Terzaghi's effective stress (see also the discussion on 'Shear strength and elastic deformation' in Section II.7.2). Various geologists have been misled into applying the stresses II.6-19 instead of Terzaghi's effective stresses in calculating the shear strength of rocks.

It may be recalled that in many rocks the compression modulus of the solid material K_s is much greater than the modulus of the (dry) porous body. In these cases 'a' approaches 1 and the σ_i^x turn into the common effective normal stresses σ_i'. It may be understood, however, that the bulk modulus K will increase and, accordingly, the parameter 'a' will decrease when the porosity becomes small. Naturally, in the limit of vanishing porosity one has $K = K_s$ and therefore $a = 0$.

$$e_1 = \frac{1}{E}\left[\sigma_1^x - \nu(\sigma_2^x + \sigma_3^x)\right]$$

$$e_{12} = \frac{1}{2G}\tau_{12}$$

$$\sigma^x = \sigma - ap$$

$$a = \frac{E}{3(1-2\nu)H} = \frac{K}{H} = 1 - \frac{K}{K_S}$$

$$\frac{\Delta V}{V} = \frac{1}{K}\,\frac{\sigma_1^x + \sigma_2^x + \sigma_3^x}{3} = \frac{1}{K}\left[\frac{\sigma_1' + \sigma_2' + \sigma_3'}{3} + (1-a)p\right]$$

$$\sigma' = \sigma - p$$

Terzaghi's effective stress

$$\frac{\Delta V_{por}}{V} = \left(\frac{1}{K} - \frac{1}{K_S}\right)\frac{\sigma_1' + \sigma_2' + \sigma_3'}{3} + \frac{\Phi}{K_S}p =$$

$$= \frac{a}{K}\,\frac{\sigma_1' + \sigma_2' + \sigma_3'}{3} + \frac{(1-a)\Phi p}{K} = \frac{1}{K}\left[a\Delta\sigma^x - (1-a)(1-\Phi)\Delta p\right]$$

Independent moduli: E, ν, Φ, K_S $\quad K_S \gg K \rightarrow a \approx 1: \quad \sigma^x = \sigma'$ *effective stress*

Table II.6 RE-FORMULATION OF STRESS/STRAIN EQUATIONS

4. THE APPLICATION OF THE PORO-ELASTIC STRESS/STRAIN RELATIONSHIPS

Limitations. Application of the 'linear poro-elastic relationships to geological problems is, in fact, very *limited*, notwithstanding the frequent and often quite uncritical use of linear elasticity in the geological literature. Although rocks can undoubtedly store some elastic energy, tectonic folding and faulting are predominantly inelastic, i.e. energy-dissipating processes. An approximately linear elastic behaviour can be expected only for small strains, where longitudinal strains do not exceed say 1–3%. Moreover, brittle rocks in an unstressed state contain numerous flaws and micro cracks, which have to be closed by some confining pressure before further loading can produce an elastic response.

It is therefore evident that, in geological applications, the elastic and poro-elastic equations relate only *increments* of stresses and strains in a rock body that is already in a state of considerable ***prestress***. The elastic moduli then refer, of course, to this state of prestress. Such incremental elastic straining takes place in the propagation of acoustic waves, the relatively rapid unloading by erosion, the adjustment of sliding fault blocks to the fault geometry, the rapid rise of pore fluid pressure caused by fluid injection from adjacent layers, and in rock fracturing by fluid injection.

It should be emphasised that the origin of the prestresses is irrelevant. They will, in general, be caused by *inelastic* deformation (e.g. compaction) and diagenetic processes. It is only essential for the *incremental changes* to be of an elastic nature. However, the existence of large prestresses poses another serious limitation when the incremental deformation implies not only pure straining but also *local rotations* (see Section II.5.1; Figs. II.5-1, 2). Then the prestresses will rotate with the material. For a co-rotating 'observer', this rotation in itself would not change the state of stress; only the incremental stresses associated with an actual straining of the rotated material element would be felt. Hence, when referred to a co-rotated coordinate system, the poro-elastic stress/strain equations of this section would apply.

In general, however, the stress/strain relationships have to be combined with the equations that describe equilibrium of stresses (Chapter II.2; eq. II.2-2), which, in analytical solution procedures, are commonly referred to *fixed* coordinate axes. (This does not apply to numerical procedures in which the rock body is replaced by an ensemble of deformable finite elements.) Naturally, the stress/strain relationships then also have to be referred to this fixed frame of reference. There, however, the rotation of large prestresses with the material will show up in the form of extra components of incremental stresses. For small body rotations (expressed by the local rotation angle) these extra components will be products of the rotation angle and the prestress components before rotation. Being of purely geometrical origin, these parts of the incremental stresses have to be subtracted from the total incremental stresses to obtain the stresses that actually cause the straining of the rotated material and which, therefore, may be related to the incremental strains. This has been discussed in detail by M. Biot (1965) in his theory of incremental deformations of bodies under large prestress.

In geological applications of poro-elasticity, rocks are usually under high prestresses – with the possible exception of formations under abnormally high pore pressures. When applying the poro-elastic equations of this section in combination with a fixed Cartesian coordinate system, one should make sure that the products of local rotation angles and initial prestress components remain small with respect to the total incremental stresses. Naturally, the poro-elastic relationships described in this section may be used legitimately when the incremental deformations are *irrotational* (see Section II.5.1; Figs. II.5-2B, C, D) and, therefore, do not rotate the prestress system.

This is the case when, for instance, a uniform incremental shortening is imposed along a principal prestress axis. Such a situation occurs in hydraulic extension fracturing, in which opening of the fracture in the direction of the least principal stress imposes shortening on the wall rock. In the following, we shall discuss examples of irrotational incremental deformations, two of which are associated with the occurrence of abnormally high fluid pressures in the crust.

Example No. 1: ***Natural hydraulic cross-fracturing of low-permeability beds.***

It has long been recognised that fluids in the crust can produce fissures through which fluid can escape. Secor, N.J. Price, Phillips and others have discussed the mechanism of natural hydraulic fracturing. Some 25 years ago Snarsky suggested hydraulic fracturing as a mechanism for oil migration and, more recently, du Rouchet has dealt with the vertical transfer of hydrocarbons from source beds through cap rocks by the opening of vertical fractures.

These natural hydraulic fractures are extension fractures. This means that the rock is separated by the fluid pressure, which therefore has to overcome the tensile rock strength T and to push aside the rock walls against the action of total rock stress. The fractures may be concordant or discordant with the bedding and may be referred to as fluid dykelets or sills – in accordance with the nomenclature for igneous intrusions.

Natural hydraulic fractures may start from micro fractures or flaws inside a bed, or from irregularities at bed boundaries, and be propagated into the bed by the intrusion of fluids from outside the bed. We are interested here in the second type – the ***'intrusion fractures'*** – and shall consider the conditions under which a bed under effective compression is wedged open by fluid injection from outside and traversed by dyke-type intrusion fractures (Fig. II.6-3). We assume that the bed is horizontal and of uniform material properties.

It is well known (Secor, 1968) that in order to form fluid dykes, the pressure p_f of the fracture fluid has to satisfy the basic condition

$$p_f \geqslant \sigma_h^o + T \qquad \text{(II.6-20)}$$

where σ_h^o is the smallest principal stress, supposed here to act horizontally, and T is the tensile rock strength parallel to the bedding and counted as positive. The tensile strength perpendicular to the bedding is usually of negligible magnitude.

One may reasonably expect the fractures to develop perpendicular to the smallest principal stress σ_h^o, which, for the sake of simplicity, is assumed to act along one horizontal direction only, thus giving rise to the formation of a single set of parallel vertical fractures. We also recall that before the onset of intrusion fracturing the beds are under effective compression, their original pore pressure p^o therefore being smaller than the smallest principal (total) stress:

$$p^o < \sigma_h^o \qquad \text{(II.6-21)}$$

It seems to be a tacit assumption in the geological literature on fluid-induced rock fractures that condition II.6-20 is the *only* requirement for vertical intrusion fracturing. This need *not* be true, however, when the rock has some permeability, which allows the fracture pressure and pore pressure in the wall rock to communicate. Then the pore pressure in the wall rock will be raised by the fracture pressure, and this rise in pore pressure will oppose the increase in horizontal effective stress necessary to compress the wall rock in order to allow the opening of the fractures.

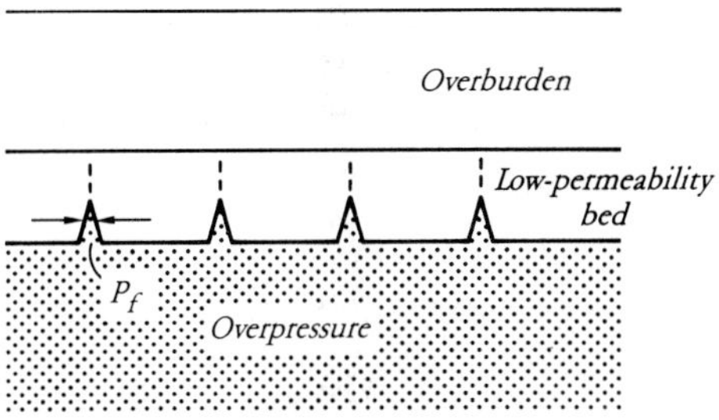

Fig. II.6-3
VERTICAL HYDRAULIC INTRUSION FRACTURES IN LOW-PERMEABILITY BED

In the following, we shall therefore examine the precise *conditions for the existence of open vertical intrusion fractures* in rocks under effective compression (Mandl & Harkness, 1987).

Compression of the rock in the direction of σ_h^o between parallel fractures that are opening implies a positive strain increment

$$\Delta e_h > 0 \tag{II.6-22}$$

while the bed as a whole is not allowed to extend in any horizontal direction. In the absence of tectonic extension, friction between the beds will usually be sufficient to provide these lateral constraints. Therefore, there will be no change of horizontal strain in the rock parallel to the fracture walls (indicated by the subscript H):

$$\Delta e_H = 0 \tag{II.6-23}$$

If we now insert the incremental strains Δe_H and Δe_h for e_1 and e_2 respectively, in the poro-elastic relationships shown in the Table (p. 270), and the horizontal and vertical stress increments $\Delta\sigma_h^x$ and $\Delta\sigma_v^x$ for σ_1^x, σ_3^x, we arrive at the following *additional condition* for the vertical intrusion fracturing of a horizontal, laterally confined layer:

$$\Delta\sigma_h^x > \frac{\nu}{1-\nu}\Delta\sigma_v^x \qquad \text{II.6-24a)}$$

It reduces to the relationship between incremental effective (Terzaghi) stresses when $K_s \gg K$:

$$\Delta\sigma_h' > \frac{\nu}{1-\nu}\Delta\sigma_v' \tag{II.6-24b}$$

where σ_v' is the effective overburden stress.

The last relationship may be put in a somewhat more convenient form by introducing the rise in pore water pressure Δp, which inside the wall rock is caused by the high-pressured fracture fluid. Since

$$\Delta\sigma_h' = \Delta\sigma_h - \Delta p, \qquad \Delta\sigma_h = p_f - \sigma_h^o \tag{II.6-25a}$$

and the total overburden stress σ_v remains unchanged, i.e.

$$\Delta\sigma_v' = -\Delta p \tag{II.6-25b}$$

equation II.6-24b may now be written as

$$p_f - \sigma_h^o > \frac{1-2\nu}{1-\nu} \Delta p \tag{II.6-26}$$

This is the condition that has to be satisfied in addition to eq. II.6-20 if the fracture pressure p_f is to separate the fracture walls. It should be noted that the rise in pore water pressure is a transient phenomenon Δp (t), which depends on the drainage conditions of the wall rock (permeabilities and conditions at bed surface) and on the capacity of the rock to take up – by virtue of the compressibilities of the rock skeleton, the skeleton material and the pore fluid – fluid that has penetrated the fracture walls.

When fluid is forced into a bed from an underlying or overlying 'reservoir' at pressure $p_f > \sigma_h^o > p^o$ by injection at distinct points of the bed surfaces (for example, at irregularities of the bedding plane) rather than by invading the bed in a uniform way, condition II.6-26 will always be satisfied *initially*, because $\Delta p\ (t=0) = 0$.

Therefore, initially intrusion fractures will form, but inside the wall rock a back-pressure will build up at a rate depending on rock type and drainage conditions. In other words, Δp will rise and, in an undrained, i.e. sealed-off, bed it will ultimately approach the maximum value

$$\Delta p_{Max} = p_f - p^o \tag{II.6-27a}$$

as the pore pressure attains the value of the fracture pressure.

If the fracture fluid is not water, but oil or gas,. the *capillary pressure difference* p_c between the wetting water phase and the non-wetting hydrocarbons diminishes Δp_{Max}:

$$\Delta p_{Max} = p_f - p_c - p^o \tag{II.6-27b}$$

The question now arises whether the fracture pressure p_f can maintain separation of the fracture walls as Δp increases. It may suffice here to consider condition II.6-26 for the limiting case that Δp has reached its maximum value (eq. II.6-27b) throughout the wall rock. For this purpose, we introduce into eq. II.6-26 expression II.6-27b for Δp and the dimensionless ratios

$$K_o = \frac{\sigma_h^{o'}}{\sigma_v^{o'}} \quad \text{and} \quad \lambda = \frac{p^o}{\sigma_v} \tag{II.6-28}$$

As the reader may easily verify, condition II.6-26 now attains the form

$$\sigma_v \geqslant p_f > A \cdot \sigma_v - \frac{1-2\nu}{\nu} p_c \tag{II.6-29a}$$

where

$$A = \lambda + K_o\ (1-\lambda) \frac{1-\nu}{\nu} \tag{II.6-29b}$$

It should be borne in mind that the injection pressure p_f cannot exceed the total overburden stress, since on reaching that value it would separate bedding planes and produce overburden-carrying fluid 'sills'. Maintaining separation of the fracture walls under the fracture pressure p_f would therefore certainly require

$$A - \frac{1-2\nu}{\nu}\frac{p_c}{\sigma_v} < 1 \qquad \text{(II.6-30)}$$

If the fracture fluid is water ($p_c = 0$) this condition reduces to $A < 1$, which implies, in view of eq. II.6-29b:

$$\nu > \frac{K_o}{1+K_o} \text{ or equivalently } K_o < \frac{\nu}{1-\nu} \qquad \text{(II.6-31)}$$

It should be recalled that Poisson's ratio ν in these formulas refers to the bulk behaviour of the porous rock under completely drained conditions (pore pressure being kept constant). Its value will therefore hardly ever lie outside the range $0.1 < \nu < 0.3$. However, the parameter K_o for sediment that has been deposited in a tectonically quiescent environment and has not subsequently been subjected to extensional tectonics will rarely have a value smaller than, say, 0.45. From soil mechanical experience and in-situ stress measurements at depth, one may infer that typical K_o-values for tectonically undisturbed beds would lie, say, between 0.45 and 0.75.

It should be also emphasised that K_o is *not related to the Poisson ratio*, as is often erroneously assumed, since the geological processes of compaction and diagenesis, which control the development of horizontal rock stresses in response to the overburden load, are of an inelastic nature.

The reader will immediately notice that ν and K_o-values, which lie in the above ranges, cannot satisfy condition II.6-31. Consequently, *water* dykelets in tectonically undisturbed beds cannot stay open when seepage drainage is impeded by low permeability or a sealing top layer. *A tectonic extension* would be required that reduces K_o sufficiently to satisfy condition II.6-31 and to allow the dykelets to stay open.

If the fracture fluid is non-wetting with respect to the solid rock skeleton ($p_c > 0$), we may rearrange eq. II.6-30 by expressing the total overburden stress in terms of an average overburden density of 2.5 g/cm³ and the overburden thickness z in km:

$$\frac{z\ (\text{km})}{p_c\ (\text{MPa})} < 0.04\ \frac{1-2\nu}{1-\lambda}\ \frac{1}{K_o\ (1-\nu)\ -\nu} \qquad \text{(II.6-32)}$$

Thus, this condition for tectonically undisturbed beds imposes an *upper limit on the overburden thickness*. In the case of a greater thickness of the overburden, the hydrocarbon pressure p_f in the fractures will not be able to maintain the separation of the fracture walls. It is of note that this limit thickness of the overburden increases with the original amount of overpressure ($\lambda > 0.45$) in the pore water of the fractured bed and is proportional to the capillary pressure difference between hydrocarbon and water inside the wall rock. If we assume, for instance, $\nu = 0.25$, $K_o = 0.5$ and $p_c = 1$ MPa, a normal pore pressure ($\lambda = 0.45$) before onset of intrusion fracturing gives a critical overburden thickness of only 0.29 km (!), while in the case of high overpressurising with $\lambda = 0.9$, the critical thickness of the overburden amounts to 1.6 km.

When condition II.6-32 (or II.6-31 for $p_c = 0$) is not satisfied, intrusion fractures, which may have formed, will close as the fluid pressure in the undrained wall rock is building up. It may be of practical interest that in strong brittle rocks this fracture closure will not be perfect and misfits between the contacting walls may even allow such 'closed' fractures to act as paths for fluid migration.

Finally, when condition II.6-32 is satisfied and intrusion fractures, formed by the injection of high-pressure non-wetting fluids (e.g. hydrocarbons), remain open, it is of interest to estimate the average *width* of such fractures. For a crude estimate it is sufficient to consider a single fracture of length L that has developed unhindered by neighbouring fractures. According to Geertsma and de Klerk (1969) the width W(0) of the fracture at the injection end may be approximated by

$$W(0) \approx 3(p_f - \sigma_h^o)\, L\, (1+\nu)/E \qquad \text{(II.6-33)}$$

Assuming for the average width $\overline{W} \approx 3W(0)/4$, one arrives at the further approximation

$$\overline{W} \approx 3(p_f - \sigma_h^o)\, L/E \qquad \text{(II.6-33a)}$$

Inserting some typical values, say $E = 3 \times 10^4$ MPa, $L = 10$ m, and a fracture fluid pressure that exceeds the total horizontal rock stresses σ_h^o, just prior to the fracture formation by 1 MPa, we obtain a fracture width of 1 mm.

Example No. 2: ***Overpressure generation by removal of overburden ('fossile pore pressure')***

When horizontal rock layers are unloaded by the removal of overburden by erosion or by the melting of an ice cap, stresses in the rock will be reduced. We consider this stress release as a poro-elastic process and ask, specifically, how the pore pressure will respond to this type of unloading. We assume that, because of a very low permeability or the presence of impermeable top and bottom seals, fluid cannot enter or leave the rock layer under consideration. In addition, the undrained rock layer is laterally confined (Fig. II.6-4). Horizontal strain increments therefore vanish ($\Delta e_1 = \Delta e_2 = 0$).

Assuming material isotropy in all horizontal directions, the changes in horizontal normal stresses will be equal ($\Delta\sigma_1^x = \Delta\sigma_2^x$). From the stress/strain equations (see p. 270), then it immediately follows that the relationship between vertical and horizontal stress changes is

$$\Delta\sigma_h^x = \frac{\nu}{1-\nu}\,\Delta\sigma_v^x \qquad \text{(II.6-34)}$$

Consequently, the mean stress becomes

$$\Delta\sigma^x = \frac{1}{3}(2\Delta\sigma_h^x + \Delta\sigma_v^x) = \frac{1}{3}\,\frac{1+\nu}{1-\nu}\,\Delta\sigma_v^x \qquad \text{(II.6-35)}$$

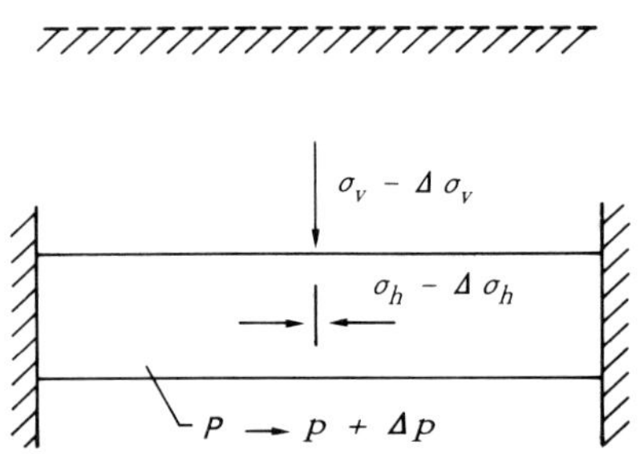

Fig. II.6-4
CHANGE OF PORE PRESSURE IN UNDRAINED, LATERALLY CONFINED ROCK LAYER BY REMOVAL OF OVERBURDEN

Inserting this expression into the last equation of the Table (p. 270), we obtain for the change in pore volume:

$$\frac{\Delta V_{por}}{V} = \frac{1}{K}\left[\frac{a}{3}\,\frac{1+\nu}{1-\nu}\,\Delta\sigma_v^x - (1-a)\,(1-\Phi)\,\Delta p\right] \qquad \text{(II.6-36)}$$

where

$$\Delta\sigma_v^x = \Delta\sigma_v - a\Delta p \qquad \text{(II.6-37)}$$

and

$$\Delta\sigma_v = [\gamma_s\,(1-\Phi) + \gamma_f\Phi]\,\Delta z$$

is the weight of a column of removed overburden with unit cross-section and height Δz; γ_s and γ_f denote the specific weights of grain material and pore fluid.

Under undrained conditions, the change in pore volume is also equal to the change in volume of pore fluid. If K_f is the compression modulus of the pore fluid, one therefore also has

$$\frac{\Delta V_{por}}{V_{por}} = \frac{\Delta V_{por}}{\Phi \cdot V} = \frac{1}{K_f}\,\Delta p \qquad \text{(II.6-38)}$$

Combining this equation with eq. II.6-36 and inserting $a=1-K/K_s$ and eq. II.6-37 in $\Delta\sigma_v^x$, after some rearranging we finally arrive at a relationship between the changes in pore pressure and total overburden stress

$$\Delta p = \frac{\left[\frac{\gamma_s}{\gamma_f}(1-\Phi) + \Phi\right]\gamma_f \cdot \Delta z}{1 + \alpha\cdot\Phi\cdot\frac{K}{K_f}\cdot\frac{K_s-K_f}{K_s-K}} \qquad \text{(II.6-39)}$$

where

$$\alpha = \frac{3(1-\nu)}{1+\nu} \qquad \text{(II.6-40)}$$

In this formula $\gamma_f\Delta z$ represents the hydrostatic head of the pore water removed with the overburden layer of thickness Δz. If $\Delta p/\gamma_f\Delta z < 1$, the pore pressure in the unloaded and undrained rock layers will be reduced less than it would by the removal of the water column Δz. The unloaded and sealed rock layer would then have acquired a pore pressure in excess of the normal hydrostatic pressure, or, if the layer was already under overpressure before unloading, the degree of overpressurising would be increased.

In view of eq. II.6-39, the *condition for overpressurising*

$$\frac{\Delta p}{\gamma_f\Delta z} < 1 \qquad \text{(II.6-41)}$$

may be expressed in terms of the elastic moduli as

$$\alpha\cdot\frac{K}{K_f}\cdot\frac{K_s-K_f}{K_s-K} > \left(\frac{\gamma_s}{\gamma_f} - 1\right)\frac{1-\Phi}{\Phi} \qquad \text{(II.6-42)}$$

Naturally, the opposite inequality sign would imply 'underpressurising' of the pore fluid in the unloaded layer.

The overpressurising condition II.6-42 depends very critically on the ratio K/K_f. This may be illustrated by inserting typical values, say for a water-saturated shale with a porosity $\Phi = 20\%$. Assuming a Poisson's ratio of 0.2 ($\alpha = 2$), a compression modulus for the skeleton material $K_s = 3 \times 10^4$ MPa, K_f (for water) $= 0.2 \times 10^4$ MPa, a compression modulus for the skeleton as a whole $K = 0.7 \times 10^4$ MPa, and the ratio of specific weights $\gamma_s/\gamma_f = 2.6$, we find that eq. II.6-42 will be satisfied ($8.52 > 6.4$). However, if we reduce the bulk modulus K by as little as 22%, condition II.6-42 will no longer be satisfied.

It may also be noted that in our example with $K = 0.7 \times 10^4$ MPa, the actual overpressurising of the pore water remains very modest. The pressure ratio (eq. II.6-41) attains the value 0.84. This would imply, for instance, that the removal of 1500 m of overburden and the associated hydrostatic head of 15 MPa resulted in a pore pressure reduction of 12.6 MPa in the unloaded rock. An initially normal hydrostatic pore pressure would, therefore, have gained an overpressure of merely 2.4 MPa, which at a remaining burial depth of 2000 m would be hardly noticeable, but at the shallow depth of 500 m would amount to the pore pressure/overburden ratio $\lambda = (5+2.4)/12 = 0.62$ compared with a 'normal' ratio of 0.43.

Essentially following the same line of reasoning as used in this example, A.J. Seriff (1979) arrived at the conclusion that, in general, *overpressurising of fluid-saturated rocks by undrained uplifting will remain of very modest degree*, if it takes place at all. It may be added that overpressurising may be somewhat higher for oil or high-pressure gas filling the pores of the unloaded rock, because of the considerably lower value of K_f.

It should also be noted that our analysis does not account for cooling of fluids and rock during uplift, which would tend to decrease the pore pressure since the pore fluid would shrink more than the rock.

Finally, for rocks with $K_s \gg K$ – as is frequently the case for highly porous rocks under low confining stress – condition II.6-42 simplifies since

$$\frac{K_s - K_f}{K_s - K} \approx 1 \qquad \text{(II.6-43)}$$

Example No. 3: ***High horizontal compressive stresses of non-tectonic origin***

Horizontal stresses equal to or higher than overburden stress are frequently encountered at fairly shallow depth in tectonically quiescent areas. The previous example of unloading by overburden removal is related to this phenomenon which has been explained by various authors (Seager, 1964; Müller, L., 1969; Baudendistel, 1974) in terms of the different mechanical behaviour of rocks during accretion and removal of overburden. As the overburden load increases, rock deformation is essentially inelastic since it is controlled by irreversible processes of consolidation and diagenesis. (In general, these processes will alter the elastic and strength moduli of the material, very likely increasing the E modulus and decreasing the Poisson ratio.) In contrast, the consolidated rock will respond in a predominantly elastic mode to unloading by removal of overburden.

The horizontal effective stresses σ_h' that are generated by the overburden load in a laterally constrained horizontal layer will be higher than in a comparable elastic process, but naturally lower than the effective overburden stress σ_v'. In the Mohr diagram (see Section II.4.2) of Fig. II.6-5, the circle 1 represents such a state of non-tectonic stress at a depth of, say, 2 km, assuming that the effective horizontal stress is 60% of the vertical stress and the fluid pressure is normally hydrostatic at the time unloading begins.

Removal of overburden of thickness Δz will alter this state of stress according to eq. II.6-34, which, in terms of Terzaghi's effective stresses,

$$\sigma' = \sigma^x - (1-a)p \quad \text{with } 1-a = \frac{K_s}{K} \tag{II.6-19}$$

may be written

$$\Delta\sigma_h' = \frac{\nu}{1-\nu}\Delta\sigma_v' - (1-a)\frac{1-2\nu}{1-\nu}\Delta p \tag{II.6-44}$$

where, by definition,

$$\Delta\sigma_v' = \Delta\sigma_v - \Delta p = 10\,\gamma\Delta z - \Delta p\,(\text{MPa}) \tag{II.6-45}$$

with γ being the average specific weight of the overburden in gf/cm³, stresses being measured in MPa and Δz in kilometres.

If the rock layer considered is sealed off – the unloading therefore taking place under *undrained* conditions – Δp has to be inserted from eq. II.6-39. For the purposes of illustration, we again use the material parameters of the previous example and determine from eqs. II.6-44 and II.6-45 the following stress reductions caused by removing an overburden of thickness Δz and $\gamma = 2.3$ gf/cm³:

$$\begin{aligned} \Delta p \ (\text{MPa}) &= 8.4\,\Delta z\ (\text{km}) \\ \Delta\sigma_v' &= 14.6\,\Delta z \\ \Delta\sigma_h' &= 2.2\,\Delta z \end{aligned} \tag{II.6-46a}$$

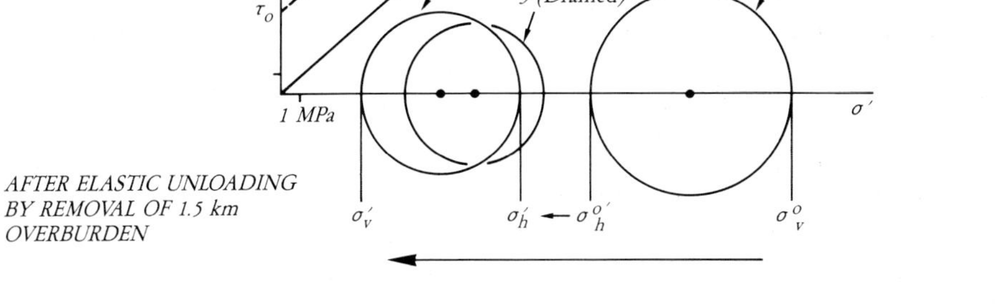

Fig. II.6-5 STRESS CHANGES DUE TO REMOVAL OF OVERBURDEN UNDER DRAINED AND UNDRAINED CONDITIONS

The stress circle 2 in Fig. II.6-5 represents the new state of effective stress at the new depth of 0.5 km after the removal of 1.5 km overburden. The diagram demonstrates the striking reversal of the role of principal stresses, σ_h' now being 8.2 MPa greater than σ_v' – provided, of course, that the material can withstand the maximum shear stress of 4.1 MPa, i.e. half the difference between the greatest and smallest principal stresses. The limiting shear stress is prescribed by Coulomb's yield or limit condition (see Section II.7.3) which maps as a straight line in the Mohr diagram (Fig. II.6-5). In the present case, the new stress state in the rock layer at 0.5 km depth below the erosion surface will still be possible (subcritical) even in cohesionless material, when the angle of internal friction Φ is greater than 30°.

If unloading takes place under *well-drained* conditions, the normal hydrostatic pressure distribution will be maintained, and the reduction in pore fluid pressure and effective stresses will therefore be

$$\begin{aligned} \Delta p \ \text{(MPa)} &= 10\,\gamma_f\,(\text{gf/cm}^3)\,\Delta z\ (\text{km}) \\ \Delta\sigma_v' &= 13\ \Delta z \\ \Delta\sigma_h' &= 1.53\ \Delta z \end{aligned} \qquad \text{(II.6-46b)}$$

The corresponding new state of stress at 0.5 km depth after drained removal of 1.5 km of overburden is depicted by circle 3 in Fig. II.6-5. We notice that the new horizontal effective stress is even somewhat higher than in the comparable case of undrained unloading.

Although the stress circles in Fig. II.6-5 do not touch the Coulomb failure line, it is fairly obvious that the construction of the figure, when carried out for rock layers at shallower depths and possessing a low cohesive strength τ_0, will predict the generation of critical stress states in a near-surface zone. The thickness of this zone will depend on the lithology of the rock, the state of stress at the beginning of the unloading process, the drainage conditions and, in particular, on the thickness of the overburden removed.

Interestingly, one has to expect that in this zone the excess of horizontal stress above the vertical one will be limited by stress release as a result of the formation of a pervasive system of *low-angle shear joints* or even thrust faults of modest displacement.

NOTES

1) A change in pore pressure can only induce shear stresses when boundary constraints enforce unequal changes in normal strains. Such boundary effects are, however, excluded from the formulation of stress/strain relationships; they are accounted for by the equilibrium equations and boundary conditions.

2) THE SKELETAL PRESSURE

Our aim is to prove that the mean skeletal normal stress $\tilde{\sigma}^s$ (eq. II.6-14) is the same as the arithmetic mean II.6-16 of the normal stresses $\tilde{\sigma}_i^s$ which are transmitted through the solid parts of three mutually perpendicular macroscopic cross-sections.

We refer again to the cubical volume element of Fig. II.6-2, the edges of which are parallel to the axes of a Cartesian coordinate system with its origin at the centre of the cube. The planar faces of the cube intersect the solid skeleton of the porous material at random, and we recall that, regarding porosity, the material of the macro element is homogeneous and isotropic.

Instrumental in the proof is Gauss' integral theorem which enables us to transform the integral over a *closed* boundary surface S into an integral over the enclosed volume V:

$$\oint_S A\, n_i\, dS = \int_V \frac{\partial A}{\partial x_i}\, dV \qquad (i = 1, 2, 3) \tag{I}$$

where n_i are the components (directional cosines) of the outwards-directed unit normal of the surface element dS. We apply this theorem to the solid skeleton of the cubical macro-element. The surface integral then extends over the entire complicated surface of the skeleton and the volume integration is carried out over the volume V_s of the skeleton material.

For convenience, the components of the stress tensor that act in the interior of the skeleton material are denoted by s_{ij} (i, j = 1, 2, 3), where, compared with the notation used elsewhere in this book (Section II.1.2), $s_{11} = \sigma_1$, $s_{12} = \tau_{12}$ etc. We then form the products

$$A_i = s_{ij} x_j \qquad \text{(summation over identical subscripts from 1 to 3)}$$

and substitute these expressions for A in (I). After the summation of the three equations with respect to i, we obtain

$$\oint_S s_{ij} x_j n_i dS = \int_{V_s} \frac{\partial}{\partial x_i} (s_{ij} x_j)\, dV_s \tag{II}$$

We now recall (Section II.1.1; eq. II.1-1a) that the Cartesian components of the stress vector $\vec{t}$ that acts on a surface element dS are

$$t_j = s_{ij} n_i \qquad (j = 1, 2, 3;\ \text{again summation over identical subscripts};\ n_i = \cos\alpha_i) \tag{III}$$

Equation (II) may, therefore, also be written in the form

$$\oint_S t_j x_j\, dS = \int_{V_s} x_j \frac{\partial}{\partial x_i} s_{ij} dV_s + \int_{V_s} s_{ij} \frac{\partial}{\partial x_i} x_j dV_s \tag{IV}$$

It should be emphasised that the closed surface S completely enwraps the skeleton volume V_s of the cubical element and, therefore, consists of all the pore walls pw and the solid parts s of the outer boundary of the cube. The stresses applied to the skeleton on pw and s are in equilibrium and do not have to balance the weight of the body. Therefore (see Section II.2.1; eq. II.2-2) the sum of stress derivatives in the first integral on the right-hand side of eq. IV vanishes. Since the derivatives in the second integral on the right-hand side of eq. IV attain the values 1 or 0, then the equation simplifies to

$$\oint_{pw+s} t_j x_j\, dS = \int_{V_s} (s_{11} + s_{22} + s_{33})\, dV_s \tag{V}$$

Comparing this expression with eq. II.6-14, we recognise the average skeletal 'pressure'

$$\tilde{\sigma}^{s} = \frac{1}{V_s} \int_{V_s} \frac{s_1 + s_2 + s_3}{3} dV_s = \frac{1}{3V_s} \oint_{pw+s} t_j x_j \, dS \tag{VI}$$

The crucial point now is proving that the contribution of the pore walls (pw) to the closed surface integral on the right-hand side of this equation vanishes.

Hydrostatic pore pressure exerts upon the pore walls the stress vector with the components

$$t_j = (p(0) - \gamma_f x_3) n_j \tag{VII}$$

where p(0) is the pore pressure at the centre of the cubical element and γ_f the specific weight of the fluid. We now assume, for the moment only, that this fluid stress is defined not only on the pore walls and the fluid part f of the outer boundary of the cube, but also on the solid part s of that boundary. A simple application of the integral theorem (I) then yields for the closed surface integral

$$\oint_{pw+s} t_j x_j dS = 3\, p(0)\, V_s - 4\gamma_f \int_{V_s} x_3 \, dV_s \tag{VIII}$$

where the integral on the right-hand side vanishes since positive and negative values of x_3 cancel.

One obtains exactly the same result when extending the integrals over the solid part s of the *outer* boundary only:

$$\begin{aligned} \int_s t_j x_j \, dS &= \int_s (p(0) - \gamma_f x_3)\, x_j n_j dS = \\ &= (1-\Phi) \oint_{s+f} (p(0) - \gamma_f x_3)\, x_j n_j dS = \\ &= 3\, p(0)\, (1-\Phi) \cdot V = 3\, p(0)\, V_s \end{aligned} \tag{IX}$$

Note that the integral over s+f extends over the total outer boundary of the cubical element. The transition from the integration over the solid part s of the outer boundary to the entire outer surface s+f would even be permitted for a quite general shape of our macroscopic volume element since the integrands are slowly varying functions on the surface s+f and completely independent of the actual microscopic pore geometry of the surface.

Comparing (VIII) and (IX) shows that indeed

$$\oint_{pw} t_j x_j \, dS = 0 \tag{X}$$

Consequently, (VI) reduces to

$$\tilde{\sigma}^{s} = \frac{1}{3V_s} \int_s t_j x_j dS = \frac{1}{3V_s} \int_s s_{ij} x_j n_i dS \tag{XI}$$

when t_j is inserted from (III). This implies that the average skeletal pressure and the associated relative change in skeletal volume are determined only by the skeletal stresses that act on the *outer* boundary of the macroscopic volume element.

The components s_{ij} of the stresses acting *locally* on the solid part of the outer boundary (Fig. II.6-6) may be replaced in (XI) by the *average* values they attain on the faces of the cubical element, for instance,

$$\tilde{\sigma}^{s}_{1} = \frac{1}{(1-\Phi)d^2} \int_s s_{11} x_1 dS, \tag{XII}$$

since the products of the random deviations of the local stresses from their average values with the components of the position vector cancel in the mean. It may also be readily seen that the skeletal shear stresses s_{ij} $(i \neq j)$ do not contribute to the integral in (XI) because of the associated x-values attaining positive and negative signs. We are therefore left with (Fig. II.6-6):

$$\int_S s_{ij}x_jn_i dS = d^2(1-\Phi)\cdot\{\ \tilde{\sigma}_1^s\left(+\frac{d}{2}\right)\cdot\frac{d}{2} - \tilde{\sigma}_1^s\left(-\frac{d}{2}\right)\cdot\frac{d}{2} + \tilde{\sigma}_2^s\left(+\frac{d}{2}\right)\cdot\frac{d}{2} - \tilde{\sigma}_2^s\left(-\frac{d}{2}\right)\cdot\frac{d}{2} + \tilde{\sigma}_3^s\left(+\frac{d}{2}\right)\cdot\frac{d}{2} - \tilde{\sigma}_3^s\left(-\frac{d}{2}\right)\cdot\frac{d}{2}\} = d^3(1-\Phi)\left[\tilde{\sigma}_1^s(0)+\tilde{\sigma}_2^s(0)+\tilde{\sigma}_3^s(0)\right] \tag{XIII}$$

Together with (VI) and (XI) this gives the final result:

$$\tilde{\sigma}^s = \frac{1}{V_s}\int_{V_s}\frac{s_1+s_2+s_3}{3}dV_s = \frac{\tilde{\sigma}_1^s(0)+\tilde{\sigma}_2^s(0)+\tilde{\sigma}_3^s(0)}{3} \qquad \text{(XIV)} \quad \text{(II.6-16)}$$

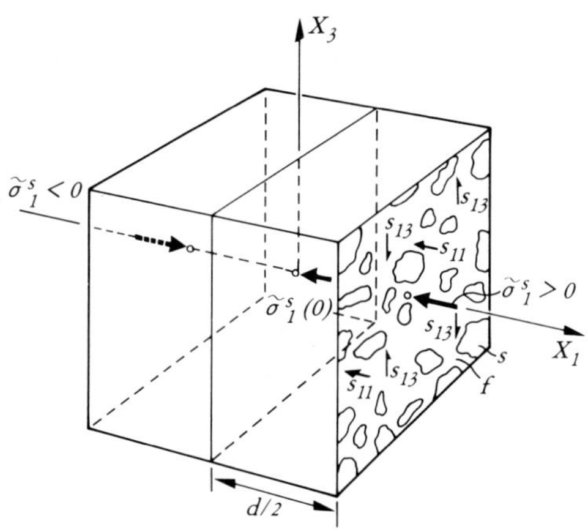

$$\tilde{\sigma}_1^s(0)\cdot d =$$

$$\frac{d}{2}\cdot\left[\tilde{\sigma}_1^s\left(+\frac{d}{2}\right) + \left|\tilde{\sigma}_1^s\left(-\frac{d}{2}\right)\right|\right] =$$

$$\frac{d}{2}\cdot\tilde{\sigma}_1^s\left(+\frac{d}{2}\right) - \frac{d}{2}\cdot\tilde{\sigma}_1^s\left(-\frac{d}{2}\right)$$

Fig. II.6-6
AVERAGE SKELETAL STRESSES ON OUTER BOUNDARY OF MACROSCOPIC VOLUME ELEMENT

II.7. PLASTICITY OF ROCKS

As remarked earlier, in the geological literature the term 'plastic' is frequently used as being synonymous with 'ductile'. It then characterises a type of rock deformation – or a material capable of such deformation – where large permanent strains can accumulate without a drastic loss of material strength. This type of behaviour is in contrast with the 'brittle' behaviour, which is characterised by relatively small and predominantly elastic deformations followed by violent fracturing. In this book, and in this chapter in particular, we deal with 'plasticity' in a more restricted and precise sense, in accord with the mechanical theory of plasticity.

Right at the beginning, it should be emphasised that this theory is concerned with the deformation of a solid on a macroscopic scale; it treats the solid material as a continuum without referring explicitly to the mode of deformation of the 'microscopic' constituents of the solid, such as the crystalline particles of a rock or soil. In fact, individual particles or interparticle bonds may break in a brittle way while, on a macroscopic scale, the deformation behaviour of the rock or rock mass may still be regarded as plastic. The concepts of plasticity, which will be presented in the following with special reference to rocks and soils, are therefore perfectly compatible with brittle deformation on a microscale.

1. *IDEALISED MATERIALS*

Rocks and soils, like any other materials, deform under load in a way that may be regarded as a combination of more elementary and idealised modes of material behaviour – known as elastic, viscous and plastic. The ideal materials that exhibit purely elastic, viscous or plastic behaviour, or a combination of these, can be visualised with the aid of the elementary models shown in Fig. II.7-1. This is possible at least as long as the ideal materials are subject to simple and uniform loading conditions (e.g. uniaxial loading or simple shearing).

It should be understood, however, that these simple models merely serve as analogies for the phenomenological description of the deformation behaviour and do not in any way represent the internal processes (e.g. intercrystalline gliding, recrystallisation, grain boundary movements, grain fracturing and crushing) that actually bring about the deformation of geological materials.

Elastic. The model of Fig. II.7-1A serves as an analogue for the perfectly elastic body. The force applied to the helical spring is the analogue of a stress, and the extension or contraction of the spring corresponds with the elastic strain. When the force is removed the spring returns to its original length. If the spring contracts or elongates in strict proportion to the force applied, it is a model of the linear-elastic material (see Chapter II.6) whose stress/strain relationship is known as Hooke's law. Fig. II.7-1A shows this relationship for a simple-shear test, which is represented schematically in Fig. II.7-2, in which the definitions of shear strain and shear rate are also given. An increase in shear stress τ is accompanied by a proportional increase in elastic shear strain (shear angle) γ. The constant of proportionality is the shear modulus G.

Viscous. The 'dash pot' in Fig. II.7-1B serves as a model for the (Newtonian) viscous liquid. It is a loosely fitting piston in a liquid-filled cylinder. Any slow displacement of the piston requires fluid flow between piston and cylinder walls. As a consequence of this, the force (i.e. stress) needed to move the piston increases with the rate of piston movement. Assuming a linear relationship between force and piston speed, the 'dash pot' becomes an analogue of the Newtonian viscous fluid[1)]. Applied to simple shearing (Fig. II.7-2), the force acting on the piston becomes the analogue of the shear stress and the piston speed represents the shear rate $d\gamma/dt$. The constant of proportionality η between shear stress and shear rate is called (dynamic) *viscosity*. It is usually measured in poise: 1 poise = 1 g cm^{-1} s^{-1} = 0.1 N.s m^{-2} = 0.1 Pa.s $=10^{-6}$ bar.s. For geological purposes viscosity can also be expressed in units of bars and million years (my): 1 my bar = 3.10^{19} poise.

Ideal plastic. The model shown in Fig. II.7-1C consists of a solid block in frictional contact with a solid plate. The block is pressed onto the plate by its own weight or by an additional vertical force. To slide the block over the plate requires that the force F be raised to a certain critical value. We consider only 'slow' motions, in which the force required to overcome the friction between block and plate is high compared with the force associated with the (small) initial acceleration of the block mass. Then it is obvious that, once sliding has started, the 'driving' force F cannot be increased further. Interpreted in terms of the simple shearing test, the model indicates that shearing does not start before the shear stress has reached a critical value ($\tau = \tau_y$ in Fig. II.7-1C). From that moment on, simple shearing can continue, i.e. the driving plate of the shearing equipment can be displaced in the direction of the applied shear stress, without increasing the shear stress. When the shear stress drops below the critical or 'yield' value, the shearing motion ceases immediately. The material 'solidifies' again – its elements maintaining the shape they had at the moment when the shear stress was reduced. This behaviour characterises the *rigid/perfectly plastic* material.

A somewhat more realistic model is obtained by allowing the conceptual plastic material to deform elastically under subcritical loads. The corresponding model of the *elastic/perfectly plastic* material (Fig. II.7-1D; Prandtl body) consists of a linearly elastic spring connected in series with a frictional block. As the force F is increased, a proportional elongation of the spring is produced until the force attains the critical value, whereupon sliding of the block starts without further change of the elastic spring. Interpreted in terms of simple shearing, the stress/strain relationship of the model is expressed by two straight lines representing, respectively, the elastic and plastic parts of the deformation (Fig. II.7-1D). The transition from elastic to plastic behaviour is idealised as a sharp separation ('yield' point). Unlike the rigid/perfectly plastic material, the elastic/perfectly plastic body shows some change of shape when it is unloaded ('unloading' curve in Fig. II.7-1D): The elastic part of the shear strain disappears, whereas the plastic strain remains unchanged.

Elastic vs Inelastic. This unloading reveals the essential difference between the elastic solid on the one hand and the viscous and plastic (i.e. inelastic) materials on the other. When unloaded, the elastic body returns to its original ('natural') state. Moreover, the perfectly elastic body unstrains without delay and returns to its environment all the mechanical energy that was expended in deforming the body. This is illustrated in Fig. II.7-3, which shows schematically the compression of an elastic spring under incremental loading. In each loading step mechanical energy is fed into the system. During unloading, the procedure is

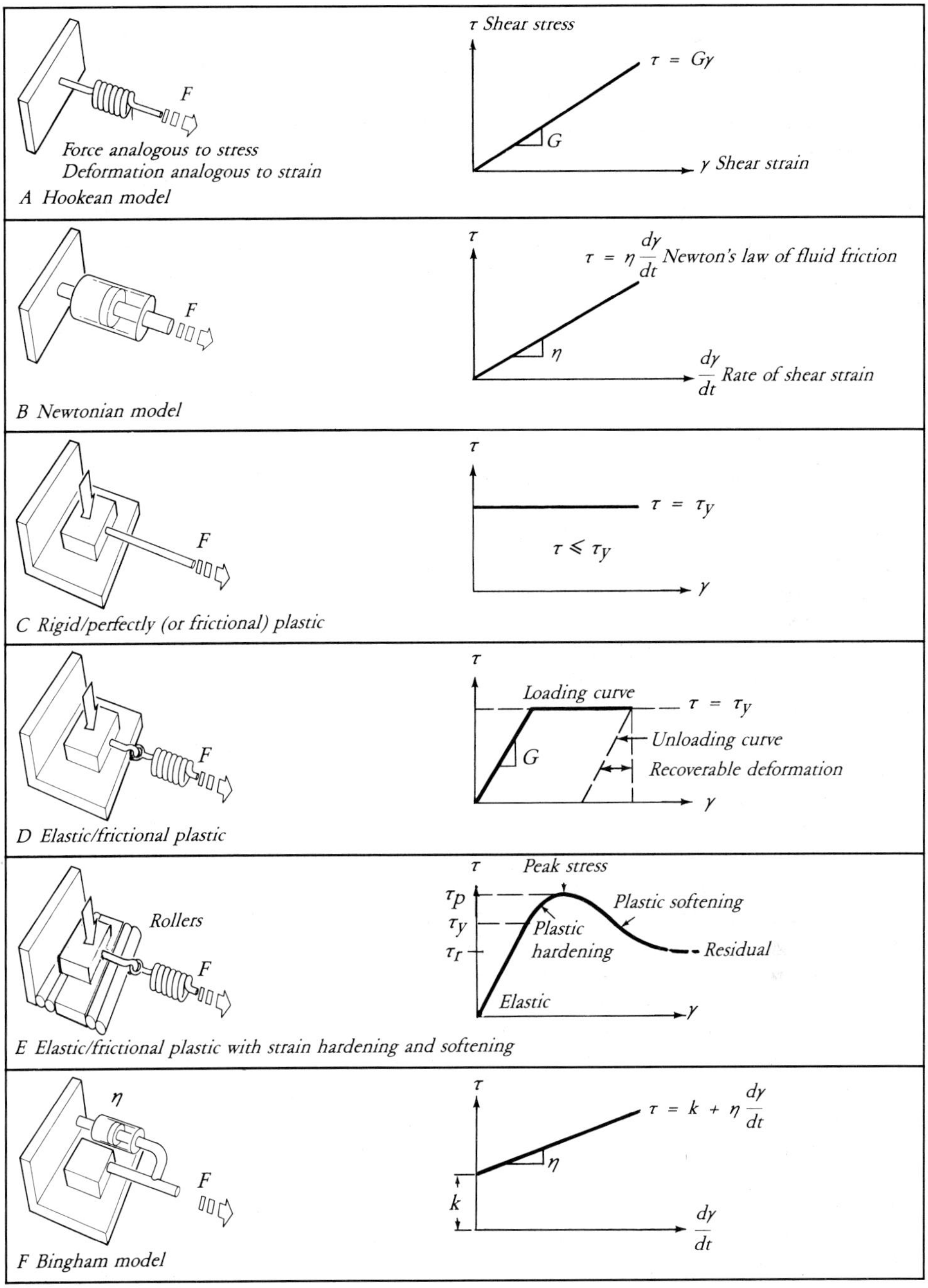

Fig. II.7-1 IDEALISED MATERIALS

reversed and the elongating spring performs mechanical work by lifting the loads. In contrast with this full recovery of mechanical strain energy, during unloading of an inelastic material (viscous or ideally plastic) none of the mechanical deformation energy is regained; all this energy is dissipated, i.e. transformed into heat or consumed in producing irreversible changes in the internal structure of the material (e.g. fracturing and reorientation of grains).

Plastic vs Viscous. If both viscous and ideally plastic materials dissipate all mechanical deformation work, what then is the essential difference between the two materials? To understand this, we turn again to simple shearing (Fig. II.7-2). As properly modelled by the 'dash pot' of Fig. II.7-1B, a viscous fluid will flow under any shear stress – no matter how small – whereas a plastic material will do so only when the shear stress attains a certain 'yield' value (cf. Fig. II.7-1C). Mutatis mutandis this holds also for more complex deformation modes. Several stress components then characterise the 'yield state'. They are not defined by yield constants but are subject to a single relationship – the 'yield condition', which will be dealt with later. However, the mere existence of a yield stress is not the only, and not even the most important, difference between the two ideal materials. In fact one may endow the viscous fluid model with a yield condition and still retain the essential difference between viscous and plastic. This modified fluid is called *'Bingham fluid'* and its model (see Fig. II.7-1F) is a combination of the dash pot and the frictional block. In simple shearing the Bingham fluid will not flow before the shear stress has reached the critical value ($\tau = k$). Until that moment it behaves in the same way as the plastic material.

However, even in flow the Newtonian viscous fluid and the ideal plastic material share an important property: since, by definition, the two ideal materials do not undergo changes in mechanical properties during inelastic shearing which, moreover, can be continued without changing the limit state of stress, the two materials have no memory of past inelastic shearing. Therefore, the stresses cannot be related to total inelastic shear strains but only to infinitesimal strain increments, or to strain rates when referred to an infinitesimal time interval. The essential difference between the viscous and the ideal plastic behaviour now lies in the fact that, in contrast with the viscous fluid, in perfectly plastic deformation there is *no one-to-one correspondence between stress and strain rate.* Arbitrary plastic shear rates can be produced by one and the same shear load. Needless to say, this is an idealisation and a complete rate independence will hardly ever be observed in nature.

The lack of a one-to-one correspondence between stress and increase of strain might make the perfectly plastic model appear incapable of linking deformations and stresses and therefore useless. This is not the case. In perfect plasticity it is not the magnitudes of stresses and strain increments that are related in a unique way, but it is the characteristic *directions of stresses and strain increments* – for instance, the direction of maximum shear stress and the direction in which the shear strain (shear angle) increases more than in any other direction. We shall deal with this relationship later (Section II.7.4).

As a consequence of the lack of a one-to-one correspondence between stress and incremental strain in perfectly plastic deformation, the full determination of this deformation requires the specification of both the boundary loads and the boundary movements. This is much more than the determination of a viscous flow would require. In viscous shear flow it is sufficient to specify either the boundary stresses or the boundary movements to obtain a full determination of flow and stresses. In particular, the motion of the driving wall is uniquely determined by the shear stresses applied and cannot be prescribed independently.

Shear flow on 'micro' scale. The practical importance of the difference between plastic and viscous flow warrants some digression on the mechanisms that, on a microscale, cause the differences in macroscopic flow behaviour. Again, we consider simple shearing between two parallel rigid plates, one plate being moved at speed U, the other being fixed (Fig. II.7-4A). As the movement of the driving plate continues, the material is forced to 'slide off' in parallel layers on top of each other – in a similar way to shearing a deck of cards.

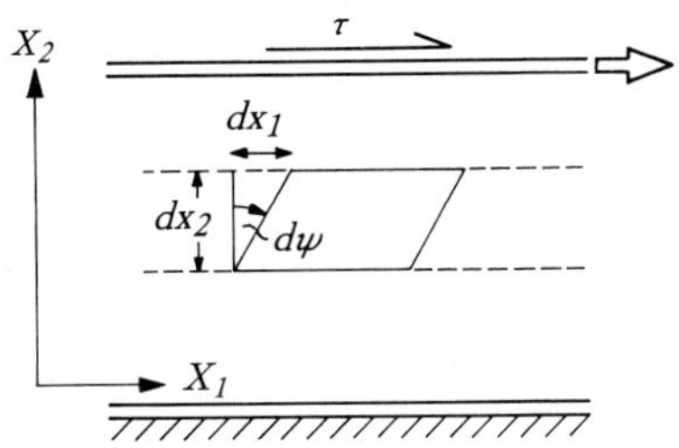

Shear strain ('engineering' strain):

The change in angle between two 'material' lines that intersect under 90° in the unstrained state; shear strain γ associated with these directions:

$$\gamma \equiv \gamma_{12} = \frac{dx_1}{dx_2} = \tan d\psi = d\psi \text{ (for small angles } d\psi)$$

Rate of shearing:

If the velocity of particle motion in the x_1-direction is denoted by ($u = dx_1/dt$) the time rate of simple shearing motion becomes

$$\frac{d\gamma}{dt} \equiv \dot{\gamma} = \frac{du}{dx_2} = \frac{1}{\cos^2 d\psi} \cdot \frac{d\psi}{dt} = \frac{d\psi}{dt}$$

(In $\cos^2 d\psi$ the angle $d\psi$ tends to zero since in determining the rate of deformation only states of deformation that are infinitesimally close are compared.)

Fig. II.7-2 SIMPLE SHEARING

Energy increments

$\Delta W_1 = F\,\Delta h$
$\Delta W_2 = 2F\,\Delta h$
$\Delta W_3 = 3F\,\Delta h$

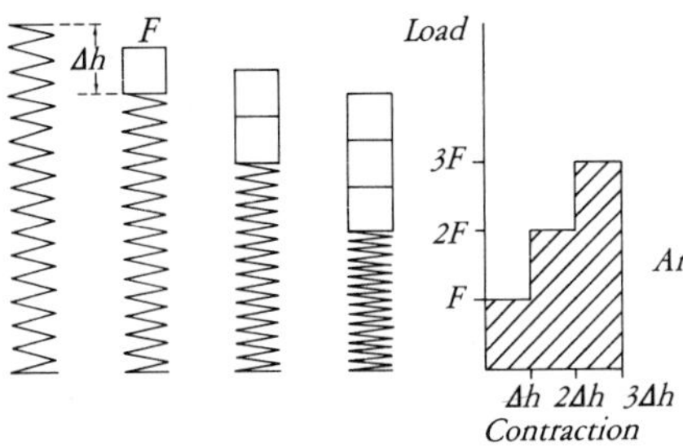

Fig. II.7-3 INCREMENTAL LOADING OF ELASTIC SPRING

In Fig. II.7-4A two neighbouring layers are sketched which move parallel to the walls at different velocities $u^{(1)}$ and $u^{(2)}$. Obviously, the layers will interact by 'internal friction', the faster layer tending to accelerate the slower layer and vice versa. In other words, between the neighbouring layers a force, or a shear stress (τ), acts parallel to the layers. What produces this frictional stress?

In a viscous fluid the internal shear stress arises from the exchange of molecules between adjacent layers. This situation is shown schematically in Fig. II.7-4B. On average, equal numbers of molecules are exchanged per unit time in both directions between the two layers. However, the molecules arriving from layer (1) at layer (2) carry a higher wall-parallel velocity component u than the molecules moving in the opposite direction. By collision with the molecules in layer (2), the arriving molecules tend to increase the average velocity $u^{(2)}$ of that layer. This molecular action is equivalent to a shear stress (τ) when the fluid is regarded as a continuous material by disregarding the molecular structure[2)]. It should be clear from this simplifying consideration that the effect of the molecular exchange, and hence the magnitude of the shear stress, is directly proportional to the gradient du/dy of the 'layer velocity' u (see Fig. II.7-4B).

Next we consider a *granular material* – such as sand or brecciated rock – under shear. We first assume that the grains do not break under the loads applied and are not bonded at the contacts ('non-cohesive frictional' material). On a microscale, shearing takes place by grains overriding each other. In Fig. II.7-4C two adjacent 'layers' of grains are shown

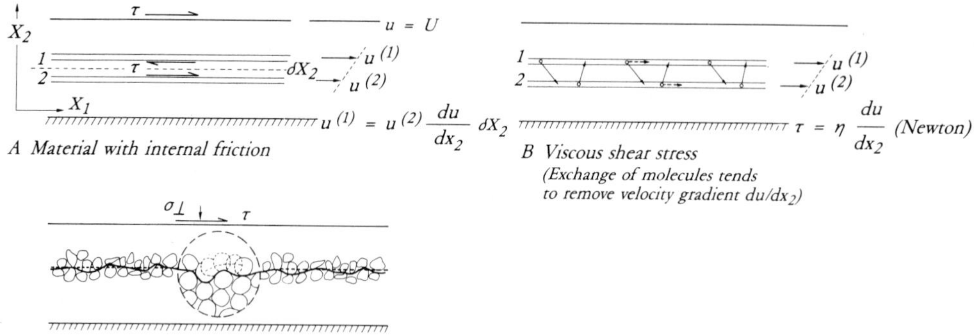

C Shear flow of granular material
(Work expended in grain 'dislocation' does not depend on dislocation speed)

Fig. II-7-4
SHEAR STRESS IN VISCOUS AND PLASTIC MATERIALS

that are separated by a 'wavy' surface SS, along which the grains of the upper layer override the grains of the lower one. As a grain is pushed up the crest of an underlying grain – as sketched in the central part of Fig. II.7-4C – mechanical work is irreversibly expended to overcome the solid friction at the intergranular contacts and in loosening the 'interlocking' of the grains. Neither the solid friction at the contacts nor the interlocking resistance are rate-dependent – at least not in the range of slow motions we are concerned with. When we now disregard the granular structure of the material and consider it a continuum (Chapter II.3, 'Micro- and macro-scale') the action of the intergranular forces is accounted for by the shear stress that is transmitted across a (macroscopic) cross-section parallel to the walls (see the dotted line in Fig. II.7-4C). As is the case with the intergranular forces, this *shear stress will not depend on the magnitude of the shearing rate.* We may, therefore, conclude that under shear a frictional granular material behaves in accordance with the concept of a perfectly plastic material rather than a viscous one.

Rate independence of the shear stress and full dissipation of mechanical energy will still hold in true *'cataclastic'* shear flow, i.e. in shearing accompanied by brittle fracturing or crushing of grains. Although fracturing of an individual grain releases elastic strain energy stored in that grain, most of this energy is dissipated at the 'spot' and cannot contribute to an elastic rebound of the sheared granular rock or soil. We think that even the presence of intergranular bonds will have little or no effect on this behaviour, as long as the bonds are 'weak' in the sense that bond rupturing requires less shear stress than is necessary to overcome intergranular friction and interlocking. An increase in load will then result in the breakage of intergranular bonds and in a gradual mobilisation of intergranular friction and interlocking resistance.

2. *FRICTIONAL OR 'CATACLASTIC' PLASTICITY*

Shear strength. In addition to dissipation of mechanical energy and rate independence of the shear stress, granular materials also exhibit shear strength – the third essential property of ideally plastic materials. However, the shear strength of a granular material, i.e. the shear stress required to produce plastic shear flow, is not a constant but depends on normal stresses. If in the shearing of the granular material between rigid plates (Fig. II.7-4C) a normal load ($\sigma_\perp$) is applied to the plates, this load will be transmitted across the contacts between the individual grains and increase the frictional and interlocking forces between the grains. It will not affect cohesive bonds between the grains. If we disregard changes in area, orientation and number of contacts that might be caused by an increase in normal stress $\sigma_\perp$, we may expect the frictional strength of the granular material to increase in strict proportion to the normal stress. The onset of plastic flow of a frictional cohesive material (i.e. a granular material with cohesive bonds between the individual grains) under simple shearing conditions may therefore be characterised by the linear stress relationship:

$$|\tau| = \tau_o + \sigma_o \tan \varphi_s \qquad \text{(II.7-1)}$$

This equation contains two material constants characteristic of the shear flow of a particular frictional material. The *cohesive strength* τ_o is the shear stress the material can withstand in the absence of any confining normal stress $\sigma_\perp$, and φ_s is the *'angle of internal friction'* in simple shearing. The absolute value $|\tau|$ of the shear stress appears in the equation, since under the same normal stress shearing can be produced in the opposite direction by merely reversing the direction of the shear stress, which implies a change of sign of τ.

The 'Effective Stress Principle'. In geological applications porous rocks are always saturated with fluid. Part of the total normal stress acting on a cross-section in such a rock is therefore always transmitted through the pore fluid (Chapter II.3). Consequently, only part of the total normal stress $\sigma_\perp$ in eq. II.7-1 will actually become effective at the grain contacts to increase the frictional strength of the fluid-saturated material. It was explained in Chapter II.3 that when the intergranular contact areas are very small ('point contacts') the average force per unit area transmitted across the contacts ('intergranular stress') becomes identical with the effective stress defined by Terzaghi (1923) as $\sigma' = \sigma - p$, where p is the fluid pressure. Since, in general, the presence of fluid has a negligible effect on the solid friction between the grains, it appears plausible for fluid-saturated materials with very small contact areas, to replace the total normal stress $\sigma_\perp$ in equation II.7-1 by the effective normal stress $\sigma'_\perp$. This has indeed been confirmed by a vast number of experiments on fluid-saturated soils and granular rocks, and these have been carried out in laboratories all over the world. But, surprisingly, this effective stress control of the shear strength was not only verified for cohesionless granular material with very small intergranular contact areas, but it was also found to apply – to a high degree of accuracy – to a great variety of rocks with 'large' inter-granular contacts, cohesive intergranular bonds, and even for rocks whose matrix hardly showed any granular structure at all.

In particular, here we should refer to the very careful triaxial testing of lead shot and quartz sand which Bishop and Skinner (1977) carried out using various pore pressures under cell pressures up to 27 MPa (270 bar). For consolidated sedimentary rocks we refer to the classic paper by Handin, Hager, Friedman and Feather (1963) who reported the results of triaxial compression tests on samples of various sedimentary rocks carried out under conditions that realistically simulated the natural conditions of overburden, fluid pressure and temperature down to a depth of about 9 km. The results indicate that the shear strength and the amount of strain preceding fracture ('ductility') are functions of Terzaghi's effective stress, provided certain conditions are satisfied.

Accordingly, the shear strength equation II.7-1 which describes the onset of plastic shear flow, is replaced by

$$|\tau| = \tau_o + \sigma'_\perp \tan \varphi_s \qquad \text{(II.7-1a)}$$

which is the linear expression of the 'effective stress principle' for the shear strength of porous rocks and soils, as it was first stated by K. Terzaghi for soils in 1923.

Although equation II.7-1a was found to apply with sufficient accuracy to many rocks and over a considerable range of the confining pressure, it has its limitations. First of all, linearity cannot hold indefinitely with increasing effective normal stresses. Since rock strength cannot become infinite, the contribution of the effective normal stress to the shear strength of the rock must eventually approach a limit value when the effective confining stress of the rock is raised sufficiently. Secondly, Terzaghi's effective stress concept itself has its limitations when used to account for the effect of pore pressure on the shearing strength of fluid-saturated rocks.

Limitations of the effective stress principle. Unfortunately, a truly conclusive and quantitative definition of the limiting factors of Terzaghi's effective stress concept cannot be given since, owing to the complexities of the failure mechanisms in cohesive or cohesionless granular materials and porous rocks in general, a satisfactory theoretical basis of the effective stress principle for the shear strength has not yet been established. Nevertheless, a reasonable idea of the limiting factors may be obtained from experimental results and mechanical considerations.

Three primary conditions for the validity of the principle have already been stated by Handin et al. (1963):

- First, the pore pressure effects should be of a purely mechanical nature; physico-chemical interactions between pore fluid and pore walls might cause deviations from Terzaghi's principle.
- Secondly, the pore space must be fully interconnected and the permeability must be sufficient to ensure a smooth distribution of fluid pressure throughout the pore space during deformation of the rock.
- Thirdly, the configuration of the pore space must be such that a uniform change in pore fluid pressure is felt throughout the solid rock matrix as the same uniform change in the local rock pressure. This condition will not, for instance, be satisfied if the pores of the rocks are too widely spaced narrow tubes.

Furthermore, it is important to distinguish between rocks with interparticle bonds that are weaker than the grain material itself and rocks with *intergranular bonds* that have the same or greater mechanical strength than the majority of the granular constituents of the rock. With weak intergranular bonds, loading of the rock will first result in the fracturing of intergranular bonds and will concomitantly mobilise frictional and interlocking resistance between individual grains or clusters of grains that are still bonded together. In this instance we may expect Terzaghi's principle of effective stress to apply just as well as in a cohesionless, frictional granular material such as sand.

When, however, the intergranular bonds are stronger than the grain material itself, fracturing will preferentially take place in the mineral grains by the rupturing of the cohesive bonding between atoms of the solid material. As far as the strength of this bonding is concerned, superposition of an isotropic pressure in the solid material will hardly have any noticeable effect. But it might perhaps affect the local concentrations of the tensile stresses on a microscale and actually cause cracks to grow. Primarily, such tensile stresses in a rock under effective compression are induced by shearing deformation: when the associated difference in principal macrostresses is raised, pre-existing cracks inside or at the surface of grains may experience additional shearing along the fracture walls. This will produce tensile stresses at the crack tips. (Actually, in an elliptical crack the greatest tensile stress will be generated slightly off the crack tip.) Under sufficient shearing, cracks will branch off the old fractures or flaws and may grow in a stable manner as long as the shear load on the rock is increasing (Brace and Bombolakis, 1963; Hoek and Bieniawski, 1966). If we now, in addition to the shear load, raise the mean normal stress (pressure) everywhere in the solid rock skeleton by the same amount Δp, by applying this extra pressure to the outer rock boundary and to the pore fluid, then the shear stresses that gave rise to the fracture-propagating local tensions will not be affected. In that sense the pressure change Δp is then truly a 'neutral' change in stress. Since it also raises any normal component σ_i of the total macrostress by Δp, the effective stress components σ_i' remain unchanged:

$$\sigma'_i = \sigma_i + \Delta p - (p + \Delta p) \qquad \text{(II.7-2)}$$

From this we conclude that even when there is severe grain fracturing the effective stress principle applies – the process of microcracking being controlled by the applied shear stresses and effective normal stresses – provided the rise in fluid pressure is fully transmitted to the walls of the grain fractures. If, however, such fractures remain 'dry', i.e. separated from the pore-fluid system, deviations from the effective stress control are to be expected. Under rising pressure in the rock matrix the walls of dry fractures will be more firmly pressed together, which increases the frictional resistance of the rock to the imposed shear stress.

Ultimate shear strength. With increasing shear deformation of the rock, damage by intra- and transgranular cracking will progress until the cohesive resistance of the rock is exhausted, while internal friction is progressively mobilised to provide a second line of resistance to continued deformation. Depending on rock type and confining pressure, this transition may be fairly sharp and characterised by a decline in the shear stress which the rock can sustain under continued shearing. The maximum shear stress that the rock can withstand is referred to as *'ultimate' shear strength*.

Such stress/strain behaviour is represented in Fig. II.7-1E by a graph and an analogue model. The latter consists of a solid block that rests partly on frictionless rollers and partly on a solid plate in frictional contact. As the block is pulled to the right, the area of frictional contact increases until maximum shear resistance is reached. In the corresponding shear stress/shear strain diagram of the same figure the maximum or 'peak' stress is denoted by τ_p, while the shear stress at which plastic deformation commences ('yield') – in the present case caused by brittle microcracking – is denoted by τ_y. As shearing continues beyond 'peak', the stress/strain curve declines and eventually approaches a residual stress level τ_r.

This more or less gradual reduction in shear strength is commonly referred to as *strain softening*. In the analogue model of Fig. II.7-1E it is accounted for by a second set of frictionless rollers which reduce the frictional resistance of the moving block. Besides shear dilatancy of granular rocks and the pervasive microfracturing considered above, various other softening mechanisms exist which, under various pressure, temperature and chemical conditions, can produce a reduction in shear strength in many – if not most – geological materials. Although such softening did not become apparent in the triaxial tests carried out by Handin and Hager (1957) and other researchers on various rock types under effective confining pressures of several hundred MPa, the very limited amount of straining in the experiment and strain rates that are many orders of magnitude higher than in corresponding tectonic processes, would not warrant the conclusion that softening mechanisms do not operate in the high-pressure regimes. We shall revert to strain softening in Chapter II.7.7 when discussing the extremely important role it plays in the generation of tectonic faults. Here our concern is still with the effective stress control of the ultimate shear strength τ_p.

Since the damage process of growing fluid-filled cracks is controlled by effective stresses, effective stress control should also apply to the ultimate shear strength of rocks, even if they have strong intergranular bonds. In view of the greatly increased number of microcracks when the shear stress approaches its maximum, and the expected increase in crack interference, the relative importance of any 'dry' cracks which may be present, is very likely to diminish. Nevertheless, a word of caution should be added. It concerns the assumption that we have tacitly made that the grains of strongly bonded grain complexes consist of mechanically uniform and isotropic material. This allowed us to conclude that a uniform all-around pressure change would not affect the growth of fluid-filled microcracks. Contrasting mechanical properties in the mineral composition of the grain material will, however, give rise to extra shear stresses on a microscale when a uniform all-around pressure is applied to the rock matrix. These shear stresses on grain size or sub-grain size scale are likely to affect the growth of individual microcracks. To what extent this will express itself in the macroscopic strength of the rock, we do not know.

Shear strength and elastic deformation. On the matter of the effective stress principle, one further point needs clarification. In Chapter II.6 on poro-elasticity we discussed how the stress/strain relationship for a linearly elastic isotropic porous body can be obtained from the corresponding relationship for the compact linearly elastic material by replacing the total normal stresses by Terzaghi's effective stresses. However, this application of the effective stress principle was subject to the restriction that the bulk of the (dry) rock matrix must be much more compressible than the material of the matrix (quartz, feldspar, etc.) itself. This well-established limitation is often assumed to apply to the shear strength of porous rocks as well. This would indeed be correct if the rock could reach its limit state (peak stress) by purely elastic deformations, as was assumed in the elastic/plastic model of Fig. II.7-1D. In reality, however, inelastic microprocesses commence at a lower, somewhat ill-defined stress level (τ_y) and progressively dominate the continuing deformation. Besides the growth of flaws and microcracks, which have been already mentioned, other processes such as the unlocking, shifting and reorienting of particles, or plastic deformation of intergranular contacts, may contribute to the inelastic deformation behaviour in this subcritical range (between τ_y and τ_p in the stress/strain diagram of Fig. II.7-1E). This material behaviour is referred to as *plastic strain hardening*, since continuation of the deformation, say by simple shearing, requires an increase in shear stress in this subcritical range.

Only when yield and peak (ultimate) stress approximately coincide will the effective stress principle applied to shear strength have the same limitations, as in poro-elastic straining. But we believe such cases of extreme brittleness to be rather rare under subsurface conditions; in general, the effective stress principle for the shear strength of rocks will not be subject to the poro-elastic limitations.

Dependence of strength parameters on confining pressure and strain. The shear stress/normal stress relationship (II.7-1a) was meant to characterise the state of limiting (ultimate) strength of a rock in simple shear. Accordingly, the angle of internal friction φ_s and the cohesive strength τ_o were associated with the shear stress peak. We may apply the same type of relation to characterise the onset of plastic hardening at the yield stress τ_y or the residual state τ_r after shear softening (Fig. II.7-1E). In both cases φ_s will be smaller than at peak. Incidentally, at yield stress τ_y, the effective stress relationship II.7-1a will be subject to the same limitations as the effective stress formulation of the poro-elastic stress/strain relations. The cohesive strength parameter τ_o will progressively decrease as plastic strain accumulates.

It will be discussed later how this general trend in the variation of strength parameters with increasing strain may be utilised in the study of tectonic faulting based on elastic/frictional plastic rock behaviour with strain hardening and softening.

Equation II.7-1a expresses a linear dependence of ultimate shear strength on the effective normal stresses as long as φ_s is a constant. However, depending on the type of rock, deviations from linearity will become conspicuous in the range of elevated effective confining pressures. In other words, the friction angle φ_S itself becomes a function of the effective confining stress. Tests on a great variety of rocks indicate that this deviation from linearity always implies a decrease in the peak friction angle with increasing normal effective stress.

Among the various causes of this decrease that have been identified, suppression of *dilatancy* under the higher confining stresses is the most prominent one for brittle rock deformation. Dilatancy is the well-known tendency of rocks under shear to increase in volume by the action of inelastic processes. When densely packed granular material, brecciated or heavily jointed rock is being sheared, the assemblages of grains or rock blocks tend to loosen and thereby to increase in volume. Dilatancy also plays a significant role in strong, low-porosity crystalline rocks under shear, where it is caused by the formation of microfractures, which begin to grow at a shear load that commonly lies between one- and two-thirds of the shear load required to produce macroscopic shear fractures (Brace, Paulding and Scholz, 1966).

The processes that cause dilatancy may be suppressed by raising the effective confining pressure sufficiently. Shear deformation will then proceed by breakage and crushing of grains, by rupturing of load-bearing rock bridges between fractures, or by the macroscopic fracturing of intact rock blocks. For a given increment in effective confining stress, these brittle destruction processes then require a lower increase in external shear stress than would be required by the dilational processes.

Whereas in low-porosity crystalline rocks the suppression of dilatancy may require several hundreds of MPa, a few tens of MPa may be sufficient to suppress dilatancy completely in a porous sandstone (Jamison and Teufel, 1979). Therefore, the associated reduction of the friction angle will become significant at quite different confining pressures. A particularly drastic reduction is to be expected when the confining pressure required to suppress dilatancy causes a transition from brittle microdeformations to intracrystalline gliding.

Frictional vs Crystal Plasticity. In the foregoing we have paid attention to brittle microprocesses such as the fracturing of grains or intergranular bonds. From metal plasticity it is well known that polycrystalline materials, such as cold-working metals, can deform permanently by completely ductile deformation of the crystal grains, in which one part of the crystal lattice slips or glides over the other by means of the propagation of dislocations (misfits in the lattice). In contrast with solid friction, the critical shear stress required for such intra-crystalline gliding is approximately independent of any normal stress component. But, in general, it will not be independent of strain rate, temperature or chemical environment. In the polycrystalline aggregate the individual crystals are randomly oriented. On a microscale the aggregate is therefore heterogeneous with respect to mechanical anisotropy. Consequently, the application of an all-around pressure – including the pore pressure in the case of a fluid-filled porous body – will induce shear stresses on the microscale. In this respect the situation is virtually the same as for grain fracturing in a brittle rock. The essential difference, however, is that in the crystal-plastic case the induced stresses trigger dislocation movements that leave the cohesive strength of the rock intact, while in the brittle rock the induced shear stresses promote the growth of microcracks which damage the rock material.

Hence, as long as the plastic deformation of a polycrystalline body is the sole result of intracrystalline gliding, and no intergranular friction is involved, the shear strength of this material is pressure independent. For a porous, fluid-filled rock that deforms in this way, the effective stress principle becomes irrelevant because equation II.7-1a reduces to

$$|\tau| = \tau_o \tag{II.7-1b}$$

To the oil geologist crystal plasticity may be of less interest, since most rock-forming minerals deform plastically at temperatures outside the range he is dealing with. Quartz, for instance, begins to deform by intracrystalline gliding around the 300 °C isotherm, provided the strain rate is low and there is sufficient water present to allow hydrolytic weakening, while feldspar requires temperatures of 550-650 °C [3].

Tectonic Significance of the Effective Stress Principle. Shear strength control by effective stresses in porous rocks is of primary tectonic significance. Abnormally high fluid pressures, i.e. fluid pressures considerably in excess of the pressures due to the weight of the static water column, reduce the shear strength of the overpressured strata and may allow such strata to plastically deform and flow under relatively low shear loads. This offers, for example, a 'lubrication' mechanism for some large overthrusts, as Hubbert and Rubey (1959) have pointed out. M.K. Hubbert deserves the credit for having introduced the effective stress concept into Structural Geology. Similar high-pressure lubrication mechanisms operate in gravitational sliding on gentle delta slopes and in flexural folding by bedding plane slips.

In tectonic overthrusting, material at the base of a thrust block is heavily sheared. Somewhat idealised, this deformation may be considered as simple shear flow – the special deformation mode to which our introduction of basic plasticity concepts has been related so far. Before we turn to conceptual generalisations, which are necessary when dealing with more general modes of plastic deformation, we shall briefly discuss some aspects of simple shearing at the base of an overthrust block.

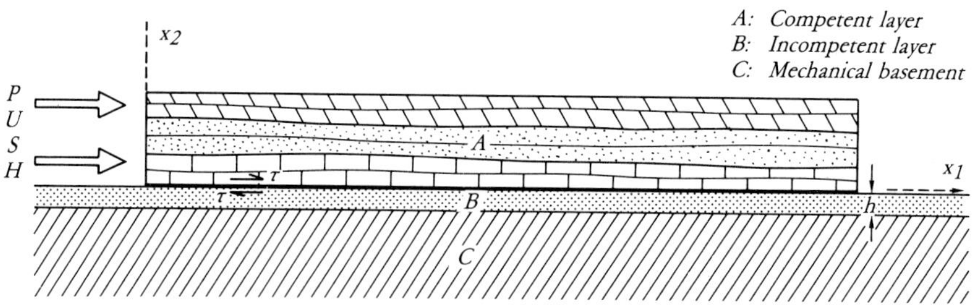

Fig. II.7-5 CROSS-SECTION OF OVERTHRUST SHEET

Example No. 1: Overthrust faulting

Consider a long 'competent' thrust block of rectangular cross-section (Fig. II.7-5) which is pushed over an 'incompetent' substratum. The latter is sandwiched between the thrust block and a 'rigid' basement. With the exception of the regions near the ends of the thrust sheet, the substratum will be subject to simple shearing. Firstly, we assume that the substratum consists of viscous material and, secondly, that it moves without acceleration ('stationary' flow) [4]. The second assumption implies that the stresses acting on any volume element inside the substratum are in static equilibrium (Chapter II.2). In particular, we may consider the stresses that act upon the element in the flow direction (see Fig. II.7-6A). If we restrict ourselves to the main case in which substratal flow is induced solely by the shearing action of the overriding thrust block, the normal stress σ_1' will not have a gradient in the x_1-direction and, consequently, the horizontal shear stresses acting on the upper and lower faces of the element must cancel. This means that the horizontal shear stress τ does not vary in the vertical direction. Because of the one-to-one correspondence between shear stresses and shear rates in viscous fluids, the shear rate du/dx_2 (see Fig. II.7-2) associated with τ will not vary in the vertical direction either. As indicated in Fig. II.7-6A, this implies $du/dx_2 = U/h$, where U is the rate at which the whole overthrust sheet moves and h is the thickness of the substratum. It therefore follows from the one-to-one correspondence between shear stress and shear rate in a viscous fluid that the shear stress along the thrust base decreases with decreasing thrusting rate (U) and increasing thickness of the viscous substratum (h). The same holds for the total push force that is required to overcome the viscous shear resistance along the base of the thrust sheet (i.e. $\tau \times$ basal area).

The situation is quite different when the incompetent substratum behaves as an ideal frictional plastic material. According to equation II.7-1a the shear stress necessary to produce shearing depends on the effective overburden stress (σ_1') but not on the rate of overthrusting. Since the shear rate du/dx_2 does not depend on the shear stress τ, the flow rate u may assume any value between the value U at the top of the substratum and zero at the bottom. In particular, a 'jump-like' change in velocity u becomes possible, as is shown in Fig. II.7-6B. This implies that the plastic substratum may shear in 'block gliding', the upper block attached to the thrust sheet sliding over the lower part of the substratum. This type of plastic shearing, in which shearing concentrates along a 'plane', is frequently observed in nature. It may be caused by the presence of a thin layer of weaker material or, more generally, by shear softening of the material.

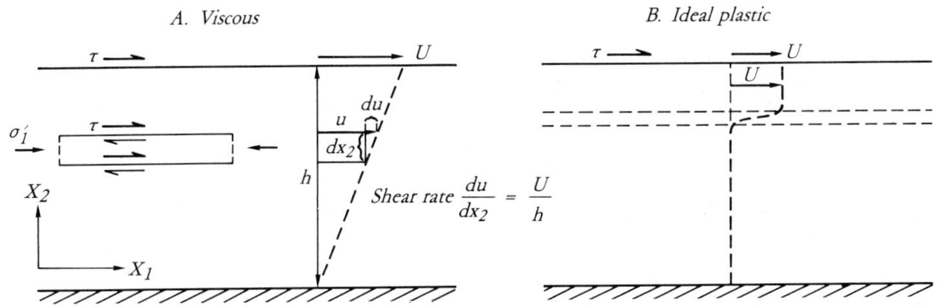

Fig. II.7-6 STATIONARY SHEARING OF SUBSTRATUM

3. COULOMB'S LIMIT CONDITION

Usually, deformations are more complex than the simple shearing considered so far, and may result from combined uniaxial contractions or extensions, shears, rigid rotations and translations of elements of the deforming body, as summarised in Figs. II.5-1, 2 of Chapter II.5. Obviously, the onset of plastic flow at peak stress will then be controlled by a more general limit condition than the criterion II.7-1a. Such a limit condition was first formulated for frictional plastic materials by C.A. Coulomb as early as 1773. We shall discuss Coulomb's concept in two steps, since it actually contains two statements. The first, concerning the state of stress at the onset of plastic deformation, belongs to the well-established part of the theory of frictional plasticity. The second part of Coulomb's limit concept deals with the mode of plastic deformation itself and postulates a link between peak stresses and deformation. It belongs to that part of the fundamentals of the theory which, unfortunately, has not yet been as firmly established as one would wish. We shall present it in a modern version ('Double-gliding model') that seems to agree very well with observed tectonic phenomena.

We consider all planar elements through a point P inside a body under stress (Fig. II.7-7). On each element the effective stress vector will have a well-defined normal stress component $\sigma'_\perp$ and a tangential component τ. According to Coulomb's criterion, plastic deformation becomes possible at P once the combination of shear stress and effective normal stress ($|\tau| - \sigma'_\perp \tan\varphi$) on an element through P attains a critical value τ_o:

$$|\tau| = \tau_o + \sigma'_\perp \tan\varphi \qquad \text{(II.7-1c)}$$

Note that rigid boundaries might prevent actual flow, even though the stresses would allow it ('contained' plastic state). The critical value τ_o – the 'cohesive shear strength' – is, at the same time, the maximum value that $|\tau| - \sigma'_\perp \tan\varphi$ can attain. In Coulomb's theory it will be maintained throughout plastic deformation. This is in line with the previously described simple shear behaviour of an ideal frictional material whose shear resistance neither increases (no 'strain hardening') nor decreases (no 'strain softening') during shearing.

Just like equation II.7-1a, the limit condition II.7-1c contains two material constants: the cohesive shear strength τ_o and the angle of internal friction φ. For densely packed sand, for instance, the value of φ will usually lie between 30° and 45°. The subscript of φ_s in equation II.7-1a has been dropped, since Coulomb's criterion implies that φ is a material parameter, independent of the mode of plastic deformation. It should be noted, however, that this is not quite in agreement with experimental findings, which indicate that for densely packed sand the friction angle φ_s measured in a shear apparatus (or some other device for plane straining) may exceed the values found in triaxial compression of cylindrical or cuboidal samples by up to 10%. Intuitively, one might expect that in three-dimensional straining of an aggregate the relative displacement of individual grains has slightly more 'freedom' than in the overriding motion sketched in Fig. II.7-4C for plane shear flow between rigid plates.

The further discussion of Coulomb's condition conveniently makes use of *geometrical representations* in the σ', τ-plane (Fig. II.7-8A). Condition II.7-1c maps into two straight lines (Fig. II.7-8A) since at a given normal stress $\sigma'_\perp$ the critical shear stress may be reckoned positive or negative, depending on its orientation (cf. Fig. II.7-7). The stress components $\sigma'_\perp$, τ that act on the elements at P (Fig. II.7-7) are represented by points in the (σ', τ)-plane which either lie on the periphery or in the interior of a circle – Mohr's stress circle (Section II.4.4, Fig. II.4-9). Among the elements at P there are three mutually perpendicular ones upon which the shear stress disappears and the normal stress attains a maximum, minimum or intermediate value, denoted by σ'_{Max}, σ'_{Min} and σ'_{Int}, respectively,

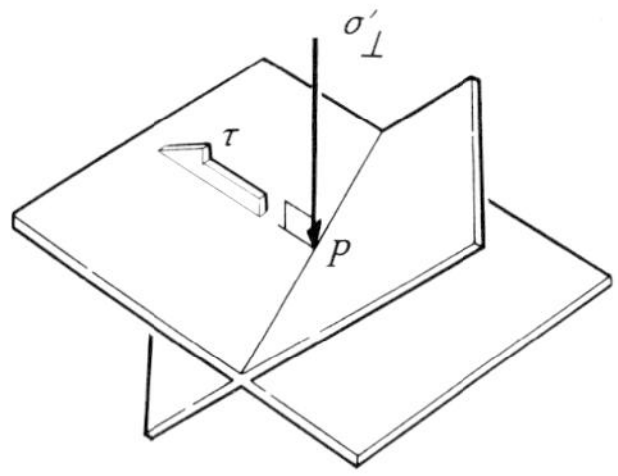

Fig. II-7-7
PLANAR SURFACE ELEMENTS IN BODY UNDER STRESS

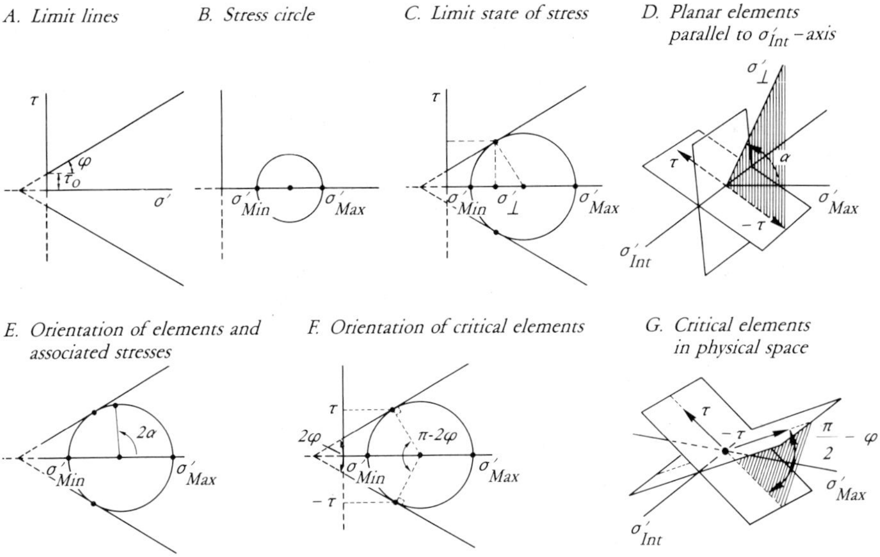

Fig. II.7-8 REPRESENTATIONS IN σ', τ-PLANE

in this chapter. The stress points associated with these 'principal stresses' (Chapter II.1; 'Principal stresses' *) in the σ', τ-plane lie on the σ'-axis and Mohr's stress circle passes through the points σ'_{Max}, 0 and σ'_{Min}, 0 (Fig. II.7-8B). To satisfy the limit condition, the stress circle has to be tangent to the straight 'limit' lines (Fig. II.7-8C). It must not intersect the limit lines, since this would violate condition II.7-1c by allowing for maximum values of $|\tau| - \sigma'_{\perp} \tan \varphi$ larger than τ_0.

As discussed in Section II.4.1, the points on the periphery of the stress circle correspond with those elements in physical space that are parallel to the direction of the intermediate principal stress σ'_{Int} (Fig. II.7-8D). Therefore, only on elements of this type can the critical stress combination II.7-1c occur. The correspondence between the individual elements and the points of the stress circle is illustrated in Figs. II.7-8D, E. When the orientation of such an element is defined by the angle (α) included by its normal with the σ'_{Max}-direction, the associated stress point on the stress circle is defined by the radius of the circle that includes twice that angle with the σ'-axis (Fig. II.7-8E). Since there are two points of tangency, condition II.7-1c will be simultaneously satisfied on two elements. They include the angle $\pi/2-\varphi$, since the corresponding radii of the stress circle (Fig. II.7-8F) include the angle $\pi-2\varphi$. Moreover, the two elements lie symmetrically with respect to the σ'_{Max} direction (Fig. II.7-8G).

The limit condition II.7-1c can be given different forms. A very useful one is obtained by expressing it in terms of stress components in a *Cartesian reference frame.* In establishing this relationship, we may now restrict ourselves to the set of potentially critical elements, i.e. the elements parallel to the σ'_{Int}-direction (Fig. II.7-8D). We note that all normal stresses (including the principal stresses σ'_{Max}, σ'_{Min}) and tangential stresses τ that act on these elements will be directed perpendicularly to the σ'_{Int}-direction[5)]. By choosing, accordingly, a Cartesian frame with x_1-axis parallel to the σ'_{Int} direction (Fig. II.7-9A), it

* The notation used for the principal stresses in Chapter II.1 on STRESS was σ_I, σ_{II}, σ_{III} with $\sigma_I > \sigma_{II} > \sigma_{III}$. Here we use different notations to avoid confusion with graphical representations in other parts.

should be possible to express condition II.7-1c in terms of the stress components σ_2', σ_3' and $\tau_{23} = \tau_{32}$ that act upon elements perpendicular to the x_2- and x_3-directions. To establish the new expression in a very convenient way, we recall that the sum of the normal stresses that act on three mutually perpendicular elements at a given point is independent of the orientation of these elements (Section II.1.3). In particular, the cubical element whose intersection with the x_2, x_3 plane is shown in Fig. II.7-9A can be rotated around the x_1-direction without changing the sum of the normal stresses that act on the faces (indicated by solid lines). This only means, of course, that the mean pressure on the cubical element is independent of its orientation in space. Rotating the element shown in Fig. II.7-9A into a position parallel to the σ_{Max}'- and σ_{Min}'-directions, we see that $\sigma_{Max}' + \sigma_{Min}' = \sigma_2' + \sigma_3'$. If we turn now to the critical stress circle in Fig. II.7-9B, this formula enables the bases of the triangles A and B to be expressed in terms of σ_2' and σ_3' and it takes little more than the Pythagorean theorem to arrive (Fig. II.7-9B) at:

$$(\sigma_2' - \sigma_3')^2 + 4\tau_{23}^2 = (\sigma_2' + \sigma_3' + 2\tau_o \cot \varphi)^2 \sin^2 \varphi \qquad \text{(II.7-3)}$$

the formulation of Coulomb-Mohr's limit condition in terms of Cartesian stress components that act in the plane perpendicular to the intermediate principal stress.

With reference to Fig. II.7-9B we may also express this condition in terms that are invariant with respect to the choice of the Cartesian coordinate axes:

$$q = p \cdot \sin \varphi + \tau_o \cdot \cos \varphi \qquad \text{(II.7-3a)}$$

where q and p stand for the maximum shearing stress $|\tau_{Max}|$ and the 'mean' stress p in the plane, respectively:

$$q = |\tau_{Max}| = \frac{\sigma_{Max}' - \sigma_{Min}'}{2} = \left[\frac{(\sigma_2' - \sigma_3')^2}{4} + \tau_{23}^2\right]^{1/2} \qquad \text{(II.7-3b)}$$

$$p = \frac{\sigma_{Max}' + \sigma_{Min}'}{2} = \frac{\sigma_2' + \sigma_3'}{2}$$

The appearance of the mean effective normal stress p in combination with the angle of internal friction on the right-hand side of equation II.7-3a accounts for the 'strengthening' effect of the effective normal stresses.

According to equation II.7-3a the limit condition may be alternatively represented in the stress plane as the locus of critical maximum shear stresses (Fig. II.7-9C). The limiting stress states are then represented by stress circles whose vertices lie on a line that is referred to as the 'shear strength' line. In the range of confining pressures where this line is straight it is simply related to the Coulomb-Mohr limit line, which is tangential to the limiting stress circles, through the relation between the respective inclination angles:

$$\tan \chi = \sin \varphi \qquad \text{(II.7-4)}$$

Typical for Coulomb's limit condition is the absence of the intermediate principal stress σ_{Int}', which supposedly has no influence on the onset of plastic flow. We shall comment on this assumption in Section II.7.4. A second striking feature of condition II.7-3 is its quadratic form in the Cartesian stress components. This has important consequences for tectonic processes, which we shall discuss next.

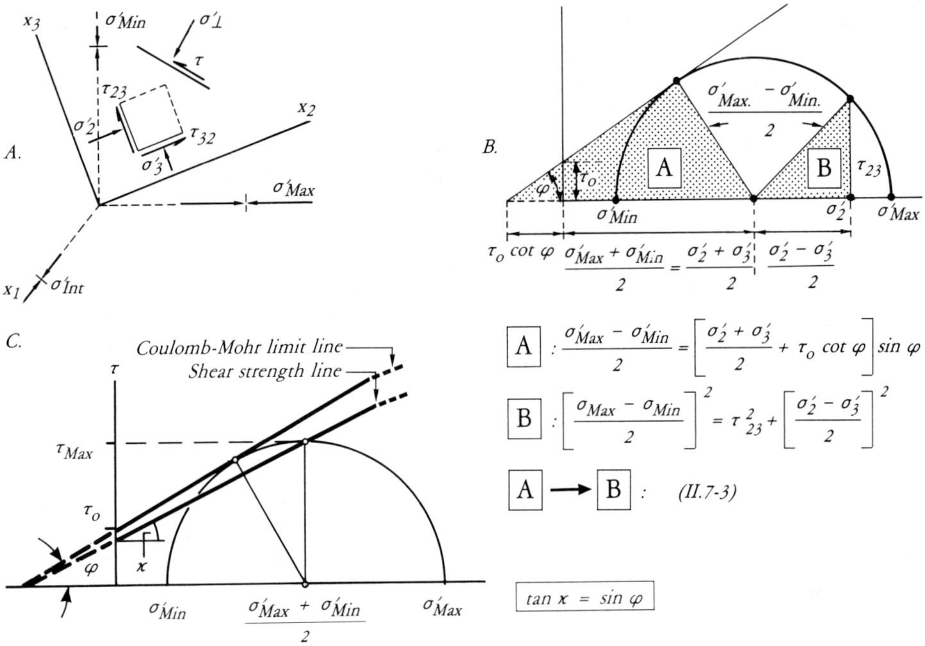

Fig. II.7-9 COULOMB'S LIMIT CONDITION IN PLANE STRAIN OR STRESS

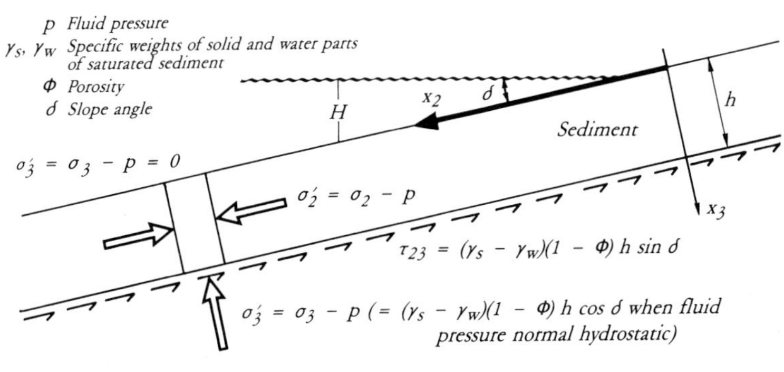

Fig. II.7-10 SUBMARINE SEDIMENT SLAB ON PLANAR SLOPE

Active and Passive Plastic States.
Example No. 2: Submarine slope

We first consider the example of a uniform submarine sediment layer on an 'infinite' slope as sketched in Fig. II.7-10. Assuming that any deformation takes place in a dip-parallel section (x_2, x_3-plane), the problem becomes of the 'plane' type. It is dealt with in Chapter II.2 (Example No. 5), where it is shown that the conditions of stress equilibrium alone determine two of the three stress components, namely the slope-parallel shear stress τ_{23} and the effective 'overburden' stress σ_3' (Fig. II.7-10). Hence, in the plastic limit state, condition II.7-3 contributes the value of the third stress component, the slope-parallel effective pressure σ_2'. Since equation II.7-3 is quadratic in the stresses, it provides, two different values for σ_2':

$$\sigma_2' + \tau_o \cot\varphi = (\sigma_3' + \tau_o \cot\varphi)(1+2\tan^2\varphi) \pm \frac{2}{\cos\varphi}\sqrt{(\sigma_3' + \tau_o \cot\varphi)^2 \tan^2\varphi - \tau_{23}^2} \tag{II.7-5}$$

The smaller value for σ_2', corresponding with the negative sign of the square root, is associated with the 'active' plastic state ('weak' solution), while the larger value, corresponding with the positive sign, is associated with the 'passive' plastic state ('strong' solution). In the σ',τ-plane the two plastic solutions map as two tangential stress circles that have the point σ_3', τ_{23} in common. These active and passive limit circles are shown in Fig. II.7-11, where it is also shown how easily the active and passive values ($\sigma_2'^a$ and $\sigma_2'^P$) of σ_2' can be determined graphically.

This duality of possible plastic states may seem strange, but actually it conforms admirably with reality. In the active state the sediment on the slope is in a 'hanging' state, whereas in the passive state it is pushing downslope against some buttress ('buttressed' state).

The following simple model may be useful in acquiring a better understanding of the active and passive state concepts. Figure II.7-12A shows an ordinary spring balance. The vertical displacement of the balance pan is made difficult by extra friction in the guides. Let us assume that the balance is at rest under the load P. It is then clear that we can – normally – increase or decrease this load in small steps until the balance starts moving – the analogue for the onset of plastic flow. When we increase the load, the internal friction in the guides will support the extra load until a certain critical value of the load is reached. Then the slightest further increase in load will cause motion. In this critical state the total load is still carried by the spring pressure F and the fully activated friction in the guides. Similarly, by decreasing the load P below its initial value (Fig. II.7-12B), friction will be acitivated in the opposite direction, now balancing the excess spring pressure until again a

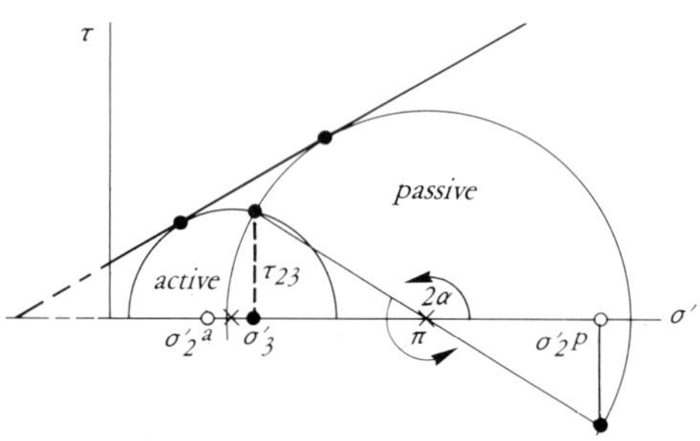

Stress points representing stresses on mutually perpendicular planes lie symmetrically with respect to centre of stress circle since the associated radii include the angle π.

Fig. II.7-11
'ACTIVE' AND 'PASSIVE' PLASTIC STATE

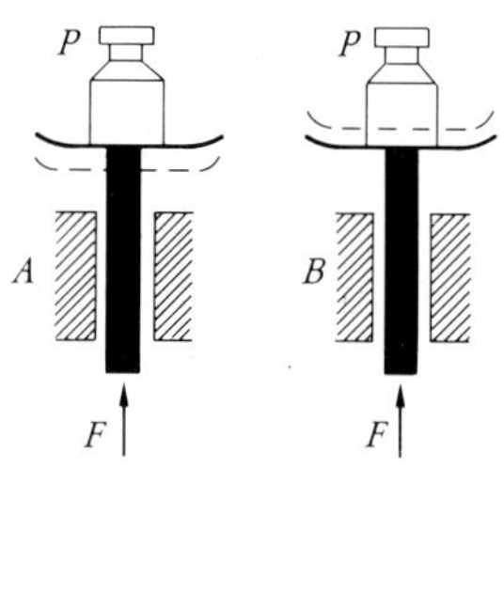

Fig. II.7-12
SPRING BALANCE WITH EXTRA FRICTION IN GUIDES ILLUSTRATING ACTIVE AND PASSIVE STATES
(after Sokolovski, V.V; 'Statics of soil media', London, 1960)

critical state is reached when any further reduction of load will cause the balance pan to move in an upwards direction. In both cases the spring pressure F was the same and it was only the action of the internal frictional resistance that brought about two different critical loads. It should also be noted that deformation in the high load pressure ('passive') state A produces a shortening of the system, while the low load ('active') state B leads to extension. Similarly, in the sloping sediment layer considered above, one may expect extensional or shortening deformations to be associated with the active and passive state, respectively.

Example No. 3: Active and passive states separated by plane of weakness; stress jumps

In Example No. 2 the whole sediment layer was either in the active or passive plastic state. Under certain conditions the two states can coexist. In the following (geologically important) case the two states may even exist in adjacent regions.

We consider a frictional material that has uniform strength properties φ and τ_o with the exception of a very narrow planar zone – termed a 'plane of weakness' (Γ in Fig. II.7-13A) – where the shear strength is smaller (φ', τ_o'). The corresponding limit lines for the material outside and at the plane of weakness are shown in Fig. II.7-13B. In nature, bedding planes, faults and joints may form such planes of weakness. Now we assume that loading has produced a normal stress (σ_3' in Fig.-II.7-13A) and a shear stress (τ_{23}) on the plane of weakness Γ, which satisfy condition II.7-1c when the reduced values φ', τ_o' are substituted for φ and τ_o. The stress point σ_3', τ_{23} then lies on the limit line for the weak material (Fig. II.7-13B). What can be said about the stresses in the 'strong' material in the vicinity of Γ? First of all, it follows from the requirement of stress equilibrium that, in the absence of pore pressure changes across Γ, σ_3' and τ_{23} have the same magnitudes at both sides of Γ as is indicated in Fig. II.7-13A (Section II.2.1; Example No. 1, 'stresses along fault planes', Fig. II.2-6). Therefore, as far as equilibrium is concerned, admissible stress states in the immediate vicinity of Γ will differ only in the normal stress σ_2' that acts parallel to the plane Γ. The corresponding stress circles in the σ', τ-plane will all have in common the stress point σ_3', τ_{23} on the limit line for the weak material. The circles may lie below the limit line for the 'strong' material, this indicating that outside the plane of weakness the material is not in the limit state and is still behaving elastically, or a stress circle may touch the upper limit line (Fig. II.7-13B), this representing a plastic state.

We note (Fig. II.7-13C) that two critical circles can be drawn through the stress point σ_3', τ_{23}, representing the active and the passive plastic state, respectively. In the active state the value of $\sigma_2' = \sigma_2'^{a}$ is smaller than σ_3', in the passive state the corresponding value $\sigma_2'^{p}$ is considerably larger than σ_3'. It may well be possible for the active state to exist on one side of Γ and the passive state on the other, or the plastic state may be confined to one side of the plane of weakness, the other side being in the elastic state. This implies that the normal stress σ_2' undergoes a jump-like change across the plane of weakness. Such a discontinuous change in normal stress is statically feasible since it does not disturb static equilibrium. However, it also has to be compatible with the associated deformations along the plane of stress discontinuity. In the present case the plane of weakness, which is itself in the limit state, allows the adjacent materials to glide along and to accommodate in this way the differences in shortening or extension that may have been caused by the difference in σ_2'. Without any gliding along the plane of stress discontinuity a discontinuous change in straining could not exist since it would be accompanied by infinite shear stresses on the plane of discontinuity. Practical implications of stress discontinuities for the theory of tectonic faulting are discussed in Part I (Section I.2.7 and close of Section I.2.8).

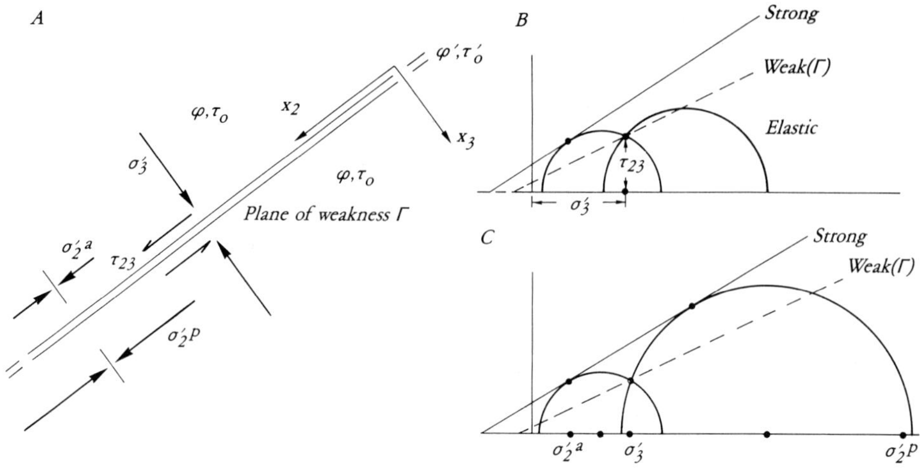

Fig. II.7-13 PLANE OF WEAKNESS IN CRITICAL STATE SEPARATING ACTIVE AND PASSIVE PLASTIC STATES

Example No. 4: Horizontal compression and extension

The existence of active and passive states can be demonstrated in a rectangular narrow container filled with dry sand (see photograph in Fig. II.7-14A) when one side wall can be moved to increase or reduce the horizontal normal stress in the sand pack. When the 'sand box' is filled, the vertical normal stress in the pack equals the overburden weight and is the maximum principal stress. When the bottom and side walls of the sand box are well lubricated, a small horizontal displacement of a side wall will change the magnitude of the horizontal normal stress but will leave the magnitude and the direction of the vertical principal stress practically unchanged.

If the side wall is allowed to recede a little, we soon notice signs of plastic yielding (Fig. II.7-14B). Starting again with a fresh pack and moving the side wall in the opposite direction, we have to raise the horizontal stress considerably before plastic deformation becomes noticeable again (Fig. II.7-14C). In both yield states the vertical normal stress is the same and the horizontal shear stress is practically zero. Only the horizontal normal stress differs considerably. In the first limit state – the active one – it is the minimum principal stress, while it becomes the maximum principal stress in the second yield state – the passive state.

The plastic deformations produced in the experiment show what may be considered the most typical feature of plastic deformation – *slip*. In the active plastic state a sand wedge is sliding down a steeply inclined plane (Fig. II.7-14B), while in the passive state a much larger wedge is pushed over a gently inclined plane (Fig. II.7-14C). Obviously, we must suspect that the inclination of the slip planes and the direction of the maximum compressive stress are directly related. This relationship is indeed the essence of the second part of Coulomb's concept of frictional plasticity, which will be discussed next.

Example No. 4: Horizontal compression and extension

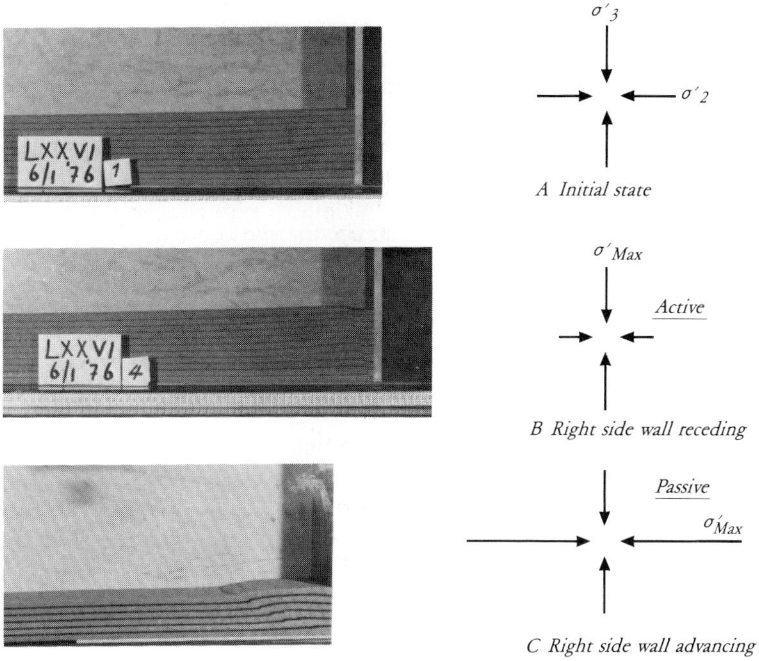

Fig. II.7-14 CHANGE IN HORIZONTAL NORMAL STRESS IN UNIFORM SANDPACK
(Side walls and bottom of sand box lubricated)

4. 'COULOMB SLIPS' AND THE 'DOUBLE-GLIDING' MODEL

According to Coulomb's concept, which was more explicitly formulated by O. Mohr in 1882, the plastic deformation and failure of cohesive or cohesionless frictional materials take place by discrete slips in well-defined directions. At any point where a material is in the limit state a pair of potential slip elements exists, namely the two elements on which Coulomb's yield condition II.7-1c is satisfied. These two conjugate elements, which we have already discussed (see Figs. II.7-8 F,G), intersect each other along the direction of the intermediate principal stress (σ'_{Int}) and their traces in the plane of the greatest and smallest principal stresses include the acute angles $\pm$ ($\pi/4$ $-$ $\varphi/2$) with the direction of σ'_{Max} (Fig. II.7-15). The sense of possible motion along the elements is indicated in Fig. II.7-15. It coincides with the orientation of the shear stress components acting on these elements and, therefore, by

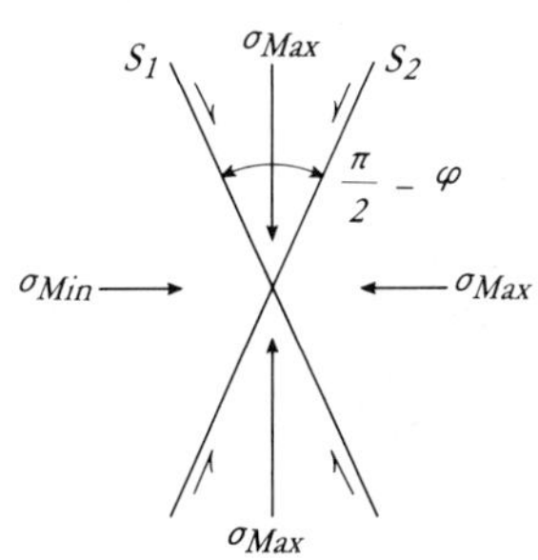

Fig. II.7-15
LOCAL COULOMB SLIP ALONG CONJUGATE PLANES

the definition of mechanical work (force × displacement), the shear stresses perform work in producing slips. Slip directions opposite to those shown in Fig. II.7-15 are impossible, as they would imply gain of mechanical energy which would contradict the energy dissipative character of plastic deformations (Section II.8.3, Examples no. 1, 2).

Typical of slip in frictional materials is that the slip directions are determined by the principal stress directions rather than by material anisotropy, as is the case in intracrystalline gliding. The simple angular relationship between slip and σ'_{Max} directions has found many successful applications in soil engineering and has been supported by numerous experiments and by observations of natural slip phenomena. The simple sandbox experiments of Fig. II.7-14 provide a geologically interesting illustration of the slip concept: The 'normal fault' in Fig. II.7-14B includes the same acute angle with the vertically directed σ'_{Max} axis as the 'low-angle thrust' with the horizontal direction of σ'_{Max} in Fig. II.7-14C (Hubbert, 1951). The simultaneous formation of two conjugate slips in a horizontally compressed sandpack is shown in Fig. II.7-16. (A slight asymmetry in dip angles is caused by a small bottom friction which causes a slight deviation of the direction of maximum compressive stress from the horizontal.)

It hardly needs mentioning that the concept of plastic slip and the mechanical theory based on it are not limited by scale. The simple sandbox experiments of Figs. II.7-14 and II.7-16 may therefore be interpreted in terms of large-scale tectonic events, thereby indicating the importance of Coulomb's slip concept for the mechanical interpretation of tectonic faults.

Fig. II.7-16 CONJUGATE SLIPS IN SANDPACK UNDER LATERAL COMPRESSION (PASSIVE STATE)

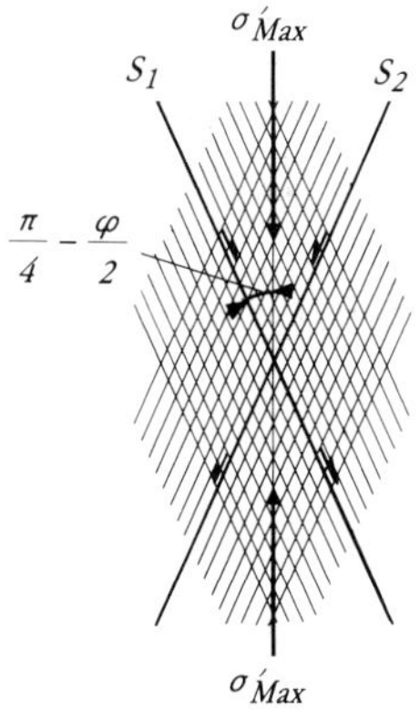

Fig. II.7-17
PLASTIC DEFORMATION BY SLIDING ALONG SLIP LINES
– Discontinuous deformation (faulting): sliding along single slip lines (S_1, S_2)
– Continuous deformation: successive sliding along two sets of closely spaced slip lines

Plastic deformation of frictional material is not restricted to deformations by one or a few discrete slips. Many slips along wider or closer spaced slip surfaces can operate at the same time. Considering the two sets of conjugate slip planes S_1 and S_2 (Fig. II.7-17) which are associated with a certain σ'_{Max}-direction, sliding may be distributed over individual elements of the two sets such that it produces any discontinuous or continuous plane deformation. Figure II.7-18 illustrates how this mechanism of '*double gliding*', i.e. gliding along both sets of conjugate slip elements, operates. Obviously, *any plane deformation* can be produced by this mechanism.

In Figure II.7-18 gliding takes place along planes. This implies constant directions of principal stresses and a constant angle of internal friction throughout the material. In a more general situation, the directions of the principal stresses and/or the frictional shear strength may vary in a rock mass that has been loaded up to the limit state. Accordingly, deformation will be accomplished by gliding along curved surfaces as sketched in Fig. II.7-19. It is important to note that curved *conjugate shear surfaces will have opposite curvature.*

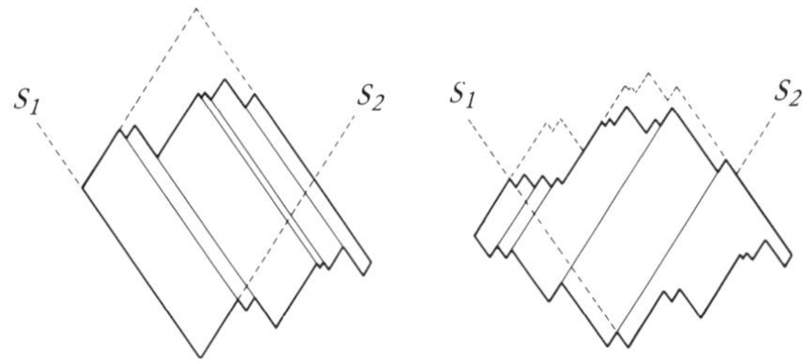

Fig. II.7-18 PLANE DEFORMATION BY 'DOUBLE GLIDING'
(after de Josselin de Jong, 1959)

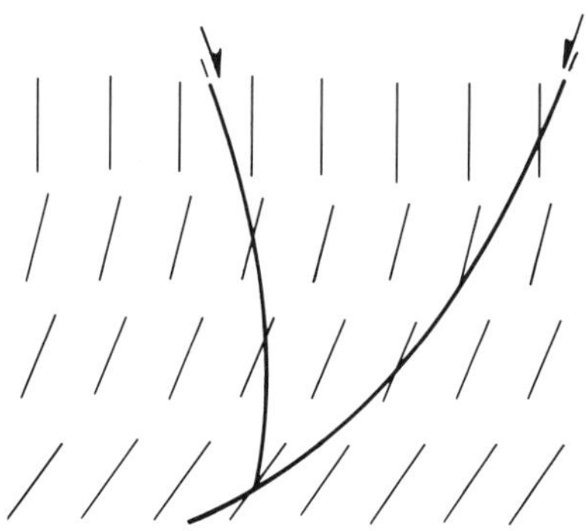

Fig. II.7-19
NON–UNIFORM FIELD OF σ_{Max} DIRECTIONS AND ASSOCIATED CONJUGATE SHEAR SURFACES

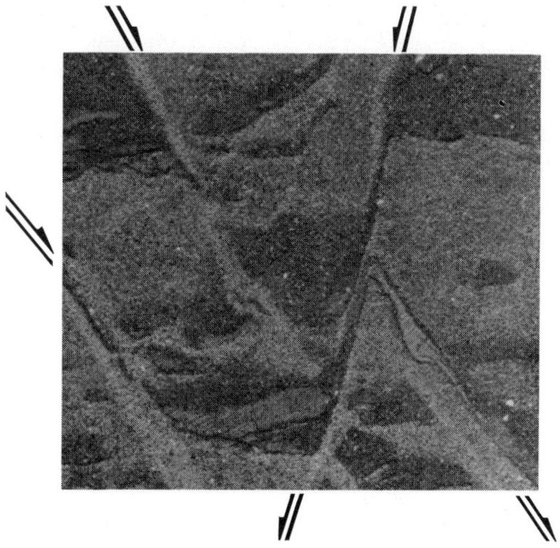

Fig. II.7-20
INTERSECTING SETS OF CONJUGATE SLIPS IN SAND/CLAY SEQUENCE
(Frechen, W. Germany; Half natural size.)
Note the continuity of clay 'smears' (dark).

In nature any type of plastic slip deformation can be found, ranging from a simple slip to 'continuous' deformations where closely spaced slips can only be detected under magnification. The illustration in Fig. II.7-20 shows a sample of slightly consolidated sediment that was deformed by double gliding, while the photograph of Fig. II.7-21 shows sets of slips in the limestone breccia of a major tectonic fault. A single slip in a plastic material can 'die out' by becoming distributed over a large number of slip surfaces. A typical example of slip (or fault) termination inside the plastic material is shown in Fig. II.7-22, in which a single slip displacement is taken up by a fan of slip surfaces.

For a good understanding of the double-gliding mechanism in plastic deformations some further aspects have to be considered.

Two-Dimensional vs Three-dimensional Strain. First, we recall that the slip deformations take place in planes perpendicular to the direction of σ'_{Int}. There is no slip component parallel to this axis and the strain produced by slip is therefore plane strain. The two-dimensional character of slip-accommodated strain does not, of course, exclude truly three-dimensional straining before peak stress is first attained. Further, one has to note that the double-gliding model applies *locally* and the orientation of slip elements will vary with location in an inhomogeneous stress field with varying principal axes. Slip elements may or may not then coalesce into sets of smooth slip surfaces (discussed in Section II.7.6). In any case, conjugate slip elements have to intersect along the intermediate principal stress axis. However, when the intermediate principal stress in the limit state equals one of the other principal stresses, the orientation of the slip planes gains a degree of freedom. For example, in the axial compression of a rock sample that is laterally supported by uniform fluid pressure, one or two conjugate slip planes will develop that make an acute angle with the direction of axial compression, but have a quite arbitrary strike direction. On a larger scale such a situation may result in a mosaic-like pattern of differently striking slip planes.

Furthermore, it is not uncommon that in the field *more than two sets* of slip planes can be observed (see, for example, Aydin and Reches, 1982). In many cases this can be explained by a change of the regional stress field after a first tectonic episode or by local readjustments of stresses accompanying faulting. Various examples of this have been given in the chapters on tectonic faulting. For example, in fully developed wrench zones at least four different fault types (R, R′, P, P′-types) can be distinguished which have accommodated in *successive* operations truly three-dimensional straining and can be very well explained in terms of Coulomb's slip concept.

Moreover, when studying a case of multiple sets of slip surfaces (faults) one always should bear in mind that an evident *difference in principal strains does not necessarily imply a difference in corresponding stresses*. Consider, for example, horizontal strata under a constant overburden ($\sigma_{ov} = \sigma_{Max}$) that are subjected to a biaxial horizontal extension with different amounts of stretching. Let us assume that stretching in both directions is contemporaneous and the strength of the rock is the same for all horizontal directions. Then, as illustrated in the Mohr diagram of Figure II.7-23A, σ'_{Min} which acts parallel to the direction of greatest stretching will decrease until the associated stress circle touches the limit line. At that instant conjugate normal faults will form that strike perpendicular to the direction of greatest stretch. It is important now to note that further stretching in that direction is accommodated by these faults without further decrease in σ'_{Min}. While this extension continues (Fig. II.7-23B), stretching in the other direction causes σ'_{Int}, which acts along that direction, to decrease until it has also attained the value of σ'_{Min}. New normal faults may then form that strike in any direction. An analogous process may be expected when unequal horizontal shortenings are applied. Such a succession of distinct events of shear fracturing has been

Fig. II.7-21 FAULT ZONE IN CARBONIFEROUS LIMESTONE SHOWING BRECCIATION AND SLIP PLANES
(Sand Point near Weston-super-Mare, Avon, Great Britain)

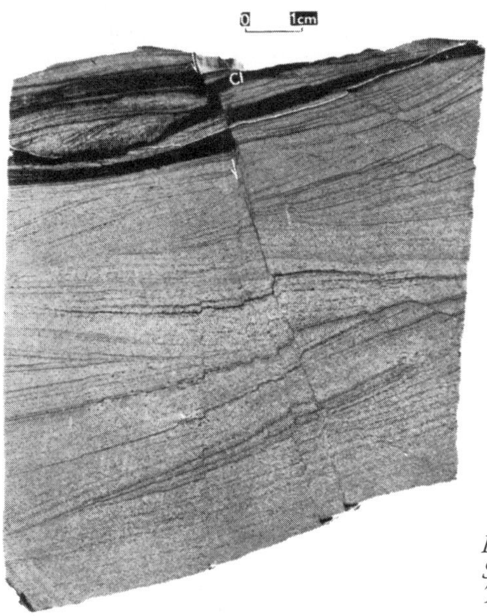

Fig. II.7-22
SMALL-SCALE EXAMPLE OF FAN-TYPE FAULT TERMINATION IN SILT/CLAY SEQUENCE

convincingly demonstrated by Reches and Dieterich (1983), who recorded the changes in the normal stresses associated with the shortening or extension imposed on various rock samples under constant strain rates in a stiff 'true' triaxial testing apparatus.

Independence from Intermediate Principal Stress. In the Coulomb theory and the associated double-gliding model both the critical state of stress and the magnitude of slip deformations are independent of the intermediate principal stress. To what extent is this a realistic assumption? Measurements on dense sand (Bishop, 1971) have shown that the friction angle at peak stress may increase by about 10% when the intermediate principal stress is raised from σ'_{Min} to σ'_{Max}. Interestingly, most of this modest increase in the friction angle at peak takes place while σ'_{Int} is still close to σ'_{Min}, with the difference between the two stresses not being more than about 15% of the limiting stress difference $\sigma'_{Max} - \sigma'_{Min}$. Similar results have been reported by Mogi (1972) for some carbonate rocks. He found a distinct increase in the peak friction angle with increasing σ'_{Int} when the effective confining pressure (σ'_{Min}) was not much higher than 25 MPa. Although for a dolomite this increase amounted to almost 20%, most of it took place while the value of the intermediate stress differed from the minimum stress by not more than about one third of the limiting stress difference. For higher effective confining pressures the effect of the intermediate principal stress was found to be of much less significance. Mogi suggested that the increase in frictional strength under an increase of the intermediate principal stress may be explained by the extra frictional resistance imposed on microcracks that are not parallel to the intermediate principal axis. In the range of relatively low or vanishing values of σ'_{Min} various authors have observed that in the extension test where $\sigma'_{Int} = \sigma'_{Min}$ the fracture angles $(45° - \varphi/2)$ were about 10% smaller than when the same material was tested under the condition $\sigma'_{Int} = \sigma'_{Max}$.

Although reliable data on the effect of the intermediate principal stress, i.e. data obtained by 'true' triaxial testing with stiff equipment, are still scarce, it would indeed seem very likely that in the upper few kilometres of the brittle crust, the angle of fault planes with the direction of the maximum compressive stress will be smaller when the minimum and intermediate principal stresses are about equal, than it is when these stresses differ considerably.

Rate-Independence of Plastic Deformation. A very important aspect of the double-gliding model is its rate independence. This major characteristic of plasticity sensu stricto, has already been discussed in Section II.7.1 ('Shear flow on microscale'). The stresses inside

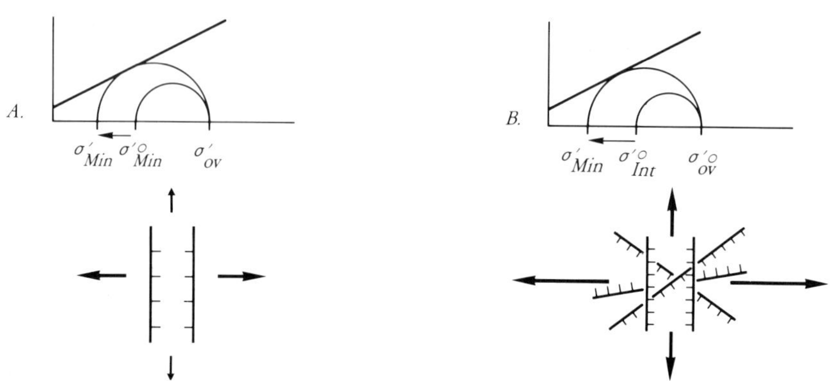

Greater horizontal extension leading first to limit state and normal faulting.

Smaller horizontal extension produces equality of horizontal normal stresses. New normal faults may have arbitrary strike direction.

Fig. II.7-23 LIMIT STRESSES IN BIAXIAL HORIZONTAL EXTENSION AND ASSOCIATED NORMAL FAULTS (Schematically shown in plan view).

the material determine the directions of the two gliding or shear motions, but leave the magnitude and rate of the actual slips or shears along S_1 and S_2 in Fig. II.7-18 completely undetermined. Instead these quantities are determined by the movements of the outer boundaries of the plastic body.

To illustrate this, we consider again a sandpack (Fig. II.7-14) which is subjected to horizontal compression in a rectangular box by moving the right hand wall. When the compression increases beyond the state shown in Fig. II.7-14C, at a certain stage of shortening the typical imbrication structure of Fig. II.7-24 will have formed. In sand, a material the behaviour of which very closely resembles that of an ideal frictional plastic material, exactly the same structure will form when, within reasonable limits, the same total shortening is produced at quite different rates. Therefore, we see that different states of internal deformation are associated with different states of the moving boundary, but that they are independent of the rate of these movements. Moreover, the internal deformation 'freezes' the moment the boundary movements stop. Hence, *time* enters only as a parameter that 'labels' subsequent positions of the moving boundary. If, instead of imposing boundary displacements, we had deformed the sandpack by a controlled loading procedure and measured the stresses inside the pack, we would have observed that the internal stresses change simultaneously with the boundary stresses. In other words, the material properties of a plastic solid sensu stricto do not affect the time history of deformations and internal stresses. The plastic solid is a material without a 'natural time'; its 'clock' is the succession of changes in the boundary conditions. In line with this, neither the double-gliding model nor any other model that relates stresses and plastic deformations will utilise material constants the physical dimensions of which contain the time.

This is in contrast with the behaviour of a viscous fluid, as has been discussed earlier (Section II.7.1), and it may be recalled that defining or measuring the material constant 'viscosity' makes reference to time. The reader may readily imagine how our sandbox experiment would develop if the sandpack were replaced by a viscous liquid. To start with, the fluid would be in the hydrostatic state, i.e. no shear stresses would exist (Chapter II.1; 'Hydrostatic state'). Moving the side wall would then disturb this state and set up a viscous resistance, i.e. shear stresses inside the fluid. These stresses are proportional to the rate at which the side wall moves and to the viscosity of the liquid. The shear stresses tend to prevent a uniform thickening of the shortening liquid layer and to produce some 'buckling' of the surface. Gravity, however, counteracts the formation of surface 'buckles' and in the course

Fig. II.7-24 IMBRICATION OF FRICTIONAL MATERIAL BY SLOW HORIZONTAL COMPRESSION
Structure is independent of compression rate.

of time will flatten any surface elevations produced. Since the shear stresses increase with the rate of boundary movements but gravity does not, the same shortening can be accompanied by strong surface 'buckling' or may even be achieved without visible disturbance of the surface at all, depending on the rate of the boundary movements.

The absence of real time as a parameter in the characterisation of the deformational response of the plastic material implies the complete neglect of typical time effects, such as creep flow and stress relaxation. Naturally, this is a simplification of real rock behaviour which will prevent a full description of deformation processes in a high-ductility environment, i.e. under conditions where large straining precedes the onset of slip. We shall discuss this point further in Section II.7.9 when assessing the range of application of plastic models to tectonic faulting.

Non-coaxiality of Stress and Incremental Strain. Next we have to draw attention to a peculiar feature of the Coulomb-plastic material, which becomes apparent when a deformation is accomplished by slips along a single set of parallel slip planes as illustrated in Fig. II.7-25A (monoclinic or 'card deck'-type deformation). Obviously, the orientation of these slip planes coincides with the direction of maximum shear rate which, by definition, is at an angle of 45° to the direction of greatest shortening rate $\dot{e}_I$. At the same time, the Coulomb slip direction has to make an angle of $45° - \varphi/2$ with the direction of σ_{Max}. Hence, in this special case the principal axes of strain rate (or incremental strain) and stress deviate from each other by the angle of $\varphi/2$. In the general case of plane deformation by the double-gliding mechanism, illustrated in Fig. II.7-25B, the $\dot{e}_I$ axis may deviate from the direction of σ_{Max} by an angle which lies between $+\varphi/2$ and $-\varphi/2$. Note that this deviation between the principal axes of stress and infinitesimal strain increments which is referred to in soil mechanics as *non-coaxiality*, is supposed to hold even for mechanically perfectly isotropic material. The existence of the phenomenon has been demonstrated experimentally by Drescher (1976). Mandl and Fernández Luque (1970) have shown theoretically that this almost paradoxical non-coincidence of corresponding principal axes of the two correlated tensors of strain rate and stress – no other tensor being involved – is perfectly compatible with material isotropy as long as the deformation is planar. Such plane deformation, be it only locally in a three-dimensional strain field, is inherent in Coulomb's slip concept, as we have discussed earlier.

It should be understood from Fig. II.7-25B that the angle of non-coaxiality is not a constant as has been assumed in some frictional plastic theories. Rather, it will depend on the respective contributions of the conjugate slip sets to the total deformation. In pure shear (Chapter II.5, Fig. II.5-3), the contributions of the two slip sets are equal and stresses and strain rates will be coaxial. In *simple shear*, however, we meet with the curious complication that non-coaxiality may be either a maximum, as illustrated in Fig. II.7-25A, or it may be found to vanish. In view of its far-reaching consequences for the theory of tectonic faults, this somewhat enigmatic point needs further comment.

Simple Shear: Kinematics vs Stress Control. Undoubtedly, the 'card-deck' mode simple shearing of Fig. II.7-25A is found to operate in nature. A typical example is sketched in Fig. II.7-26A which represents the stretching of a long sediment layer on a slope. The sediment layer is in the (active) limit state, induced by steepening of the slope, and rests on weaker material which easily accommodates the saw-tooth deformation of the base of the layer by a set of parallel ('normal') faults. Slip along these parallel fault planes produces a quasi-continuous simple shearing of a stack of fault blocks. The 'driving' stress σ'_{Max} is the result of the sediment weight and – with the exception of some disturbance of the stress field near the serrated boundaries of the sediment layer – is unaffected by the fault kinematics. No unyielding boundaries interfere with this process of simple shearing, which is *controlled solely by the stresses* (static control).

Quite different to this is the situation depicted in Fig. II.7-26B, in which simple shearing is imposed by the movement of parallel 'rigid' walls. Such *kinematic control* is applied in shearing experiments ('direct shear test') and does operate in nature on a wide range of scale (see Section I.2.8). For this process, the initial orientation of σ'_{Max} will normally be altered by the shearing action of the moving walls. In Fig. II.7-26B it is assumed that this stress acted at a right angle to the driving wall before shearing started – the common situation in direct shear tests. The imposed shearing will force the σ'_{Max} direction to rotate as indicated in the figure. The shearing will increase the difference between the maximum and minimum principal stresses until a limiting state of stress (eq. II.7-3a) is reached. At that moment first Coulomb slips tend to form as far as boundary constraints, local compaction of the material, reorientation of particles, etc. allow for this. The occurrence of such 'precursory' Coulomb slips or faults is well-established by experiments and field observations (Morgenstern & Tchalenko, 1967; Tchalenko, 1970; Mandl, de Jong & Maltha, 1977; Bartlett, Friedman & Logan,

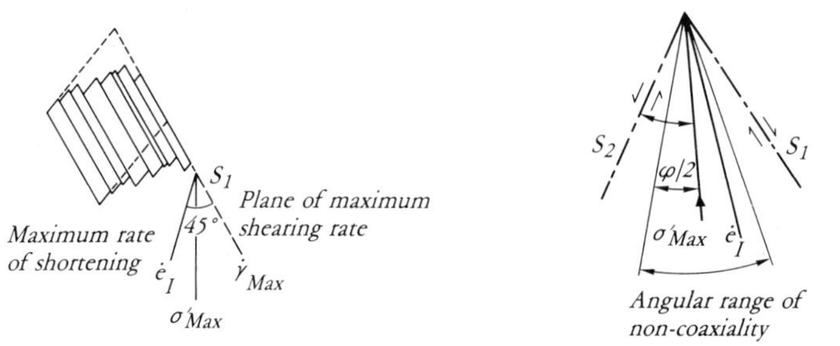

A. Single set of Coulomb slips

B. Conjugate Coulomb slips

Fig. II.7-25 NON-COAXIALITY OF THE PRINCIPAL DIRECTIONS OF STRESS AND STRAIN RATE

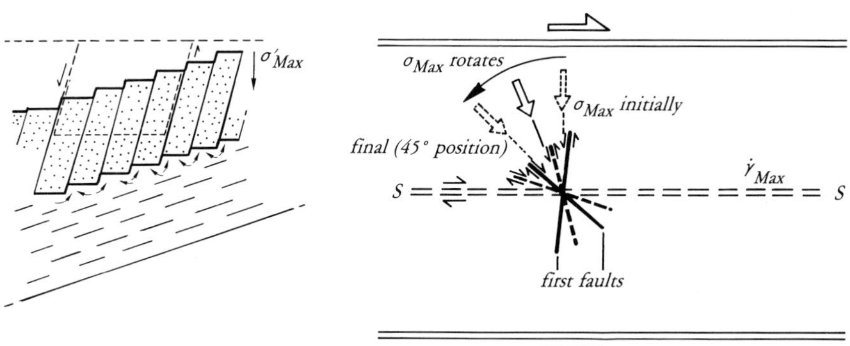

A. Normal faulting on steepening slope – 'Stress-controlled' simple shear

B. 'Kinematically controlled' simple shear

Fig. II.7-26 STRESSES IN SIMPLE SHEAR

1981). Figures II.7-27 and II.7-28 should give an impression of experimental evidence that was obtained in quite different shearing devices. Fig. II.7-27 shows in longitudinal cross-section a slab of plasticine that was sheared between saw-toothed platens, but not up to shear failure of the sample. Coulomb slips, synthetic with the imposed shear direction, are clearly visible. Figure II.7-28 shows en échelon slips which preceded the formation of a through-going shear zone in a sugar pack that was sheared in a ring-shear apparatus. The photograph shows the interior of the annulus after removal of the top plate and part of the pack. Shearing of the pack between top and bottom plates was terminated before peak stress was reached. Obviously, such precursory slip planes cannot accommodate continued shearing by the boundary walls, and σ'_{Max} has to rotate further (Fig. II.7-26B). One would now expect this rotation to continue until the inclination of $45° - \varphi/2$ with respect to the boundary is attained, but, surprisingly, this is not the case.

In fact, careful measurements in a ring-shear apparatus (Mandl, de Jong & Maltha, 1977) leave little doubt that the final inclination of σ'_{Max} towards the moving boundary is 45°, and therefore coinciding with the direction of greatest rate of shortening between the shearing plates. This has been observed for quite different frictional materials, both cohesionless and slightly cohesive, in dry and fluid-saturated states, at different loads (up to 1 MPa) and shearing rates[6)]. This result is supported by the observation made by Bartlett et al. (1981) in high-pressure shearing of limestone that the extension microfractures produced in shearing are inclined at $45 \pm 5°$ to the imposed shear direction. Since extension fractures grow parallel to the direction of greatest compressive stress, the observed orientation of the microfractures is indicative of the orientation of that stress.

When σ'_{Max} has attained the 45° orientation and shearing continues, the shear deformation concentrates in a shear band parallel to the moving boundary. In shearing experiments the shear band forms preferentially at the midplane of the shear apparatus where fixed and moving parts of the apparatus are in contact. In nature, strength inhomogeneities will play a major role in determining the location of shear bands. Now that the shear band (SS in Fig. II.7-26B) is bounded by planes of maximum shear stress rather than by Coulomb-type slip planes, Coulomb's slip concept may seem to be of no avail in the case of kinematically controlled simple shear. The contrary is true, however. Various experimental studies have shown that the kinematically controlled shear bands are actually formed by several sets of Coulomb slips which successively develop in a progressively narrowing zone. Particularly illuminating in this respect was Morgenstern and Tchalenko's study of micro-shear features in clay sheared in a conventional shear box (1967). A more detailed discussion of the fairly complex processes of kinematically controlled shear band formation is given in Part I of this book in connection with 'simple-shear tectonics' (Section I.2.8) and wrench faulting (Chapter I.4) – geological manifestations of kinematically controlled shear 'band' formations. Even in fully developed shear bands in experiments and in the field, traces of the Coulomb-slip history may be observed, as the two examples in Fig. II.7-29 illustrate.

Application to Tectonic Faulting. From our discussion of the Coulomb theory and its double-gliding generalisation, it will be clear that Coulomb slip surfaces may serve as models of actual tectonic faults of various types, excluding kinematically controlled simple shear zones. The various applications of the double-gliding model in previous parts of this book should provide sufficient proof of the usefulness of this mechanical model in explaining the processes of tectonic faulting. Here we summarise the *limitations* of the model.

First of all, application of the double gliding model requires that the rock mass or a known part of it is in the plastic limit state.

Secondly, the stress field, or at least, the field of principal stress directions has to be known. To determine the stress field, one has to prescribe the stresses along the entire boundary of the plastic region considered. We know that in *plane problems* the number of unknown stress components exceeds the number of equations that describe stress equilibrium by only one (Chapter II.2, eq. II.2-2a). Therefore, with one additional stress equation – the limit condition equation II.7-3 – the set of required stress equations is complete. Unfortunately, problems of practical interest where the boundary conditions can be exclusively described in terms of stresses are rare. (The submarine sediment layer on an infinite slope, example no. 5 in Chapter II.2 is such a 'statically determinate' problem.) Usually, stresses are only known along parts of a boundary, while conditions on the remaining parts are known in terms of displacements. Within this framework, as in any model of rigid/fric-

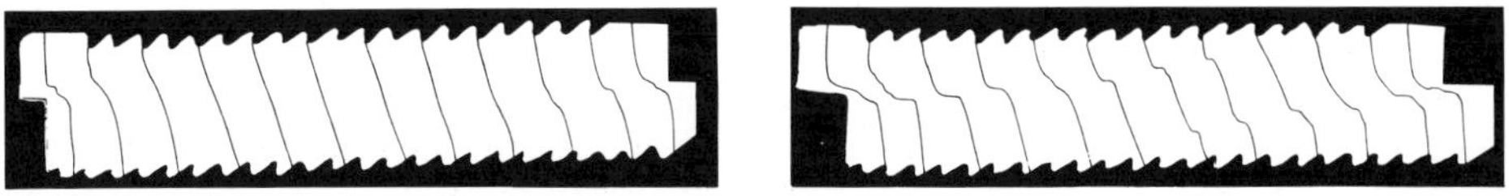

Fig. II.7-27 INTERNAL DEFORMATIONS IN DIRECT SHEAR TEST
(plasticine; after Hvorslev, 1960; reprinted from L. Šuklje (1969) by permission of John Wiley & Sons, Ltd.)

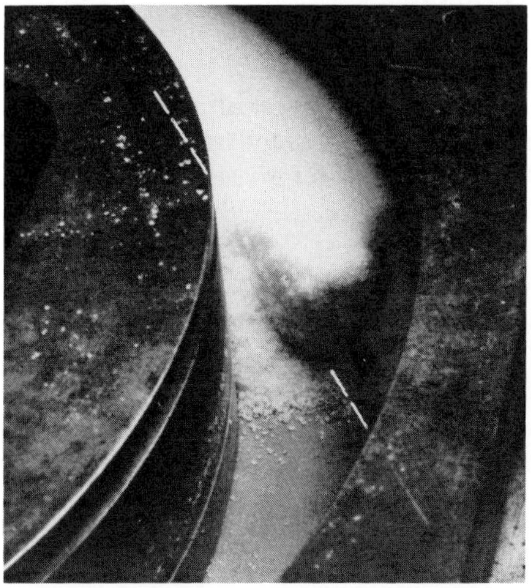

Fig. II.7-28
'PRECURSORY SHEARS' IN RING-SHEAR APPARATUS. Slips in sugar pack precede formation of shear zone.
(Shearing was stopped before shear load reached peak. Part of pack was sucked off to expose slip lines.)

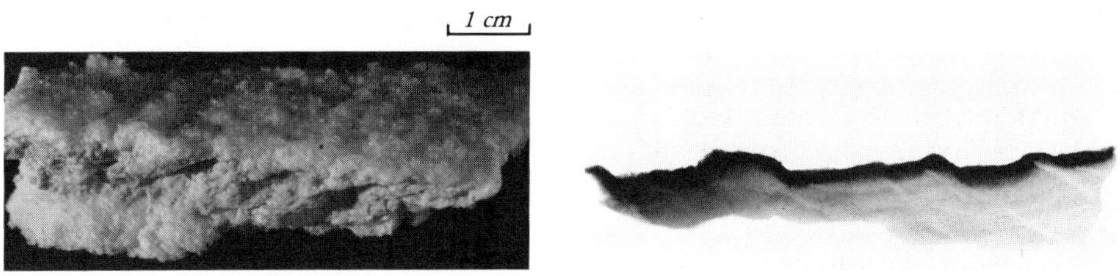

A. Sinistral shear zone in sugar pack (side view)

B. X-ray photograph of tectonic ripples on fault plane in sandstone (Hastings, Sussex, England). Dextral shearing produced slips transverse to fault plane.

Fig. II.7-29 SECONDARY SLIP PLANES.

tional plastic material behaviour (Fig. II.7-1C), kinematics and stresses are only loosely linked via a relationship between the directions of principal stresses and infinitesimal strain increments (Fig. II.7-25B). Within this framework, displacement boundary conditions are therefore of no use in determining the stress field, and a rigorous application of the model will be restricted to plane problems of the statically determinate type. (Note that even in the case of axi-symmetric deformation this symmetry has to be expressed in terms of stresses through stress/strain relationships.) However, this does not exclude many useful semi-quantitative applications of the model in cases where reasonable assumptions about the field of principal stress directions can be made.

Although the model enables the shape of potential faults in dip section to be determined, it does not provide information on which potential slip surface will actually be mobilised or on what the distance between neighbouring faults may be. Here, informations on boundary displacements may be helpful in determining locations of fault initiation.

The model also assumes that the shear strength is *isotropic*. In other words, the friction angle may change with position but should not vary with the orientation of potential slip elements. In Section I.2.3 such effects of strength anisotropy are discussed; these no longer enable the slip orientation to be determined from the simple Mohr diagram of Fig. II.7-8 F,G).

Finally, we wish to mention the very questionable procedure of determining potential Coulomb-type faults, which may be found in most textbooks on Structural Geology or Tectonics. Following this procedure, one first calculates a *purely elastic* stress field and then fits Coulomb slip lines into the field of principal stress trajectories without regard for the existence of a limiting stress state. Where the stress field has not yet reached a limiting state, no slip surfaces can exist. Where it has reached the limiting state, its further development is not governed by elastic stress/strain laws. However, we do not have to discard this approach completely, if we exercise some discretion, for the following reason: although the values of the stress components determined under the assumption of elastic behaviour may be quite different from what they should actually be in a plastically deformed region, the directions of the principal stresses may be reasonably correct when the elastic deformation started from a stress state with small or even vanishing difference between maximum and minimum stress (small or vanishing Mohr circle). In this case, a considerable build-up of stress difference in the elastic range will be required before the limit state can be reached. It would seem very likely that most of principal stress reorientation would have already occurred during the long elastic part of the total deformation.

Flow rules. Active slip lines may be so closely spaced and the individual slip movements may be so small that the resulting plane deformation may be considered to be *continuous*. In this case the model could be better referred to as a double-shearing model (Spencer 1964). The associated mathematical formalism for the description of the displacement field has been presented by Spencer (1964). The equations, which are presented in Fig. II.7-30, relate velocity components and their gradients to the direction of the maximum principal stress. These differential equations replace the relationships between strain rates and stresses that hold in the conventional theory of rigid/plastic materials (e.g. metals) in which corresponding axes of principal strain rates and stresses coincide (coaxiality) [7)].

Spencer's equations reduce to the well-known (Geiringer) equations for a frictionless ideal plastic solid when $\varphi = 0$.

When the stress field is known, the velocity equations enable the velocity and deformation fields to be calculated by numerical methods, provided the boundary conditions are properly prescribed in terms of velocity components. Three types of boundary conditions are possible: (1) Both velocity components are prescribed along a line that does not coincide with a slip line [8)] ('Cauchy problem'); (2) One velocity component only is given along each of two conjugate slip lines ('Goursat problem'); (3) A condition for the velocity is given along a non-slip line, while a velocity component is prescribed along a slip line. So far, we know of only one application of the velocity equations to a geological problem; this has dealt with gravitational faulting and deformations on a delta slope (Crans et al. 1980, 1981).

In concluding this section on the double-gliding or double-shearing model, we recall the *basic assumption* of the model that any deformations, including internal rotations, result solely from gliding or simple shearing along the conjugate slip lines. Naturally, rigid body motions imposed from outside on the whole of the plastically deformed region will have to be added to the internal deformations [9].

V_α V_β Velocity components (parallel projections of velocity vector $\vec{V}$ on local slip directions which make obtuse angle)

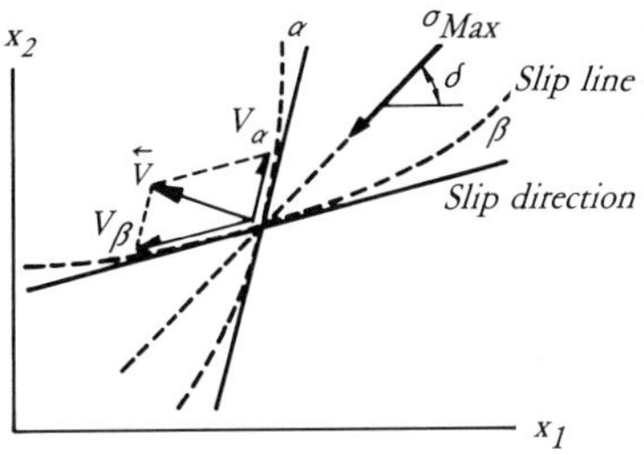

Velocity equations (A.J.M. Spencer, 1964):

$$\frac{\partial V_\alpha}{\partial \alpha}\cos\varphi + (V_\alpha \sin\varphi - V_\beta)\frac{\partial \delta}{\partial \alpha} - \sin\varphi \cdot \Omega = 0$$

$$\frac{\partial V_\beta}{\partial \beta}\cos\varphi + (V_\alpha - V_\beta \sin\varphi)\frac{\partial \delta}{\partial \beta} + \sin\varphi \cdot \Omega = 0$$

where $\frac{\partial}{\partial \alpha}$, $\frac{\partial}{\partial \beta}$ *mean directional differentiations along α- and β-directions, and*

$$\Omega = \frac{\partial \delta}{\partial t} + V_\alpha \frac{\partial \delta}{\partial \alpha} + V_\beta \frac{\partial \delta}{\partial \beta}$$

Fig. II.7-30 PLANE DEFORMATION OF INCOMPRESSIBLE PERFECTLY FRICTIONAL-PLASTIC MATERIAL

5. THE ELASTIC/FRICTIONAL PLASTIC MATERIAL

We have seen that a complete analysis of a plastic deformation field by means of the double-shearing model requires a knowledge of the extent of the plastic region and the field of the principal stress directions. However, such information is available in only a few very special cases. Normally, tectonic loading will not induce the plastic limit state simultaneously at every point of the loaded rock mass, but rather it will generate plastic regions which gradually grow and spread out as the limit of the elastic deformability of the rock is reached. Obviously, therefore, modelling of *growing plastic zones* and calculating the stresses and deformations inside and outside these zones requires that elastic deformation behaviour be incorporated in the material model.

In its simplest form, the Prandtl model, illustrated by the analogue of Fig. II.7-1D, such an elastic/plastic model assumes linear elastic deformation up to the limit state. Thus below the limit state the stresses are related to the strains by Hooke's law (see, for example Fig. II.6-1). In the limit state any further increment in strain is the sum of a purely plastic strain increment δe^p and a purely elastic increment δe^e. Unloading from the limit state is purely elastic and is controlled by the same elastic moduli as loading. In addition, it is assumed here that volume changes are of a purely elastic nature. The sum of the plastic increments of normal strains (eq. II.5-2) therefore vanishes:

$$\delta e_1{}^p + \delta e_2{}^p + \delta e_3{}^p = 0 \qquad \text{(II.7-6)}$$

We have already stated (Section II.7.1) that a characteristic of ideal plastic behaviour is that one-to-one correspondence between the stress components and the strain increments does not exist. The relationship between stresses and kinematical quantities therefore reduces to a one-to-one correspondence between *directions* of principal stresses and principal plastic strain increments.

To derive these relationships for the *plastic* strain increments in a frictional plastic material, it would seem most natural to invoke the double-shearing model. Unfortunately, the corresponding relationships derived by Spencer (1982, eq. 6.13) are complicated by the occurrence of various non-linear terms. So far, it seems, Spencer's equation have not been used in computer programs for the solution of elastic/frictional plastic problems. We therefore follow here the conventional and simpler approach of utilising the *St. Venant/Levy/von Mises flow rules* for perfectly plastic materials; these were originally introduced to characterise the plastic deformation of metals. For plane strain these plastic flow rules are exactly equivalent to the assumption of coaxiality of corresponding principal directions of stress and plastic strain increments. Admittedly this disagrees with our previous statement about the possible non-coaxiality of strain increments and stresses, inherent in the double-gliding mechanism. However, this inconsistency appears less serious when we note from Fig. II.7-25 that non-coaxiality is a maximum when only one set of slip lines is activated, and completely disappears when the two conjugate sets contribute equally to the deformation. The latter mode we expect to be approximated by the actual deformation that takes place in a growing plastic zone, which, for the greatest part of its boundary, is still surrounded by elastic material. It is difficult to see how in such a contained plastic region one set of slip lines could predominate.

From the assumption of plane strain, plastic incompressibility (eq. II.7-6), and coaxiality of strain increments and stresses follow the plastic flow rules [10] for the normal and shear components:

$$\begin{aligned} \delta e_i{}^p &= \lambda \cdot s_i \\ \delta e_{ij}{}^p &= \lambda \cdot \tau_{ij} \qquad (i \neq j) \end{aligned} \tag{II.7-7}$$

where s_i are deviatory stress components. For example:

$$\begin{aligned} s_2 &= \sigma_2 - (\sigma_2 + \sigma_3)/2 \\ s_3 &= \sigma_3 - (\sigma_2 + \sigma_3)/2 \end{aligned} \tag{II.7-8}$$

where the subscripts denote the co-ordinate axes in the plane of deformation. In a straightforward generalisation to general modes of straining (including non-planar ones) the same flow rules are applied with subscripts referring to all three co-ordinate axes (i,j=1,2,3).

The parameter λ is an arbitrary coefficient of proportionality rather than a material constant. It is always positive, since in plastic deformation mechanical work is always expended and never gained [11].

When, during plastic deformation, the stress components in the limit state are kept constant, as, for instance, in the case of simple shearing under constant confining pressure, the state of elastic strain will not change. Generally, however, stresses in the limit state may change and such changes will also be accompanied by elastic strain increments. For example, plastic simple shearing of a frictional material can be carried out under a changing confining pressure. The associated limit shear stress (τ_y in Fig. II.7-1D) will change accordingly. The elastic strain increments induced in an isotropic material in the limit state by changes in effective normal stresses and shear stresses are determined by Hooke's law:

$$\delta e_i^e = \frac{1+\nu}{E}\,\delta\sigma_i' - \frac{\nu}{E}\,\delta(\sigma_1' + \sigma_2' + \sigma_3') \qquad (i=1,2,3)$$

$$\delta e_{ij}^e = \frac{1}{2G}\,\delta\tau_{ij} \qquad (i\neq j) \qquad \text{(II.7-9)}$$

By adding the elastic strain increments to the plastic ones (eq. II.7-7) the Prandtl-Reuss equations are obtained for the isotropic, plastically incompressible, linear elastic/perfectly plastic material in the limit state:

$$\delta e_i = \delta e_i^e + \delta e_i^p = \frac{1+\nu}{E}\,\delta\sigma_i' - \frac{3\nu}{E}\,\delta\sigma' + \lambda s_i \qquad (i=1,2,3)$$

$$\delta e_{ij}^e = \frac{1}{2G}\,\delta\tau_{ij} + \lambda\tau_{ij}\ ;\ (\sigma' = (\sigma_1' + \sigma_2' + \sigma_3')/3) \qquad (i\neq j) \qquad \text{(II.7-10)}$$

The Plastic Potential. At this point it may be convenient to introduce the concept of 'plastic potential', which was first introduced by von Mises in 1928 and which has played a crucial role in the development of plasticity theories and, in particular, of theories of frictional granular materials.

It may be easily verified that the plastic strain increments II.7-7 in the x_2, x_3-plane are obtained by differentiating the stress function

$$G \equiv \tfrac{1}{4}(\sigma_2' - \sigma_3')^2 + \tau_{23}^2 - K^2 = 0 \qquad \text{(II.7-11)}$$

with respect to the corresponding effective stress components and multiplying the result by λ. (In this the shear stresses τ_{23} and τ_{32} have to be treated as formally different [12]).

The relationship II.7-11 is von Mises' yield condition for cohesive, perfectly plastic material in plane deformation. In contrast with Coulomb-Mohr's limit condition II.7-3, it does not contain the mean effective stress, since the shear strength of a purely cohesive material is assumed to be unaffected by a change in the mean effective normal stress. Accordingly, the theory for a purely cohesive material is considerably simpler than the plasticity theory for a frictional (i.e. pressure-sensitive) material.

Following von Mises' procedure, Drucker and Prager (1952) have derived plastic flow rules for frictional materials by taking Coulomb-Mohr's limit condition as plastic potential [13]. Because of this association with the limit condition these flow rules are commonly referred to as *'associated' flow rules*. In contrast, the plastic flow rules II.7-7, when used in combination with Coulomb-Mohr's limit condition, are *'non-associated' flow rules*.

Example No. 5: Differential uplift

Drucker-Prager's plastic flow rules have been used tentatively by J.N. Thomas and J.V. Walters in a finite-element analysis of stresses and plastic zones that develop in a uniform sedimentary overburden when a (rigid) basement block is vertically uplifted. The geometry of the problem and the initial stress conditions are shown in Fig. II.7-31A. One may expect that uplift of the basement block will initiate plastic zones at the corners of the rising block from where they will grow in an upwards direction (Fig. II.7-31B). Because of the symmetry of the configuration, only one-half of the geometry of the problem needs to be considered (Fig. II.7-32). The uplift itself was simulated by 50 incremental vertical displacements.

Different stages in the development of the plastic zones are presented in Figs. II.7-32A-D. In addition to the plastic zone that emanates from the corner of the uplifted block, a plastic extension region was formed at the free surface in response to the upwards-bending effect of the basement uplift, indicating the likely situation of a shallow graben in a comparable situation in nature. The two zones coalesce (Fig. II.7-32C) and the plastic region finally attains the shape shown in Fig. II.7-32D. Typically for this problem, the plastic region encloses an elastic 'core' attached to the uplift block. Figure II.7-32E displays in magnitude and direction the field of σ_{Max} at the moment when the plastic zone has been fully developed. Into this stress field one may fit Coulomb slips as far as the associated slip motions are compatible with the kinematic boundary conditions. In the present case only slip lines of the family shown in Fig. II.7-32F could represent potential faults. They would be steep reverse and upwards-convex faults. Such faults are actually observed both in the field and

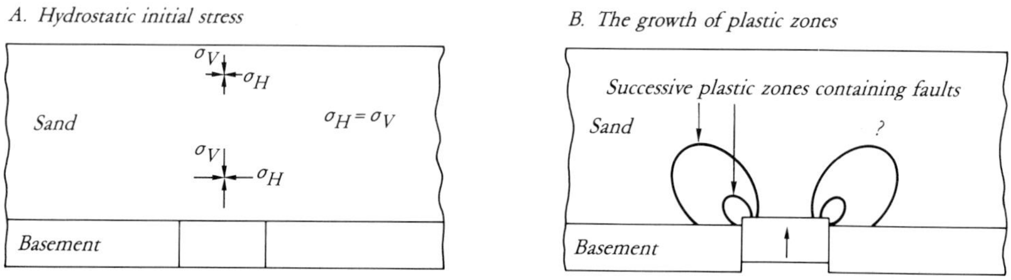

Fig. II.7-31 DEVELOPMENT OF PLASTIC ZONES IN SEDIMENT ABOVE BASEMENT UPLIFT

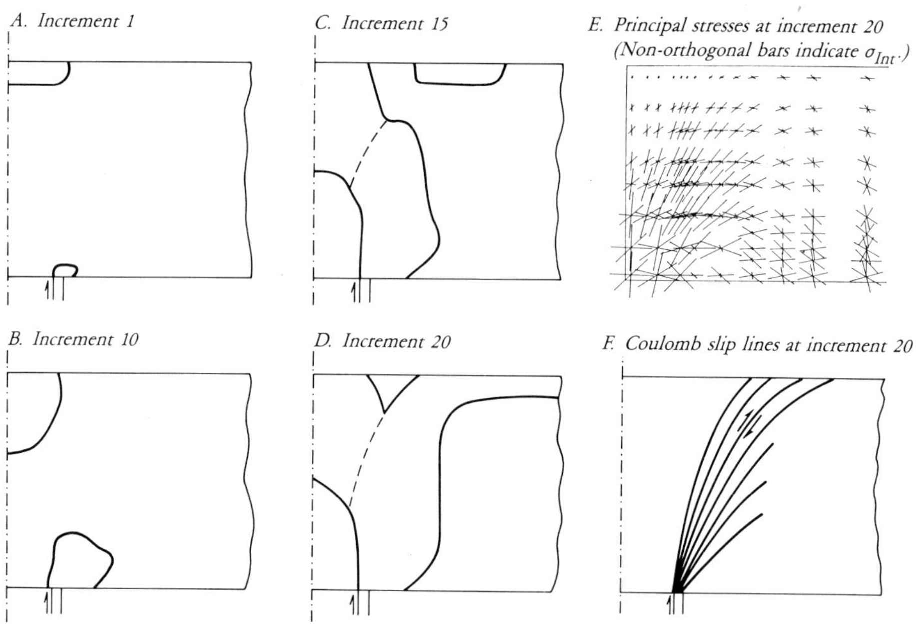

Fig. II.7-32 PLASTIC ZONE GROWTH ABOVE BASEMENT UPLIFT
(One half of region shown); Finite element simulation by Thomas and Walters, 1979 (unpubl.).

Configuration of sandbox.

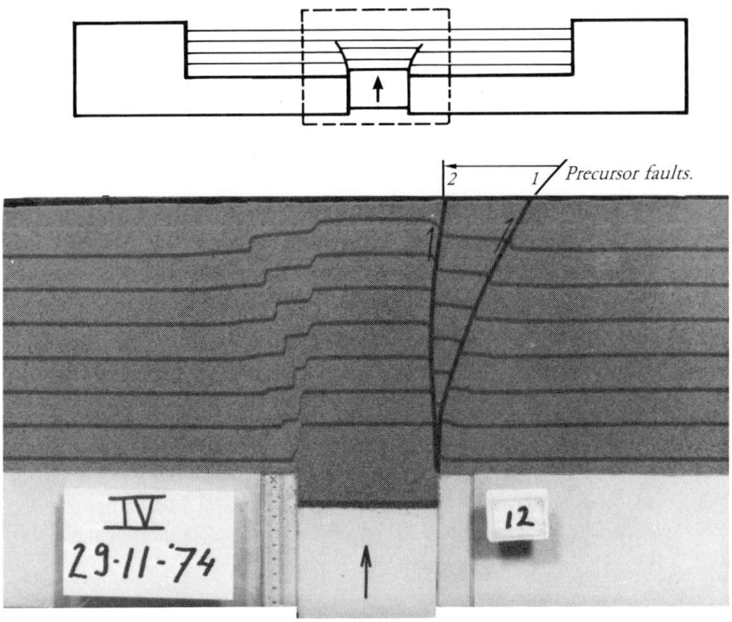

Fig. II.7-33 SANDBOX EXPERIMENT SHOWING DEVELOPMENT OF PRECURSOR FAULTS DURING UPLIFT

in sandbox experiments (Fig. II.7-33) in which they precede the formation of a final near-vertical main fault.

Incidentally, the orientation of σ_{Max} inside and outside the fully developed plastic region is almost identical with the σ_{Max}-orientation determined by Sanford (1959) in a purely elastic uplift analysis. Although the two solutions differ considerably with respect to stress magnitudes the agreement in orientation of the principal stresses is somewhat surprising. We therefore recall (see Section II.7.4) that such agreement may be explained by the fact that even in the elastic/plastic analysis most of the drastic redistribution of principal stress orientations, caused by the uplift, takes place before the material attains the limit state. This is a consequence of the isotropic state of stress that was chosen as the initial state, which in Mohr's stress plane maps as a point on the axis of normal stresses. The build-up of a limiting stress circle therefore requires extreme stress changes in the elastic range, accompanied by drastic reorientations of principal stress directions.

The dilatancy problem. In spite of the reasonable result obtained in the uplift example, it has to be pointed out that the associated flow rules have serious shortcomings. In particular, they predict that any increment of plastic shearing will be accompanied by a proportional increase in specific volume of the sheared material: [13)]

$$\delta V^p / V = -\delta\gamma^p_{Max} \cdot \sin\varphi \tag{II.7-12}$$

where the relative volume change $\delta V^p/V$ is counted as negative for increasing volume and $\delta\gamma^p_{Max}$ is the maximum change in shear angle (measured in radians), i.e. twice the corresponding tensorial shear strain (see eq. II.5-3c and Fig. II.5-2D). While for very small plastic shearing deformations it might be argued that this dilatation may have little effect on the growth of plastic zones and the stress system, such a persistent increase in volume cannot be tolerated in the case of continued shearing. The non-associated plastic stress/strain equations II.7-7, however, imply volume constancy under plastic shearing, which is not realistic either in view of the general tendency of rocks to dilate under shearing in the brittle range (see Section II.7.2, 'Dependence of strength parameters...'). It is therefore not surprising that the issue of associated vs. non-associated relationships has played a central role in the development of plastic theories for frictional materials.

This issue is of more than historical interest only and deserves some further comments since it is at the base of various modifications that have been proposed to achieve a better agreement with the true material behaviour in a given situation, without unduly complicating the mathematical apparatus. In dealing with a given tectonic problem, one therefore has to choose, for use in a computer program (e.g. of the finite-element type), from a number of alternative frictional/plastic stress/strain relationships, the one most apposite to the given situation. This requires some understanding of the physical-mechanical background of the flow-rule problem for frictional plastic materials.

Consider a limit state of stress represented in Mohr's stress plane by a stress circle the crestal point P of which lies on the straight shear strength line with inclination angle χ (Fig. II.7-34). We recall that $\tan \chi = \sin \varphi$, as was shown in Fig. II.7-9C. Next, we identify the stress plane with the corresponding strain plane by using the same co-ordinate axes for normal stresses and plastic normal strains, and shear stresses and shear strains. Then we associate the vector $\vec{\delta e}^P$ of incremental plastic strain with the stress vector $\vec{\sigma}'$ that defines the crestal stress point P in Fig. II.7-34. As explained in the figure, this association is motivated by the fact that the inner product of the two vectors defines the increment of plastically dissipated mechanical work δW^P when the incremental strain and the limit stress are coaxial. Figure II.7-34 also shows that the vector $\vec{\delta e}^P$ has the normal strain component $\delta e^P_{Max} + \delta e^P_{Min} = \delta V^P/V$. Considering in the same figure the angle ψ which $\vec{\delta e}^P$ makes with the ordinate, one immediately notices that

$$-\delta V^P/V = \delta \gamma^P_{Max} \cdot \tan \psi \qquad \text{(II.7-13)}$$

The factor $\tan \psi$ is called 'dilatancy factor' and ψ is the *'angle of dilatancy'*.

If we now compare II.7-13 with the corresponding expression II.7-12 that followed from Drucker-Prager's associated flow rules and take note of $\tan \chi = \sin \varphi$ (eq. II.7-4), we have

$$\psi = \chi \qquad \text{(II.7-14)}$$

which implies that the incremental plastic strain vector in Fig. II.7-34 is directed normal to the shear strength line.

This *condition of normality* or, more generally, the concept of a constant angle $\psi > 0$, *has to be abandoned* if an unrealistic volume increase under continued shearing is to be avoided.

Moreover, normality of the vector of plastic strain increment has another unrealistic consequence: for a cohesionless frictional material the stress vector in the diagram of Fig. II.7-34 coincides with the shear strength line and, therefore, becomes orthogonal to the strain-increment vector. Consequently, the inner product of the two vectors, which is the incremental plastic work, vanishes. In other words, without cohesion the frictional material would not dissipate the mechanical work expended in plastic deformation.

Drucker's stability postulate. Unfortunately, any attempt to abandon the normality condition seems to conflict with a postulate of material stability that was formulated by Drucker (1951, 1959). In plain words, it states: "When a deformable body is acted upon by an arbitrary equilibrium system of stresses (and is kept under isothermal conditions), additional static loading as distinct from the initial loading, cannot extract mechanical work from the system 'body + initial load' ." This postulate entails the normality condition for $\vec{\delta e}^P$, as can be easily demonstrated (Fig. II.7-35). The normality, in turn, justifies the use of Coulomb-Mohr's limit function as plastic potential since $\vec{\delta e}^P$ in Fig. II.7-34 has the direction of the gradient of the function II.7-3a that describes the shear strength line in Mohr's stress plane.

The way out of the normality dilemma was first shown by Drucker himself (1954) pointing out that his stability postulate does not generally apply to frictional systems. Following Drucker (1954) and Brown (1964), we may illustrate this important point by a solid block in frictional contact on an inclined plane (Fig. II.7-36). In the initial equilibrium situation (I) a force F, greater than the slope-parallel weight component W_t of the block, tends to pull the block upslope, but its magnitude is not yet sufficient to overcome the frictional resistance. This resistance may not yet be at its maximum value and is therefore characterised by a friction angle φ^* smaller than the friction angle φ at fully mobilised friction. Then an additional force $\vec{f}$ is applied (II) with a component that opposes the normal component N of the block weight, while the slope-parallel component of $\vec{f}$ tends to pull the block downslope. As discussed in the figure, such a force $\vec{f}$, when properly chosen, may reduce the friction at the base of the block sufficiently to allow the initial force F to pull the block upslope against the downslope-directed component of the additional force $\vec{f}$. Thus, the additional force does extract mechanical work from the system (I) and Drucker's postulate of material stability does not apply to this system.

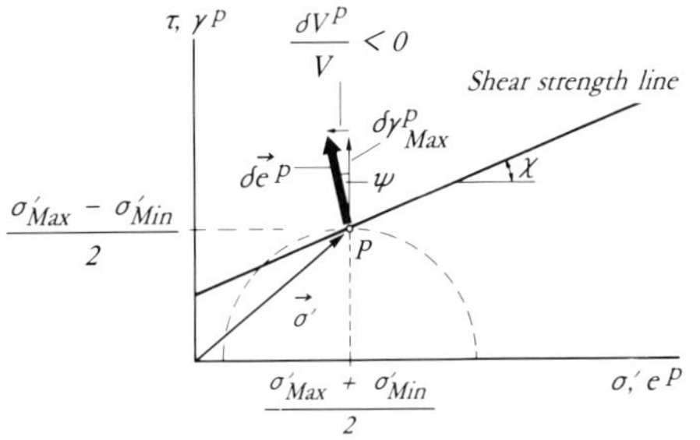

Plastic work per unit volume in plane straining (eq. II.8-11a) (co-axiality of stress and strain increments)

$$dW^P = \sigma'_{Max}\, de^P_{Max} + \sigma'_{Min}\, de^P_{Min} = \frac{\sigma'_{Max} + \sigma'_{Min}}{2}\left(de^P_{Max} + de^P_{Min}\right) + \frac{\sigma'_{Max} - \sigma'_{Min}}{2} \cdot \left(de^P_{Max} - de^P_{Min}\right) = \vec{\sigma}' \cdot \vec{de}^P$$

i.e. the scalar product of vectors with components

$$\vec{\sigma}' = \left(\frac{\sigma'_{Max} + \sigma'_{Min}}{2},\ \tau_{Max}\right) \text{ and } \vec{de}^P = \left(\frac{dV^P}{V},\ d\gamma^P_{Max}\right)$$

Dilatancy relationship:

$$\frac{dV^P}{V} = -\, d\gamma^P_{Max} \cdot \tan\psi$$

Fig. II.7-34 LIMIT STRESS AND PLASTIC STRAIN INCREMENTS IN THE COMBINED STRESS/PLASTIC STRAIN PLANE. THE DILATANCY RELATIONSHIP

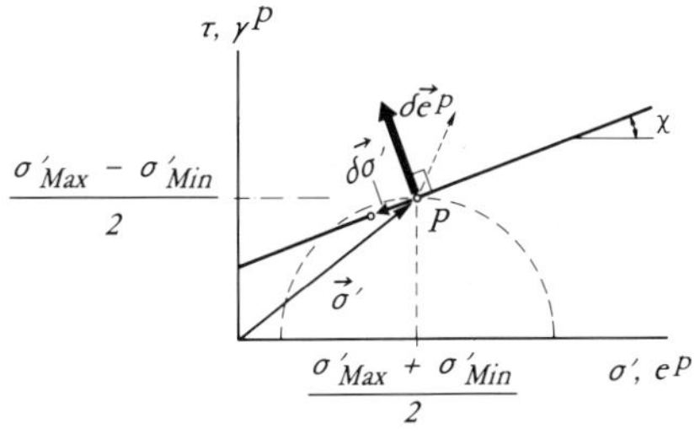

A volume element is brought into static equilibrium at limit state P by stress $\vec{\sigma}'$. Then an incremental stress system $\vec{d\sigma}'$ is added, which maintains equilibrium and leaves the element in a limit state. The work performed by the total stress in producing the plastic strain increment $\vec{de}^P$ is fully dissipated, i.e.

$$dW^P = (\vec{\sigma}' + \vec{d\sigma}') \cdot \vec{de}^P > 0$$

If, in addition, it is assumed that the incremental stress $\vec{d\sigma}'$ cannot extract mechanical work from the system 'volume element + initial stress $\vec{\sigma}'$', i.e.

$$\vec{d\sigma}' \cdot \vec{de}^P \geqslant 0,$$

*$\vec{de}^P$ must be **normal** to the shear strength line. Otherwise one could always find a $\vec{d\sigma}'$ which would render the plastic work negative.*

Fig. II.7-35 MATERIAL STABILITY AND 'NORMALITY'

The example of Fig. II.7-36 also indicates the cause of the discrepancy: Drucker's postulate is a *work principle*, linking plastic strain increments to the stresses that actually do work on these strains. The component f_n of the additional force $\vec{f}$, which reduces the frictional resistance of the system, does *not* perform work. If the added force $\vec{f}$ did not have such a component, which, because of internal kinematic constraints of the system, cannot perform mechanical work, Drucker's principle would apply. Various authors have therefore suggested that, when applying Drucker's principle to frictional materials, workless stresses associated with internal constraints should be excluded. For the concept of a plastic flow potential, this implies that the differentiation with respect to the stress components has to be carried out at constant constraining stresses. In frictional materials, internal constraining stresses will be associated with shear dilatancy of the material. In the simplest case when dilatancy is completely suppressed by an effective confining pressure, it is this effective mean stress which becomes the constraining stress that has to be kept constant when differentiating the limit condition.

In plane strain, with dilatancy being suppressed, the mean effective stress $p = (\sigma'_{Max} + \sigma'_{Min})/2$ is to be considered as a constraining, workless stress and therefore has to be treated as a constant when differentiating Coulomb-Mohr's limit condition

$$G \equiv \tau_{Max} - p \cdot \sin\varphi - \tau_o \cos\varphi = 0 \qquad \text{(II.7-3a)}$$

The associated components of the incremental plastic strain vector then become:

$$\delta\gamma^p{}_{Max} = \lambda \cdot \partial G / \partial \tau_{Max} = \lambda, \qquad \delta V^p/V = \lambda \cdot \partial G/\partial p = 0 \qquad \text{(II.7-15a)}$$

Thus, in the case of ***suppressed dilatancy***, the incremental plastic strain vector is orthogonal to the line τ_{Max} = const., which, in the stress/strain plane of Fig. II.7-37 is tangential to the crestal point P of the limit stress circle. This line therefore represents the plastic flow potential. It is important to note that every limit state of the frictional material has its own flow potential, represented by the crestal tangent to the respective stress circle, as indicated by the points P and P′ in Fig. II.7-37.

If dilatancy is not or only partly suppressed, the constraining stress is no longer the full effective mean stress and is difficult to assess [14)]. We may circumvent this difficulty by postulating a straightforward generalisation of the flow potential τ_{Max} = const., which contains the mean normal stress p. This parameter must then no longer be treated as a constant when dilatancy is not fully suppressed.

The simplest form of a plastic flow potential that may account for various degrees of plastic shear dilatancy is:

$$G \equiv \tau_{Max} - p \cdot \tan\psi = \text{const.} \qquad \text{(II.7-16)}$$

In the diagram of Fig. II.7-37 it plots as a set of parallel straight lines (dotted) that run

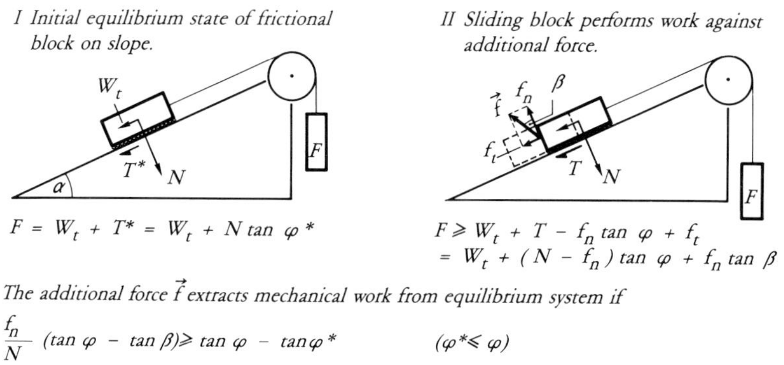

Fig. II.7-36 THE EXTRACTION OF MECHANICAL WORK FROM A FRICTIONAL SYSTEM

through the crests of the limit circles and are inclined at the dilatancy angle ψ to the axis of normal stresses and strains. Differentiation of this flow potential with respect to stresses, and multiplication by the undetermined factor λ, produces strain relationships for plane flow:

$$\delta\gamma^{P}_{Max} = \lambda \cdot \partial G / \partial \tau_{Max} = \lambda, \qquad \delta V^{P}/V = \lambda \cdot \partial G/\partial p = -\lambda \tan \psi \qquad \text{(II-7.15b)}$$

It may be immediately noticed that by changing the dilatancy angle ψ between 0 and χ (the inclination angle of the shear strength line in Fig. II.7-37, which is related to the friction angle by equation II.7-4), the plastic volume strain is varied between 0 and the value predicted by equations II.7-12.

Although this plastic flow potential enables us to prescribe the amount of plastic dilatancy at will and thus to approach real material behaviour more closely, it should be noted that the dilatancy factor chosen remains constant during plastic deformation, whereas in reality the dilatancy rate should decrease and eventually vanish when plastic shearing is continued, as is shown schematically by the plot of accumulated dilatancy in Fig. II.7-42B.

Application of a plasticity theory based on flow rules which are derived from the flow potential II.7-16 will therefore remain restricted to an initial phase of plastic deformation in which, in a dilatant frictional material, the dilatancy rate may be considered as approximately constant. Within such limits the theory may enable the numerical simulation – particularly by the finite-element method – of the growth of plastic zones and the geometry of potential faults. An example of such an application was shown in Fig. II.7-32 for the plastic zones developing above the basement uplift. Coulomb-slips fitted into the stress field of the plastic regions (Fig. II.7-32F) represent potential faults. The model itself, however, does not specify which of the potential slip surfaces may actually be mobilised. Although in some cases of geological interest, sufficient extra information may be available to single out the slip surfaces actually mobilised, further refinements of the plasticity model will normally be required when aiming at the numerical simulation of the development of individual faults.

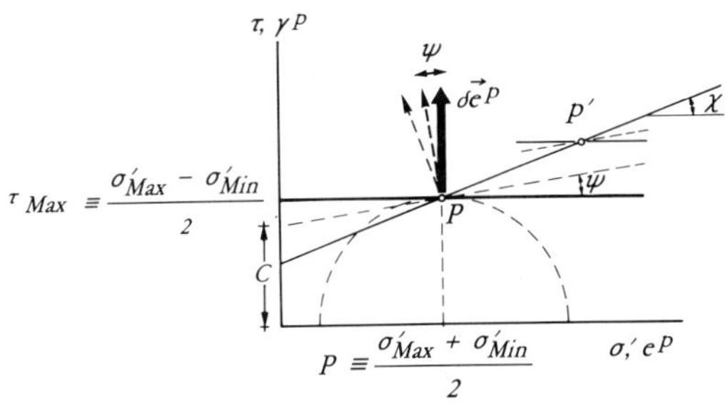

Plastic flow potentials:

$$G \equiv \frac{\sigma'_{Max} - \sigma'_{Min}}{2} - \frac{\sigma'_{Max} + \sigma'_{Min}}{2} \cdot \tan\psi = C$$

Notice that the plastic potential differs for different limit states through the value of the constant C.

The slope angle ψ may be varied between 0 and χ. Accordingly, the plastic volume increase varies from 0 to

$\delta\gamma^{P}_{Max} \cdot \tan\chi$, *where* $\tan\chi = \sin\varphi$.

Fig. II-7-37 PLASTIC FLOW POTENTIAL AND PLASTIC VOLUME INCREASE (DILATATION)

Before turning to such further refinement of the elastic/plastic model, we shall discuss an important, though little recognised, implication of Coulomb's slip concept for the formation of tectonic faults under *non-planar* strain or stress conditions. Such conditions may, for instance, be induced in the overburden of a faulted basement, when basement blocks are horizontally displaced along basement faults.

6. CONTINUOUS AND DISCONTINUOUS SLIP SURFACES

In the Coulomb-Mohr theory of frictional plasticity, plastic straining takes place by slip displacements along slip elements. These displacements are also orthogonal to the direction of the intermediate principal stress. The strain produced is therefore of the plane type locally. This is in accord with the fact that the intermediate principal stress neither enters the limit condition nor the plastic potential (eq. II.7-16). Therefore, the flow rules derived from that potential also predict zero plastic straining in the direction of the intermediate principal stress. As stated earlier (Section II.7.4 – 'Two-dimensional vs. three-dimensional strain') this locally two-dimensional character of Coulomb-plastic straining does not exclude truly three-dimensional deformation, since in an inhomogeneous stress field the principal axes and associated orientations of slip elements may vary with position.

The important question now arises: Will the slip elements in a plastic region integrate into smooth, coherent slip surfaces? In other words, will the planar Coulomb-slip elements form the tangential elements of sets of smooth surfaces? Or, applied to tectonic faulting, will the stress field allow or impede the inception of smooth, coherent fault surfaces?

To investigate this question we have applied (Mandl, 1987a) the following theorem for the integrability of vector fields:

A vector field $\vec{v}$ (with all the required differentiability, etc.) will be everywhere normal to a set of smooth surfaces if, and only if,

$$\vec{v} \cdot \text{curl}\, \vec{v} = 0 \tag{II.7-17}$$

i.e. the scalar product of $\vec{v}$ in its own curl vanishes (Fig. II.7-38A). Written in extenso for a Cartesian x, y, z-system this condition becomes:

$$v_x \cdot (\text{curl}\, \vec{v})_x + v_y \cdot (\text{curl}\, \vec{v})_y + v_z \cdot (\text{curl}\, \vec{v})_z \equiv \tag{II.7–17a}$$

$$v_x \cdot (\partial v_z/\partial y - \partial v_y/\partial z) + v_y \cdot (\partial v_x/\partial z - \partial v_z/\partial x) + v_z \cdot (\partial v_y/\partial x - \partial v_x/\partial y) = 0$$

For a relatively simple proof of the integrability theorem the reader is referred to, for example, Kestin's textbook of thermodynamics (Kestin, 1966).

Now we identify the vector $\vec{v}$ with the unit normal $\vec{n}$ of a set of Coulomb-slip elements and consider unit vectors $\vec{e}_I$ and $\vec{e}_{III}$ in the direction of the maximum and minimum principal stresses. Denoting the angle between $\vec{n}$ and $\vec{e}_I$ by μ ($\mu = \pi/4 + \varphi/2$; see Fig. II.7-15) the three coplanar vectors $\vec{n}$, $\vec{e}_I$, $\vec{e}_{III}$ are related by

$$\vec{n} = \vec{e}_I \cos\mu + \vec{e}_{III} \sin\mu \tag{II.7-18}$$

If the material has the same angle of internal friction everywhere, then the application of eq. II.7-17 to this vector $\vec{n}$ yields the following condition for the existence of continuous and smooth slip surfaces:

$$\cos^2\mu\,\vec{e}_I\cdot\text{curl}\,\vec{e}_{III}+\sin^2\mu\,\vec{e}_{III}\cdot\text{curl}\,\vec{e}_{III}$$
$$+\sin\mu\cos\mu\,[\vec{e}_I\cdot\text{curl}\,\vec{e}_{III}+\vec{e}_{III}\cdot\text{curl}\,\vec{e}_I]=0 \qquad \text{(II.7-19)}$$

This quasilinear partial differential equation of first order in the derivatives must thus be satisfied by the vector fields $\vec{e}_I$ and $\vec{e}_{III}$ if Coulomb-slip elements are to define continuous, smooth slip surfaces.

It is now easily seen that this condition is always satisfied if the directions of the greatest and smallest principal stresses are everywhere parallel to a given plane, as in problems of plane stress or strain, and do not vary in a direction orthogonal to this plane. By identifying this plane with the x, y-plane, all derivatives in the z direction vanish and, from the definition of curl $\vec{v}$ (see eq. II.7-17a), it follows that the only non-vanishing component of curl $\vec{e}_I$ and curl $\vec{e}_{III}$ is the z component. Since, moreover, the z components of $\vec{e}_I$ and $\vec{e}_{III}$ vanish, each of the inner products in equation II.7-19 vanishes and the integrability condition is satisfied. Hence, in an isotropic, uniform material in a state of *plane stress or plane strain* (with σ_I and σ_{III} remaining parallel to the plane of deformation), there will always be a set of surfaces whose normals coincide with the normals (eq. II.7-18) of Coulomb-slip elements (Fig. II.7-38B). In other words, the stress field will not impede the formation of *continuous, smooth fault surfaces*.

However, the situation is drastically changed when the *stress field is truly three-dimensional* in the sense that no Cartesian reference frame can be found such that the z component and/or z derivatives of the unit vectors $\vec{e}_I$ and $\vec{e}_{III}$ vanish. The products in condition II.7-19 will then no longer vanish separately, and the condition will impose a genuine and severe constraint on direction fields of the maximum and minimum principal stresses compatible with the existence of coherent, smooth slip surfaces. Therefore, in general, *incipient faults are not expected to form continuous, smooth surfaces in three-dimensional stress fields*. This may be illustrated by a few examples of three-dimensional stress fields.

A.

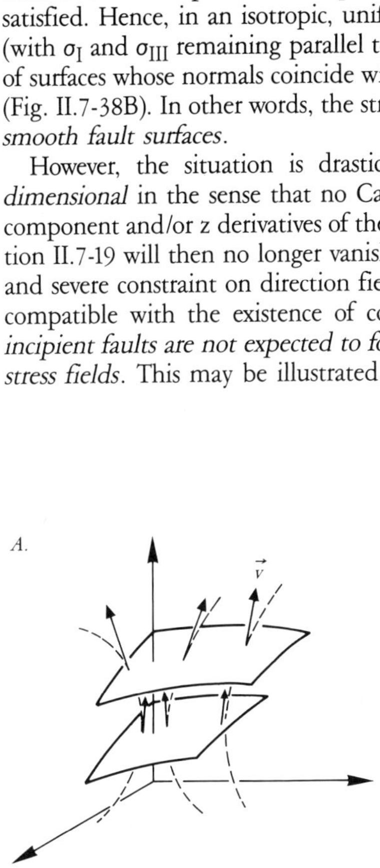

Three-dimensional vector field $\vec{v}$ with trajectories: Necessary and sufficient for the existence of a set of orthogonal surfaces is:
$\vec{v}\cdot\text{curl}\,\vec{v}=0$

B.

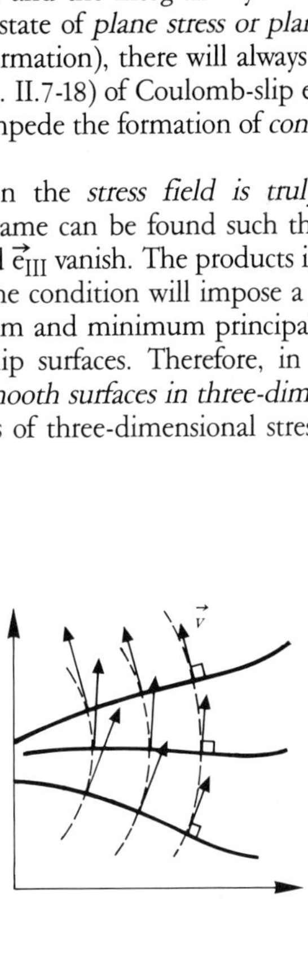

Two-dimensional vector field $\vec{v}$ with trajectories: There always exists a set of lines which are orthogonal to vector field.

Fig. II.7-38 SURFACES AND ORTHOGONAL VECTOR FIELD

Examples: Discontinuous Fault Surfaces in Three-Dimensional Stress Fields

Figure II.7-39A shows the case of pure (small) *torsion* of a circular cylindrical bar whose axis coincides with the z axis. The planar cross-sections perpendicular to the cylinder axis are subject to the maximum shear stress which acts in the circumferential direction. Consequently, the greatest (compressive) and smallest (tensile) principal stresses act tangential to the curved cylinder surface at an inclination of ± 45° to the cylinder axis. One may easily verify that the associated unit vectors, $\vec{e}_I$, $\vec{e}_{III}$ have the same x and y components, but z components of opposite signs:

$$v_x = y/r\sqrt{2}\ ,\ v_y = -x/r\sqrt{2}\ ,\ v_z = \pm\ 1/\sqrt{2} \qquad \text{(II.7-20)}$$

The inner product of the two vectors therefore vanishes, which is in agreement with their orthogonality. With reference to equation II.7-17a, one notices that only the z components of the curls of $\vec{e}_I$ and $\vec{e}_{III}$ do not vanish and attain the same value $-1/r\sqrt{2}$. Inserting the components II.7-20 in II.7-19, one then finds that the left-hand side becomes $(\cos^2\mu - \sin^2\mu)/2\,r$. Since $\mu = 45° + \varphi/2$, this value differs from zero, and condition II.7-19 is not satisfied. In fact, the deviation from zero increases in magnitude as the distance from the cylinder axis decreases.

Similar to this is the case of a stationary screw dislocation shown in Fig. II.7-39B. The similarity is apparent when the elements of a hollow cylinder are considered, the axis of which coincides with the dislocation line. The dislocation slip has occurred on a vertical plane over the triangular area with base b. If we interpret this dislocation as slip on a basement wrench fault and assume basement and overburden to have the same elastic properties, the stress field associated with the dislocation will represent the overburden response to *incipient wrench faulting* in the basement. Again, as in the similar torsion case, condition II.7-19 is violated, which strongly suggests that Coulomb slip surfaces ('Riedel' faults), which form when the stresses reach the limit state, should – at least initially – consist of incoherent segments in the vicinity of the basement fault.

A further example, which is inspired by a paper by McGarr (1980), is shown in Fig. II.7-40. Here the two horizontal principal stresses σ_I and σ_{III} of a regionally uniform tectonic stress system vary with depth, both in magnitude and direction. This state of stress may be viewed genetically as resulting from the superposition of two elementary regional states of stress that reflect two distinct tectonic phases. The first phase has established a state of stress (marked by a superscript (1)) such that the principal stress difference $\sigma_I^{(1)} - \sigma_{III}^{(1)} = \sigma_x^{(1)} - \sigma_y^{(1)} = a - bz$ decreases linearly with depth. This may, for instance, have taken

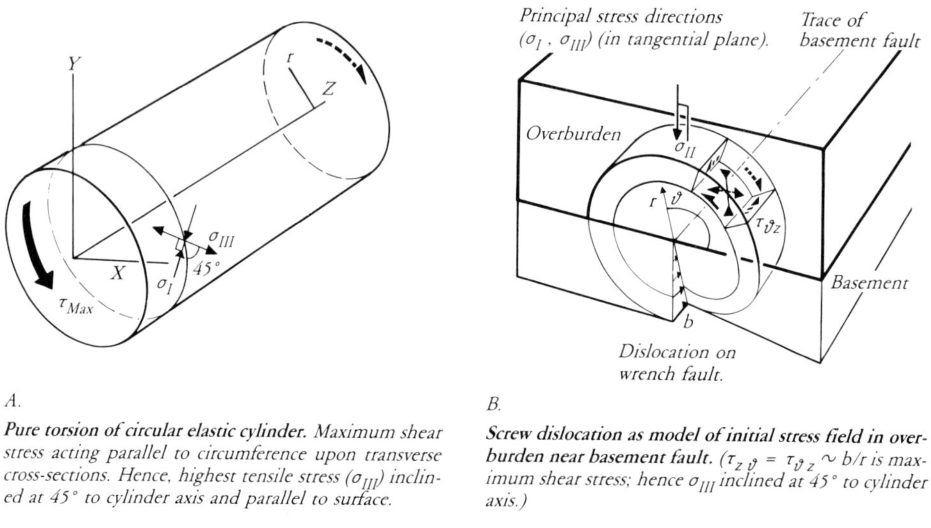

A.

Pure torsion of circular elastic cylinder. *Maximum shear stress acting parallel to circumference upon transverse cross-sections. Hence, highest tensile stress (σ_{III}) inclined at 45° to cylinder axis and parallel to surface.*

B.

Screw dislocation as model of initial stress field in overburden near basement fault. *($\tau_{z\vartheta} = \tau_{\vartheta z} \sim b/r$ is maximum shear stress; hence σ_{III} inclined at 45° to cylinder axis.)*

Fig. II.7-39 THREE-DIMENSIONAL FIELDS OF PRINCIPAL STRESS VECTORS

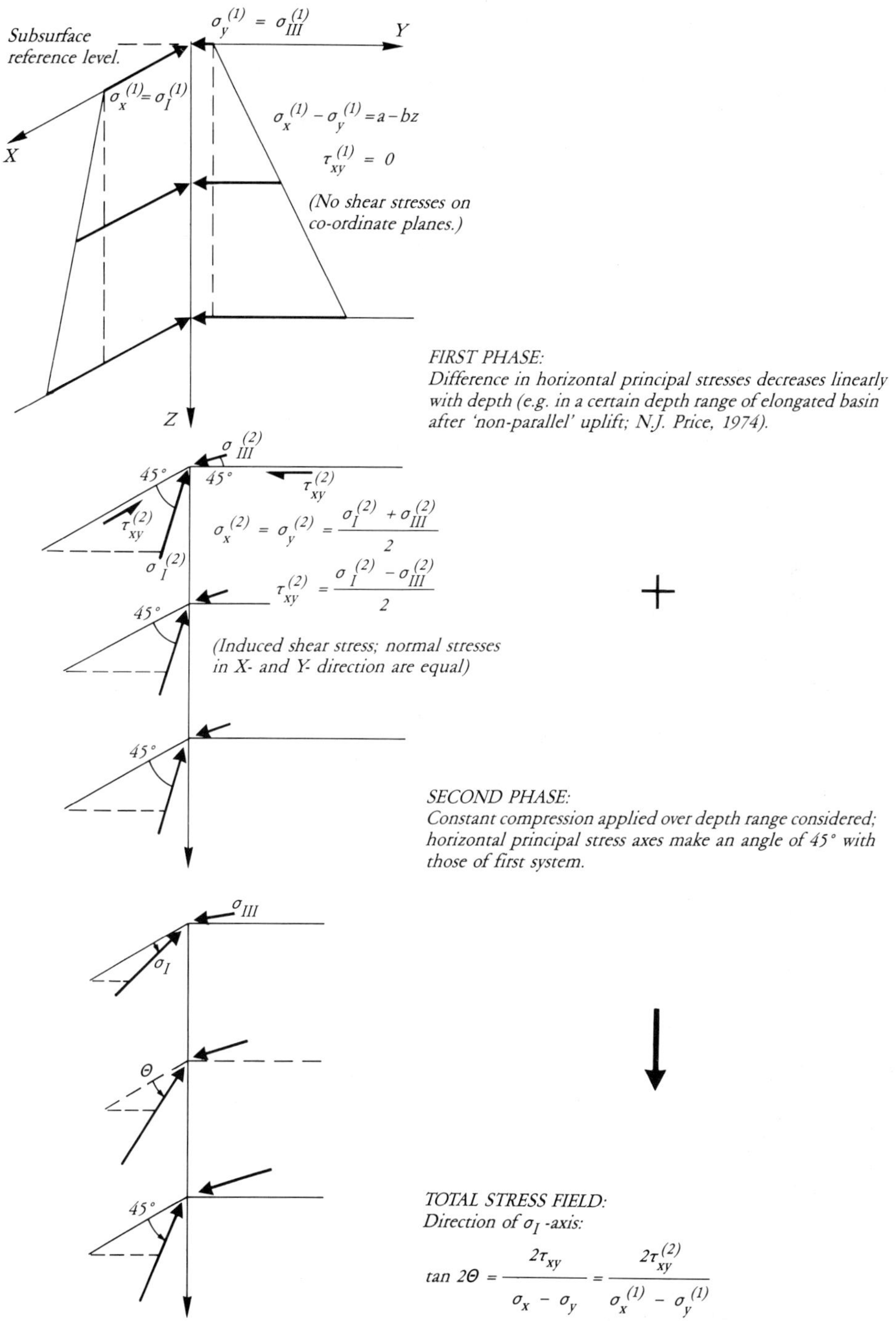

Fig. II.7-40 SUPERPOSITION OF TWO REGIONAL STRESS REGIMES WHOSE PRINCIPAL STRESSES DO NOT COINCIDE (One principal stress is vertical.)
(First Phase + Second Phase → Total Stress Field)

place during the regional uplift of an elongated basin. In a second tectonic phase (indicated by the superscript (2) in Fig. II.7-40) a horizontal compression is produced with horizontal principal stresses $\sigma_I^{(2)}$ and $\sigma_{III}^{(2)}$ constant over the depth range considered and acting at 45° to the principal stress directions of the first tectonic phase. The x- and y-planes are then planes of maximum shear stress of the second system. The shear stress $\tau_{xy}^{(2)}$ and the normal stresses $\sigma_x^{(2)} = \sigma_y^{(2)}$ on these planes can then immediately be expressed in terms of the principal stresses of system (2) (see, for example, eqs. II.1-4, 5) as stated in the figure. Assuming linear elastic behaviour of the rock mass during the compression phase (2), we may add corresponding components of the two stress fields. When the total stress components σ_x, σ_y, τ_{xy} have been obtained, the angle Θ the σ_I axis makes with the positive x-axis may be determined by an elementary formula (see, for example, eq. II.4-4) as indicated in the figure. Its value becomes the following function of the depth coordinate z:

$$\Theta = \frac{1}{2} \arctan \frac{2\,\tau_{xy}}{\sigma_x - \sigma_y} = \frac{1}{2} \arctan \frac{2\,\tau_{xy}}{\sigma_x^{(1)} - \sigma_y^{(1)}} = \frac{1}{2} \arctan \frac{\sigma_I^{(2)} - \sigma_{III}^{(2)}}{a - bz} \qquad \text{(II.7-21)}$$

Note that the relevant principal stress axes of the combined system are again horizontal, the intermediate principal stress remaining vertical, and that therefore the z components of the unit vectors $\vec{e}_I$, $\vec{e}_{III}$ vanish. However, these vectors *rotate with depth* and, consequently, the z derivatives in the curl components (eq. II.7-17a) will not vanish. The components of the two unit vectors are:

$$\begin{aligned} e_{Ix} &= \cos\Theta\,, & e_{Iy} &= \sin\Theta\,, & e_{Iz} &= 0 \\ e_{IIIx} &= -\sin\Theta\,, & e_{IIIy} &= \cos\Theta\,, & e_{IIIz} &= 0 \end{aligned} \qquad \text{(II.7-22)}$$

With these components inserted, the left-hand side of condition II.7-19 turns out to be equal to the magnitude of the gradient $d\Theta/dz$ of the rotation angle Θ, which, in view of equation II.7-21, does not vanish as long as $b \neq 0$ and the principal stresses of the compressive tectonic phase (2) differ in magnitude.

Within the framework of plasticity theory, based on Coulomb's slip concept, incipient faulting in the *rotating stress fields* of our example, or in similar cases, should therefore not be expected to produce smooth, coherent fault surfaces but rather separate segments which, only in a later phase of continued faulting, may be interconnected, but not integrated into smooth surfaces.

It should also be noted that, even if a certain truly three-dimensional stress field were found to satisfy the integrability condition II.7-19, any slight variation of the field would cause violation of that condition.

In stress fields with non-planar character, or not associated with plane deformation, it will therefore be the rule, rather than the exception, that an incipient fault consists of separate segments that do not lie on a smooth surface.

7. *THE ELASTIC/FRICTIONAL PLASTIC MATERIAL WITH STRAIN HARDENING AND SOFTENING*

The plasticity models discussed so far were of the 'ideal' plastic type (Fig. II.7-1 C, D). This means they had in common that plastic deformation did not start before the stresses satisfied a limit condition that was a function of stresses only and was held unchanged throughout the process of plastic deformation. At the end of Section II.7.5, we pointed out the need for further refinement of this model for the mathematical modelling of individual faults. Such an improvement of the model is achieved by approximating real material behaviour more closely by allowing the shear strength to change during plastic deformation.

When dealing with the shear strength of rocks and its control by effective stresses (Section II.7.2), we adverted to the growth of brittle microcracks which commences in a rock under shear at some shear stress τ_y well below the 'ultimate' shear stress τ_p. As shearing continues beyond this state, the shear stress has to rise – the material is said to 'strain harden' – until the limit stress τ_p is reached and shear failure becomes possible.

In the plastic hardening of frictional materials it is mainly the frictional shear strength that will change and monotonically increase with the amount of plastic shearing that the material has undergone. This has its analogy in the block model of Fig. II.7-1E when the frictional block is being moved from a bed of frictionless rollers onto a frictional base. Assuming that frictional hardening is an isotropic process, i.e. a process which affects any cross-sectional element regardless of its orientation, it may be accounted for by a *variable friction angle* φ_h which increases monotonically with the accumulated shear strain

$$k = \int d\gamma^{p}_{Max} \qquad \text{(II.7-23)}$$

and attains its maximum value φ at peak. Concomitantly, the cohesive strength, i.e. the stress-independent part of the shear strength, may also undergo some changes. In a brittle rock it is likely to attain its smallest value at peak. Since, however, virtually nothing is known about the strain dependence of the cohesive strength, one might as well disregard such changes and maintain a constant cohesion value τ_o instead. It should be recalled that this cohesion value is determined in the Mohr diagram (Fig. II.7-9B) by the intersection of the limit line with the τ-axis. Because the limit line itself is the envelope of stress circles associated with failure states, the cohesion value τ_o is also associated with the limit state.

The introduction of frictional hardening requires a redefinition of the limit condition II.7-3a in the form of a *'current yield' function*

$$q - p \cdot \sin \varphi_h = \tau_o \cdot \cos \varphi_h \qquad \text{(II.7-24)}$$

where φ_h attains its smallest value at 'initial yield' and its maximum at peak stress where the 'current yield' function turns into the limit condition (II.7-3a). This is illustrated in Fig. II.7-41A.

Shear Softening. Eventually, as shearing continues, in most geological materials the processes of plastic hardening become dominated by processes that weaken the material and cause a reduction in shear strength. This process of 'shear softening' is illustrated in Fig. II.7-42 by a schematic stress/strain curve describing the material behaviour when the effective confining stress is held constant. This phenomenon has been already mentioned (Section II.7.2; 'Limitations of the effective stress principle'). Again, it should be noted that

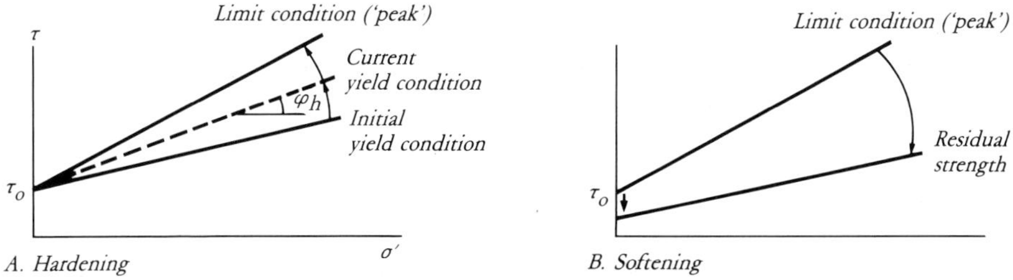

Fig. II.7-41 ISOTROPIC STRAIN HARDENING AND SOFTENING

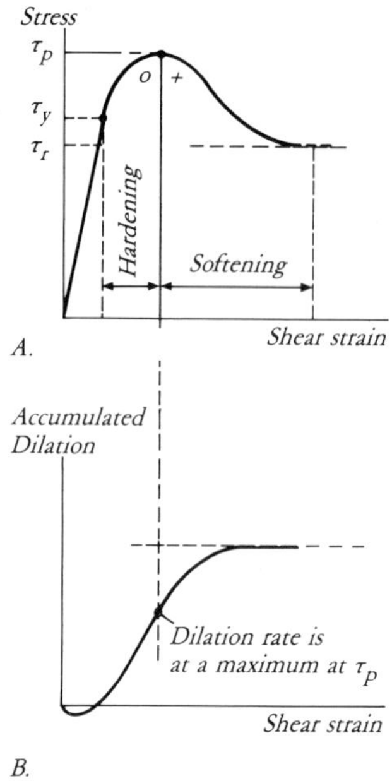

Fig. II.7-42
TYPICAL STRESS/STRAIN RESPONSE AND RELATED DILATION

the fact that in certain high-pressure/high-temperature experiments on rocks a reduction in shear strength has not been observed may not be considered proof for the absence of shear softening in the rocks tested under the pressure/temperature conditions of the experiments. It might more likely be attributed to insufficient straining and/or too high strain rates in the experiments. In fact, a gradual, but marked drop in shear strength at elevated temperatures and pressures has been reported for triaxial creep tests on olivine (Post, 1977) and dry granite (Tullis & Yund, 1977).

Various softening mechanisms have been proposed for rocks and soils. Shear dilatancy, cataclastic diminution and rounding of grains, reorientation of grains, the growth of microcracks or macroscopic joints are *brittle softening* mechanisms, i.e. mechanisms which produce a shear strength reduction that is only slightly affected by the rate of applied straining. It should be noted, however, that softening by the growth of brittle microcracks in the presence of water may, depending on the rock type, be more pronounced at lower strain rates because of stress corrosion at the crack tips (Atkinson, 1984). It should be added, however, that this rate dependence may become negligible at 'normal' tectonic strain rates (Costin & Mecholsky, 1983), since at those rates stress corrosion has sufficient time to become fully effective. Assuming that, in contrast with creep or slip along an existing fault, the formation of a new fault zone proceeds at 'normal' tectonic strain rates, strain-rate dependence due to stress corrosion may be disregarded when considering the elasto-plastic process of fault zone formation.

At *high temperatures and pressures* shear-strength reducing processes, such as pressure solution and dynamic recrystallisation, become strongly rate-dependent, since these processes are controlled by some type of diffusion. As the strain rate is decreased, the peak strength decreases and the pre-peak part of the stress/strain curve becomes less steep. Unfortunately, little seems to be known about the decline of shear resistance (post-peak) with increasing shear strain at constant strain rate. As mentioned earlier, the decline in shear strength may become noticeable only after relatively large straining.

Although time-dependent processes are alien to plasticity theory proper, the rate dependence of maximum shear strength is not incompatible with the theory as long as the strain rate remains approximately constant throughout the inelastically deforming region, and therefore does not enter as an extra variable. One may envisage many tectonic situations in which, prior to the onset of softening, in a widespread deformation field the variations in space and time of the strain rate are sufficiently small to enable their effect on yield function and limit condition to be neglected. This requires that, in the ductile regime, the boundary displacements take place at constant rate. If, however, the deformations are mainly controlled by boundary stresses, constant or varying in time, the strain rates will usually vary and their effect on yield and limit condition and on the flow rules cannot be ignored. In other words, under such loading conditions the elastic/frictional plastic strain hardening and softening model has to be extended to include strain rate for it to cope with the development of strains and stresses at the elevated temperatures and pressures prevailing in the ductile crust. Although it is outside the scope of this book, we shall later sketch the essence of such visco-plastic generalisation.

For the discussion of macroscopic plasticity models and their application it is not necessary to delve into the difficult field of the processes mentioned above or others that may possibly cause hardening and softening of rocks and soils in the brittle and ductile range. Instead we can be content with accounting for a gradual post-peak decline of shear strength to some residual value τ_r, by defining in the stress/strain diagram (as shown schematically in Fig. II.7-42A) a smooth transition from the peak to the residual strength at a prescribed maximum rate of decline. The stress components themselves have to satisfy a current yield condition. Similarly to isotropic strain hardening, isotropic strain softening may be accounted for by allowing the strength parameters – friction angle and cohesive strength – to decrease monotonically beyond peak as a function of the integrated plastic strain k (II.7-23). This is illustrated in Fig. II.7-41B.

The plastic flow potential (II.7-16) is affected by isotropic hardening/softening only by the changes in the 'dilatancy factor' tan ψ (II.7-13). If dilatancy is not completely suppressed by high effective confining pressures, tan ψ may be prescribed to increase monotonically with the strain parameter k (II.7-23) until peak stress is reached, and thereafter to decrease gradually to zero, as shown in the schematic diagram of Fig. II.7-42B.

The yield condition in three-dimensional stress space. Current yield condition II.7-24, limit condition II.7-3 and plastic flow potential II.7-15 are stress relationships that do not depend on the intermediate principal stress. Accordingly, these conditions may be graphically represented in the plane of the principal stresses σ_I' and σ_{III}', where they map as pairs of lines. One such line representing a current yield condition for a cohesionless, frictional material ($\tau_o = 0$) is drawn in Fig. II.7-43A in the σ_I', σ_{III}'-plane of a Cartesian reference frame, the coordinates of which are the values of the principal stresses. Let us now consider a point C on the yield line and the possible values the intermediate principal stress σ_{II}' may attain. Since this stress is bounded by the two other principal stresses, it is easily seen that the possible triplets of principal stress values associated with the stress point C define the segment AB on the straight line that is parallel to the σ_{II}'-axis and runs through C. The line segment is bounded by the plane containing the line $\sigma_{II}' = \sigma_I'$ and the σ_{III}'-axis and by the plane containing the line $\sigma_{II}' = \sigma_{III}'$ and the σ_I'-axis. Hence, all stress states compatible with a state of current plastic yielding map on a triangle with vertex at 0 and with two sides running though A and B in Fig. II.7-43A. If viewed along the space diagonal ($\sigma_I' = \sigma_{II}' = \sigma_{III}'$), this triangle intersects an octahedral plane (i.e. a plane with normal parallel to the space diagonal) along a segment AB as shown in Fig. II.7-43B.

The triangular planar surface in principal stress space represents the *complete* (current) yield condition when the stress axes are *ranked* in a definite way according to the magnitude of the principal stresses. In

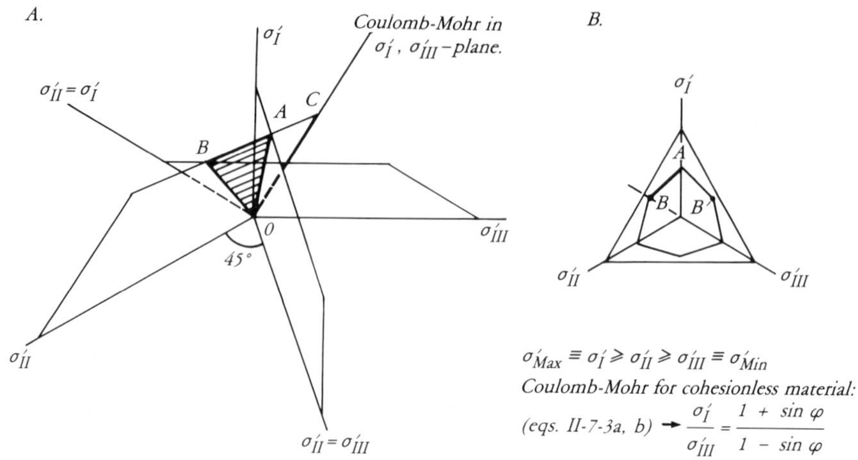

Fig. II.7-43 MOHR–COULOMB YIELD CRITERION IN PRINCIPAL STRESS SPACE

analysing concrete problems, however, it is often convenient, or perhaps even necessary, to drop this ranking. For example, in plane deformation with two principal stress axes lying in the plane of deformation, the roles of the out-of-plane principal stress and one of the in-plane principal stresses may change along the plane of deformation. By *abandoning ranking* of the principal stress axes in stress space one may allow, for instance, an interchange of the roles of σ_{II}' and σ_{III}' as intermediate and minimum principal stresses. The associated triangular yield surface will then intersect the octahedral plane in Fig. II.7-43B along AB'. When all possible permutations of ranking are being performed, the corresponding representations of the (current) Coulomb-Mohr-type yield condition constitute in principal stress space a hexagonal pyramid, the axis of which coincides with the space diagonal. The hexagonal intersection of this yield pyramid with an octahedral plane is also shown in Fig. II.7-43B.

In experiments with dry sand in cubical triaxial test apparatus, it has been found (Lade & Duncan, 1973; Goldscheider, 1976) that the corners of the composed yield surface in principal stress space are rounded off. For rocks one may reasonably expect some similar rounding of the corners of the yield pyramid. Therefore, and because of the less stringent accuracy requirements in geological applications, it seems justified to approximate the pyramidal yield surface in principal stress space by a circular cone (Fig. II.7-44) – the type of yield surface first introduced by Drucker and Prager (1952 [15])).

For cohesive materials the apex of the cone lies in the tension octant of the principal stress space at the distance $\tau_0 \cdot \cot\varphi$ from the origin. The half apex angle γ of the cone inscribed in the yield pyramid is related to the friction angle φ by the simple relationship given in Fig. II.7-44. Frictional plastic hardening and softening may then be accounted for by increasing or decreasing the apex angle as illustrated in Fig. II.7-45. The variation of the apex angle as a function of the accumulated shear strain k (II.7-23) may be chosen so as to match any isotropic hardening/softening behaviour; in particular, to define the friction angle at the onset of plastic yielding, at peak and after full softening [16]). Whereas in purely frictional hardening/softening the apex of the current yield cone remains at a fixed position, an additional cohesive softening may easily be accounted for by shifting the apex toward the origin.

Shear Concentration. The phenomenon of *shear softening* is of fundamental importance for the mechanics of tectonic faulting since it promotes the formation of narrow shear zones or discrete slip movements. The slightest inhomogeneity in shear strength or in the stress field may trigger a local concentration of shearing deformation, since softening is allowed to start earlier or to progress faster than in the surrounding rock. As this permits more shear deformation to be accommodated locally, and this, in turn, causes further softening, the process is self-accelerating.

Secondly, the reduced strength of the shear bands or slip surfaces generated will enable them to remain active even when – within certain limits – the values and directions of the principal stresses change in time. This is important since it enables incipient shear zones to develop into major tectonic faults in spite of the stress changes that usually accompany the formation of faults.

Naturally, pre-existing zones (or planes) of weakness may favour shear localisation, and shear softening may be restricted to certain layers or interfaces.

Although we believe the reduction in shear strength to be the most prominent mechanism in the formation of tectonic faults, it has to be admitted that *other mechanisms*, such as elastic buckling, strong stress concentrations, or geometrical factors such as heterogeneities in bed thickness, may control the initiation and localisation of shear bands even in the absence of shear softening or any pre-existing contrasts in shear strength. It should be emphasised, however, that without subsequent shear softening such 'embryonic' shear bands are very likely to be immobilised by minute changes in the directions of boundary displacements or external tectonic forces.

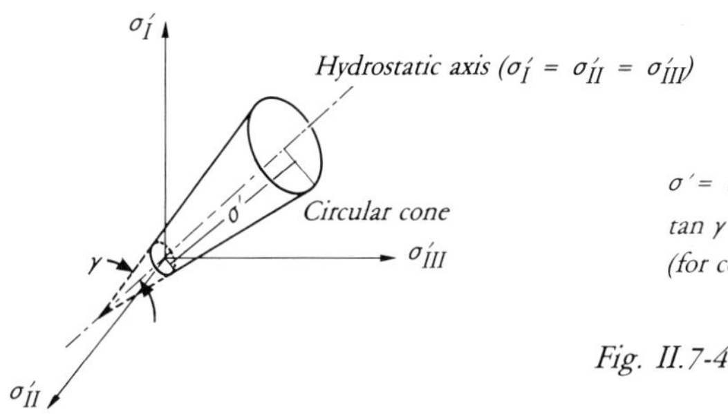

$\sigma' = (\sigma'_I + \sigma'_{II} + \sigma'_{III})/3$

$\tan\gamma = \sqrt{3}\sin\varphi\,(3 + \sin^2\varphi)^{-1/2}$

(for cone inscribed in Mohr-Coulomb pyramid)

Fig. II.7-44 DRUCKER-PRAGER YIELD CONE IN THREE-DIMENSIONAL STRESS SPACE

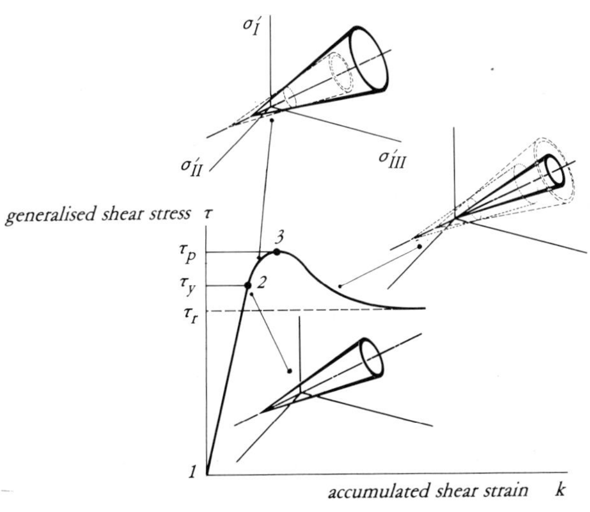

Fig. II.7-45 CURRENT YIELD SURFACE FOR ISOTROPIC STRAIN HARDENING AND SOFTENING

Example No. 7: Differential uplift

It is of interest to return to the development of the plastic zone in the overburden above a vertically uplifted basement block, which has been dealt with in Example No. 5 of Section II.7.5, and to re-examine the problem with the frictional hardening/softening model (Walters and Thomas, 1982). The geometry of the problem and the initial stress conditions were shown in Fig. II.7-31. The finite-element calculation [17)], based on the elastic/frictional plastic model without shear hardening/softening, produced broad plastic regions in the overburden, the development of which was shown in Figs. II.7-32A-D. In Fig. II.7-46 a few stages of the plastic zone growth are shown; this was modelled for exactly the same problem by using the frictional hardening/softening model. Because of the symmetry of the configuration, again only one-half of the geometry needs to be considered. The small dots at the integration points of the finite elements indicate elastic response of the material; the large dots represent plastic hardening, while the crosses indicate that peak stress has been passed and softening is taking place. Fig. II.7-46A shows the onset of a plastic hardening zone both at the edge of the uplifted basement block and at the overburden surface (extension regime). As uplifting continues (Fig. II.7-46B), a narrow softening zone develops within a hardening region already formed and extends to the surface. This softening zone represents a first shear band, closely simulating the precursor fault observed in the corresponding sandbox experiment of Fig. II.7-33. Typically, during the growth of the softening zone much of an initially broad hardening zone (not shown in the figure) has unloaded back to an elastic state of stress. As uplift continues further (Figs. II.7-46C, D), steeper shear zones form, with final movement along a vertical shear zone – in excellent agreement with the experimental result of Fig. II.7-33.

This example illustrates very well the concentration of shear deformation by plastic softening. Although the width of the calculated shear bands is affected by the size of the finite elements, it is no longer necessary to fit Coulomb-slip surfaces into the plastic regions. The *narrow plastic regions themselves represent the faults* – at least at an incipient stage, because the program used was designed for small deformations only.

In spite of the successful fault simulations achieved with the frictional plastic hardening/softening model, it should be noted that the implemented hardening/softening rules represent isotropic processes and, therefore, may be somewhat at variance with the actual hardening/softening behaviour of rocks, which is likely to show some degree of *anisotropy*. Hardening anisotropy may be induced, for instance, by the pre-peak growth of statistically oriented microcracks, while shear softening might be more realistically associated with Coulomb-slip elements determined by the state of stress at peak. A first attempt to incorporate anisotropic shear softening along Coulomb-slip directions into an elastic/frictional plastic finite-element program was undertaken by Malina (1969), whose program has been used by Bock (1971) in analysing the growth of a fault zone and the associated internal secondary shears under plane strain conditions and biaxial loading. Unfortunately, it seems that this line of investigation has not been pursued further.

On the matter of softening, it should also be understood that the softening, introduced in this section, prepares the formation of a new shear zone by affecting a whole rock volume. Softening may also be confined to single surfaces, such as bedding planes, joints or pre-existing fault planes. Then, softening will not depend on shear strain but on the relative displacement of the adjacent rock blocks ('Slip softening').

Finally, it should be recalled that softening as a fault-promoting mechanism is not restricted to brittle rock behaviour, but may also operate in the *ductile range*. When, under an applied differential stress, rock deforms by *creep*, the flow may either stop after a primary deceleration phase or accelerate again and eventually lead to failure if the applied differential stress exceeds a critical value, which is referred to as the *'fundamental shear strength'*. In general, both strain-dependent and strain-rate-dependent softening mechanisms will contribute in localising the shear during the final (tertiary) creep phase. Orowan (1960), in an attempt to explain deep-focus earthquakes, proposed some kind of rock thixotropy as the cause of fault formation in the ductile range, implying a decrease in shear stress with increasing shear rate. Such behaviour would be caused by flow-induced structural changes in the material. Where the creep rate was initially somewhat higher, it will increase concomitantly with a further reduction in flow resistance, while the associated drop in shear stress will slow down the flow elsewhere.

As the strain rate increases, the predominant flow mechanism may change. For example, intergranular slip may give way to cataclasis, as can often be observed on overthrusts, where sharp fault planes of a cataclastic nature are embedded in a wider zone of ductile deformation.

8. STABILITY AND INSTABILITY OF ELASTIC/PLASTIC DEFORMATION

Shear softening is not only the primary mechanism in converting continuous deformations into discrete faults, it can also cause the growth of plastic regions to become unstable. This means the reduction in shear resistance may disturb the static equilibrium of the internal forces in systems under external load, giving rise to acceleration forces. Thus the growth of a softening zone may become an accelerating process, only restrained by the inertia of the

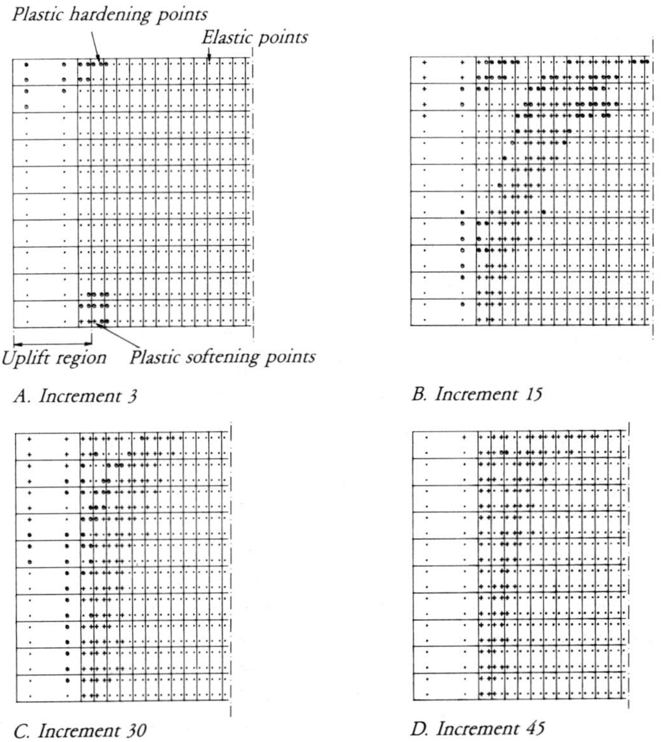

A. Increment 3

B. Increment 15

C. Increment 30

D. Increment 45

Fig. II.7-46 DEVELOPMENT OF PLASTIC ZONES IN STRAIN–SOFTENING ELASTIC/FRICTIONAL PLASTIC MATERIAL ABOVE BASEMENT UPLIFT OF FIG. II.7-32 (After Walters and Thomas, 1982)

system. Although, intuitively, this may seem quite clear, the problem of defining the conditions for unstable or stable (i.e. statically controlled) growth of plastic shear zones is by no means an easy one and in need of further research. In the following we shall content ourselves with discussing the problem in qualitative terms.

First of all, when speaking of instability of plastic deformation, one has to distinguish between the behaviour of an homogeneous, uniformly strained material element and the behaviour of a material system the components of which may differ in material properties, stress state or strain history. Secondly, one has to distinguish between the behaviour of the material under imposed boundary displacements (strain control) and the response to imposed load increments.

Material Stability. To begin with, we consider a uniform material element subject to homogeneous straining, say, a simple shear. In the Figures II.7-47A-D a schematised stress/strain diagram for an elastic/strain softening plastic material is presented and discussed in terms of mechanical energies stored or dissipated in a unit volume. During small elastic deformation the stresses increase in proportion to the associated strains. The elastic energy stored in a unit volume at a given state of strain is then determined as the sum of products of strains and the associated stress components at the final state of strain (equation II.8-10b). In particular, the elastic energy stored in a unit volume element under simple shear is the product of shear strain and associated shear stress (equation II.8-9c and Fig. II.8-9C). Since the shear strain is half the shear angle measured in radians – the 'engineering' strain (equation II.5-3c) – the elastic energy stored in the unit volume in simple shear is that represented in Fig. II.7-47A by a triangular area under the stress strain line when the shear angle is used as strain measure. Area I in Fig. II-47A represents the elastic strain energy stored at peak stress. When shearing continues beyond peak, strain softening will reduce the recoverable elastic energy which, for a given state of plastic shearing, is represented by the shaded area II in the Figure. The difference between recoverable peak and post-peak energies (I – II), represented by the shaded areas in Figs. II.7-47B, C, is the elastic energy released by the stress drop during strain softening. If this excess energy is dissipated by plastic straining, the deformation of the homogeneous material element is stable; otherwise it is unstable.

In looking into this further, we consider the mechanical work performed on the element in producing the post-peak plastic shearing. This work, which is represented in Fig. II.7-47D by the shaded area III, has to be expended, independently of any elastic pre-peak strain, and is fully dissipated by the plastic deformation. In other words, this is the amount of mechanical work the plastic post-peak deformation can absorb. Therefore, *a stable mode of deformation requires that the absorbed energy exceeds the amount of elastic energy released by softening* during the post-peak straining episode. With reference to Fig. II.7-47D, this means that the rectangular area III has to be greater than the area I-II, which represents the released elastic energy, to allow stable plastic shearing of the material element. Obviously, this limits the admissible steepness of the post-peak descent of the stress/strain curve.

If this condition of stability is not satisfied for a material in *strain-controlled* plastic deformation, the softening element will feed mechanical energy into its surrounding. In simple shearing, this would mean that during plastic shearing the shear stress that acts upon the element will have to reverse its direction to prevent acceleration and to maintain quasi-static shearing. In contrast, however, when the deformation is *load-controlled*, the material element will always deform in an unstable, i.e. accelerating, mode while softening. For example, increasing the shear load in a simple shear test up to peak and then maintaining peak load would invariably lead to accelerative deformation, restrained only by inertial forces. The softening material could no longer provide the resistance required for static equilibrium.

Drucker's stability postulate, dealt with in Section II.7.5, applies to the load-controlled case. It is easily seen that an additional load, referred to in Drucker's postulate, could in fact extract mechanical work from a frictional material that softens in simple shear. Consider, for instance, in close analogy with the sliding block example of Fig. II.7-36, a frictional material in a pre-peak state of simple shear, acted upon by a certain shear stress and a normal stress perpendicular to the shear direction. Now an additional load is imposed, which reduces both shear stress and normal stress. When properly chosen, this additional load will reduce the strength of the material sufficiently to bring it into a post-peak state and thus allow shearing to continue in the direction of the originally applied (pre-peak) shear stress. Since the corresponding component of the additional load acts in the opposite direction, the softening material performs mechanical work on the additional load system.

Stability and Instability of a System. What has been said about the stable behaviour of the strain-controlled uniform deformation of a softening homogeneous material element and the instability of such an element under load-controlled deformation should not be confused with the *deformation behaviour of a system*. Such systems, generally heterogeneous in stress/strain history or material properties, may exhibit stable or unstable deformation behaviour both under strain and load control.

This may be illustrated by the two models shown in Fig. II.7-48. The models have only one degree of freedom and are both composed of two bars of the same elastic/plastic material; the two bars have the same length but different thickness. In the first model the bars

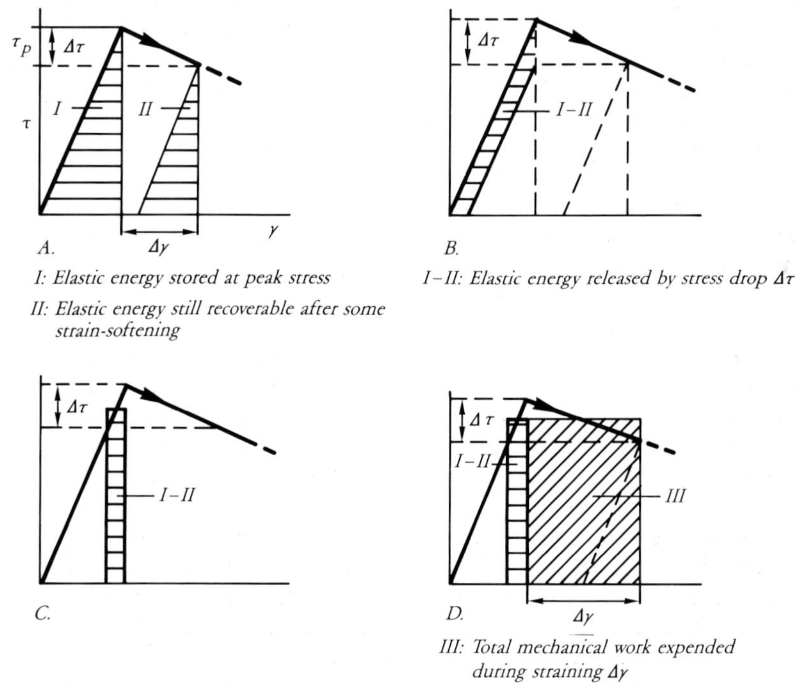

Fig. II.7-47 THE ROLE OF ELASTIC STRAIN ENERGY IN PLASTIC STRAIN SOFTENING (SCHEMATISED STRESS/STRAIN DIAGRAM)

are arranged in parallel, in the second model in series. In the first model, one bar (PL), originally somewhat longer than the other one (EL), was pre-compressed to the length of bar EL. Then the two bars were rigidly linked together and subject to an external load Q which was just sufficient to induce peak stress in bar PL while leaving bar EL still at some elastic pre-peak state of stress. These states (1) are also indicated in the schematised stress/strain diagram of Fig. II.7-48C. Next, an additional load ΔQ is applied to the system, shortening both rigidly linked bars by the same amount $\Delta e = -\Delta u/l$. Under this post-peak straining the material of bar PL suffers a reduction in load-carrying capacity by an amount $\Delta Q'$ which is found by multiplying the stress drop $\Delta\sigma_{ep}$ in Fig. II.7-48C by the cross-section d_{PL} of bar PL:

$$\Delta Q' = d_{PL}.\Delta e \cdot |\tan\alpha| \qquad \text{(II.7-25)}$$

Assuming that under the additional loading of the system bar EL still remains in the elastic state, it will carry an additional load which, with reference to Fig. II.7-48C, is

$$\Delta Q'' = d_{EL} \cdot \Delta e \cdot E \qquad \text{(II.7-26)}$$

where E is Young's modulus. As long as

$$\Delta Q'' - \Delta Q' > 0 \qquad \text{(II.7-27)}$$

the system is stable since it is capable of statically supporting an extra load by internal load transfer from the softening to the elastically deforming part. When the above expressions for $\Delta Q'$ and $\Delta Q''$ are inserted, condition II.7-27 becomes

$$|\tan\alpha| /E < d_{EL}/d_{PL} \qquad \text{(II.7-28)}$$

If this condition is not satisfied, the system will collapse under the additional load.

This simple model illustrates that plastic softening of a subregion does not have to unbalance the static equilibrium of the whole rock mass under an imposed external load or to cause unstable, i.e. self-accelerating, growth of the softening plastic zone. In spite of the unstable deformation behaviour of material elements when softening under an imposed load, the system may deform in a stable manner, as long as the elastic/strain hardening rock, which surrounds the plastically softening part, can support the extra load transferred from the softening material. Or, expressed in terms of energies, the *system remains stable as long as the elastic strain energy released in the softening subregion can be absorbed by pre-peak deformation of other parts of the system.*

The second model (Fig. II.7-48B) shows a quite different internal configuration of elastic and plastically softening parts of the two-bar system. The two bars in series have to carry the same load. When they are shortened together, the thinner bar PL will therefore experience a higher compressive stress and will reach the peak state when the thicker bar EL is still in the elastic deformation range. Undoubtedly, any further increase in the external load will cause the collapse of the system. However, if instead, shortening is continued beyond peak in a controlled way (strain control), while the load is free to drop concomitantly, the system may deform in a stable manner. To demonstrate this, we assume that the system has been shortened so far that the bar PL has just reached peak state. Then we apply in the axial direction a small displacement to the interface of the two bars, producing the shortening Δe of bar PL and the extension $-\Delta e$ of bar EL, without changing the overall dimensions of the system. It is easily inferred from Fig. II.7-48D that concomitantly with the shortening Δe the load-carrying capacity of the softening bar PL is reduced again by the amount $\Delta Q'$ defined in II.7-25, while the load transmitted through the elastic bar EL is decreased by extension by the amount $\Delta Q''$, defined in II.7-26. Obviously, as long as the reduction $\Delta Q''$ in transmitted load exceeds in magnitude the reduction in load-carrying capacity, i.e. $\Delta Q'' - \Delta Q' > 0$, the system will remain stable. Thus the stability condition turns out to be the same as that for the previous model with the two bars in parallel. If this condition is satisfied, any internal perturbation of the type considered will die out and the system will return to its original internal configuration, since the elastic bar tends to shrink and the plastic bar tends to extend again.

Condition II.7-28 clearly indicates the stabilising function of the elastic part in the special system considered: the stiffer and bigger the elastic component in comparison with softening rate and volume of the softening part, the better this is for the stability of the deforming system. Although derived for a very special model, this statement may hold true for various tectonic systems, and enable fault zones to grow in a stable, quasi-static manner. If the elastic energy released in the softening zone is not absorbed by the elastic surrounding or boundaries, the softening region will start growing at an accelerating rate, developing a *'runaway' instability*. The conditions for 'runaway' of a softening zone will not only include the elastic and plastic material characteristics, but in particular will depend on the shape of the softening zone and the type of the tectonic stress field.

Rudnicki (1977) has analysed the conditions for 'runaway' instabilities of elliptical inclusions of a weaker, softening material embedded in an infinite, uniformly strained matrix of homogeneous elastic material. The analysis showed that a near-spherical shape of the softening zone, high internal friction and dilatancy of the softening material tend to inhibit a 'runaway' growth [18].

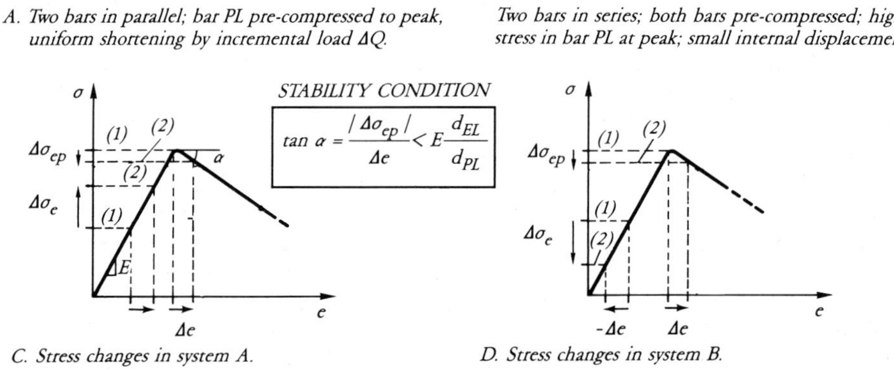

C. Stress changes in system A.

D. Stress changes in system B.

Fig. II.7-48 STABILITY OF ONE-DEGREE OF FREEDOM SYSTEM OF ELASTIC/PLASTICALLY SOFTENING MATERIAL

The simple models considered so far cannot do full justice to the complexity of real tectonic processes. In particular, it should be pointed out that tectonic settings exist in which the elastic environment, instead of stabilising the growth of softening zones, has the opposite effect and promotes 'runaway' instabilities. To illustrate the difference between the *stabilising and de-stabilising roles of the elastic environment* of a growing fault zone we consider the modification of the sliding-block model of Fig. II.7-36 which is shown in Fig. II.7-49A. We assume that the block on the slope is on the verge of being pulled up-slope by the weight at the other end of the string. An elastic spring connects the rear of the block to the immobile slope. Were it not for this spring, the additional pull of a small extra weight would be sufficient to trigger up-slope sliding of the block, which would accelerate since the reduction from static to sliding friction unbalances the static forces. The stability of this displacement-softening system is maintained by the extension of the elastic spring which provides the extra force to compensate for the loss in frictional resistance. Movement of the block may then still be envisaged as a succession of static equilibrium states produced by a gradual increase of the pulling weight. Any surplus work momentarily performed by the 'driving' force is completely stored as elastic strain energy. Without an elastic spring, movement of the block-weight system would be unstable. This is worsened by placing the elastic spring between the block and the pulling weight (Fig. II.7-49B). When the block starts sliding, release of strain energy of the extended spring will momentarily add to the acceleration of the up-slope sliding of the block. Incidentally, when a displacement is applied to the up-slope end of the spring instead of a constant pulling force, and the spring is extended until the block starts moving, the reduction in friction will cause the spring to contract again. The inertial mass of the system will make the elastic contraction 'overshoot' and will cause the force acting upon the sliding block to drop below the level that is necessary to overcome the sliding friction. The block will come to rest and the cycle will repeat itself, resulting in a jerky sliding movement of the block, commonly referred to as *'stick-slip'*.

A typical example of how the elastic strain energy of the ambient material can cause a softening plastic zone to grow in an unstable way has been analysed by Palmer and Rice (1973). A thin interlayer or interface of lower shear strength is embedded in a matrix of stronger, uniform elastic material that is subject to simple shearing between long, parallel rigid platens. The shear stress transmitted through the elastic material may initiate slip along the plane or thin zone of weakness, as shown schematically in Fig. II.7-50. The 'driving' upper plate may, for instance, represent the drag of an overburden sheet on a long slope. As a slip region forms along the weak interlayer, the shear stress will be distributed unequally in that region since the shear strength varies with the shear displacement, which itself increases from the leading edge towards the interior of the spreading slip zone [19)]. Since a small displacement, usually of the order of millimetres or a few centimetres, is required to mobilise the full peak strength, the shear stress attains its maximum value τ_p at some distance from the leading edge and decreases from there towards its residual value τ_r. In other words, a new element of the growing slip zone will first experience an increase in shear stress to peak value, followed by a gradual decrease. This reduction in shear strength causes a release of elastic strain energy and a rebound of the ambient elastic material, indicated in Fig. II.7-50 by the reorientation of the material marker line A into A′. This mechanical energy contributes to the advance of the slip zone. The thicker the elastic matrix and the greater its stiffness (elastic shear modulus), the greater is the energy release associated with a given shear displacement and softening characteristic. As Palmer and Rice have shown, this surplus energy can become high enough to compensate for the energy dissipated during the advance of the slip zone and may therefore propagate the slip zone, even when movement of the 'driving' plate is stopped. For this unstable, progressive growth of the slip region to take place, the thickness of the elastic matrix must exceed a certain critical value.

But it should be recalled that in this and similar cases of unstable slip driven by the release of elastic energy from ambient rock masses, the slip propagates along pre-existing weak interlayers or interfaces, such as bedding planes or existing fault planes.

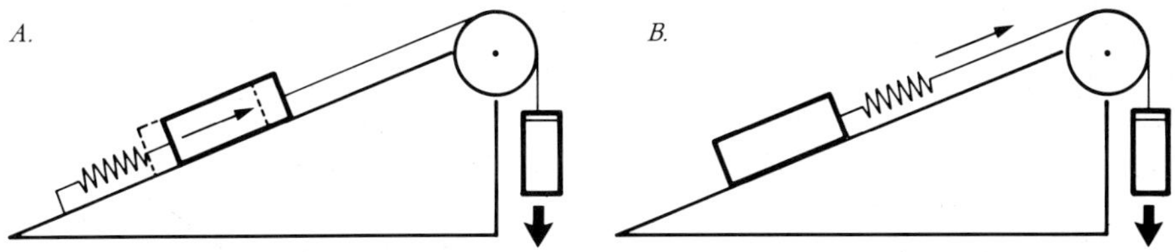

Fig. II.7-49 STABILITY OF LOAD–CONTROLLED ELASTIC/FRICTIONAL PLASTIC SYSTEM

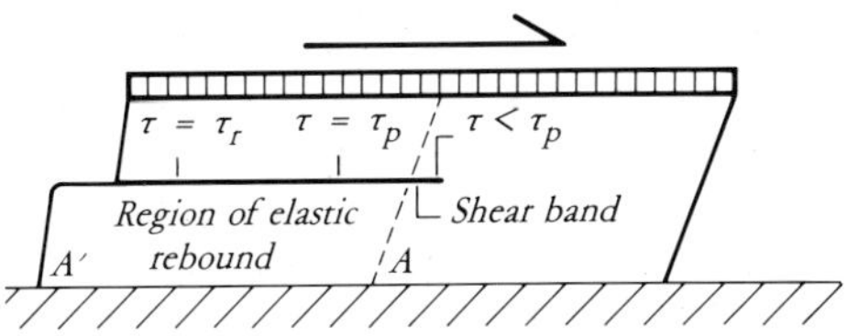

Fig. II.7-50 PLASTIC SHEAR BAND INSIDE ELASTIC MATERIAL PROPAGATING UNDER SIMPLE SHEARING

9. APPLICABILITY AND FUTURE IMPROVEMENTS

Finally, the question arises under what conditions the concepts and models of frictional plasticity, which we have discussed in a sequence of increasing complexity, may be reasonably applied to model the formation of tectonic faults. The question cannot be answered completely, mainly because the plasticity theory is a phenomenological theory and therefore treats the material on a scale that is large in comparison with the size of grains, crystals or individual joint blocks the material is made of, and ignores the underlying 'micro'processes. Assessment of the range of applicability of such a theory accrues, to a large extent, from actual applications to concrete cases – whether successful or not. Nevertheless, certain limitations in applying frictional plastic models may be related to basic assumptions made in these models, or to certain flaws that still exist within the framework of the theory.

Limitations. Beginning with the limitations implicit in the basic assumptions of the theory, the reader is reminded that in all models discussed it has been assumed that the forces are in static equilibrium. This contrasts with the possibility of an accelerated growth of a plastic softening zone discussed in the preceding section. Obviously, the static hardening/softening model cannot cope with such a growth phase, and a dynamic plastic model would be needed instead, i.e. a theoretical model in which the equations of static equilibrium are extended to include the body force associated with the acceleration of material elements. To our knowledge, numerical calculations based on a dynamic frictional plastic model with hardening/softening have not yet been undertaken to model the unstable growth of softening regions. However, the application of the static model allows us to decide whether we are dealing with a stable or an unstable fault formation, since instability would show in the numerical approach as the occurrence of uncontrollably large displacements inside the system in response to a small step in boundary displacements.

The most severe limitation of the plasticity approach to tectonic faulting stems from the rate independence of the plastic material sensu strictu, often also referred to as 'time independence'. This basic feature of plasticity, which was discussed in Section II.7.4 ('Rate independence of plastic deformation'), implies that the elastic/plastic changes in the strain and stress field occur co-instantaneously with associated changes in the boundary conditions. If the boundary geometry is changed by a displacement increment or if boundary loads are changed, the strains and stresses will change instantaneously and then remain fixed as long as the new state of boundary conditions is maintained. This excludes the phenomenon of rock *creep*, which is a flow-like continuation of deformation under a fixed load. It may die out after some deceleration phase ('primary' creep) or accelerate again after having passed through a phase of straining at approximately constant rate ('secondary' creep) finally to result in rupture ('tertiary' creep).

The Ductile Regime. These viscous phenomena are, by definition, negligible in the brittle range but, depending on the rock type, they become increasingly important at the elevated temperatures, pressures and low strain rates characteristic of what is usually referred to as the ductile range. The *transition zone between brittle and ductile crust* will therefore also mark a gradual modification in the mechanism of faulting: pressure-sensitive frictional plastic behaviour will give way to an approximately pressure-insensitive deformation with a very small stress drop that may need very large shear strains to manifest itself; faults will therefore be embedded in wide deformation zones.

The transition between the two regimes is primarily controlled by an increase in temperature that activates microscopic creep processes which, with increasing temperature, will more and more outweigh the rate-independent microscopic deformation processes and result in an increasing strain-rate sensitivity of the deformational response. Sibson (1982, 1983) has estimated the depth of the transition zone in continental crust. His estimates – based on the rheological properties of quartz-rich rocks in the laboratory, on focal depths of moderate to larger earthquakes, geothermal gradients and tectonic regimes – demonstrate the overriding effect of the crustal *heat flow*. The problem of the depth of the brittle/ductile transition zone has been further explored by Smith and Bruhn (1984) for intraplate extensional regimes. They claim only a qualitative character for their rheological/thermal model in view of the uncertainties in the calculation of temperature profiles and in the extrapolation of rheological data obtained in the laboratory [20]; their results indicate that, for the same rock type, the brittle/ductile transition zone in a cool geothermal regime may easily lie about 10 km deeper than in a hot regime. Moreover, it is understandable that the depth of the transition zone increases with the tectonic strain rate, although the predicted effect is less than that of temperature. Rock type, however, may have an important effect. Comparison of the predicted depth data shows, for instance, that for limestone the depth of the transition zone is several kilometres less than say for wet quartzite. Roughly speaking, the depth values predicted for transition zones in continental crust under various temperature and strain rate conditions and for different rock types vary between about 7 km and 30 km or more. Interestingly, Smith and Bruhn also arrive at the possible existence of vertical stacks of brittle/ductile transition zones.

Hence, in aiming for a complete description of the continuous and discontinuous deformations that accompany tectonic faulting in the rate-sensitive regime below a brittle/ductile transition zone, one must include strain rates explicitly in the stress/strain laws of plasticity. Thus the *concepts of a plastic solid and a viscous liquid have to be combined*. It should be noted, however, that such an explicit introduction of strain rates is not required for the current and limit yield conditions. As has been pointed out already (Section II.7.7, 'Shear softening'), these conditions may still be applicable in the ductile regime in unchanged form, provided the levels of strain rate, temperature and effective confining pressure are implicitly taken into account as additional parameters controlling peak strength and the strain hardening and softening moduli.

The mathematical formulation of general visco-elastic-plastic flow laws for strain hardening/softening frictional materials is outside the scope of this book. Besides, hardly any experience has been gained so far in calculating the growth of softening zones in such materials. However, we shall illustrate the necessary generalisations of the plastic hardening/softening material by modifying the analogue model of Fig. II.7-1E. Various modifications of elastic/plastic analogue models have been proposed in the past to include viscous behaviour, for example by Benioff (1951; referred to by Orowan, 1960) to explain seismic sequences, and by Price (1966) to account for time dependence of rock failure. These models have in common with the model in Fig. II.7-51 an elastic spring element (1) to account for an initial, instantaneous elastic response, a Voigt-Kelvin element (2) (spring and dashpot in parallel) to account for a delayed elastic response and a frictional (St. Venant) element (3). In the model of Fig. II.7-51 the frictional element incorporates plastic hardening and softening behaviour as in the model of Fig. II.7-1E. A viscous element (4) is coupled in parallel to the frictional element as in the Bingham model of Fig. II.7-1F. When a displacement u is applied at the rate $\dot{u}$ to both sides of the system, the spring (1) will extend instantaneously, while extension of the spring in (2) will be delayed by the viscous resistance of the dashpot ('primary' creep). As extension of the elastic elements proceeds, the force transmitted through the whole system increases and eventually the frictional element will be activated. This then represents the onset of plastic yielding with strain hardening. Further elongation of the model requires an increase in force until the maximum frictional resistance (peak stress) is mobilised in the sliding element. Note that the extension is the same for the frictional element and the dashpot in parallel. Therefore, at very low displacement rates $\dot{u}$ the frictional resistance of the dashpot is very small, but it is high at high displacement rates. Accordingly, the forces required to start movement of the sliding element and to continue it through the hardening range will increase with the displacement rates. This corresponds with the strain-rate dependence of yield and limit stresses. When the motion of the frictional element enters the softening range, the force transmitted through the system (3+4) decreases, allowing the elastic spring (1) to shrink, as indicated by the displacement u *, and to release elastic strain energy. Likewise, the spring in (2) will shrink, but with delay. If then at some softening state the force that acts upon the system is fixed, instead of proceeding with the application of a constant displacement rate, deformation will continue at increasing rate as a further decrease of the resistance in (3) will have to be compensated by an increase in the viscous resistance of (4), which implies a higher rate of deformation. When the resistance of the frictional element has dropped to its residual value, the system 3+4 will continue to deform at maximum rate ('tertiary' creep) under the constant boundary force while the elements (1) and (2) are inactive. When the external force is eventually reduced below the level required to overcome the residual frictional resistance of (3), the deformation of the whole system will stop after some delayed rebound of the elastic springs.

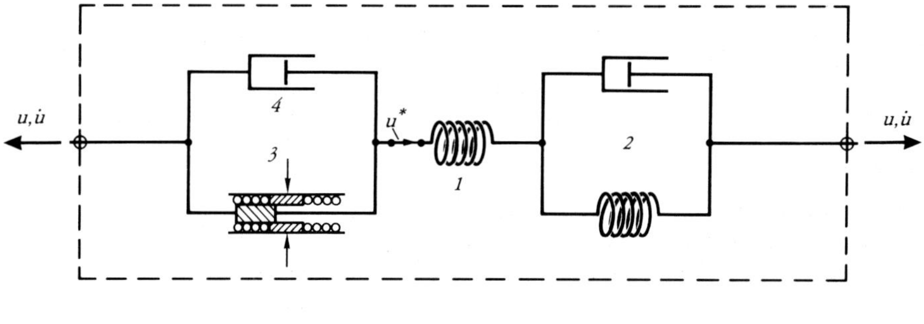

Fig. II.7-51 'BLACK BOX' ANALOGUE OF ELASTIC/FRICTIONAL PLASTIC MATERIAL WITH STRAIN RATE DEPENDENCE (Boundary displacement u and displacement rate $\dot{u}$ prescribed)

The adoption of plasticity to the ductile range is greatly simplified when *plastic strain hardening and softening is abandoned*. The generalised Bingham element 3+4 in the analogue model of Fig. II.7-51 is then replaced by the simple frictional element of the elastic/frictional plastic model of Fig. II.7-1D. The price paid for this simplification is that we can no longer model the shear concentration into individual faults but have to content ourselves with modelling the growth of plastic zones at limit state, as was discussed in Section II.7.5.

If this reduced analogue model is subject to a constant displacement rate, the stress transmitted to the frictional element will depend on the applied displacement rate. If this rate is high, the Kelvin element can accommodate only a little of the imposed displacement, since the high resistance of the dashpot blocks the deformation of the parallel spring. Most of the deformation is accommodated by the spring (1), which results in a high stress inside the system. However, if a very slow external deformation is imposed on the system, the Kelvin element will participate much more in the deformation, leaving less strain to the spring (1) and consequently enabling less stress to build up inside the system. Applying this to multidimensional deformations, we conclude that *the slower the deformation* is, which is imposed upon the system, *the more continuous strain will accumulate* before the limit state is established [21].

If boundary loads are prescribed instead of displacements, each load increment of the analogue model is transmitted to the frictional element instantaneously, since both the Kelvin-Voigt element and the elastic spring (1) are carrying the same load. Thus the rate of loading has no effect on the internal state of stress. The strains, however, are affected by the loading rate. The slower the increase in external load, the more time is provided for the Kelvin element to deform in addition to the deformation of the spring (1) which represents the instantaneous elastic response to external loading. Hence it follows again, as in the case of imposed boundary displacements, that *the slower the loading, the more continuous deformation will precede the plastic limit state*. Again we infer from this analogue of one-dimensional deformation a similar deformation behaviour in multidimensional deformation.

In accord with attaching a viscous dashpot to the elasto-plastic analogue model of Fig. II.7-1D, on the mathematical side the stress/strain equations II.7-10 have to be altered by replacing the elastic part by visco-elastic relationships between stresses, strains and strain rates. A linear relationship would correspond to a Newtonian viscous response of the dashpot in the Kelvin element, but, in general, a non-linear relationship between stresses and strain rates will be more realistic in modelling rock flow. The development of plastic regions and stress fields in the ductile range can then be calculated by numerical procedures, such as the finite-element technique.

Again, as in the example presented in Section II.7.5 (Fig. II.7-32), kinematically admissible slip lines can be fitted into the field of principal stress directions of a fully developed plastic region to represent the traces of potential faults [22]. In other words, the double-gliding model may then be applied, and it should be emphasised again that the occurrence of discrete slip surfaces and the operation of the *double-gliding mechanism are by no means restricted to the brittle domain*, but are also common phenomena in the ductile regime, as is well known from metal technology. Naturally, the angle between conjugate slip lines has to be adjusted accordingly. Since in the ductile range pressure sensitivity disappears, the angle of internal friction vanishes and conjugate slips will make an angle of 90°. The suitability of this theory for predicting the shape and pattern of potential faults can, in the end, be ascertained only by comparing the numerical results with evidence from model experiments and the field.

Here then we arrive at the *most rudimentary form of the plasticity approach to tectonic faulting*, in which the whole rock mass is assumed to be in a limit state and the slip lines are constructed after the stresses have been either determined for a plane state of stress from the equilibrium equations, the limit condition and the stresses prescribed around the entire boundary, or the field of principal stress directions has somehow been conjectured. This is essentially how plasticity theory was first applied to problems of tectonic faulting after Odé (1960) presented the concept of faulting as a velocity discontinuity in plastic deformation [22] to the geological public (Varnes, 1962; Kanizay, 1962; Rance, 1967; Cummings, 1976; Tapponier and Molnar, 1976; Mandl and Shippam, 1981). The limitations of this approach, which in fact is a special application of the double-gliding model, have already been summarised in Section II.7.4 ('Application to tectonic faulting'). In particular, it is required that the stresses be known along the entire boundary of the plastic rock mass, which restricts this approach to a few, albeit important, geological problems. However, the solutions obtained are not impaired by neglecting strain-rate effects in the ductile range, since the stress/strain relationships do not enter the calculations at all.

In fact, this *indifference with respect to brittleness or ductility* of the environment holds for any application of the double-gliding model, as long as the underlying field of principal stress directions – be it calculated, estimated or guessed – is associated with the limit state in the whole region and not with some subcritical state of deformation. Naturally, the double-gliding model itself will only describe the onset of potential slip movements without regard for viscous flow that might accompany the discontinuous movements.

Coulomb-slip Fault Representation vs Plastic Shear Concentration. The reader will have realised that in this chapter on rock plasticity, and in the tectono-mechanic applications of Part I, Coulomb's slip concept and its double-gliding generalisation play a central role. He will also have noticed a certain inconsistency in our hierarchic presentation of frictional-plastic models. While in the rigid/plastic and elastic/frictional plastic models (illustrated by the analogues of Figs. II.7-1C,D), potential faults were identified with Coulomb-slip surfaces, embedded in or bounding plastic regions, faults were modelled in a shear hardening/softening frictional material (Fig. II.7-1E) as narrow plastic zones surrounded by elastic material. Although in the example of a basement uplift (examples no. 5 and 7) the calculated softening zones agreed well with the configuration of calculated Coulomb slips, it is difficult to see why such agreement might be expected in general. It should be recalled that in the hardening/softening model shear strength was an isotropic property, while, presumably more realistically, shear softening should take place preferentially along Coulomb-slip directions associated with peak stress state. It is hoped that in the near future the incorporation of this type of anisotropic strain softening will reconcile the two concepts of plasticity-mechanical representation of tectonic faulting.

NOTES

1) Fluids with a non-linear relationship between shear stresses and shear rates are known as 'non-Newtonian'. For instance, the behaviour of ice and rock salt in simple shear flow is approximately described by the 'power law' $\tau = A(d\gamma/dt)^{1/n}$ ($n \geqslant 1$; for glacial ice $3 < n < 5.5$, for dry polycrystalline halite $n \div 6$). It should be noted, however, that Urai (1983) and Spiers, Urai et al. (1986) have demonstrated that small quantities of brine (0.1 - 1% w) in natural salt cause a significant weakening effect at low strain rates, resulting in $n \rightarrow 1$ (Newtonian) for steady-state flow.

2) A molecule of mass m carries a wall-parallel linear momentum m.u′, where u′ is the wall-parallel velocity component of the individual molecule – the average velocity of a great number of molecules being u. The net influx of linear momentum per unit time into a volume is, according to the basic laws of mechanics, equivalent to the action of boundary forces.

3) There are exceptions. Polycrystalline halite, for instance, starts flowing as a result of intracrystalline gliding at temperatures and pressures that already prevail in the upper few kilometres of the crust. This flow, however, is not of the plastic type but rather of a (non-linear) viscous type. It seems that unrealistically high shear rates would be required to approach ideal plastic behaviour more closely.

4) Naturally, during the initiation of thrusting the substratal flow will be accelerative. It can be shown, however, that during a period of changing rate of overthrusting U (t), at any instant t the substratal flow is practically identical with the 'stationary' flow shown in Figure II.7-6A, i.e. the steady flow produced by a plate moving at a constant speed equal to the instantaneous rate U(t). This conclusion is based on the rigorous solution for the 'transient' shear flow caused by a constant shear stress (Lamb, H.; 'Hydrodynamics', Dover Publ. Inc., New York, 1932, p. 592). The time required to reduce the difference between instantaneous and stationary velocity profiles to about 1/3 of its initial value is $0.4\ \varrho\ h^2/\eta$ seconds, where ϱ is the rock density, η the viscosity in poise, h the substratum thickness in cm. This time becomes practically zero even for large values of h in view of the high rock viscosities. The viscosity of rock salt, for instance, would presumably range in order of magnitude between 10^{16} and 10^{20} poise.

5) The shear stresses are directed perpendicularly to the σ_{Int} direction, since otherwise the principal plane upon which σ_{Int} is acting would not be free of shear stresses. By virtue of the symmetry of shear stresses (Chapter II.2, eq. II.2-3, 'Symmetry of shear stresses ...', 'Plane problems', Fig. II.2-5) a shear stress component on elements of Fig. II.7-8D acting parallel to the σ_{Int} direction would imply shear stresses on the principal plane.

6) The stress measurements in the ring-shear apparatus (Mandl et al., 1977) were carried out in the granular pack outside the shear band proper. Thus only for the material outside the shear band may we assert the 45° orientation of principal stress axes. Nothing is known about the stress orientation inside the shear zone proper.

7) Spencer also showed (1971), in a discussion of similar, but somewhat simpler equations derived by Mandl and Fernàndez Luque (1970) and Harkness and Butterfield (1971) from the double-gliding model, that his equations (Fig. 30) are the more general ones since they also correctly account for shearing along *curved* slip lines, while the equations derived by the other authors imply shearing only along the *tangents* to curved slip lines. The difference between the two formulations disappears when the direction of σ_{Max} remains constant both in space and time. It is now agreed that when σ_{Max} varies, Spencer's equations should be applied. Readers interested in more details may be referred to a paper by Spencer (1982).

8) In mathematical terms, the slip lines are the characteristics of the two quasi-linear hyperbolic velocity equations. As such, they are the potential carriers of discontinuities in the tangential velocity component or its tangential derivative.

9) The fact that within the framework of the double-shearing model no *internal rotations* can exist other than those associated with the action of the simple shears along the slip lines has been critisised by de Josselin de Jong on several occasions (e.g. 1971). He argues that, in addition to the double-shearing deformation, 'free rotations' may exist that vary from place to place. Accordingly, de Josselin de Jong's equations form an indeterminate system (cf. Spencer 1982, p. 622). We believe that such extra internal rotations would have to be the result of elastic deformations not accounted for in the double-shearing model of a rigid/frictional plastic material.

10) If we make use of the well-known expressions for the direction of σ_{Max} in the plane of maximum and minimum principal stresses (see, for example, eq. II.4-4) and of the direction of the maximum contractive strain (eq. II.5-15), co-axiality in the x_2, x_3-plane implies:

$$\frac{\tau_{23}}{\sigma_2 - \sigma_3} = \frac{\delta e^p_{23}}{\delta e^p_2 - \delta e^p_3}$$

This may be rewritten by replacing the stress difference by the difference of deviatory stresses and noting that the sum of deviatory stresses $s_2 + s_3 = 0$, and $\delta e^p_2 + \delta e^p_3 = 0$:

$$\frac{\delta e^p_2 - \delta e^p_3}{s_2 - s_3} = \frac{\delta e^p_2}{s_2} = \frac{\delta e^p_3}{s_3} = \frac{\delta e^p_{23}}{\tau_{23}} = \lambda$$

11) The work expended in incremental plastic plane straining of a unit volume element follows from equation II.8-11a:

$$\delta W_p = (dW_p/dt)\ \delta t = \sigma'_{Max}\delta e^p_{Max} + \sigma'_{Min}\delta e^p_{Min} = \frac{\sigma'_{Max} + \sigma'_{Min}}{2}(\delta e^p_{Max} + \delta e^p_{Min}) + \frac{\sigma'_{Max} - \sigma'_{Min}}{2}(\delta e^p_{Max} - \delta e^p_{Min})$$

Since we assume that the plastic deformations do not change the volume, the sum of the principal strains vanishes and the plastic work increment becomes:

$$\delta W_p = \tau_{Max}(\delta e^p_{Max} - \delta e^p_{Min}) = \tau_{Max} \cdot \delta\gamma^p_{Max} > 0$$

where γ is the 'engineering' shear strain, i.e. the shear angle in radians (see for example, eq. II.5-3). The work expression is positive because the work is 'fed' into the volume element. Because of the flow rule II.7-7, one has

$$\delta\gamma^p_{Max} = 2\,\delta e^p_{Max} = 2\lambda \cdot \tau_{Max}$$

and obtains

$$\delta W_p = 2\,\lambda\,\tau^2_{Max} > 0$$

and hence

$$\lambda > 0\,.$$

12) One obtains, for example,

$$\delta e^p_2 = \lambda \frac{\partial G}{\partial \sigma'_2} = \frac{\lambda}{2}(\sigma'_2 - \sigma'_3) = \frac{\lambda}{2}(s_2 - s_3) = \lambda\, s_2$$

because of $s_2 + s_3 = 0$ in plane deformation, according to eq. II.7-8.

13) Expressed in our x_2, x_3-coordinate system the Drucker-Prager flow rules read:

$$\delta e^p_2 = \lambda\left[\, s_2 - \left(\frac{\sigma'_2 + \sigma'_3}{2} + \tau_o \cot\varphi\right)\sin^2\varphi\,\right]$$

$$\delta e^p_3 = \lambda\left[\, s_3 - \left(\frac{\sigma'_2 + \sigma'_3}{2} + \tau_o \cot\varphi\right)\sin^2\varphi\,\right]$$

$$\delta e^p_{23} = \frac{\delta\gamma^p_{23}}{2} = \lambda\,\tau_{23}$$

From these flow rules follows a plastic volume increase (counted as negative in accord with the negative sign assigned to extensional strains):

$$\frac{\delta V^p}{V} = \delta e^p_2 + \delta e^p_3 = -\lambda\left[\,\sigma'_2 + \sigma'_3 + 2\,\tau_o \cot\varphi\,\right]\sin^2\varphi$$

or, after inserting from the Coulomb-Mohr limit condition II.7-3a:

$$\frac{\delta V^p}{V} = - \lambda \left(\sigma'_{Max} - \sigma'_{Min} \right) \sin \varphi$$

Since the first two of Drucker-Prager's equations above also hold when δe^p_2, δe^p_3 are replaced by δe^p_{Max}, δe^p_{Min} and s_2, s_3 by s_{Max}, s_{Min}, one obtains for λ;

$$\lambda = (\delta e^p_{Max} - \delta e^p_{Min})/(\sigma'_{Max} - \sigma'_{Min})$$

Inserting this expression into the formula for the volume change, one finally arrives at

$$\frac{\delta V^p}{V} = - \left(\delta e^p_{Max} - \delta e^p_{Min} \right) \cdot \sin \varphi = - \delta \gamma^p_{Max} \cdot \sin \varphi$$

14) The problem is further complicated by the fact that internal constraints that vary in time give rise to stresses that do perform work.

15) When the same approximation is applied to defining the plastic flow potential in principal stress space, the following difference should be noted: when the pyramid-type potential (resembling the yield pyramid in Fig. II.7.43B) is differentiated with respect to the principal stress component that happens to be the intermediate one, the associated plastic strain-rate component vanishes. In other words the plastic strain remains confined to the plane of the maximum and minimum principal stresses, in accord with Coulomb's concept. When, however, the plastic potential is represented by a circular cone in principal stress space, the derivative with respect to the intermediate principal stress component does not vanish and plastic straining is no longer planar. When dealing with plane strain problems, we then have to invoke an elastic out-of-plane strain increment to cancel the corresponding plastic component.

16) In numerical studies of faulting, Walters and Thomas (1982) applied the frictional hardening/softening relationship

$$\tan \gamma = \tan \gamma_r + [4 (\tan \gamma_p - \tan \gamma_r) \cdot B \cdot k] \cdot (B+k)^{-2}$$

where γ_r and γ_p correspond to the residual friction angle φ_r and the friction angle at peak φ_p:

$$\tan \gamma_{r,\,p} = 3^{1/2} \sin\varphi_{r,\,p} \cdot (3 + \sin^2 \varphi_{r,\,p})^{-1/2}$$

17) The finite element scheme adopted by Thomas and Walters for modelling the generation of tectonic faults as a phenomenon of elastic/frictional plastic rock deformation with strain hardening/softening utilises 8-noded isoparametric elements with 2×2 Gaussian integration points.

The scheme is based on an incremental-iterative form of the 'initial stress method'. This implies that at each loading step first a purely elastic problem is solved and strains are determined by using the *same* elastic stiffness matrix. Naturally, this will lead to stress states, which, although in static equilibrium, may lie outside the current yield cone. In an iterative procedure nodal correction loads are applied to redistribute the 'excess' stresses elastically so as to re-establish a current yield state and, at the same time, to satisfy the conditions of static equilibrium. In each iteration step the elastic excess stresses are determined by comparing the elastically determined state of stress (outside the yield surface) with a corresponding state of stress on the yield surface. The latter state is found by considering the total strain associated with the elastic state of stress and dividing it into an elastic strain and a plastic strain increment. The plastic increments are determined in each iteration step by means of the plastic flow rules.

The extensive computing effort has been significantly reduced by a new procedure (Thomas, 1984) which accelerates the convergence of the iteration cycles and improves the numerical stability of the analysis. This has proved particularly useful in the post-peak range. So far, this finite element scheme has been applied only to problems of plane and axisymmetric strain.

18) The analysis is complicated by the shear concentration that is likely to take place inside a softening inclusion and may effectively change its shape.

19) The non-uniformity of shear displacement along the base of large, internally deforming sliding masses and its effect on the onset of progressive slope failure has been already analysed by Müller and Malina (1968).

20) An excellent 'state-of-the-art' review of the rheological deformation mechanisms in the lithosphere has been presented by Kirby (1983).

21) For a Voigt-Kelvin material under multi-axial loading it holds true that, if stresses are prescribed along the entire boundary, the internal stress field is exactly the same as that calculated for a linearly elastic material under the same boundary conditions.

Similarly, when the boundary conditions are formulated in terms of displacements only, the displacement field will asymptotically approach that calculated under the assumption of linearly elastic response.

22) Sometimes the objection has been raised that a slip line *inside* a plastic region cannot carry displacement discontinuities, and, therefore, cannot represent a fault, since at the same time the tangential normal stress would have to suffer a jump-like discontinuity across the slip line. This, however, would contradict the mutual exclusion of stress and displacement jumps across a slip line in the interior of a plastic region as stated by a classical theorem of plasticity theory (Hill, 1950; Salençon, 1977). The theorem applies only to the kind of plasticity theories in which the characteristics of the displacement field (see also note 8) are identical with the stress characteristics. This is indeed the case when Spencer's flow equations (Fig. II.7-30) or flow equations derived by Shield (1953) from Drucker-Prager's flow rules (see note 13) are used.

The objection against the slip line representation of faults does not concern slip lines that form part of the boundary of a plastic region. However, it may be also discarded with regard to internal slip lines for the following reason: when a slip is mobilised, the shear strength will change inside a narrow slip band and inhomogeneity in the mechanical properties of the system does allow for a jump in the tangential normal stress component. In fact, the decoupling of different stress regimes by tectonic faults is based on such shear-induced strength inhomogeneities (see Fig. II.7-13).

II.8. WORK

1. THE CONCEPT

The mechanical interaction between a material system, e.g. a rock mass, and its surroundings is envisaged as the action of forces. Such forces are necessary to produce a mechanical process in the system, i.e. to generate motion of the system as a whole or relative motion of its parts (i.e. deformation). However, the mere occurrence of forces at the surface of a heavy material body in itself is not sufficient to produce such a process; there must be some surface forces acting across a part of the boundary material that *moves* in the direction of the applied forces. In other words, motion can only be imposed mechanically [1)] upon a material body by movement of some adjacent bodies. To illustrate this, Figure II.8-1A shows a 'rigid' block resting on an immobile base. The stresses at the base of the block support its weight but do not induce motion. In contrast, Fig. II.8-1B shows the same block being moved by a push force, while Fig. II.8-1C shows a deformable block 'glued' to an immobile base, but deformed by a tangential force $\vec{T}$ along its upper surface. (Vertical surface forces are applied to maintain the horizontal position of the top surface.) A more general case is sketched in Fig. II.8-2 which shows surface loads ($\vec{F}_1, \vec{F}_2, \ldots$) that act upon a deformable body. Part of the body surface is kept at rest, while the material of the deformable part of the surface is displaced ($\vec{u}_1, \vec{u}_2, \ldots$) at the points where the boundary loads are applied. Although, in general, the boundary displacements and boundary forces will not be collinear, the boundary displacements will usually have a component (d_1, $d_2, \ldots$) parallel to the corresponding force.

Definition. From this preliminary consideration it should be clear that, when looking for physical quantities suitable to characterise a motion-generating mechanical interaction between a material system and its surroundings, one has to take into account both boundary forces and displacements. The simplest quantity that comprises boundary forces and boundary displacements is the product of the magnitude of a boundary load $\vec{F}$ and the load-parallel boundary displacement d at the point where the load is applied. This product [2)]

$$W = |\vec{F}|\, d \qquad \text{(II.8-1)}$$

is called the mechanical *'work'* of the boundary force $\vec{F}$. The work is zero when force or displacement in the force direction vanish. If none of the applied forces perform work, they do not induce any motion or deformation. The 'work' is counted as positive or negative, depending on whether the displacement component d has the same orientation as the corresponding external boundary forces or is opposite to it. In the first case a boundary force – or, as the case may be, the surrounding – is said to perform work on the material system;

in the second case the system performs work on its surroundings. For example, in Fig. II.8-1C the boundary force $\vec{T}$ performs work in shearing the solid block, while in Fig. II.8-1D the block, approaching its original shape against an opposing boundary load, performs work ($W<0$) on its surroundings.

In general, it is a whole system of boundary forces that performs work on a material body. The total work done by this system is then simply the algebraic sum of the contributions of the individual boundary forces $\vec{F}$,... $\vec{F}_n$ (Fig. II.8-2).

$$W = |\vec{F}_1| \cdot d_1 + |\vec{F}_2| \cdot d_2 + \ldots\ldots + |\vec{F}_n| \cdot d_n \qquad \text{(II.8-1a)}$$

Usually, in geological problems the boundary forces cannot be regarded as acting at a number of points on the boundary of a rock mass; rather they will be continuously distributed over its surface Σ. The boundary force $\vec{F}$ that acts across a small element $d\Sigma$ of the boundary then has to be expressed in terms of stress $\vec{t}$ (i.e. the force per unit area of a surface element; Section II.1.1) by $\vec{F} = \vec{t} \cdot d\Sigma$. To obtain the total work of the boundary stresses, one has to sum (integrate) the work contributions $|\vec{t}| \cdot d \cdot d\Sigma$ of all the infinitesimal elements of the entire boundary Σ [3)].

Body forces – in particular the specific weight – can also perform work in displacing the volume elements of a material body. Consider, for instance, a block sliding down an inclined plane (Fig. II.8-3). On each unit volume element of the body the specific weight $\vec{\gamma}$ performs the work $\gamma \sin \delta \cdot d$, where d is the distance the block has been displaced and δ is the slope angle. If we add the work contributions of all volume elements, then the total work performed by the weight in a downslope displacement d is $V \cdot \gamma \sin \delta \cdot d$, where V is the volume of the block.

Units. According to its definition, work is a number with the physical dimensions of force times length. It may be expressed in various units:

$$1\ \text{J(oule)} = 10^7\ \text{erg (dyne.cm)} = 1\ \text{N(ewton).m} = \frac{1}{9.81}\ \text{kp (kilopound).m}$$

Variable boundary forces. Before discussing the eminent physical significance of the quantity 'work', we have to emphasise that the definition II.8-1,1a assumes that the forces remain constant while performing work. This is not always the case. Take, for instance, an elastic spring and shorten it by a total amount d^* (Fig. II.8-4). Let the force that keeps the spring at rest in this contracted state be F^*. It would then be wrong to expect that the work $F^* \cdot d^*$ had to be performed in producing this shortening. We know that keeping the spring in a less contracted state requires a smaller load. Consequently, the final state of shortening can be obtained by gradually increasing the load F from 0 to F^*, allowing the spring to attain static equilibrium after each loading step. In order to determine the total work performed on the spring during this quasi-statical operation, we therefore have to divide the total displacement d^* into a succession of incremental displacements δd, each of which is small enough to allow the loading force to be approximated as constant during the incremental displacement. The *elementary work* performed during such an incremental displacement then becomes

$$\delta W = F(d)\, \delta d$$

or, more generally, (II.8-1b)

$$\delta W = F_1(d_1,..d_n) \cdot \delta d_1 + \ldots F_n(d_1,...d_n) \cdot \delta d_n$$

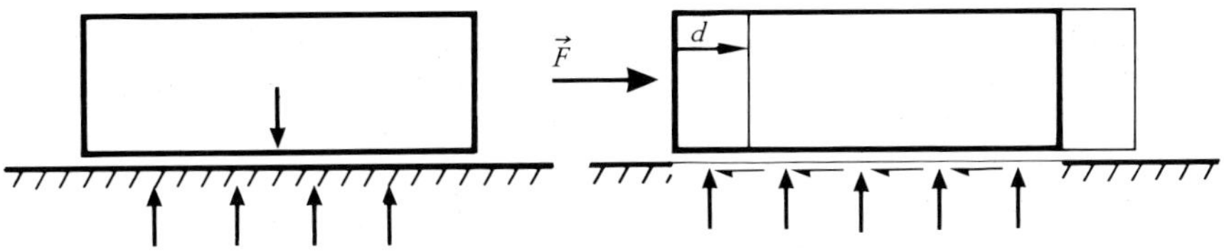

A Weight supported by normal stresses along base.

B Block pushed over frictional base.

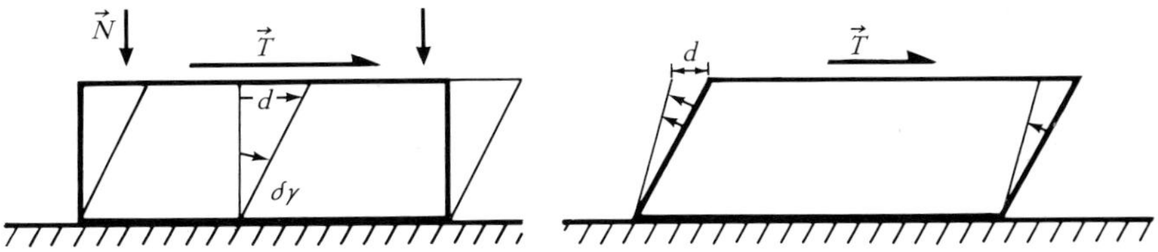

C Block adhering to base and sheared by tangential force $\vec{T}$ (Normal forces $\vec{N}$ keep surface horizontal.)

D Block approaching original shape against tangential force.

Work performed on block: $W = |\vec{T}| \cdot d > 0$

Work performed by block itself: $W = |\vec{T}| \cdot d < 0$

Fig. II.8-1 HEAVY BLOCK ON IMMOBILE BASE SUBJECT TO DIFFERENT LOADS

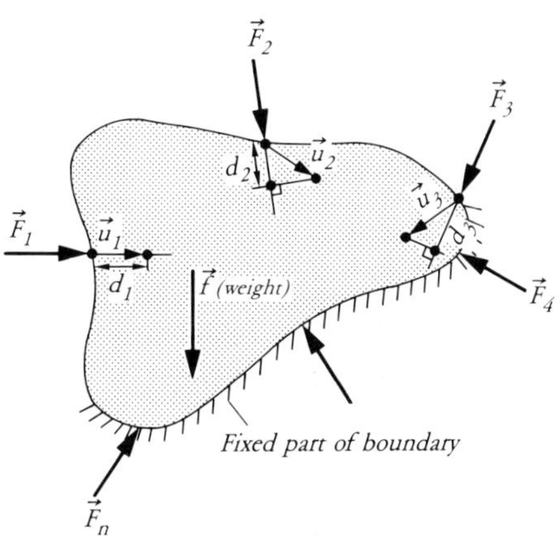

Fig. II.8-2
FORCES AND DISPLACEMENTS

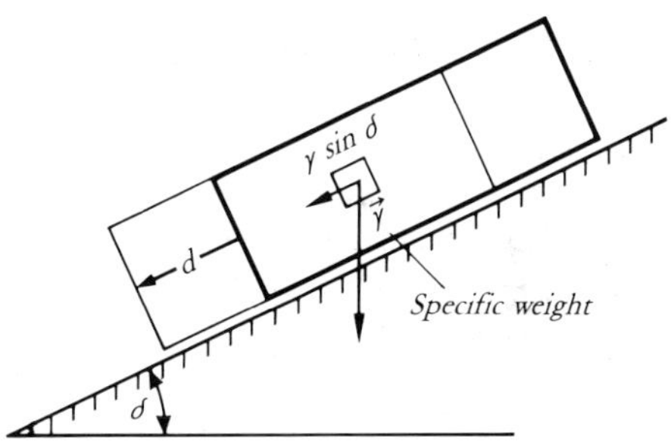

Fig. II.8-3
WORK DONE BY WEIGHT OF BLOCK ON SLOPE

when a *system of varying boundary forces* is applied. Note that for an elastic solid the loads at a given state of boundary deformation will depend on the displacement at the points of load application.

The *total work* performed in producing the final state of deformation is obtained by summing over the elementary works performed during infinitesimal loading steps (integration). For a linearly elastic spring the load is proportional to the shortening, as is indicated very schematically in Fig. II.8-4. The sum of the elementary works, i.e. the total work, appears in the load – contraction diagram as the area bounded by the load curve. It is obvious that as the displacement increments δd are made 'infinitesimally' small ($\delta d \rightarrow 0$), the total work performed in producing the linearly elastic shortening d^* becomes

$$W_{el} = \frac{1}{2} F^* \cdot d^* \tag{II.8-2}$$

rather than $F^* \cdot d^*$, the work that would have been performed by a constant load F^*.

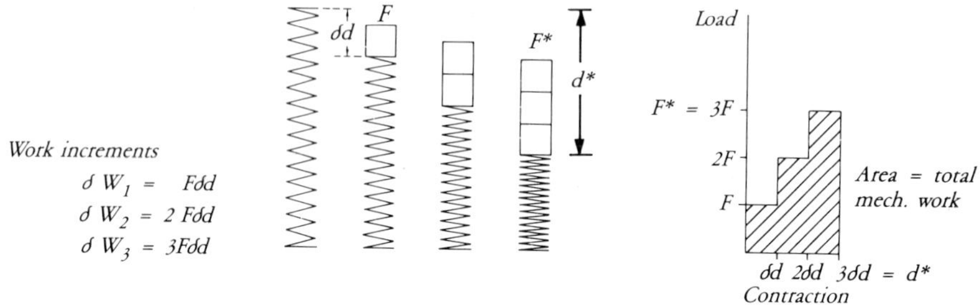

Fig. II.8-4 INCREMENTAL LOADING OF ELASTIC SPRING

2. WORK AND ENERGY

If the loading of the elastic spring in Fig. II.8-4 is reversed, the spring will perform work by lifting the loads while returning to its original shape. For an ideally elastic spring this work equals exactly the work 'fed' into the spring during the quasi-statical loading operation. By definition, this holds for any ideally elastic body. The mechanical work performed in deforming the elastic body is therefore stored as 'potential' or 'strain' energy inside the deformed body to be transformed back into mechanical work during unloading.

The reader may already have asked himself what would happen if the elastic spring were suddenly loaded by the full weight F^* instead of quasi-statically increasing the load. Obviously, the load in excess of that required for quasi-static shortening of the spring would accelerate the spring/weight system and the work performed by the constant load F^* would partly become kinetic energy of the system,[4)] feeding oscillatory motion of the system around the state of shortening d^*.

Similarly, the work done against gravity in pushing, for instance, a block up an inclined 'frictionless' plane is stored as potential energy in the elevated block and is completely recoverable as mechanical work when the block is allowed to slide down the frictionless slope.

In reality the mechanical work done in moving or deforming a body is never completely recoverable. There are always frictional processes operating at the boundary or inside the material system which dissipate at least part of the work that is fed into the system. For instance, in keeping the block in Fig. II.8-1B moving at constant speed, the work performed by the necessary push force is completely dissipated by the friction along the base of the block. Two important classes of idealised materials – the viscous fluids and the perfectly plastic solids (see Section II.7-1) – are even distinguished by the complete dissipation of any mechanical work performed on such materials.

Dissipation of mechanical work does not mean that work is 'annihilated', but rather that it is transformed into a non-mechanical form of energy – heat. That mechanical work is a form of energy transformable into heat is the essence of the First Law of Thermodynamics. The equivalence of mechanical work and heat is specified quantitatively by the *'mechanical equivalent of heat'*

$$1 \text{ kcal (kilocalorie)}^{5)} \approx 427 \text{ kp.m} \approx 4200 \text{ joule} = 4.2 \ 10^{10} \text{ erg} \qquad \text{(II.8-3)}$$

It is often necessary to specify the amount of heat produced *per unit time* from mechanical work. One therefore has to define the rate at which mechanical work is performed on a material body. This rate is called *'power'* and is defined as the elementary work δW divided by the 'infinitesimal' time span δt during which it is performed: $\delta W/\delta t$.

3. APPLICATIONS

We shall now present applications of the work concept to tectonic faulting, which, although very elementary, provide some interesting results on the general mechanics of tectonic faulting.

Example No. 1: Work of tectonic forces in faulting

Tectonic faults are narrow zones of concentrated shearing activity and most, if not all, of the work performed by the tectonic forces in producing or maintaining fault movements will be dissipated inside the narrow fault zone proper by solid friction, cataclasis, intercrystalline gliding, etc. The force that performs work in faulting is the tangential force exerted by the moving fault block upon the fault zone proper (Fig. II.8-5). If we consider 1 cm² of the fault boundary, this force becomes the shear stress τ, and the rate of work performed on this unit area is

$$\delta W/\delta t = \tau \cdot \delta d/\delta t \qquad \text{(II.8-4)}$$

where d is the displacement of the moving fault block.

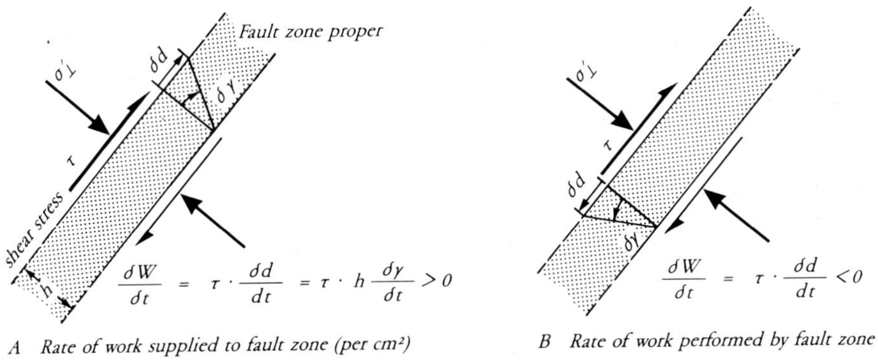

A Rate of work supplied to fault zone (per cm²) *B Rate of work performed by fault zone*

Fig. II.8-5 DISSIPATION OF MECHANICAL WORK IN FAULT ZONE

When both shear stress and displacement have the same direction (Fig. II.8-5A) the rate of work performed on the fault zone proper is positive, which means that mechanical energy is 'fed' into the fault zone in which it is dissipated. In contrast, a situation in which shear stress and displacement have opposite directions (Fig. II.8-5B) would imply that the fault zone proper performs work against the external shear stress and, consequently, mechanical energy would flow from the fault zone into the adjacent rocks. This is obviously impossible and puts certain restrictions on the sense of fault movements, as is discussed in Section I.2.1.

Example No. 2: Accommodation faulting along a curved main fault

When a rock mass, in moving along a partly curved stationary fault, has to adjust its shape to the changing curvature of this fault, this accommodation is often accomplished by subsidiary faulting of the moving mass. Figure II.8-6A shows the dip section of a stationary normal fault that flattens with depth. The necessary deformation of the downthrown block is brought about by subsidiary faults of the type shown. These subsidiary faults move along the stationary main fault with the sliding block. The sense of displacement along these faults is opposite ('antithetic') to that of the main fault. They are therefore referred to as 'antithetic' faults. In contrast, subsidiary faulting in the downthrown block of Fig. II.8-6B has to be 'synthetic' in displacement with the stationary main fault, whose dip increases with depth [6)]. Along all these faults mechanical work has to be dissipated. Let us examine this and see what further conclusions of interest may be drawn.

At any point in the material two perpendicular planes exist that are free of tangential stresses and are acted upon by, respectively, the maximum and the minimum principal stresses σ_I and σ_{III} (see Section II.1.3). Figure II.8-7 shows the intersections of these 'principal' planes with the plane parallel to which the principal stresses σ_I and σ_{III} act. The angles between the principal planes are bisected by the planes along which the shear stress attains its maximum value and the orientation indicated. On any other element the orientation of the shear stress is found by considering the symmetry with respect to the principal planes ($\tau = 0$). Along any of these planes a possible fault displacement (slip) would have to be oriented in the same way as the shear stress to ensure the dissipation of mechanical

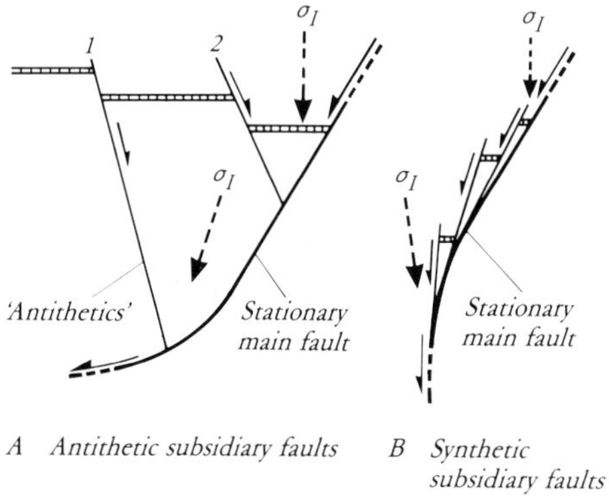

Fig. II.8-6
ACCOMMODATION OF MOVING BLOCK BY SUBSIDIARY FAULTING

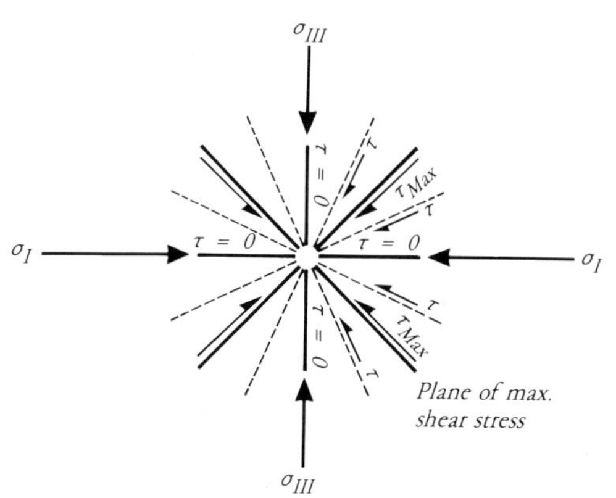

Fig. II.8-7
ORIENTATION OF SHEAR STRESSES WITH RESPECT TO PRINCIPAL STRESS DIRECTIONS

work. Therefore, when comparing Fig. II.8-7 with Fig. II.8-6A, we notice that the σ_I-direction must always lie within the angle enclosed by the main fault and subsidiary antithetic faults. In contrast, the σ_I-direction can never lie in the angle between the main fault of Fig. II.8-6B and one of its synthetic subsidiaries. As indicated, it will always lie at the downthrown side of the subsidiary faults.

Finally, the reader may note by considering Fig. II.8-7 that the angle between two active synthetic faults can never exceed 90°.

Example No. 3: Heating of fault zone material

The mechanical work transformed into heat in a fault will tend to raise the temperature of the material in the fault zone proper. Particularly in rapid faulting ('seismic faulting'), in which the rate of actual fault movement $\delta d/\delta t$ lies in the range of 10–100 cm/s, practically all the heat produced during a single 'jerk' displacement of the fault blocks will first accumulate inside the fault zone proper before it can gradually flow off into the surrounding rock. We may therefore expect a substantial increase in the temperature of the fault material and shall try to quantify this increase.

Expressed in heat units, according to eq. II.8-3, the rate of work (eq. II.8-4) dissipated in the fault zone proper amounts to heat supplied at the rate

$$q = 24 \cdot 10^{-9}\,\tau \cdot \delta d/\delta t \ \text{cal/cm}^2\text{s} \qquad \text{(II.8-5)}$$

where the shear stress τ is measured in dynes/cm², d in cm and t in seconds. Let us consider the case that during an episode of rapid faulting the heat is generated at a constant rate and, more or less, uniformly throughout a shear band of width h (Fig. II.8-8). Then the heat q.t, produced during the time span t inside a prismatic shear zone element of 1 cm² cross-section, is stored in this element as 'sensible' heat. Its amount is ϱ C.h.ΔT, where ϱ is the rock density, C the heat required to raise the temperature of 1 gram by 1°C, and ΔT the increase in temperature above that of the ambient material. Hence we have the simple 'heat balance'

$$\varrho C \cdot h \cdot \Delta T = q \cdot t \qquad \text{(II.8-6)}$$

or, after inserting from eq. II.8-5:

$$\varrho C \cdot h \cdot \Delta T = 24 \cdot 10^{-9}\,\tau \cdot d \qquad \text{(II.8-6a)}$$

If we insert average values for density and specific heat of the sheared material ϱ = 2.5 g/cm³, C = 0.2 cal/g °C, and choose a realistic value for the shear stress that acts along the fault plane, namely τ = 200 bar = 2.10^8 dynes/cm², the temperature rise becomes

$$\Delta T \cong 10\, d/h\,^\circ\text{C} \qquad \text{(II.8-6b)}$$

Since shear stresses along active faults will hardly attain values more than two or three times larger than the one assumed above, formula II.8-6b implies that it is mainly the ratio between fault displacement and width of shear zone that determines the rise in fault zone temperature, provided the fault movement is rapid.

The temperature considered here is, of course, an average temperature; the temperature

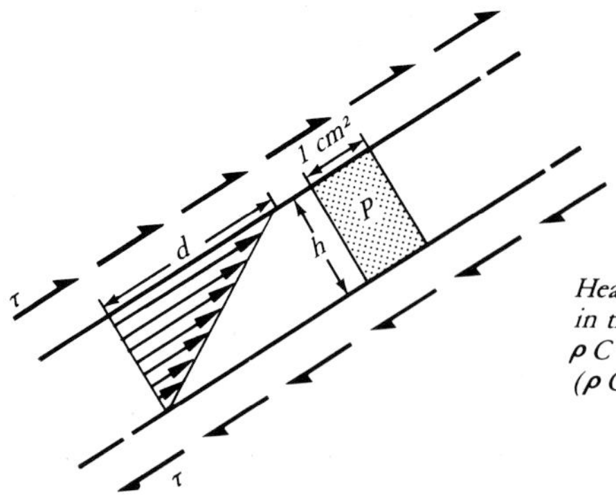

Heat stored in prismatic element P in the absence of heat losses:
$\rho C \cdot h \cdot 1 \cdot \Delta T$
(ρC specific heat capacity, ΔT increase in temperature)

Fig. II.8-8 HEAT GENERATED INSIDE SHEAR BAND

at frictional contacts between individual grains of a rock may easily attain higher values, giving rise to fusion of the material at intergranular contacts. It is of interest to know under what conditions the average temperature T would rise to a level high enough to initiate melting of fault-zone material. The melting temperatures for most rocks lie in the range of 1000° – 1500°C (see Rowley, 1974). It is obvious from eq. II.8-6b that such temperatures will not be reached in those cases where a rapid shearing event affects the full width of a cataclastic shear zone tens of centimetres or even metres wide. However, many observations in nature and in the laboratory indicate that, after a cataclastic shear zone has formed, shearing eventually concentrates on slip *planes*. Then formula II.8-6b is no longer applicable since practically no heat can be stored in the 'infinitely' thin shear zone and all the heat produced by friction must flow instantaneously into the adjacent rock. This is in contrast with the assumptions made in deriving eq. II.8-6b and, therefore, the problem has to be tackled in a different way.

If the flow of heat across the plane face of a body of semi-infinite extent is known, a well-known formula of the theory of heat conduction (see Carslaw & Jaeger, 1959; p. 75, eqs. 7,8) enables the temperature distribution inside and at the boundary of the heated body to be calculated as a function of time. Naturally, the temperature is highest at the heated surface of the body and this is the temperature we are primarily interested in. For this boundary temperature the formula mentioned above simplifies to

$$\Delta T(t) = \frac{\sqrt{\varkappa}}{K} \cdot \frac{q}{\sqrt{\pi}} \cdot \sqrt{t} \qquad \text{(II.8-7)}$$

where K is the heat conductivity, i.e. the flow of heat per second through a cross-section of 1 cm^2, 'driven' by a temperature gradient of 1°C per cm, and $\varkappa = K/\varrho C$ is the 'heat diffusivity'. Typical values of these material coefficients for rocks are $K = 5.10^{-3}$ cal/cm.s. °C and $\varkappa = 10^{-2}$ cm^2/s. With these values inserted, and substituting for q from eq. II.8-5, equation II.8-7 may be written as

$$\Delta T \cong 2{,}7 \cdot 10^{-7}\, \tau \sqrt{\frac{\delta d}{\delta t} \cdot d} \qquad \text{(II.8-7a)}$$

For the shear stress value of 200 bar = 2.10^8 dynes/cm² the temperature rise at the slip plane becomes

$$\Delta T = 54 \sqrt{\frac{\delta d}{\delta t} \cdot d} \qquad \text{(II.8-7b)}$$

and we see that the product of displacement rate and total fault displacement has to attain values above 370 to enable the fault plane temperature to rise by more than 1000°C. Thus, at a slip rate of say 50 cm/s, the total displacement of one 'jerky' movement would have to be at least 7.4 cm. These are very reasonable values and one should therefore expect melting of fault plane material to be quite commonly associated with rapid faulting events.[7] Surprisingly enough, *melted fault gauges* ('Pseudotachylytes') have been observed on only a few occasions, and several researchers have tried to explain the unexpected scarcity of rock melts in fault zones. Jeffreys (1942) concluded that the individual fault slips were generally too short in length to produce the heat required for melting. N.J. Price (1970) inferred that the shear stresses along faults are smaller than generally assumed, and Sibson (1973) pointed out that in low-permeability rocks the rise in fault plane temperature will be accompanied by a local increase in pore water pressure which lowers the shear stress τ, thereby impeding further heat generation. According to Sibson, melting of fault material is only likely to take place along faults in high-strength (i.e. requiring high τ-values) initially dry rocks where seismic faulting takes place on discrete planes.

We would like to add here that, once a skin of melted material has formed on the slip plane, it takes only a modest further rise in temperature to reduce the viscosity of the melted rock drastically.[8] The melted fault lining itself would then act as a lubricant and cause a drastic drop in shear stress τ. Similar to an increase in pore pressure, the melted skin layer would impede further heating and so prevent the production of conspicuous layers of melted rock.

4. WORK IN STRAINING DEFORMABLE MATERIAL

We have discussed the work performed by the boundary forces in deforming a material body. What has been stated with respect to the body as a whole also applies to each of its deforming elements. This means that in straining a volume element inside the body the adjacent material performs work on this element, or vice versa, via the stresses that act across the surface of the element. Naturally, the work contributions of all the volume elements together equal the deformation work performed by the boundary forces on the body as a whole. When material and deformation are uniform throughout the deformed body the work expended in deforming an element of given volume is a well-known part of the work expended in deforming the whole body. Often, however, the deformation is inhomogeneous and the work is not uniformly distributed over the interior of the body. In that case it is often desirable to have the work that is *locally* expended in straining a small volume element expressed in terms of the local stresses and strains (or strain increments). It is this work that is stored in the volume element as elastic strain energy or is transformed inside the volume element into heat. We shall give an elementary derivation of the work expended in straining a small volume element, restricting ourselves to 'small' strains (i.e. strains where the relative changes in lengths or angles (measured in radians) are $\ll 1$).

We shall first examine two simple cases in which the expression for the work is easily found. We consider the *uniaxial* compression or extension of a small cubical element with faces perpendicular to the axes of a Cartesian coordinate system (Fig. II.8-9A). The only stress that acts upon this element is the normal stress σ_1, so that the faces perpendicular to the x_1-axis are loaded by the force $\sigma_1 \cdot dx_2 \cdot dx_3$. It causes the distance between these faces to change by $d = e_1 dx_1$. Let us first assume that σ_1 remains constant during the uniaxial straining (e.g. in deforming viscous or plastic material). Then, according to eq. II.8-1, the work performed on the element is $W = \sigma_1 dx_2 dx_3 \cdot d$, which can be further written

$$W = \sigma_1 \cdot e_1 \cdot dx_1 dx_2 dx_3 \qquad \text{(II.8-8a)}$$

According to the sign conventions used (compressive stresses and contractions positive), the product $\sigma_1 e_1$ remains positive, irrespective of whether the positive work performed on the element is associated with compression or extension. (If the element performs work on its surrounding, σ_1 and e_1 have opposite signs and $W < 0$.) It should be noted that whatever the changes in distance between the other pairs of faces of the cubical element, they take place without expenditure of work.

Expressions analogous to expression II.8-8a hold for uniaxial compressions or extensions in the other two coordinate directions, and the work performed by an arbitrary system of normal stresses (in equilibrium) in producing small straining of the cubical element is obtained by adding the three work contributions:

$$W = (\sigma_1 e_1 + \sigma_2 e_2 + \sigma_3 e_3) dx_1 dx_2 dx_3 \qquad \text{(II.8-8b)}$$

Dividing by the volume of the cubical element, we obtain for the local work performed per unit volume by *constant* stresses:

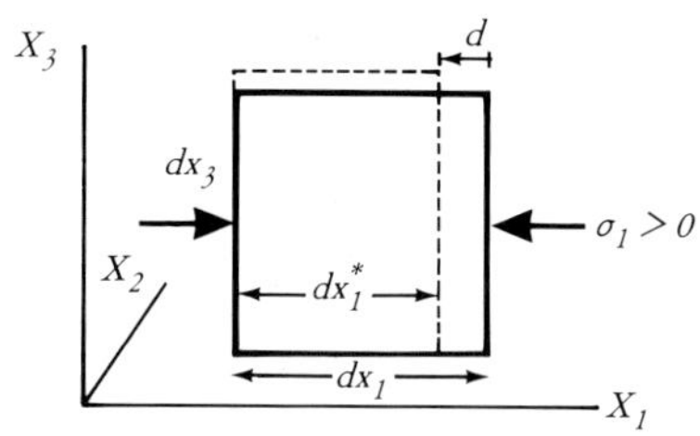

A Uniaxial compression of cubical element

$$d = \frac{dx_1 - dx_1^*}{dx_1} dx_1 = e_1 dx_1 > 0$$

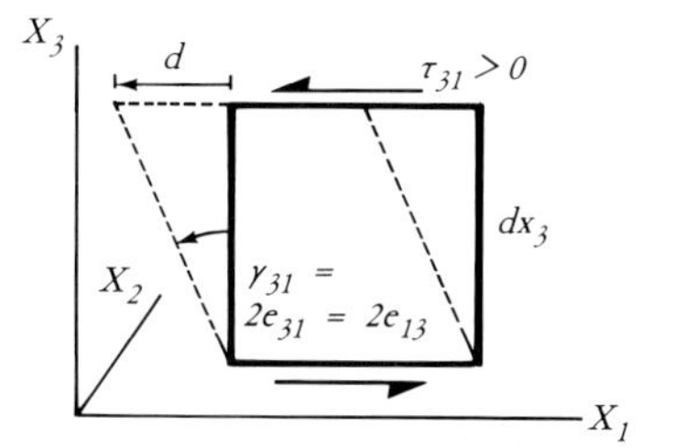

B Simple shearing of cubical element

$d \approx \gamma_{31} dx_3 = 2e_{31} dx_3 > 0$

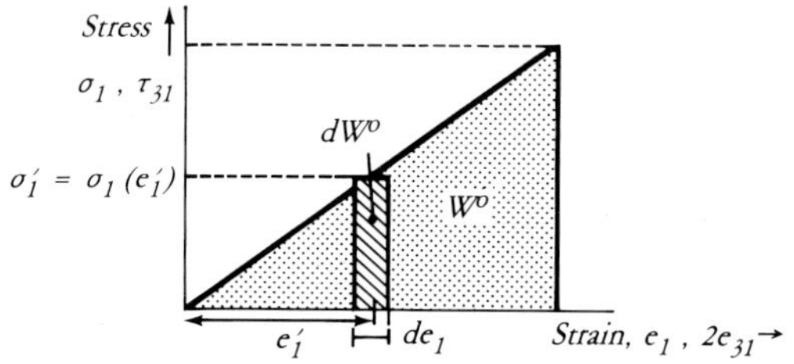

C Work per unit volume performed by stresses that vary in proportion to strains.

Fig. II.8-9
WORK PERFORMED IN STRAINING
('Small' strains)

$$W^o = \sigma_1 e_1 + \sigma_2 e_2 + \sigma_3 e_3 \qquad \text{(II.8-8c)}$$

It has already been discussed ('Variable boundary forces', see also Fig. II.8-4) that in quasi-statical deformations of an elastic material the stresses applied will gradually increase from 0 to the final values associated with the deformed end state. In Fig. II.8-9C the argument that led to eq. II.8-2 is repeated for the cubical element. For stresses that *vary* in proportion to the strains the work performed per unit volume is found to be half of the work performed by constant stresses:

$$W^o = \frac{1}{2}(\sigma_1 e_1 + \sigma_2 e_2 + \sigma_3 e_3) \qquad \text{(II.8-8d)}$$

Next, we subject the cubical element to simple shearing parallel to the x_1, x_2-plane under a constant shear stress τ_{31}. As is pointed out in Fig. II.8-9B, the work performed by the *constant* shear stress is

$$W = |\tau_{31}|\, dx_1 dx_2 \cdot d = 2\tau_{31} e_{31} dx_1 dx_2 dx_3 \qquad \text{(II.8-9a)}$$

or per unit volume

$$W^o = 2\tau_{31} e_{31} \qquad \text{(II.8-9b)}$$

For the work performed per unit volume by a shear stress that *varies* in proportion to the shear strain, we find with reference to Fig. II.8-9C:

$$W^o = \tau_{31} e_{31} \qquad \text{(II.8-9c)}$$

i.e., again half the work performed by a constant shear stress in producing the same shear strain.

The reader may easily convince himself that the most general angular change of the cubical element, which leaves the length dimensions unchanged, is obtained by the combination of three simple shears parallel to, respectively, the x_1, x_2-, x_2, x_3- and the x_1, x_3-planes. The associated shear stresses are τ_{31}, τ_{12}, τ_{23}. They act parallel to the x_1-, x_2- and x_3-axes, respectively. Accordingly, the work performed per unit volume in this combined shearing operation is

$$W^o = 2(\tau_{31} e_{31} + \tau_{23} e_{23} + \tau_{12} e_{12}) \qquad \text{(II.8-9d)}$$

if the shear stresses are constant, or half of this if the shear stresses vary in proportion to the shear strains:

$$W^o = \tau_{31} e_{31} + \tau_{23} e_{23} + \tau_{12} e_{12} \qquad \text{(II.8-9e)}$$

Finally, if we add the work performed by the normal stresses (eq. II.8-8c) and that performed by the shear stresses (eq. II.8-9d), we obtain the work performed per unit volume by *constant stresses* in an arbitrary process of 'small' straining, expressed in stresses and strains:

$$W^o = \sigma_1 e_1 + \sigma_2 e_2 + \sigma_3 e_3 + 2\,(\tau_{12} e_{12} + \tau_{23} e_{23} + \tau_{31} e_{31}) \qquad \text{(II.8.10a)}$$

If the *stresses vary* in proportion to the associated strains, adding eqs. II.8-8d and II.8-9e gives:

$$W^o = \frac{1}{2}(\sigma_1 e_1 + \sigma_2 e_2 + \sigma_3 e_3) + \tau_{12} e_{12} + \tau_{23} e_{23} + \tau_{31} e_{31} \qquad \text{(II.8-10b)}$$

This is the strain energy stored in a unit volume of a linearly elastic material subject to 'small' strains. The stress components in this expression are associated with the final state of strain.

We conclude this chapter by formulating the general expression for the rate at which work ('power') is expended in deforming a unit volume. During an infinitesimal time span dt a state of strain is changed by the infinitesimal strain increments $de_1, ... de_{31}$. Since during an infintesimal time span the stress components may be regarded as constants, we may apply eq. II.8-10a to obtain the 'elementary' work $dW^o = \sigma_1 de_1 + \ ... \ + 2\tau_{31} de_{31}$ performed during the time interval dt. For the *rate of work* performed per unit volume it therefore follows that

$$\frac{dW^o}{dt} = \sigma_1 \dot{e}_1 + \sigma_2 \dot{e}_2 + \sigma_3 \dot{e}_3 + 2(\tau_{12} \dot{e}_{12} + \tau_{23} \dot{e}_{23} + \tau_{31} \dot{e}_{31}) \qquad \text{(II.8-11)}$$

where $\dot{e}$ stands for de/dt.

Of special interest is the rate of work performed in *plane strain*, say in the x_1, x_3-plane ($\dot{e}_2 = \dot{e}_{12} = \dot{e}_{23} = 0$). If, moreover, the principal axes of stresses and strain increments coincide, as for instance in isotropic elastic or viscous materials, these axes may be chosen as reference frame x_1, x_3. Since the shear components vanish in this reference system, equation II.8-11 reduces to the expression in maximum and minimum principal components:

$$\frac{dW^o}{dt} = \sigma_{Max} \cdot \dot{e}_{Max} + \sigma_{Min} \cdot \vec{e}_{Min} \qquad \text{(II.8-11a)}$$

NOTES

1) A second – non-mechanical – way to deform a material body is to supply heat.

2) Work is the 'scalar product' of the two vectors $\vec{F}$ and $\vec{u}$:
$W = \vec{F} \cdot \vec{u} = |\vec{F}|\,|\vec{u}| \cos \alpha = |\vec{F}| d$ (α being the angle between $\vec{F}$ and $\vec{u}$).

3) The work becomes the boundary integral $W = \int_\Sigma |\vec{t}| \cdot d \cdot d\,\Sigma$.

4) The part $F^* - F(d)$ of the constant load F^* that is in excess of the load F, which is required for quasi-static shortening d of the spring, will accelerate the weight/spring system according to Newton's Second Law. Hence, if m is the mass of the loading weight, v its velocity, and the spring mass is neglected: $F^* - F(d) = m\delta v/\delta t$. During an incremental shortening $\delta d (= v\delta t)$ the excess force performs the work $\delta W = [F^* - F(d)]\delta d = m(\delta v/\delta t)\, v\delta t = \delta(mv^2/2)$. This means that the extra work performed is equivalent to an increase in kinetic energy $mv^2/2$ of the system.

5) 1 kilocalorie (kcal) is defined as the energy change which is produced in 1 kg of water when its temperature is raised from 14.4°C to 15.5°C at atmospheric pressure. Note also that in the 'English system' the unit of heat is the British thermal unit (Btu): 1 Btu = 0.251996 kcal.

6) Within the framework of rock plasticity theory (Section II.7.4; Figs. II.7-15 to 21) the main fault would appear as a Coulomb slip and the 'antithetics' of Fig. II.8-6A as belonging to a second set of slips conjugate to the first slip. In contrast, the synthetic subsidiary faults of Fig. II.8-6B belong to the same set of Coulomb slips that envelope the curved main fault.

It should be emphasised, however, that the present discussion is based solely on the general concept of dissipation of work along faults and is not restricted to any particular model of material failure behaviour.

7) Naturally, one can hardly expect a slip event to produce more than a thin skin of melted ('glassy') material lining the fault plane, since the temperature drops very sharply in the immediate vicinity of the active slip plane. In this connection, see Friedman, Logan and Rigert (1974).

8) Miscellaneous glasses have viscosities of about 1000 poise at temperatures between 1000 and 1200°C. The viscosity of an Olivine basalt decreases in the range between 1150 and 1300°C from 38 000 to 300 (!) poise. (Clark, S.P., Handbook of Physical Constants, Geol. Soc. Am. Memoir 97, p. 229, New York 1966). From Newton's law of fluid friction it is easily found that it would take a shear stress of only 50 bar to produce a fault displacement at a rate of 50 cm/s on a melted layer only 1 mm thick and still having a viscosity of 10^5 poise. The shear stress decreases in proportion to a decrease in viscosity and an increase in layer thickness.

II.9. BUOYANCY IN POROUS ROCKS

1. PRELIMINARIES

Fluid-saturated rocks are multi-component systems whose components (solid, water, oil, gas) differ in specific weight. These differences create forces acting between the components, and these will be discussed in this chapter.

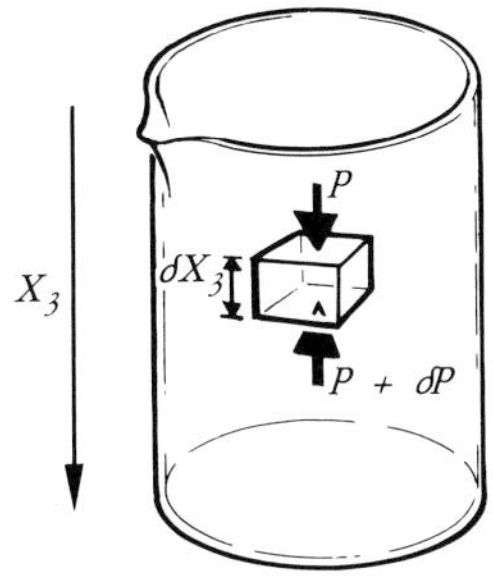

Fig. II.9-1
SMALL FLUID ELEMENT
Weight force causes pressures to be different at top and bottom of element.

Hydrostatic pressure. In a fluid at rest the pressure increases with depth. The cause of this is easily recognised if one considers a small element within a beaker filled with a liquid (Fig. II.9-1). If the normal stress p – the hydrostatic pressure – acts on the top face of the element, the pressure on its bottom face must be greater, since it has to balance not only p but also the weight of the element. An increase in depth (x_3) by dx_3 will therefore be accompanied by an increase in hydrostatic pressure $dp = \gamma_f dx_3$, where γ_f is the specific weight of the fluid. The 'hydrostatic gradient' dp/dx_3 thus equals the specific fluid weight

$$dp/dx_3 = \gamma_f \qquad \text{(II.9-1)}$$

when all quantities are referred to the same units of force and length. In terms of the international (SI) units newton (N) for force, metre (m) for length, kilogram (kg) for mass and seconds (s) for time, eq. II.9-1 then reads [1)]:

$$dp/dx_3\,(N.m^{-3}) = \gamma_f(N.m^{-3}) = \varrho_f(kg.m^{-3})\,g\,(m.s^{-2}) \qquad \text{(II.9-1a)}$$

where ϱ_f is the mass density and $g = 9.81\ m.s^{-2}$ is the gravity acceleration. Noting that the unit of pressure 1 Pa = 1 $N.m^{-2}$ and 1 bar = 10^5 Pa, we may write the equation in more practical units:

$$dp/dx_3\,(bar.m^{-1}) = 9.81.10^{-5} \cdot \varrho_f(kg.m^{-3}) = 9.81.10^{-2} \cdot \varrho_f(g.cm^{-3}) \qquad \text{(II.9-1b)}$$

The pressure gradient is often specified in pounds per square inch per foot (psi/ft), where 1 psi/ft = 0.23 kp/cm²/m. One psi/ft coincides with the average of the specific overburden weight of the Tertiary U.S. Gulf Coast deposits. It should be added that in fresh or brackish water dp/dx_3 = 0.433 psi/ft (Fig. II.9-2), while the limit gradient for brine is 0.5 psi/ft.

Archimedes' principle. Because of the hydrostatic gradient, a body completely submerged in a fluid at rest will experience a higher fluid pressure along its base than along the upper part of its surface. This pressure difference results in a vertically upwards directed force – the 'buoyancy' force – which the fluid exerts upon the submerged body.

Let us consider, for instance, a solid of cubical shape, suspended from a spring balance

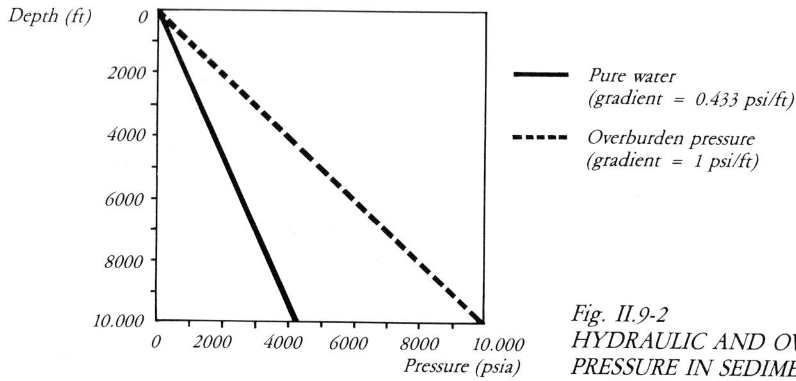

Fig. II.9-2
HYDRAULIC AND OVERBURDEN PRESSURE IN SEDIMENTS

(Fig. II.9-3A). When the solid is fully submerged in a liquid (Fig. II.9-3B), the spring balance will indicate a reduction in the 'pull' of the suspended solid. This reduction equals the weight of the displaced fluid, which can readily be established by comparing the forces exerted by the fluid on the top and base faces of the submerged cube. The hydrostatic fluid pressure – acting always perpendicular to a confining wall – exerts a force directed vertically upwards of magnitude $(p_o + \gamma_f x_3'')A$ upon the basal face of the cube (Fig. II.9-3B), while a force of magnitude $(p_o + \gamma_f x_3')A$ acts vertically downwards upon the top face. Since the resultant fluid force on the vertical faces of the cube vanishes, the fluid at rest exerts the vertical net force

$$F_f = - \gamma_f \cdot V \tag{II.9-2}$$

upon the submerged solid, where V is the solid volume and the minus sign indicates that the buoyancy force acts opposite to the positive x_3 – direction, which we chose as vertically downwards. This result also holds for completely submerged bodies of any shape, as one may recognise by considering the solid in Fig. II.9-4, which is composed of many cubical elements. Since, obviously, eq. II.9-2 will apply to each vertical column of cubical elements, it will also hold for the whole ensemble of cubical blocks. Since the division into elements can be made arbitrarily fine, the result must hold for any completely submerged body [2)].

It is important to realise that expression II.9-2 for the buoyancy force was found for a solid that was acted upon by the hydrostatic fluid pressure over its *entire surface*. While this would still be the case if the solid were supported by rods making 'point' contacts with the base of the solid (Fig. II.9-3C), the fluid pressure would have no or only part access to the base of the cubical solid when the solid rested on the bottom of the container (Fig. II.9-3D) or was supported by rods as shown in Fig. II.9-3E.

When the cubical block rests on the bottom of the container (Fig. II.9-3D) the fluid does not exert any buoyancy force upon the solid – on the contrary – it adds to the effect of the solid's weight in pressing the cube against the supporting container bottom. To lift this body from the bottom requires a force equal in magnitude to the true solid's weight plus the weight of the overlying liquid and air column [3)]. In Fig. II.9-3E the downwards-acting fluid force $(p_o + \gamma_f x_3')A$ is counteracted by the fluid pressure acting over the area ΦA of the base of the cubical body.

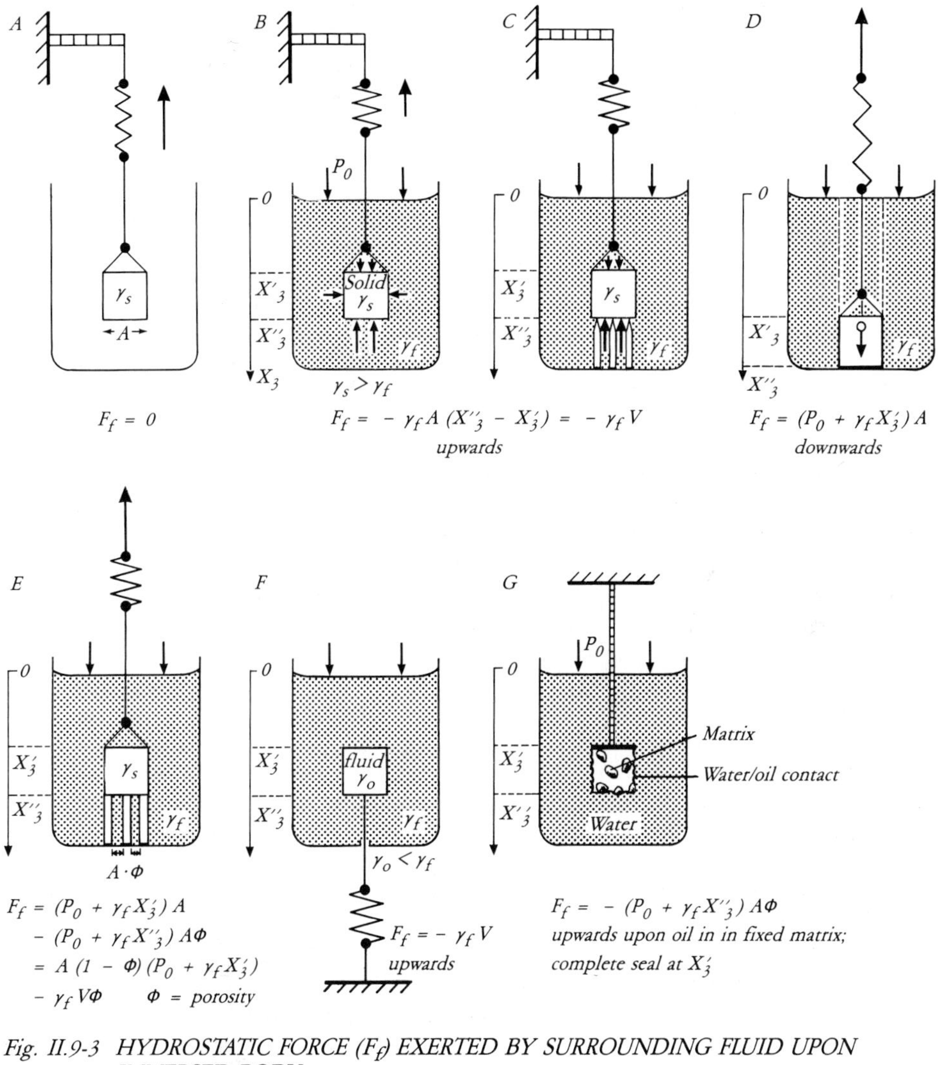

Fig. II.9-3 HYDROSTATIC FORCE (F_f) EXERTED BY SURROUNDING FLUID UPON IMMERSED BODY (Specific fluid weight γ_f; $P_0 = P(0)$ atmospheric pressure)

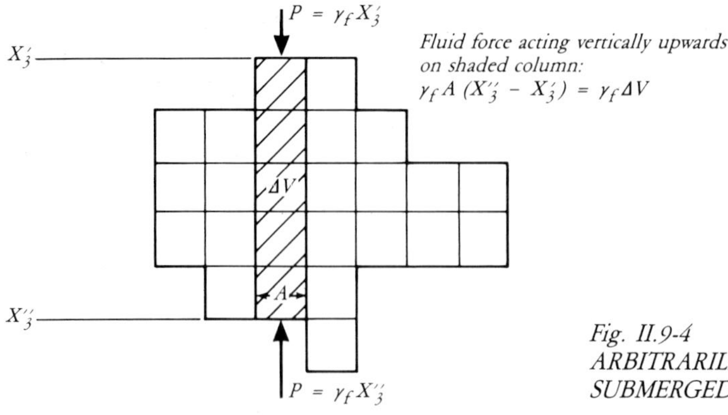

Fig. II.9-4
ARBITRARILY SHAPED SOLID SUBMERGED IN FLUID

2. FLUID-FILLED POROUS ROCKS

'Submerged' weight of rock matrix. After these preliminaries we turn to a porous material composed of grains and completely submerged in a fluid. An individual grain, when completely surrounded by fluid, will experience the full buoyancy force (eq. II.9-2). This is illustrated schematically in Fig. II.9-5A for a spherical grain. The full buoyancy will also act upon grains in a pack whose intergranular contacts have negligible areas ('point contacts'; Fig. II.9-5B). The buoyancy will, however, be reduced when the intergranular contacts have finite areas as sketched in Fig. II.9-5C. In this case, which is frequently encountered in natural sediments when, for instance, the grains are bonded together by cementing material, the fluid has access to only part of the surface of each grain and the force it exerts upon the grain will be smaller than specified by eq. II.9-2.

Since, in general, we are not concerned with individual grains but rather with rock elements containing a large number of grains, we wish to specify the buoyant force the pore fluid exerts upon the solid content of a *unit bulk element*. We shall see that this is not as trivial a matter as it might seem at first sight. Let us consider a cubical bulk element of unit bulk volume and porosity Φ (Fig. II.9-6A). How should we define exactly the outer boundary Σ of that element? There are essentially two possible ways: we may keep the faces of the cubical element perfectly plane, in which case they will intersect the solid material at random (Fig. II.9-6B), or we can choose 'wavy' surface elements that intersect the solid matrix at intergranular contacts only (Fig. II.9-6C). In the latter case the unit bulk element contains only complete grains and the buoyancies of the individual grains add up to the buoyancy per unit bulk volume. According to eq. II.9-2, the buoyancy force will have the magnitude $\gamma_f(1-\Phi)$ for point contacts ($1-\Phi$ being the solid volume of the unit bulk element). Subtracting the buoyancy from the true weight $\gamma_s(1-\Phi)$ of the solid part, the *'submerged' solid weight* γ' per unit bulk volume is obtained:

$$\gamma' = (1-\Phi)(\gamma_s - \gamma_f) \tag{II.9-3}$$

The simple interpretation of this expression as specific 'submerged' solid weight holds only for intergranular 'point' contacts[4]. Moreover, it is applicable only to bulk elements bounded by a *'wavy' surface* (Fig. II.9-6C) and not to those bounded by a surface that cuts *randomly* through the solid matrix (Fig. II.9-6B), despite the volume of solid in the two bulk elements being the same. It may be surprising that in the latter case the total buoyant force exerted upon the solid matrix in the unit bulk volume *vanishes*. This can be demonstrated by a simple force balance. Let us assume that the fluid will exert a buoyant force of magnitude F_f upon the solid material inside the unit bulk element of Fig. II.9-6B. Then – in reaction – the solid material will exert an opposite total force of equal magnitude upon the fluid inside the element considered. The fluid will therefore experience, vertically downwards, a force of magnitude F_f, its own weight $\gamma_f \Phi$ and the pressure force $p' \cdot \Phi \cdot 1$ acting on the fluid part (Σ_f) of the upper horizontal face of the cubical element. Therefore, a total force $F_f + \gamma_f \Phi + p' \Phi$ has to be balanced by the vertically upwards directed pressure force of magnitude $(p' + \gamma_f)\,\Phi$ that acts on the fluid part of the basal face of the cubical element in Fig. II.9-6B. Comparison of the two forces shows that $F_f = 0$ and that therefore the pore fluid pressure inside the bulk element, whose boundary intersects the solid matrix at random, does not exert any net force upon the solid material. A little reflection will show that the striking difference in buoyancy between the two cases considered is due to the uncompensated downwards-directed force exerted by the pore pressure upon the upper parts of the boundary grains that are cut by the basal face of the element in Fig. II.9-6B.

At this point the reader may wonder what the practical implications of the distinction between the buoyant case of Fig. II.9-6C and the 'non-buoyant' case of Fig. II.9-6B may be. It will be found stated repeatedly in this book that the stress field is the key to an understan-

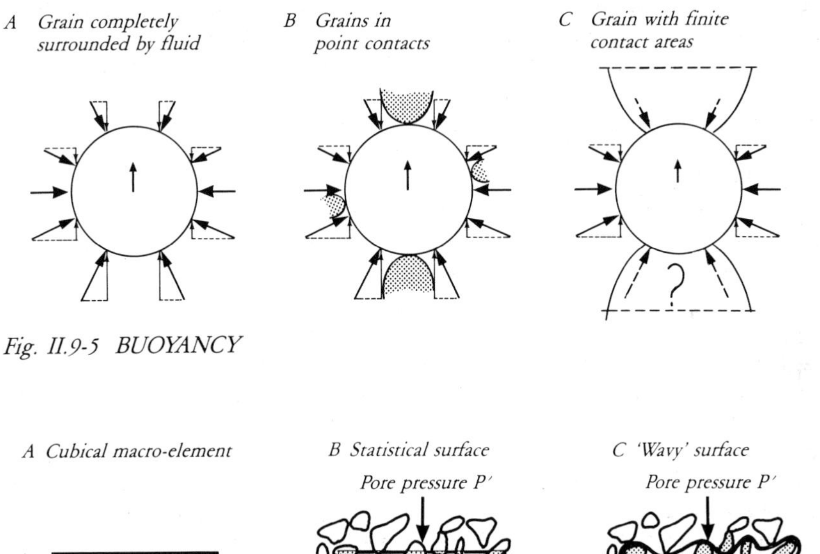

Fig. II.9-5 BUOYANCY

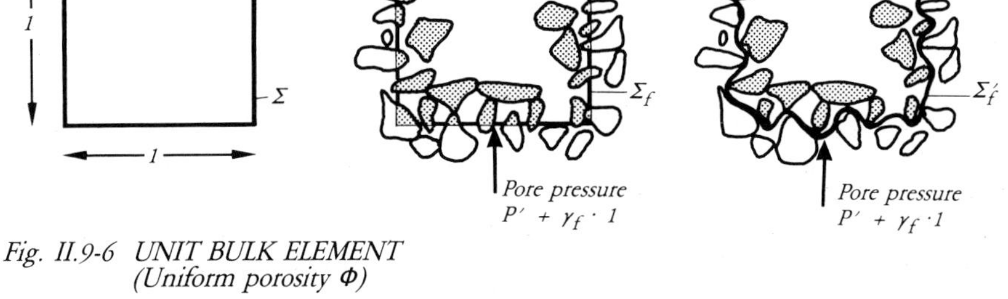

Fig. II.9-6 UNIT BULK ELEMENT (Uniform porosity Φ)

ding of tectonic features. Unfortunately, in a porous rock under a given load there is not just one type of stress: the stress transmitted through the solid matrix of a granular rock can be specified as the force transmitted across the solid part of a random cross-section – such as the plane faces of the cubical element in Fig. II.9-6B – or it may be defined as a measure of the forces acting across the intergranular contacts on a 'wavy' cross-section (Fig. II.9-6C). The different stress concepts are discussed in Section II.3.2, where it is also shown that the (numerical) difference between the two concepts is due to the pore pressure. Consequently, when dealing with 'intergranular stresses' one has to utilise volume elements of the type shown in Fig. II.9-6C and in analysing the equilibrium of these stresses buoyancy has to be taken into account (see Section II.2.2, eq. II.2-2c). When 'skeletal stresses' are being dealt with, i.e. stresses transmitted through a plane section, volume elements of the type shown in Fig. II.9-6B have to be used and buoyancy does *not* enter the conditions of stress equilibrium.

As the intergranular contacts in a granular rock become enlarged – in particular by diagenetic processes – the buoyancy force associated with Fig. II.9-6C will decrease. It will even vanish completely – no matter how the bulk element is bounded – in the extreme case that the rock has a random 'tubular' pore geometry (Fig. II.9-7A). Again this can be proved by the argument used in dealing with the bulk element type of Fig. II.9-6B, since in the 'tubular' rock, plane and 'wavy' cross-sections have the same ratio between fluid-covered area and total area [5]. The vanishing of the net buoyancy force can be visualised in yet another way. In Fig. II.9-7B, the 'tubular' pore space is idealised as an assemblage of straight vertical pipes. Obviously, in this model the pore pressure cannot exert a vertical force upon the solid matrix. To make the model more realistic, we may replace the straight pipes by tortuous and interconnected ones, where vertical force components will now act upon non-vertical elements of the pore walls (Fig. II.9-7C). Because of the assumed randomness of the pore geometry, however, the contributions of upwards- and downwards-directed vertical force components will cancel at any horizontal level (x_3).

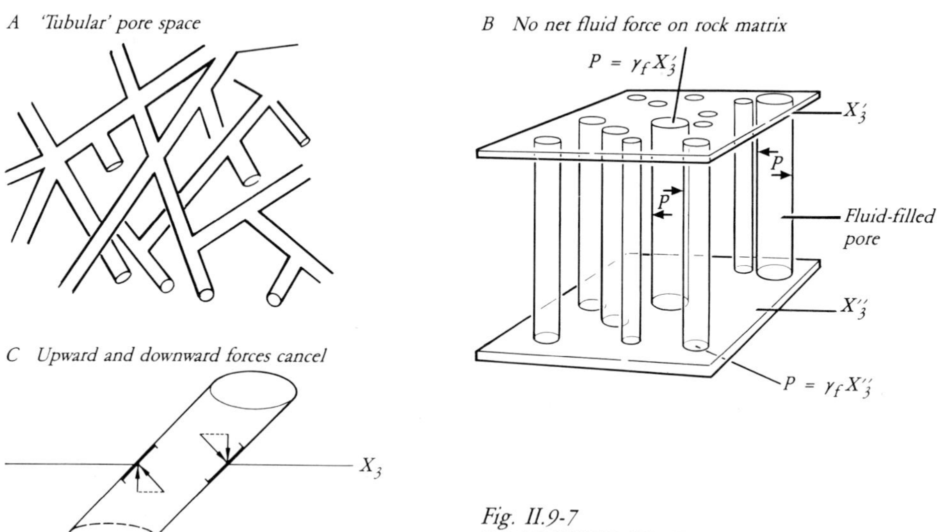

Fig. II.9-7
ROCK WITH 'TUBULAR' PORE SPACE

Different pore fluids. So far we have discussed the net force a hydrostatic[6)] pore fluid may exert upon the solid matrix of a porous rock. Now we turn to the forces that act *between pore fluids* of different specific weights. A fluid mass surrounded by a heavier fluid at rest will experience a buoyant force that tends to make the lighter fluid rise. To maintain hydrostatic equilibrium an extra downwards force has to be exerted upon the lighter fluid, as is illustrated in Fig. II.9-3F, in which the lighter fluid is completely filling a 'weightless' cubical container surrounded by the heavier fluid. Similarly, buoyancy will make hydrocarbons in bulk migrate upwards through permeable rocks when this motion is not blocked by impermeable barriers. Fig. II.9-3G shows a porous rock element filled with oil and perfectly sealed at the top. Both oil and surrounding water are at rest and the water is pushing against the horizontal oil/water contact (at x_3''). According to eq. II.9-1 both the oil pressure and the water pressure decrease in the upwards direction, but since $\gamma_o < \gamma_w$ the oil pressure decreases less. Consequently, the fluid pressure will differ across the sealing top layer (x_3'), the pressure in the oil being higher. The seal has to withstand this difference in pore pressure to prevent the oil from rising. In Fig. II.9-8 a realistic example is sketched of an oil accumulation that is hydrostatically 'trapped' by a sealing shale layer and a sealing fault plane. It is important to realise that, in order to act as a *seal against hydrocarbon migration,* a layer or a fault plane need not be perfectly impermeable to fluids. Within certain limits of differential fluid pressure across such a layer, although permeable to formation water, it may very well completely obstruct the passage of hydrocarbons in the bulk phase[7)]. This is because of the capillary forces that act at the contacts between water and hydrocarbons inside the pore space. These capillary forces enter the interplay of hydrostatic forces in a porous rock containing two or more immiscible pore fluids. In the following we shall give a brief account of the interplay between buoyant and capillary forces in the process of migration and accumulation of hydrocarbons.

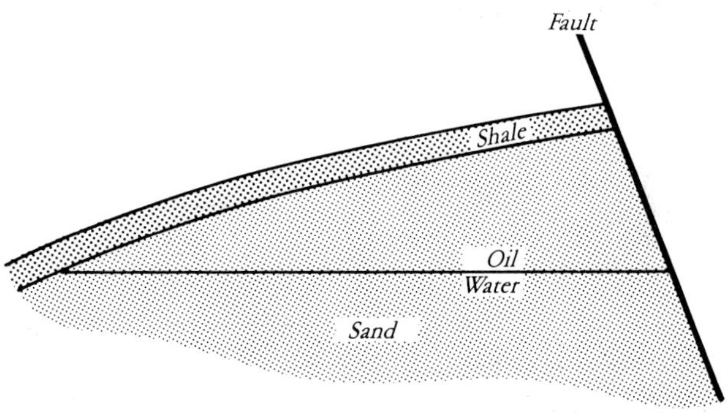

Fig. II.9-8 FAULT TRAP

3. MIGRATION AND TRAPPING OF HYDROCARBONS

Interfacial tension. The interface – or rather the very thin interlayer – between two immiscible fluids (e.g. oil and water) or between a fluid and a solid wall is the seat of special energy. This 'interfacial energy' arises from the difference in attraction which the interface molecules experience from the molecules in the bulk materials on either side of the interface. Increasing the area of the interface, e.g. in deforming a liquid droplet, requires work to be performed in bringing additional molecules to the interface, and this is regained when the interfacial area is decreased. The interfacial energy per cm^2 of interface is called interfacial 'tension' (α), or surface 'tension' when the interface separates a fluid from its vapour. The term 'tension' refers to the analogy with an elastic membrane which reacts to an extension of its area by setting up extra tensile stresses. For instance, increasing the volume of a liquid drop requires mechanical work to be performed by the fluid pressure inside the drop, rather like blowing up a balloon. (The analogy, however, fails in so far as the tensile stress in the skin of the balloon increases with the size of the balloon, whereas the surface tension of the droplet remains constant as long as the temperature is not changed.) From the foregoing it will be clear that the pressure p_i inside a drop will be higher than the pressure p_{ou} outside. The pressure difference $p_c = p_i - p_{ou}$ is called *'capillary pressure'*, since its existence shows up most strikingly in capillaries. It may be mentioned without proof that for a spherical droplet of radius ϱ and surface tension α this capillary pressure is

$$p_c = 2\alpha/\varrho \qquad \text{(II.9-4)}$$

Oil Slug in Capillary. Now let us consider an oil or gas slug in a water-filled, horizontal capillary (Fig. II.9-9A) of radius R, completely submerged in water. At constant temperature the slug attains an equilibrium shape and the total interfacial energy of the slug/capillary system is minimal. This implies that the oil/water menisci are spherical and convex towards the fluid that possesses the lower interfacial energy at the contact with the solid wall of the capillary. In the present case it is the water phase that has the smaller interfacial energy with respect to the solid wall(s), i.e. $\alpha_{ws} < \alpha_{os}$. It is accordingly called the *'wetting'* phase, since in order to achieve the minimum of total interface energy the system tends to keep as much of the solid wall covered with water as is compatible with the given volume of the oil slug and the energy required for the water/oil interface. The requirement of minimum interface energy of the system determines the radius of curvature ϱ of the spherical menisci as a unique function of the interfacial tensions between the different phases. Elementary geometry provides the simple relationship $\varrho = R/\cos\vartheta$ between the radius of curvature of the menisci, the radius R of the capillary and the 'contact angle' ϑ shown in Fig. II.9-9A [8]. Hence, according to eq. II.9-4 the pressure in the oil slug will exceed that in the surrounding water by the capillary pressure:

$$p_c = 2\alpha_{ow}/\varrho = 2\alpha_{ow}\cos\vartheta/R \qquad \text{(II.9-5)}$$

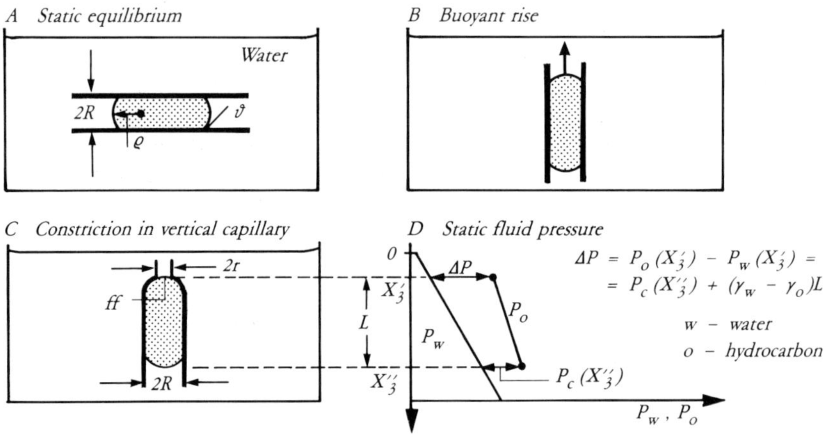

Fig. II.9-9 HYDROCARBON SLUG IN CAPILLARY

Critical length of oil slug. Figure II.9-9B shows the same oil slug, but now in a vertical capillary. Obviously, the slug cannot be in static equilibrium, since it is acted upon by the full buoyancy force and can move in the capillary without a change in interfacial energy [9]) and hence without capillary forces opposing the rising of the slug. Let us now suppose that the slug has arrived at the entrance to a narrower capillary (Fig. II.9-9C) of radius r. To force its way into the narrower capillary the element ff of the oil/water interface has to decrease its radius of curvature ϱ to r/cos ϑ. According to eq. II.9-5 the higher curvature is associated with the higher capillary pressure $p_c = 2\,\alpha_{ow} \cos\vartheta / r$. Therefore, it can only be attained if the pressure in the oil phase at x_3' (see Fig. II.9-9C) exceeds the pressure in the water just ahead of the water/oil interface by this value p_c. As long as the oil slug has not entered the narrow capillary the fluids in Fig. II.9-9C are at rest, and in the upward direction the pressure p_o in the oil slug decreases less than the water pressure p_w. This results, as shown in Fig. II.9-9D, in a static pressure differential across the interfacial part ff at x_3' of $p_o(x_3') - p_w(x_3') = p_c(x_3'') + (\gamma_w - \gamma_o)L$, where $p_c(x_3'') = 2\,\alpha_{ow} \cos\vartheta/R$ and $L = x_3'' - x_3'$ is the length of the oil slug [10]). For the slug to enter the narrower capillary, this pressure differential has to attain the value $p_c(x_3') = 2\,\alpha_{ow} \cos\vartheta / r$ as mentioned earlier. This is achieved when the slug is given the critical length

$$L_c = \frac{2\,\alpha_{ow} \cos\vartheta}{(\varrho_w - \varrho_o)g} \left(\frac{1}{r} - \frac{1}{R} \right) \tag{II.9-6}$$

where ϱ_w, ϱ_o are the densities of water and oil, and g is the gravity acceleration.

To obtain an idea of the critical slug length required to make buoyancy overcome the capillary resistance at the entrance of a narrower pore, let us assume that the capillary walls are perfectly water-wet ($\vartheta = 0$), that $\varrho_w - \varrho_o = 0.1$ g/cm³, $\alpha_{ow} = 30$ dyne/cm, R = 0.005 cm, and r = 0.0015 cm. Since g = 981 cm.s^{-2} the oil slug would have to have a critical length of 285 cm to make it enter the narrower capillary.

Oil slug in porous rock. The simple capillary system discussed may serve as a model for both the migration of hydrocarbons through porous and permeable rocks and for the trapping of hydrocarbons by the interface between different rocks. The application of formula II.9-6 to hydrostatic subsurface conditions has been discussed by Berg (1975). In applying the capillary model to hydrocarbon *migration* the narrower capillary represents the pore 'throats' through which the oil/water interface has to be pushed. From the numerical example discussed above, one must expect that the migration of hydrocarbons will take place in the form of 'stringers' rather than 'globules'. When the model is applied to capillary trapping, r in eq. II.9-6 becomes the 'effective' pore radius of the sealing rock layer or fault gouge. The crucial problem in applying formula II.9-6 to (hydrostatic) subsurface conditions is, of course, the estimate of the 'effective pore size', which in the absence of data from direct capillary measurements can only be obtained in a very crude approximation. We may mention here that water-saturated fault gouges may provide considerable capillary barriers to hydrocarbons, even if the gouge has been produced by cataclasis only and has not been affected by diagenetic effects [11]).

Downward Flow of Pore Water. It should be recalled that formula II.9-6 was derived assuming that the water outside the capillary was at rest. If the capillary were embedded in a water-saturated porous material as sketched in Fig. II.9-10, vertical percolation of water would produce a pressure difference $p_w(x_3'') - p_w(x_3')$ over the length of the oil slug which would differ from the hydrostatic one $[\gamma_w(x_3''-x_3')]$. As shown in Fig. II.9-10, a downward flow, driven by the fluid weight, is possible where the fluid pressure p_w still increases with depth, although this will be less than in the hydrostatic case (see also Section II.10.1). In such a case the water pressure still exerts an upward-directed net force upon the oil slug, but less than buoyancy would do in the static case (i.e. $dp_w/dx_3 < \gamma_w$). Accordingly, the oil slug would have to attain a greater length than determined by eq. II.9-6 to enter the narrower capillary. It should be understood that this only applies to downward flow with pressure increasing in the flow direction. The reader will notice that by raising the fluid pressure at the top of the porous pack in Fig. II.9-10 a downward flow can be produced with fluid pressure decreasing in the downward direction. Then no upwards-directed net force will act upon the oil slug and, irrespective of its length, it cannot be made to pass a capillary barrier in the upward direction. Figure II.9-11 gives a slight variant of Fig. II.9-10, in which part of the permeable pack is given a lower permeability. Closer inspection reveals that in such a case it becomes possible for the fluid pressure to decrease over the low-permeability interval while still increasing with depth in the high-permeability part. Embedded in the low-permeability part, our capillary 'throttle' would not allow the oil slug to pass in an upwards direction, except if the slug were made to continue far into the underlying high-permeability part.

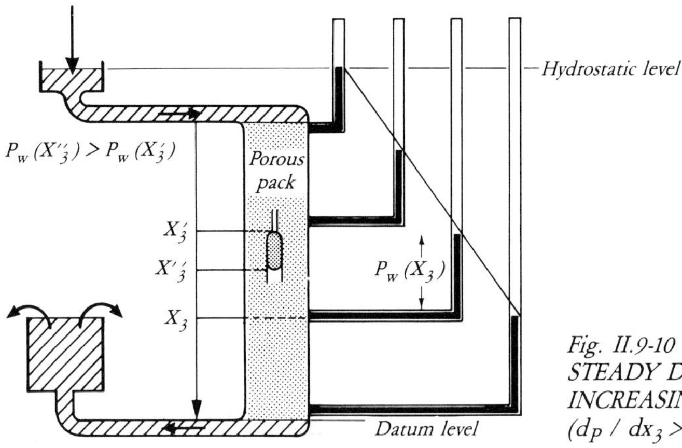

Fig. II.9-10
STEADY DOWNWARD FLOW WITH INCREASING WATER PRESSURE ($dp/dx_3 > 0$)

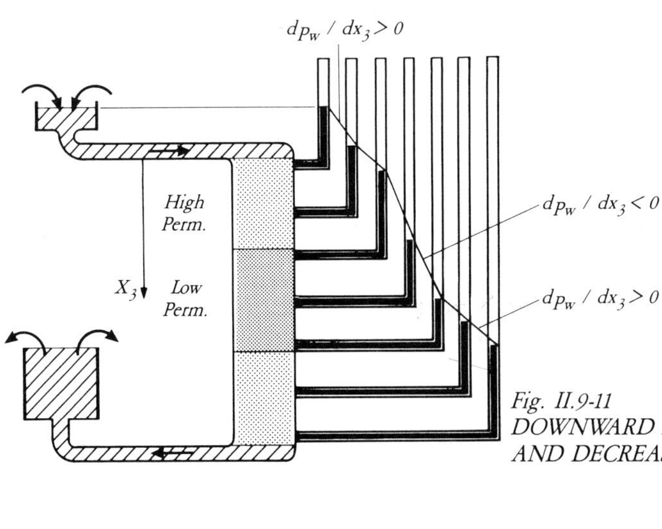

Fig. II.9-11
DOWNWARD FLOW WITH INCREASING AND DECREASING WATER PRESSURE

Unconventional trapping of hydrocarbons. Artesian water flow from the margin of a basin downdip towards its central parts is an example of the downward water flow considered in Fig. II.9-11. In intervals of low permeability the water pressure may decrease with depth and preclude the buoyant rise of hydrocarbons that have possibly been generated in these low-permeability formations (e.g. biogenic or thermally generated gas). This trapping mechanism is unconventional in the sense that it does not require the presence of a lithological top seal. It may have operated in commercial gas accumulations trapped by updip water in the absence of a permeability seal at the top, as reported by Masters (1979) and Gies (1981, 1982).

NOTES

1) $1\ N = 1\ m.kg\ (mass) \cdot s^{-2}$; $1\ Pa = 1\ N.m^{-2} = 10^{-5}$ bar.

2) The analytical proof of eq. II.9-2 for arbitrarily shaped bodies is accomplished by making use of Gauss' integral theorem. Let Σ be the closed surface of the submerged body, V its volume, and $\vec{t} = -p \cdot \vec{n}$ the stress vector on the surface, with $\vec{n}$ being the outward unit-normal (compressive stress counted as positive). The total surface force $\vec{F}_f$ exerted by the hydrostatic fluid upon the solid is then

$$\vec{F}_f = \oint_\Sigma \vec{t}\, d\Sigma = - \oint_\Sigma p\vec{n}\, d\Sigma$$

with the Cartesian components

$$F_{fi} = - \oint_\Sigma p \cos \alpha_i\, d\Sigma \qquad (i = 1, 2, 3)$$

where α_i are the directional cosines of $\vec{n}$. Applying Gauss' theorem and inserting $p = p(0) + \gamma_f x_3$ yields for the vertical component F_{f3}:

$$F_{f3} = - \oint_\Sigma p \cos \alpha_3\, d\Sigma = - \int_V \frac{dP}{dx_3} dV = - \gamma_f V.$$

3) We have emphasised the limitations of eq. II.9-2 to avoid confusion still persisting in the literature regarding the application of Archimedes' principle to porous rocks. For instance, in their text book, Rieke and Chilingarian, ('Compaction of argillaceous sediments' p. 6, 1974) state that 'the fluid pressure p is effective over the entire area of any surface passed through the system, irrespective of the fraction of the area lying within the solid-filled part of the space'. This would, of course, imply that the solid in our Fig. II.9-3D would experience the full buoyancy – a view which the authors even claim to support by an 'experiment' (see their Fig. 2B on p.6). Anyone who feels the need for it can easily demonstrate the incorrectness of their view by taking, for instance, a container with mercury and a piece of cork with a flat base. As the cork is pushed below the mercury surface and brought into contact with the flat bottom of the container, it will remain in that position, kept there by the weight of the overlying liquid and atmospheric pressure.

4) With finite contact areas the 'submerged' weight of the solid content will be greater than eq. II.9-3; the exact expression for γ' will then become

$$\gamma' = (1-\Phi)(\gamma_s - \gamma_f) + (1-\psi)\gamma_f \qquad \text{(IIIa)}$$

where ψ is the 'porosity' of the wavy surface as defined in section II.3.2 (cf. Note 2 ibid.). With point contacts the wavy surface is completely occupied by fluid and $\psi = 1$. Expression IIIa then reduces to II.9-3. However, the 'porosity' ψ cannot become smaller than the bulk porosity Φ. In the limit case $\psi = \Phi$ the 'submerged' weight becomes equal to the true solid weight per unit bulk volume $\gamma = \gamma_s(1-\Phi)$.

5) Cf. the case $\Phi = \psi$ in Note 4.

6) The fluid force exerted upon the rock matrix in fluid *flow* is discussed extensively in Chapter II.10 on flow through porous rocks, eqs. II.10-2,3.

7) The continuous character of the oil phase is emphasised to distinguish this type of oil migration from other possible transport mechanisms, such as molecular diffusion.

8) Cos $\vartheta = R/\varrho = (\alpha_{so} - \alpha_{sw})/\alpha_{ow}$. If $\alpha_{so} - \alpha_{sw} = \alpha_{ow}$ the contact angle $\vartheta = 0$ and the menisci in Fig. II.9-9A become hemispheres of radius R. The water is then said to be perfectly wetting the pore wall. If $\alpha_{so} - \alpha_{sw} > \alpha_{ow}$ the three phases cannot have a common contact and a water layer will then separate the oil slug from the capillary wall.

9) Except for some hysteresis effects which may make the contact angle at the lower end of the moving slug somewhat larger.

10) Naturally, all these considerations also apply to a *gas* slug when interfacial and specific weight data for the gas are inserted.

11) This has also been demonstrated by Mandl, de Jong and Maltha (1977) by shearing coarse-grained sugar packs in a ring-shear apparatus. The crushed shear zones provided capillary barriers to the non-wetting fluid which – in terms of oil and water – would have allowed the trapping of oil accumulations of 10 to 25 m thickness.

It should be noted that interfaces (e.g. bedding or fault planes), even when offering only a very modest capillary resistance, insufficient for sealing hydrocarbon accumulations of commercial interest, may play an important role in guiding buoyancy-driven hydrocarbon migration.

II.10. FLOW THROUGH POROUS ROCKS

1. PRESSURE AND FLOW

'Pressure head'. When a fluid is at rest in an arbitrarily shaped container (Fig. II.10-1A) the level to which the fluid rises in open standpipes immersed in the fluid will be the same for any standpipe – irrespective of the position of its lower orifice. The fluid pressure at the orifice of a standpipe is equal to the weight of the vertical fluid column h_p with unit cross-section. The height of the fluid column in the standpipe is accordingly called the 'pressure head'. The elevation h_e of its orifice above (or below) an arbitrarily chosen 'datum level'

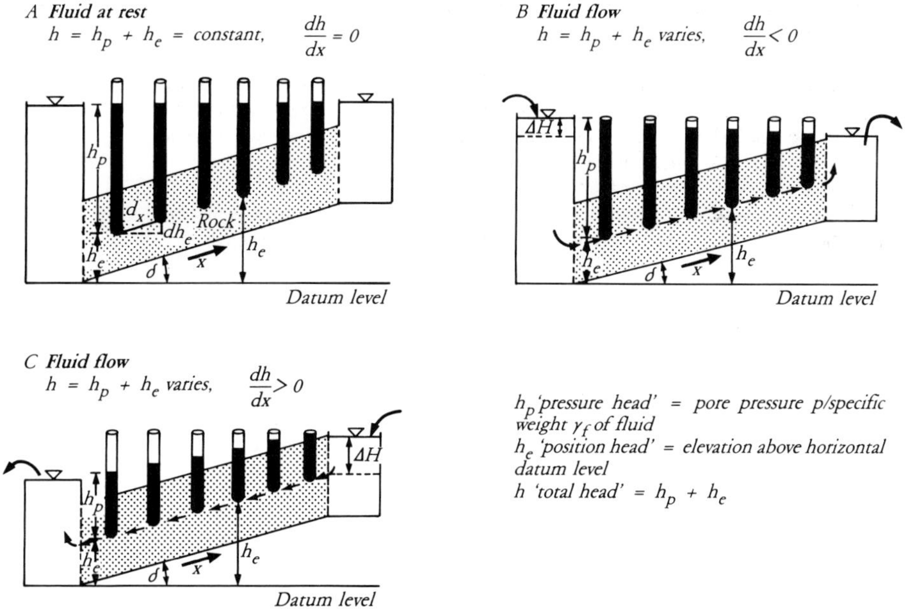

Fig. II.10-1 PORE PRESSURE AND 'PRESSURE HEADS' IN POROUS ROCKS
The fluid flow is determined by the totel head. If it is constant (A), the fluid is at rest. If it varies, the fluid flows (B,C) the flow direction being always that in wich the total head decreases. As shown in case (C), this may even imply flow from low to higher fluid pressure (if $dh_p/dx < 0$, although $dh/dx > 0$).

is referred to as 'elevation' or 'position head' and $h = h_e + h_p$ is the 'total' or 'hydraulic' head. The total head is constant throughout a fluid at rest. All this also holds when the fluid fills the pore space of a porous rock. Then the orifice of standpipes used (or the sensitive part of any other pressure gauge) will be large compared with the dimensions of the individual pores and the 'pressure' measured will approximately equal the local mean value of the (micro-)pressures that act inside the pores. (This mean pressure p has been defined in Section II.3.2; Fig. II.3-2.)

'Filter velocity'. If the total head varies from place to place, the fluid is bound to move [1]. As illustrated in Figs. II.10-1B, C the fluid always flows in the direction of decreasing total head [2]. It is important to note that this does not imply that the fluid always moves in the direction of decreasing pore pressure, as is clearly demonstrated in Fig. II.10-1C. The flow itself is characterised by the 'filter' or 'seepage' velocity. To define this quantity, we consider a cross-sectional element of unit bulk area (solid + pores) inside the porous rock and 'measure' the total volume of fluid that passes this element per unit time. Obviously, for a given flow, the magnitude of the volume flux through the unit element considered will depend on the orientation of the element. If the unit normal of the element is perpendicular to the (macro-)flow direction, net flow through this element will be zero; if the normal of the element is parallel to the flow direction the flux will attain its maximum value. The magnitude of this maximum flux and the direction of (macro-)flow determine the vector of the filter velocity $\vec{V}$ (Fig. II.10-2A). The rate of net flow through any other unit element is obtained by orthogonally projecting $\vec{V}$ upon the normal to that element. In particular, by projecting $\vec{V}$ upon the axes of a Cartesian reference frame (x_1, x_2, x_3), one obtains the components of the vector $\vec{V}$ with respect to the chosen frame (Fig. II.10-2B). The components V_1, V_2, V_3 therefore represent the rates of net flow through elements of unit bulk area with normals pointing along the respective axes of the Cartesian reference system.

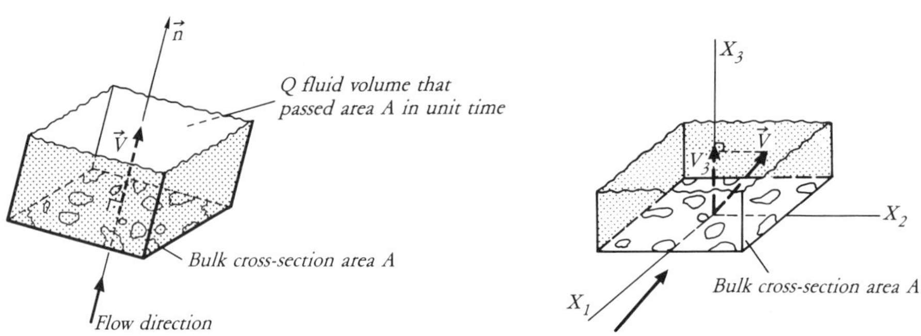

A Filter velocity $\vec{V}$ has direction of macro-flow and magnitude $|\vec{V}| = Q/A$.

B Cartesian flow components

Fig. II.10-2 FILTER VELOCITY

2. DARCY'S FLOW LAW

Many experiments – the first carried out by H. Darcy in 1856 – have shown that in 'slow' and stationary (i.e. time-independent) percolation of an incompressible fluid through a porous material, the filter velocity is proportional to the negative gradient of the total head h – the *'hydraulic gradient'.* If one separates the specific weight γ_f of the fluid and its viscosity η_f from the general constant of proportionality, the empirical Darcy law can be written in terms of the three filter-velocity components and associated hydraulic gradients in the x_1-, x_2-, x_3-directions:

$$\left.\begin{matrix} V_1 \\ V_2 \\ V_3 \end{matrix}\right\rangle = -\frac{k}{\eta_f}\gamma_f \cdot \left\langle\begin{matrix} \partial h/\partial x_1 \\ \partial h/\partial x_2 \\ \partial h/\partial x_3 \end{matrix}\right. \qquad (h = h_e + \frac{P}{\gamma_f}) \qquad \text{(II.10-1)}$$

Permeability. The constant of proportionality k in Darcy's law represents the reciprocal of the specific resistance of the porous material to fluid flow and is called *'permeability'.* Its physical dimension follows from the dimensions of the other quantities in eq. (II.10-1) as a length squared. Therefore, when using centimetre-gram-second units (c.g.s.), permeability is measured in cm^2. Values of permeabilities cover a wide range. For instance, the permeability of a cigarette is about 10^{-5} cm^2; for sandy materials ranging from silts to well-sorted coarse sand, permeability ranges between 10^{-11} and 10^{-8} cm^2 and for overpressured shales the in-situ permeability will lie between 10^{-14} and 10^{-19} cm^2.

The practical unit used in the oil industry is the 'darcy'. It was introduced for horizontal flow, where according to eq. (II.10-1) $\gamma_f dh/dx_i = dp/dx_i$. The permeability was defined to be 1 *darcy* when the filter velocity was 1 $(cm^3/cm^2)/s$, the viscosity η_f = 1 cP = 0.01 poise (= 0.01 dyne $cm^{-2}s$; e.g. η (water at 20°C) = 1 cP) and the pressure gradient was measured in atmospheres (≈ bars) per cm. In the general flow case the gravitational contribution $\gamma_f h_e$ has to be taken into account as well and, for consistency, expressed in atmospheres (≈ bars) – which is the pressure at the bottom of an open fluid-filled standpipe of height h_e. The permeability expressed in darcies is converted into cm^2 units by

1 darcy $\cong 10^{-8}$ cm^2.

In groundwater hydrology the coefficient $(k/\eta_f)\gamma_f$ is referred to as permeability and is expressed in terms of the filter velocity associated with unit drop of hydraulic head. The unit of hydrological permeability is therefore 1 cm/s, which for water corresponds with 1.02×10^{-5} cm^2 for k.

Anistropic material. Equation (II.10-1) holds for seepage through an isotropic material. In stratified sediments permeability is often direction-dependent and may be considerably lower in the direction perpendicular to the bedding plane. If, in an anisotropic medium the permeability is measured in all directions and on the radii emanating from a point the square roots of the corresponding permeability values are plotted, the end points lie on an ellipsoid. If the coordinate axes x_I and x_{III} are chosen so as to coincide with the directions of the longest and shortest axes of the ellipsoid, i.e. with the directions of the greatest and smallest permeabilities ($k_I \leqslant k_{II} \leqslant k_{III}$), Darcy's law attains the generalised form for anisotropic media:

$$V_I = -\frac{k_I}{\eta_f}\gamma_f\frac{\partial h}{\partial x_I}$$
$$V_{II} = -\frac{k_{II}}{\eta_f}\gamma_f\frac{\partial h}{\partial x_{II}}$$
$$V_{III} = -\frac{k_{III}}{\eta_f}\gamma_f\frac{\partial h}{\partial x_{III}}$$

In the anisotropic case $\vec{V}$ and the gradient of h are no longer parallel because the permeabilities differ. To find the rate of flow V_n across an element whose normal makes the directional angles α_I, α_{II}, α_{III} with the x_I-, x_{II}-, x_{III}-axes, one has to form the product $V_n = V_I \cos\alpha_I + V_{II}\cos\alpha_{II} + V_{III}\cos\alpha_{III}$.

'Driving' and Resistive Forces. Although Darcy's law has been established from experiments, it can also be derived from basic laws of hydromechanics. Such a derivation has the advantage of providing an understanding of the law and its range of validity. For application in rock and soil mechanics it is of particular importance to have a clear notion of the exact mechanical meaning of the terms in Darcy's law. We shall therefore demonstrate in an elementary way how Darcy's law can be understood as the balance between forces that 'drive' the fluid through a porous material and forces that resist that flow.

A volume element inside the porous material is delineated as shown in Fig. II.10-3. Fluid flow, when viewed on a macro scale, is parallel to the x-axis and the faces perpendicular to this direction have unit bulk area (solid + pores). The faces of the volume element are planar and cut randomly through the solid skeleton of the porous material. The fluid inside the volume element is 'driven' by two forces: firstly, the x-component of the fluid weight W_f and, secondly, the resultant of the hydrodynamic pressures that act on the fluid parts of the two bulk sections 1 and 2 [3]. When we denote the average values of the pore pressures that act on the fluid parts of bulk sections 1 and 2, by p_1 and p_2 respectively, the resulting 'driving' force of the pore pressure is $(p_1 - p_2) \cdot \Phi$. The porosity Φ, assumed as constant, enters this expression because the pore pressure acts on the pore part of the unit bulk section only. As shown in Fig. II.10-3, the sum of the two driving forces, when divided by the pore volume they act upon, becomes $-\gamma_f dh/dx$.

The macro flow considered is uniform and steady, which means that the filter velocity is constant in time and space. Consequently, the fluid leaves cross-section 2 in Fig. II.10-3 at the same average velocity as it enters through cross-section 1. This implies that the fluid does not experience a net acceleration between the two cross-sections and, therefore, all forces that act upon the fluid volume considered must be in static equilibrium, i.e. their resultant must vanish. The total driving force $-\gamma_f dh/dx$, derived above, must therefore be balanced by the resistive force R_x exerted by the solid skeleton upon the moving fluid inside a unit bulk element. Since the driving force was referred to unit pore volume, the resistive force has to be taken per unit pore volume as well. According to Darcy's law eq. II.10-1, for isotropic porous materials the resistive force R_x in the case of Fig. II.10-3 becomes $-(\eta_f/k)V_x$ or, for arbitrarily oriented Cartesian axes,

$$\vec{R} = -\frac{\eta_f}{k}\vec{V} \qquad \text{(II.10-2)}$$

The opposite force $-\vec{R}$ is often referred to as *'drag force'*, since it is the resultant force exerted by the moving fluid upon the solid content of the unit bulk element [4].

We have emphasised that expression II.10-2 represents the resistive force per *unit pore volume.* In the literature, however, it is often interpreted as force per unit bulk volume. This has created confusion, particulary with regard to a proper evaluation of the 'drag' which, for instance, percolating water may exert upon unconsolidated sediments. The following remarks should clarify the apparent contradiction between the two statements.

Fluid drag on matrix. Our simple equilibrium consideration concerned the fluid inside a volume element bounded by planar faces (Fig. II.10-3) that cut randomly through the solid material. As in the preceding chapter (Fig. II.9-6C), it also makes sense, if the skeleton has a granular structure, to delineate the volume element in a slightly different way by means of a *'wavy'* boundary that intersects the skeleton of the rock at intergranular contacts only. The enlarged part of such a boundary is sketched in Fig. II.10-4. Considering now the driving force that acts upon the fluid inside the 'wavy' boundary, we notice that the driving weight component remains the same as in the case of planar boundaries (enclosing the same fluid volume), but that the driving pressure force and, consequently, the resistive force (or, respectively, the drag force) have changed. If the intergranular contact areas occupy only a negligible fraction of the total area of the wavy boundary ('point contacts'), it is easily seen from Fig. II.10-4 that the total pressure force that acts in the x-direction upon a wavy cross-section is equivalent to the force that would result if the pressures were to act on all elements – fluid and solid – of the associated *planar* cross-section. The respective pressure forces on 'wavy' cross-sections 1 and 2 in Fig. II.10-3 therefore become p_1 and p_2 (times unit cross-section area) instead of $p_1\Phi$ and $p_2\Phi$. The corresponding driving pressure is $-(dp/dx)\,dx \cdot 1$. Adding the unchanged weight component (Fig. II.10-3) and dividing by the bulk volume dx.1, we obtain the driving force *per unit bulk volume* $-(dp/dx + \gamma_f \Phi\, dh_e/dx)$. Comparison with Darcy's law (II.10-1) immediately shows:

$$-\left(\frac{dp}{dx} + \gamma_f\;\Phi\;\frac{dh_e}{dx}\right) = \frac{\eta_f}{k}V_x + \gamma_f\;(1-\Phi)\;\frac{dh_e}{dx} \qquad \text{(II.10-3)}$$

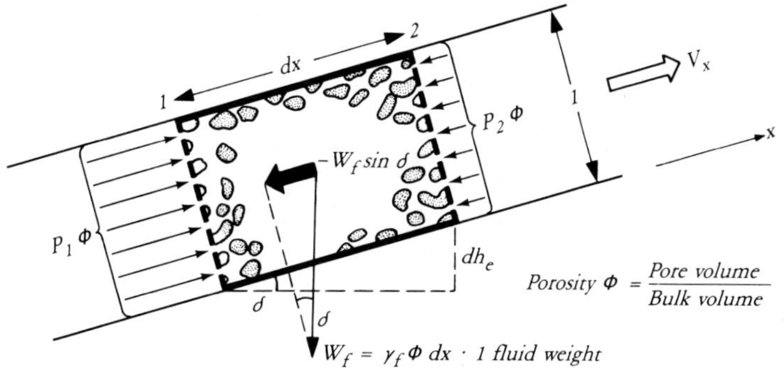

Fluid weight component in x-direction: $-\gamma_f\Phi\cdot dx\cdot\frac{dh_e}{dx}$

Net fluid pressure on fluid part of unit bulk cross-sections: $-(p_2-p_1)\cdot\Phi\cdot 1 = -\frac{dp}{dx}dx\cdot\Phi\cdot 1$

Total 'driving' force per unit of pore volume ($\Phi\cdot 1\cdot dx$): $-\gamma_f\frac{dh}{dx} = -\frac{d}{dx}(p+\gamma_f h_e)$

Fig. II.10-3 'DRIVING' FORCE IN FLOW THROUGH POROUS ROCKS

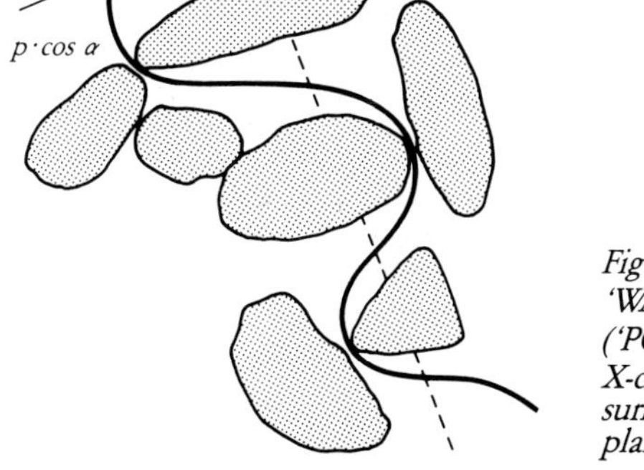

Fig. II.10-4
'WAVY' CROSS-SECTIONS IN GRANULAR MEDIA ('POINT CONTACTS')
X-component of pressure p times element dS of wavy surface = pressure p times projected area $dS\cdot\cos\alpha$ on planar surface.

The right-hand side of eq. II.10-3 now represents the total drag force per unit bulk volume that acts upon the granular material enclosed by the 'wavy' boundaries. The second term in the right-hand part of eq. II.10-3 represents the buoyancy force exerted by the fluid upon the completely immersed grains. In horizontal flow the buoyancy term vanishes and $(\eta_f/k)V_x$ becomes the total drag force per unit bulk volume on the completely fluid-surrounded material. It exceeds the drag force on the material between planar cross-sections by the amount $(\eta_f/k)V_x(1-\Phi)$, since the latter, when also referred to unit bulk volume, becomes $(\eta_f/k)V_x \cdot \Phi$. Necessarily, the excess drag $(\eta_f/k)V_x(1-\Phi)$ is exerted upon the boundary grains of the fluid-enwrapped element and indeed equals, according to Darcy's law eq. (II.10-1), the extra driving force $-(dp/dx)(1-\Phi)$ that is associated with the choice of a 'wavy' boundary.

From the foregoing it follows that, surprisingly enough, in a granular material with intergranular *'point' contacts* the force $\vec{R}$ in eq. II.10-2 can be interpreted as resistance to fluid flow in *two* ways: (1) as total resistance per unit *pore volume* when the boundary grains and pores of the volume considered are randomly intersected by the outer boundary; (2) as total resistance per unit *bulk volume* when the boundary grains are fully part of the bulk element considered.

Example: Flow through overpressured sediment on infinite slope

To illustrate the application of Darcy's law we consider a uniform submarine sediment layer on a very long slope (Fig. II.10-5). We assume that rapid sedimentation of material of low permeability has prevented normal compaction and has caused the pore pressure to attain values in excess of the corresponding hydrostatic one, i.e. $p > \gamma_f h_p^o$ as illustrated in Fig. II.10-5. This excess pressure and, hence, the total head h may increase with 'depth' x_3 as indicated by the water level in the standpipes of Fig. II.10-5. If the compaction behaviour of the sediment is the same all along the slope, the fluid will flow in the negative x_3-direction only ($V_3 < 0$, $V_2 = 0$). The filter velocity V_3 is determined by Darcy's law.

It should be recalled that the filter velocity is referred to a unit bulk cross-section of the solid material. In a compacting sediment this cross-section moves in the x_3-direction, say at the rate W. The filter velocity V_3 referred to this moving cross-section will be equal to $-W$, i.e. independent of the porosity Φ of the material, provided the compacting sediment rests on an impermeable substratum. This can be seen as follows: if the fluid were at rest, the moving cross-section would pass in unit time the fluid volume ΦW. Since, in addition, subsidence at the rate W reduces the pore volume of the underlying sediment by $(1-\Phi)W$ per unit time, an equal volume of fluid has to be squeezed through the moving unit cross-section of porosity Φ. (Compressibility of fluid and solid material may be considered negligible with respect to the much higher compressibility of the bulk material.) Adding the two flows and using the negative sign to indicate the upward direction of the flow gives $V_3 = -W$.

Obviously, the upward flow will reduce the weight of the sediment column resting on a cross-section at x_3. In particular, one may ask about the reduction of the weight that is carried by the (very small) grain contacts at x_3, since it is the compressive stress transmitted through these contacts that determines the shear strength of the sediment which, in turn, controls the stability of the slope. To calculate this weight reduction per unit bulk volume, one has to apply the right-hand part of eq. II.10-3, which represents the drag force on the solid content of a unit bulk element that is bounded by a 'wavy' surface. With reference to Fig. II.10-5 and reading x_3 instead of x in eq. II.10-3, the reduction of the x_3-component of the weight of the solid material in a unit bulk element becomes

$$-\frac{\eta_f}{k}W - \gamma_f(1-\Phi)\cos\delta \qquad \text{(II.10-3a)}$$

leaving the reduced solid weight component

$$(\gamma_s - \gamma_f)(1 - \Phi) \cos \delta - \frac{\eta_f}{k} W \qquad \text{(II.10-4)}$$

per unit bulk volume to be carried by the grain contacts.

The first term in eq. II.10-4 is the weight of the solid material reduced by the full buoyancy (see Section II.9). This 'submerged' solid weight is further reduced by the drag of the fluid leaving the compacting sediment. It is of interest to ask under what conditions will the reduced solid weight, eq. II.10-4, vanish completely. In c.g.s. units the numerical value of the first term in eq. II.10-4 is, in practice, close to 10^3. Assuming for the viscosity of the pore water at 100°C the value of 0.005 poise and introducing as a measure of the subsidence rate mm/year instead of cm/s, one finds that the two terms in eq. II.10-4 balance when W/k $\cong 6.10^{13}$. Thus at a subsidence rate of 0.6 mm/year the permeability of the overpressured sediment has to be 10^{-14} cm² to make the fluid carry the whole solid weight. This is, as was mentioned earlier ('Permeability'), certainly not a low permeability value for overpressured shales in situ.

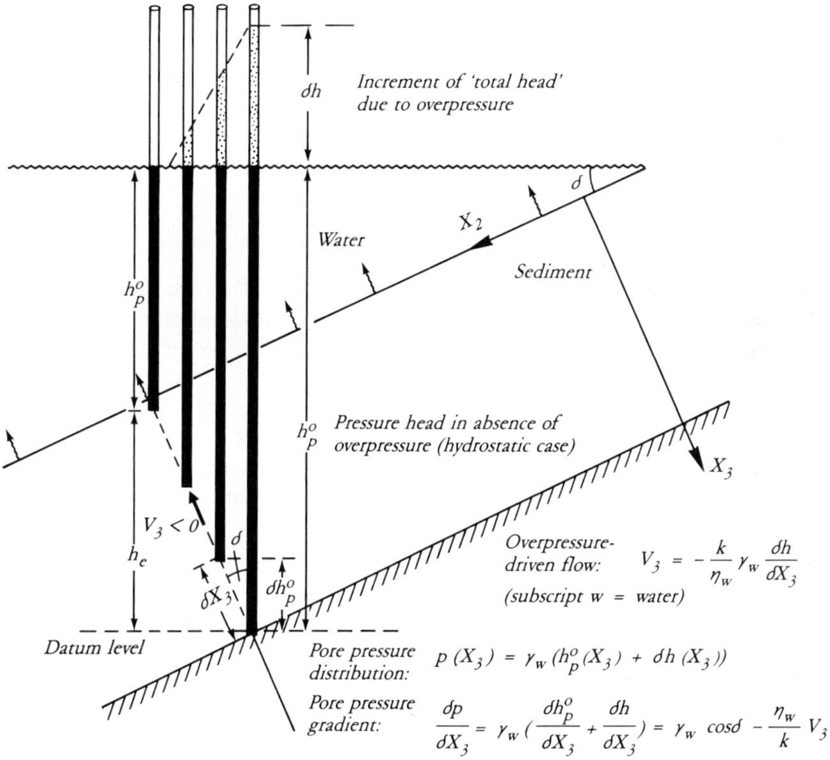

Fig. II.10-5 FLUID FLOW THROUGH OVERPRESSURED SUBMARINE SEDIMENT SLAB ON SLOPE

Limitation of Darcy's Law. We have interpreted Darcy's law as a balance between the forces 'driving' the fluid flow and the force resisting this flow. A complete theoretical foundation of Darcy's law still requires a derivation of expression II.10-2 for the resistive force from basic hydrodynamics. Such a derivation is of considerable practical value, since it reveals the limitations of the law first obtained in an empirical way. The derivation clearly demonstrates that Darcy's law is a 'limit' law holding for *'slow' steady flow* only. Characteristic of this type of flow is a steady laminar flow pattern inside the pore space and a flow rate so small that inertia forces (accelerative forces) do not noticeably affect the micro-flow pattern, which is controlled by the pressure and the viscous friction stresses [5]). In other words, if we were to mark the flow paths inside the pore space by a suitable tracer material, in 'slow' flow the micro-streamline pattern would not allow distinction between upstream and downstream sides of grains or pore restrictions, since any systematic asymmetry in the flow pattern (e.g. wakes on downstream sides) would mean an inertial effect. Consequently, in 'slow' flow a reversal of the macro-flow direction would not affect the micro-streamline pattern.

In general, flow through porous rocks is sufficiently slow to allow inertial forces, i.e. the forces associated with local acceleration or deceleration of the fluid, to be safely disregared with respect to the frictional forces imposed upon the fluid by the huge internal surface of the porous rock [6]). For the interested reader we show below how, for such a flow, expression II.10-2 for the resistive force can be derived.

3. THEORETICAL FOUNDATION

The resistive force $\vec{R}$ exerted upon the moving fluid by the solid skeleton is the reaction to the drag force $-\vec{R}$ exerted by the fluid upon the skeleton of the rock. This fluid drag is caused by two types of fluid stresses acting upon the pore walls: the hydrodynamic pressure [7]) p (Fig. II.10-6A) and the viscous friction stresses (Fig. II.10-6B). The fluid drag is the resultant of the components of these boundary stresses that are taken parallel to the macro-flow direction (see the dotted arrows in Fig. II.10-6). Naturally, the microscopic stress components will vary strongly along the pore walls, and it is our task to express the resultant of these unknown micro-stress components in terms of measurable macro-quantities. In the case of 'slow' steady flow we may proceed as follows.

We consider an arbitrary fluid volume inside a pore. Since accelerative forces are negligible, the weight of the fluid element and the stresses acting upon its boundary are in equilibrium. Mathematically, this is expressed by three equilibrium equations (II.2-2; read γ_f for γ) that relate the spatial gradients of the stress components to the specific fluid weight. The specific fluid weight may be formally replaced in these equations by the gradient of an 'auxiliary' pressure $\sigma = \gamma_f h_e$ where h_e is the local elevation head. This simple procedure is explained in Fig. II.10-7. Now the equilibrium equations contain only terms that are spatial derivatives of the fluid stresses and $\gamma_f h_e$. Every normal component σ_i (i=1, 2, 3) of the fluid stresses is the sum of the hydrodynamic pressure p and a viscous normal stress σ_i^v, while the shear stresses are of a purely viscous nature. The hydrodynamic pressure terms can be combined with the 'auxiliary' pressure terms into spatial derivatives of $\gamma_f h = p + \gamma_f h_e$ (h is the 'local head'). Consequently, the equilibrium equations now contain only terms that are spatial (first-order) derivatives of $\gamma_f h$ and of the viscous stresses. Since, further, the viscous stresses are linear functions of the spatial derivatives of the velocity components [8]), the equilibrium equations for slow steady flow represent three *linear relationships* [8]) between the velocity components of the micro-flow and the 'driving' micro-head h. (In addition, the volume balance equation [8]) provides a fourth equation linear in the three velocity components.)

The ***linearity*** of the equations for slow flow has the following consequences: If, by some procedure, we could change the velocity field $\vec{v}$ inside the pore space in a uniform way, i.e. alter $\vec{v}$ into $m \cdot \vec{v}$, where m is a constant, then the values of the total micro-head h would also change into m.h everywhere [9]). Likewise, the viscous stresses, which are linear functions of the derivatives of the velocity components [8]), would uniformly change by the constant factor m. Therefore, in particular the viscous stress contribution (Fig. II.10-6B) to the total drag force would change by the factor m. It is not difficult to see how the hypothetical, uniform velocity change would affect the contribution of the pressure (p) to the total drag force. Let us first consider the 'auxiliary' normal stress $\gamma_f h = p + \gamma_f h_e$ which is supposed to act upon the entire internal solid surface of the rock (Fig. II.10-6D). With $\vec{v}$, also h and, consequently, the resultant force associated with this 'auxiliary' normal stress will be changed by the common factor m. In contrast, the part of this normal stress with magnitude $\gamma_f h_e$ (Fig. II.10-6C) remains unchanged since the elevation head h_e is a purely geometrical quantity and therefore is not affected by a change of the velocity field. The resultant of these normal stresses (Fig. II.10-6C) vanishes, however; that is to say, at least as long as we consider bulk elements of the porous rock that are bounded by (planar) faces that cut randomly through the skeleton. It is not difficult to demonstrate that the normal stresses of magnitude $\gamma_f h_e$ acting on the entire internal skeleton surface of such a bulk element do not produce a net force. The uniform change $h \rightarrow m \cdot h$ therefore implies, for the type of bulk elements considered, that the resultant drag force exerted by the hydrodynamic (micro-)pressure (Fig. II.10-6A) upon the internal surface of the skeleton would also change by the same constant factor as the velocity field.

The uniform change in velocity $\vec{v} \rightarrow m \cdot \vec{v}$ does not only entail the drag force change $-\vec{R} \rightarrow -m \cdot \vec{R}$, but also the change in filter velocity $\vec{V} \rightarrow m \cdot \vec{V}$. The ratio between drag force and filter velocity therefore remains the same for all 'similar' micro-velocity fields, i.e. for all fields that differ by a uniform factor only. Further, it follows from the basic hydrodynamic equations [10] that not only the viscous drag but also the 'pressure drag' is proportional to the viscosity. The proportionality between $\vec{R}$ and $\vec{V}$ may therefore be written precisely in the form of equation II.10-2.

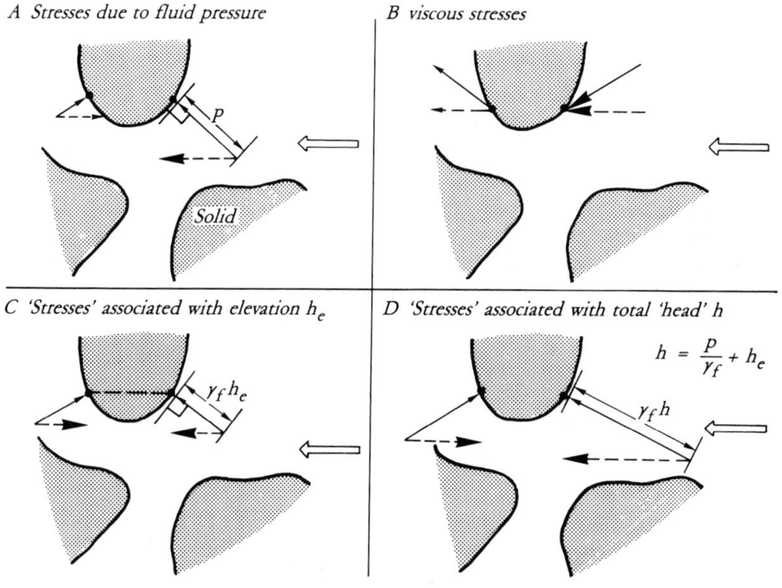

Fig. II.10-6 DRAG FORCE AS SUM OF FLUID STRESSES ON PORE WALLS

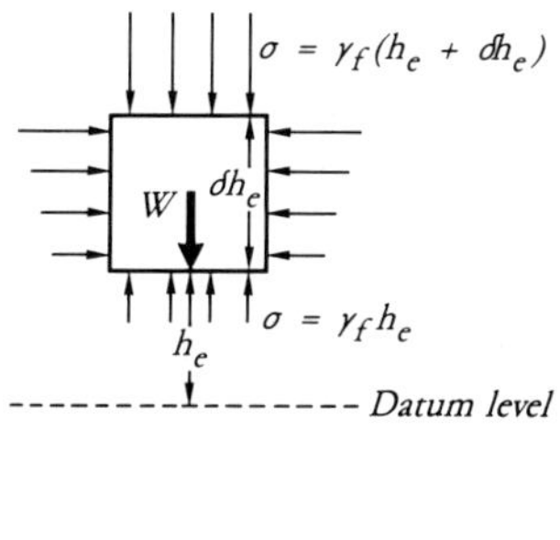

Fig. II.10-7
CUBICAL FLUID ELEMENT
Boundary stresses $\sigma = \gamma_f h_e$ produce resultant force equal to weight $W = \gamma_f dh_e^3$

Uniformly changing the micro-flow in a porous material is not merely a hypothetical operation. It is easily accomplished, as is shown in Fig. II.10-8. The fluid-filled porous material occupies the central container and fluid flow is driven by the difference in total head between two overflow vessels I and II. By raising vessel I and keeping it filled, the total flow rate through inflow and outflow cross-sections A and A′ will be increased by, say, a factor m. If, in addition, the supply tubes are long enough and the cross-sections A and A′ are located far enough from their ends, the flow at A and A′ will be of the Poiseuille type (characterised by a parabolic velocity profile) and the velocity change will be uniform on these cross-sections. Now, the fluid between A and A′ is also bounded by the tube walls and the huge internal surface of the porous material. On solid walls at rest, however, the velocity of a viscous fluid is always zero. Next we invoke a famous theorem by Helmholtz and Korteweg [11] which states for slow steady flow of incompressible, viscous fluids that the specification of the velocity along the *entire* boundary uniquely determines the whole velocity field. Applied to our case this means that, if $\vec{v}$ and h are the velocity and total head fields before increase of the driving head difference, the new fields will be $m \cdot \vec{v}$ and $m \cdot h$ plus an arbitrary constant head h_o, since they satisfy the flow equations [12] and the velocity boundary conditions at A, A′ and at the solid walls ($\vec{v} = 0$). Hence, $m \cdot \vec{v}$ is the new uniformly increased velocity field. In other words, all micro-flows produced in the system of Fig. II.10-8 differ only by a constant factor and our derivation of equation II.10-2 therefore applies [13].

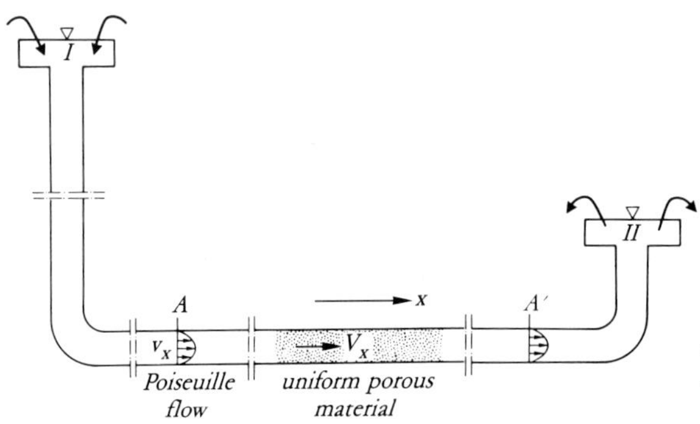

Fig. II.10-8
PARALLEL FILTER FLOW THROUGH UNIFORM POROUS MATERIAL

NOTES

1) For the sake of completeness, it should be mentioned that fluid flow can also be produced without applying a difference in total head. This can be achieved by moving material along the outer boundary of a fluid-filled porous material (e.g. by moving one of two concentric cylinders that are separated by porous material); the shear stress produced in this way will drive fluid flow, but only within a narrow boundary zone of the porous material.

2) If the resistance of the porous material to fluid flow is the same in any direction (isotropic material), the fluid will flow in the direction of highest decrease in total head h.

3) It can easily be shown that the viscous stresses that act across the fluid parts of the cross-sections cancel in the mean.

4) The identification of the filter-velocity term (II.10-2), with the resistive force $\vec{R}$ is correct for *constant porosity*, as was assumed in deriving the expression for the 'driving' force. It is not difficult to show that for an *inhomogeneous*, isotropic porous material the resistive force has the components

$$R_i = -\frac{\eta_f}{k} V_i + \gamma_f p \frac{1}{\Phi}\frac{\partial \Phi}{\partial x_i} \qquad (i=1, 2, 3) \qquad \text{(II.10-2a)}$$

a decrease in porosity adding an extra resistive force.

5) The dimensionless Reynolds' number $R_e = L\,U\,\gamma_f/\eta_f$ serves as a measure of the ratio between inertia forces and frictional forces. L is a characteristic dimension of the pore space and U the average flow rate in the pores. For 'slow' flow $Re \ll 1$.

6) To avoid any misunderstanding, what has been stated above (see 'Driving' and Resistive Forces) with respect to the *resultant* of accelerative (i.e. inertia) forces that at higher rates will act inside the pore space of a bulk element should be recalled: if the macro-flow is uniform and steady, the fluid mass occupying the pore space of a bulk element at a given instant does not experience a net acceleration. Therefore, the 'static' balance between 'driving' force (with components $-\gamma_f \partial h/\partial x_i$; $i=1, 2, 3$) and resistive force $\vec{R}$ remains valid for any (constant) flow rate $\vec{V}$. Although the local acceleration and deceleration forces that are active in the pores cancel in the mean, they do affect – at higher flow rates – the micro-flow pattern. This, in turn, affects the frictional resistance $\vec{R}$ of the fluid flow. Only when the accelerative forces are negligible everywhere inside the pore space does the resulting relative force $\vec{R}$ attain the simple expression (II.10-2).

7) It should be understood that the symbol 'p' which was used previously to denote the average pore pressure, as it may be measured in a standpipe, will in the following stand for the 'micro'-pressure that acts anywhere inside a pore.

8) When we restrict ourselves to incompressible, linearly viscous fluids, the viscous stresses are related to the spatial derivatives of the velocity field $\vec{v}\,(x_1,x_2,x_3)$ by Stokes' stress/strain rate relationships:

$$\sigma_i^v = -2\eta_f \frac{\partial v_i}{\partial x_i} \qquad (i=1, 2, 3) \qquad \text{(I)}$$

$$\tau_{ij} = \tau_{ji} = -\eta_f \left(\frac{\partial v_i}{\partial x_j} + \frac{\partial v_j}{\partial x_i} \right) \qquad (i \neq j\,;\, i,j=1, 2, 3)$$

Contrary to general usage in hydrodynamics, we put negative signs at the right-hand side of these relationships in accordance with the sign convention adopted in Section II.1 (Fig. II.1-16). In addition, the velocity field has to satisfy the volume balance equation for incompressible fluids:

$$\frac{\partial v_1}{\partial x_1} + \frac{\partial v_2}{\partial x_2} + \frac{\partial v_3}{\partial x_3} = 0 \qquad \text{(II)}$$

When one substitutes into the equilibrium equations II.1-2 for the stresses $\sigma_i = p + \sigma_i^v$ and τ_{ij} from (I) and makes use of (II), the equilibrium equations turn into the Navier-Stokes equations for slow steady flow of an incompressible fluid:

$$\gamma_f \frac{\partial h}{\partial x_i} = \eta_f \left(\frac{\partial^2 v_i}{\partial x_1^2} + \frac{\partial^2 v_i}{\partial x_2^2} + \frac{\partial^2 v_i}{\partial x_3^2} \right) \qquad (i=1, 2, 3) \qquad \text{(III)}$$

where $h = p/\gamma_f + h_e$.

9) It should be understood that h is only determined up to an additive constant h_o, since h enters the flow equations (eq. (III) of note 8) via its derivatives. In the following considerations this constant plays no role and has therefore been omitted.

10) Cf. equation (III) of note 8). Given the velocity field, the h-field will – with the exception of an additive constant – be proportional to the viscosity.

11) Lamb, H., 'Hydrodynamics', Art. 344, 6th ed., Cambridge 1932.

12) Cf. equations (II) and (III) of note 8).

13) In our argument, similarity of the micro-flow at the bounding cross-sections A and A′ was ensured by making provisions for Poiseuille flow. It is fairly obvious that this is not an essential restriction and that in the interior of the porous body – with the exception of some narrow end regions – similarity of the micro-flow will hold even if at the inflow and outflow ends of the porous body an increase in average flow rate is not accompanied by a strictly uniform increase in micro-flow.

REFERENCES

Anderson, E.M., 1951. *The Dynamics of Faulting*, Oliver & Boyd, London, 83 pp.

Angelier, J., Colletta, B., Chorowitz, J., Ortlieb, L. and Rangin, C., 1981. Fault tectonics of the Baja California Peninsula and the opening of the Sea of Cortez, Mexico. *J. Struct. Geol.*, 3 (4): 347-357 pp.

Angelier, J. and Colletta, B., 1983. Tension fractures and extensional tectonics. *Nature,* 301: 49-51 pp.

Arthaud, F. and Matte, Ph., 1977. Late Paleozoic strike slip faulting in southern Europe and northern Africa: result of a right lateral shear zone between the Appalachians and the Urals. *Bull. Geol. Soc. Am.*, 88: 1305-1320 pp.

Arthur, J.R.F. and Menzies, B.K., 1972. Inherent anisotropy in a sand. *Géotechnique,* 22 (1): 115-128 pp.

Arthur, J.R.F., Chua, K.S. and Dunstan, T., 1977. Induced anisotropy in a sand. *Géotechnique,* 27 (1): 13-30 pp.

Ashby, M.F. and Verall, R.A., 1973. Diffusion-accommodated flow and superplasticity. *Acta Metallurgica,* 21: 149-163 pp.

Atkinson, B.K., 1984. Subcritical crack growth in geological materials. *J. Geoph. Res.*, 89 (B6): 4077-4114 pp.

Aydin, A. and Reches, Z., 1983. Number and orientation of fault sets in the field and in experiment. *Geology,* 10: 107-112 pp.

Bally, A.W., 1981. Thoughts on the tectonics of folded belts. In: K.R. McClay and N.J. Price (Editors), Thrust and Nappe Tectonics. *Geol. Soc. Spec. Publ.* 9, London.

Badgley, P.C., 1965. *Structural and Tectonic Principles*. Harper & Row, New York

Bally, A.W., 1981. Thoughts on the tectonics of folded belts. In: K.R. McClay and N.J. Price (Editors), Thrust and Nappe Tectonics. *Geol. Soc. Spec. Publ.* 9, London.

Bally, A.W., Bernoulli, D., Davis, G.A. and Montadert, L., 1981. Listric normal faults. *Oceanologica Acta,* NoSP, 87-101 pp.

Bally, A.W. (Editor), 1983. *Seismic Expressions of Structural Styles: AAPG Studies in Geology* 15, vol. 2, 3.

Barber, D.W. and Sowers, G.M., 1974. A photoelastic study of the effects of surface geometry on fault movements. Advances in Rock Mechanics, II.A: 585-590 pp; *Proc. 3rd Congr. I.S.R.M.*, Denver.

Bartlett, W.L., Friedman, M. and Logan, J.M., 1981. Experimental folding and faulting of rocks under confining pressure; part IX Wrench faults in lime stone layers. *Tectonophysics,* 79: 255-277 pp.

Baudendistel, M., 1974. Abschätzung der Seitendruckziffer and deren Einfluss auf den Tunnel. *Rock Mechanics Suppl. 3:* 89-96 pp.

Beach, A., 1975. The geometry of en-échelon vein arrays. *Tectonophysics,* 28: 245-263 pp.

Beenker, F.P.M., 1981. *Mathematical study of salt creep under a non-uniform overburden.* Report, Eindhoven Univ. of Technology, The Netherlands.

Belousov, V.V., 1961. Experimental geology. *Scientific American,* 204: 97-106 pp.

Berg, R.R., 1975. Capillary pressure in stratigraphic traps. *AAPG Bull.* 59: 939-956 pp.

Bergerat, F., 1977. La fracturation de l'avant-pays Jurassien entre les fosses de la Saône et du Rhin, analyse et essai d'interpretation dynamique. *Revue de Géographie Physique et de Géologie Dynamique* (2), 19 (4): 325-338 pp.

Biot, M.A., 1941. General theory of three-dimensional consolidation. *J. Appl. Phys.*, 12: 155 pp.

Biot, M.A., 1965. *Mechanics of Incremental Deformations,* Wiley, New York.

Bishop, A.W., 1972. Shear strength parameters for undisturbed and remolded soil specimens. *In:* R.H.G.Parry (Editor), *Stress-strain behaviour of soils,* Foulis, Henley-on-Thames, Oxfordshire, 3-58 pp.

Bishop, A.W., 1973. The influence of an undrained change in stress on the pore pressure in porous media of low compressibility. *Géotechnique,* 23 (2): 435-442 pp.

Bishop, A.W. and Hight, D.W., 1977. The value of Poisson's ratio in saturated soils and rocks stressed under undrained conditions. *Géotechnique,* 27 (3): 369-384 pp.

Bishop, A.W. and Skinner, A.E., 1977. The influence of high pore-water pressure on the strength of cohesionless soils. *Phil. Trans. R. Soc. London A,* 284 (1318): 91-130 pp.

Blay, P., Cosgrove, J.W. and Summers, J.M., 1977. An experimental investigation of the development of structures in multilayers under the influence of gravity. *J. Geol. Soc. London,* 133: 329-342 pp.

Blundell, D.J., Hurich, C.A. and Smithson, S.B., 1985. A model for the MOIST seismic reflection profile, N. Scotland. *J. Geol. Soc. London,* 142: 245-258 pp.

Bock, H., 1971. Computer simulation of second order faults. *Rock Mechanics,* 3: 225-238 pp.

Bombolakis, E.G., 1968. Mohr circle analysis of infinitesimal strain. *In:* R.E. Riecker (Editor), *Rock Mechanics Seminar I,* Terrestrial Sciences Laboratory, Bedford, Mass.

Bott, M.H.P., 1956. The mechanics of oblique slip faulting. *Geol. Mag.* 96 (2): 109-117 pp.

Bott, M.H.P. and Dean, D.S., 1973. Stress diffusion from plate boundaries. *Nature,* 243 (5406): 339-341 pp.

Brace, W.F. and Bombolakis, E.G., 1963. A note on brittle crack growth in compression. *J. Geophys. Res.,* 68: 3709-13 pp.

Brace, W.F., 1964. Brittle fracture of rocks. *In:* W.R. Judd (Editor), *State of Stress in the Earth's Crust.* New York.

Brace, W.F., Paulding, B.W. and Scholtz, C.H., 1966. Dilatancy in fracture of crystalline rocks. *J. Geophys. Res.,* 71: 3939-3953 pp.

Brown, E.H., 1964. A theory for mechanical behaviour of sand. *Proc. 11th Int. Congr. Appl. Mechanics,* Munich, 183-191 pp.

Bruce, C.H., 1973. Pressured shale and related sediment deformation mechanisms for development of regional contemporaneous faults. *AAPG Bull.,* 57: 878-886 pp.

Bruce, C.H., 1983. Shale tectonics, Texas Coastal Area growth faults. *In:* A.W. Bally (Editor), *Seismic Expressions of Structural Styles Vol. 2,* AAPG, Tulsa.

Butterfield, R., Harkness, R.M. and Andrawes, K.Z., 1970. A stereo-photogrammetric method for measuring displacement fields. *Géotechnique,* 20: 308-314 pp.

Carslaw, H.S. and Jaeger, J.C., 1959. *Conduction of Heat in Solids.* Clarendon Press, Oxford.

Chapple, W.M., 1978. Mechanics of thin-skinned fold-and-thrust belts. *Geol. Soc. Am. Bull.,* 89: 1189-1198 pp.

Charpal de, O., Guennoc, P., Montadert, L. and Roberts, D.G., 1978. Rifting, crustal attenuation and subsidence in the Bay of Biscay. *Nature,* 275: 706-711 pp.

Chenevert, M.E. and Gatlin, C., 1965. Mechanical anisotropies of laminated sedimentary rocks. *Trans. AIME,* 234: 11-67 pp.

Chinnery, M.A., 1966. Secondary faulting I, II. *Can. J. Earth Sciences,* 3: 163-190 pp.

Cloos, H., 1928. Experimente zur inneren Tektonik. Centralbl. f. *Mineralogie,* Abt. B: 609-671 pp.

Cloos, H., 1930. Künstliche Gebirge II. *Natur u. Museum,* 258-270 pp.

Costin, L.S. and Mecholsky, J.J., 1983. Time dependent crack growth and failure in brittle rock. *Proc. 24th U.S. Symposium on Rock Mechanics,* Texas A & M University.

Couples, G., 1977. Stress and shear fracture (fault) patterns resulting from a suite of complicated boundary conditions with application to the Wind River Mts. *Pure and Appl. Geophys.* 115: 113-133 pp.

Crans, W., Mandl, G. and Shippam, G., 1973. Geomechanische Modelle tektonischer Strukturen. Sonderforschungsbereich 77, Felsmechanik, *Jahresber. 1973, Univ. Karlsruhe.*

Crans, W., Mandl, G. and Haremboure, J., 1980. On the theory of growth faulting: A geomechanical delta model based on gravity sliding. *J. Pet. Geol.,* 2 (3): 265-307 pp.

Crans, W. and Mandl, G., 1980. On the theory of growth faulting Part II(a). *J. Pet. Geol.*, 3 (2): 209-236 pp.
Crans, W. and Mandl, G., 1981. a. On the theory of growth faulting Part II(b). *J. Pet. Geol.*, 3 (3): 333-355 pp.
Crans, W. and Mandl, G., 1981 b. On the theory of growth faulting Part II(c). *J. Pet. Geol.*, 3 (4): 455-476 pp.
Cummings, D., 1976. Theory of plasticity applied to faulting, Mojave Desert, southern California. *Geol. Soc. Am. Bull.*, 87: 720-724 pp.
Cundall, P.A., 1976. Explicit finite-difference methods in geomechanics. *Proc. Int. Conf. Num. Meth. in Geomech.*, Blacksburg, Virginia, 1: 132-150 pp.
Cundall, P.A., Hansteen, H., Lacasse, S. and Selnes, P.B., 1980. *NESSI - soil structure interaction program for static and dynamic problems.* Norwegian Geotechnical Institute report 51509-8.

Dahlen, F.A., 1984. Noncohesive critical Coulomb wedges: an exact solution. *J. Geophys. Res.*, 89 (B12): 10.125-133 pp.
Dahlen, F.A. and Suppe, J., 1984. Mechanics of fold-and-thrust belts and accretionary wedges: cohesive Coulomb theory. *J. Geophys. Res.*, 89 (B12): 10.087-101 pp.
Dailly, G.C., 1976. A possible mechanism relating progradation, growth faulting, clay diapirism and overthrusting in a regressive sequence of sediments. *Bull. Can. Petr. Geol.*, *24* (1): 92-116 pp.
Dallmus, K.F., 1958. Mechanics of basin evolution and its relation to the habitat of oil in the basin. *In:* L.G. Weeks (Editor), *Habitat of Oil.* Am. Ass. Pet. Geol., Tulsa.
Davis, D., Suppe, J. and Dahlen, F.A., 1983. Mechanics of fold-and-thrust belts and accretionary wedges. *J. Geophys. Res.*, 88 (B2): 1153-72 pp.
Donath, F.A., 1967. Role of experimental rock deformation in dynamic structural geology. *In:* R.E. Riecker (Editor), *Rock Mechanics Seminar II.* Terrestrial Sciences Laboratory, Bedford, Mass.
Drescher, A., 1976. An experimental investigation of flow rules for granular materials using optically sensitive glass particles. *Géotechnique*, 26: 591-601 pp.
Drucker, D.C., 1951. A more fundamental approach to plastic stress-strain relations. *Proc. 1st U.S. Nat. Congr. Appl. Mechanics*, Am. Soc. Mech. Eng., 487-491 pp.
Drucker, D.C. and Prager, W., 1952. Soil mechanics and plastic analysis of limit design. *Quart. Appl. Math.*, 10: 157-165 pp.
Drucker, D.C., 1954. Coulomb friction, plasticity and limit loads. *J. Appl. Mech.*, 21: 71-74 pp.
Drucker, D.C., 1959. A definition of stable inelastic material. *J. Appl. Mech.*, 26: 101-106 pp.

Elliott, D., 1976a. The motion of thrust sheets. *J. Geophys. Res.*, 81 (5): 949-963 pp.
Elliott, D., 1976b. The energy balance and deformation mechanisms of thrust sheets. *Phil. Trans. R. Soc. London A*, 283: 289-312 pp.
Emmons, R.C., 1969. Strike-slip rupture patterns in sand models. *Tectonophysics*, 21: 93-134 pp.
Evamy, D., Haremboure, J., Kamerling, P., Knaap, W.A., Molloy, F.A. and Rowlands, P.H. 1978. Hydrocarbon habitat of the Tertiary Niger Delta. *AAPG Bull.*, 62 (1):1-39 pp.

Faugere, E. and Brun, J.P., 1984. Modelisation experimentale de la distention continentale. *C. R. Acad. Sc. Paris*, 299 (II. no. 7): 365-370 pp.
Freund, R., 1974. Kinematics of transform and transcurrent faults. *Tectonophysics*, 21: 93-134 pp.
Friedman, M., Logan, J.M., Rigert, J.A., 1974. Glass-indurated quartz gouge in sliding friction experiments on sandstone. *GSA Bull.*, 85: 937-942 pp.
Fung, Y.C., 1965. *Foundations of Solid Mechanics*, Prentice-Hall, Englewood Cliffs, New Jersey.

Gassman, F., 1951. Über die Elastizität poröser Medien. *Vierteljahrschr. d. Naturforschenden Ges.* in Zürich, 96: 1.
Geertsma, J., 1957. The effect of fluid pressure decline on volumetric changes of porous rocks. *Trans. AIME*, 210: 331-340 pp.

Geertsma, J. and de Klerk, F., 1969. A rapid method of predicting width and extent of hydraulically induced fractures. *J. Pet. Tech.*, 21: 1571-1581 pp.

Gangi, A.F., Min, K.D. and Logan, J.M., 1977. Experimental folding of rocks under confining pressure: Part IV - Theoretical analysis of faulted drape folds. *Tectonophysics*, 42: 227-260 pp.

Gerber, E.K. and Scheidegger, A.E., 1965. Probleme der Wandrückwitterung, im besonderen die Ausbildung Mohrscher Bruchflächen. *Rock Mech. & Eng. Geology*, Suppl. II: 80-87 pp.

Gies, R.M., 1981. Lateral trapping mechanisms in 'Deep Basin gas trap, western Canada'. *Abstract, AAPG Bull.*, 65 (5): 930 pp.

Gies, R.M., 1982. Origin, migration and entrapment of natural gas in Alberta Deep Basin: Part 2. *AAPG Bull.*, 66 (5): 572 pp.

Goguel, J., 1948. Introduction à l'ètude mécanique des déformations de l'écorce terrestre. *Mém. Expl. Carte Géol. France* (2nd ed.), Paris.

Goldscheider, M. and Gudehus, G., 1973. Some sectionally proportional rectilinear extension tests on dry sand. *In:* A.C. Palmer, *Proc. Symp. on the Role of Plasticity in Soil Mechanics*, Cambridge, 55-66 pp.

Gretener, P.E., 1969. Fluid pressure in porous media - its importance in geology: a review. *Bull. Can. Pet. Geology*, 17 (3): 255-295 pp.

Gretener, P.E., 1972. Thoughts on overthrust faulting in a layered sequence. *Bull. Can. Pet. Geology*, 20 (3): 583-607 pp.

Hafner, W., 1951. Stress distribution and faulting. *Bull. Geol. Soc. Am.*, 62: 373-98 pp.

Hallbauer, D.K., Wagner, H. and Cook, N.G.W., 1973. Some observations concerning the microscopic and mechanical behaviour of quartzite specimens in stiff, triaxial compression tests. *Int. J. Rock Mech. Min. Soc. & Geomech. Abstracts*, 10: 713-7-26.

Handin, J. and Hager, R.V., 1957. Experimental deformation of sedimentary rocks under confining pressure: Tests at room temperature on dry samples. *AAPG Bull.*, 41 (1): 1-50 pp.

Handin, J., Hager, R.V., Friedman, M. and Feather, J.N., 1963. Experimental deformation of sedimentary rocks under confining pressure: Pore pressure tests. *AAPG Bull.*, 47: 717-755 pp.

Hansen, B., 1965. *A theory of Plasticity for Ideal Frictionless Materials*, Teknisk Forlag, Copenhagen.

Harding, T.P., 1973. Newport-Inglewood Trend, California - an example of wrenching style deformation. *AAPG Bull.*, 57 (1): 97-116 pp.

Hill, R., 1960. *The Mathematical Theory of Plasticity*, University Press, Oxford.

Hoek, E. and Bienawski, Z.T., 1966. Fracture propagation mechanism in hard rock. *Proc. First Int. Congr. Rock Mech.*, Lisbon, 1: 243-249 pp.

Horsfield, W.T., 1977. An experimental approach to basement-controlled faulting. *Geologie en Mijnbouw*, 56, (4): 363-370 pp.

Horsfield, W.T., 1980. Contemporaneous movement along crossing conjugate normal faults. *J. Structural Geology*, 2 (3): 305-310 pp.

Hsü, K.J., 1970. Cohesive strength, toe effect, and the mechanics of imbricated overthrusts. *Proc. 2nd Congr. of I.S.R.M.*, paper 3-36, Beograd.

Hubbert, M.K., 1937. Theory of scale models as applied to the study of geological structures. *Bull. Geol. Soc. Am.*, 48: 1459-1520 pp.

Hubbert, M.K., 1951. Mechanical basis for certain familiar geological structures. *Bull. Geol. Soc. Am.*, 62: 355-372 pp.

Hubbert, M.K. and Rubey, W.W., 1954. Role of fluid pressure in mechanics of overthrust faulting. *Bull. Geol. Soc. Am.*, 70: 115-205 pp.

Humphris, C.C., 1978. Salt movement on continental slope, northern Gulf of Mexico. *Bull. AAPG*, 51 (979): 69-85 pp.

Hvorslev, H.J., 1960. Physical components of the shear strength of saturated clays. *Proc. ASCE Res. Conf. on Shear Strength of Cohesive Soils*, Univ. of Colorado, 169-273 pp.

Jackson, J. and McKenzie, D., 1983. The geometrical evolution of normal fault systems. *J. Structural Geol.*, 5 (5): 471-482 pp.
Jaeger, J.C., 1964. *Elasticity, Fracture and Flow.* Methuen, London.
Jaeger, J.C. and Cook, N.G.W., 1968. *Fundamentals of Rock Mechanics.* Methuen, London.
Jamison, W.R. and Teufel, L.W., 1979. Pore volume changes associated with failure and frictional sliding of a porous sandstone. *Proc. 20th U.S. Symp. on Rock Mech.*, 163-170 pp, Austin, Texas.
Jeffreys, H., 1942. On the mechanics of faulting. *Geol. Mag.*, 79: 291-295 pp.
de Jong, G. de Josselin, 1959. *Statics and kinematics in the failable zone of a granular material.* Thesis, Techn. University, Delft.
de Jong, G. de Josselin, 1971. The double-sliding, free-rotating model for granular assemblies. *Géotechnique,* 21: 155-163 pp.
Judson, Sh., 1947. Large-scale superficial structures - a discussion. *J. Geology,* 55: 168-175 pp.

Kanizay, S.P., 1962. Mohr's theory of strength and Prandtl's compressed cell in relation to vertical tectonics. *U.S. Geol. Survey Prof. Paper,* 414-B: 1-16 pp.
Kehle, R.O., 1970. Analysis of gravity sliding and orogenic translation. *Bull. Geol. Soc. Am.*, 81: 1641-64 pp.
Kestin, J., 1966. *A course in Thermodynamics.* Blaisdell Publ., London.
Kirby, S.H., 1983. Rheology of the lithosphere. *Rev. Geophys. Space Phys.*, 21: 1458-87 pp.
Koenigsberger, J. and Morath, O., 1914. Theoretische Grundlagen der experimentellen Tektonik. *Z. Dtsch. Geol. Ges.*, 65 (Mber.): 65-86 pp.
Kollmann, K., 1977. The oil and gas exploration in the Molasse zone of Upper Austria and Salzburg from a regional geological point of view. *Erdöl - Erdgas Z.*, 93: 36-49 pp.
Körner, H.J., 1969. Kinematische Betrachtungen zum Rankineschen Spannungzustand in der geneigten, kriechenden Schicht. *Felsmechanik u. Ingenieursgeologie, Suppl. V.*, 33-54 pp.

Lade, P.V. and Duncan, J.M., 1973. Cubical triaxial tests on cohesionless soil. *J. Soil Mech. & Found. Div.*, SM 10: 793-812 pp.
Lama, R.D. and Vutukuri, V.S., 1978. *Handbook on Mechanical Properties of Rocks,* Vol. II: 162-164 pp., Trans. Techn. Publ.
Last, N.C., 1986. Simulation of overburden deformation in prograding delta. *Proc. European Conf. Numerical Methods in Geomech.*, Sept. 1986, Stuttgart.
Laubscher, H.P., 1972. Some overall aspects of Jura dynamics. *Am. J. Science,* 272: 293-304 pp.
Laubscher, H.P., 1980. Die Entwicklung des Faltenjuras - Daten und Vorstellungen. *N. Jb. Geol. Pal. Abh.*, 160 (3): 289-230 pp.
Lawn, B.R. and Wilshaw, T.R., 1975. *Fracture of Brittle Solids.* Cambridge University Press.
Lehner, F.K., 1977. A theory of substratal creep under varying overburden with applications to tectonics. *EOS Abstr.*, 58: 508 pp.
Lehner, F.K., 1986. Comments on 'Noncohesive critical Coulomb wedges: an exact solution' by F.A. Dahlen. *J. Geophys. Res.*, 91 (B1): 793-796 pp.
Letourneur, J., 1953. Le Grand Sillon Houiller du Plateau Central Français. *Bull. Service Carte Geologique de la France,* 238, Paris.
Logan, J.M., Friedman, M. and Stearns, M.T., 1978. Experimental folding of rocks under confining pressure: Part VI. Further studies of faulted drape folds. *Geol. Soc. Am. Memoir* 151: 79-99 pp.
Logan, J.M., Friedman, M., Higgs, N.G., Dengo, C. and Shimamoto, T., 1979. Experimental studies of simulated gouge and their application to studies of natural fault zones. *In: Proc. Conf. VIII, "Analysis of Actual Fault Zones in Bedrock", U.S. Geol. Survey,* Open-File Report 79-1239, Menlo Park, Cal., 305-343 pp.
Lovering, T.S., 1928. The fracturing of incompetent beds. *J. Geology,* 36: 709-717 pp.

Malavieille, J., 1984. Modélisation expérimentale des chevauchements imbriqués: application aux chaînes de montagnes. *Bull. Soc. Géol. France,* (7), XXVI (1): 129-138 pp.
Malina, H., 1969. Berechnungen von Spannungsumlagerungen in Fels und Boden mit Hilfe der

Elementenmethode. *Veröffentl. Inst. Bodenmech. & Felsmech.*, Univ. Karlsruhe, 40: 1-89 pp., Karlsruhe.
Mandl, G., 1964. Change in skeletal volume of a fluid-filled porous body under stress. *J. Mech. Phys. Solids,* 12: 299-315 pp.
Mandl, G. and Fernandez Luque, R., 1970. Fully developed plastic shear flow of granular materials. *Géotechnique,* 20 (3): 277-307 pp.
Mandl, G., de Jong, L.N.J. and Maltha, A., 1977. Shear zones in granular material: An experimental study of their structure and mechanical genesis. *Rock Mechanics,* 9: 95-144 pp.
Mandl, G. and Crans, W., 1981. Gravitational gliding in deltas. In: K.R. McClay and N.J. Price, *Thrust and Nappe Tectonics.* Geol. Soc. Special Publ. No. 9: 41-54 pp, Geol. Soc. London.
Mandl, G. and Shippam, G.K., 1981. Mechanical model of thrust sheet gliding and imbrication. *In:* K.R. McClay and N.J. Price, *Thrust and Nappe Tectonics.* Geol. Soc. Special Publ. No. 9: 79-97 pp, Geol. Soc. London.
Mandl, G., 1987a. Discontinuous fault zones. *J. Structural Geol.,* 9(1): 105-110 pp.
Mandl, G. and Harkness, R.M., 1987b. Hydrocarbon migration by hydraulic fracturing. *In:* M.E. Jones and R.M.F. Preston (Editors), *Deformations of Sediments and Sedimentary Rocks,* Geol. Soc. Special Publ. No. 29; 39-53 pp.
Mandl, G., 1987c. Tectonic deformation by rotating parallel faults - the 'bookshelf' mechanism. *Tectonophysics*, 141:27-316 pp.
Mann, P., Hempton, M.R., Bradley, D.C. and Burke, K., 1983. Development of pull-apart basins. *J. Geology,* 91: 529-554 pp.
Marshall, E.A., 1967. The compression of a slab of ideal soil between rough plates. *Acta Mechanica III,* 2: 82-92 pp.
Masters, J.A., 1979. Deep basin gas trap, Western Canada. *AAPG Bull.,* 62 (2): 152-181 pp.
McGarr, A., 1980. Some constraints on levels of shear stress in the crust from observations and theory. *J. Geophys. Res.,* 85 (B11): 6213-38 pp.
McLamore, R. and Gray, K.E., 1967. The mechanical behaviour of anisotropic sedimentary rocks. *Trans. ASME,* 62-76 pp.
McTigue, D.F. and Mei, Ch.C., 1981. Gravity-induced stresses near topography of small slope. *J. Geophys. Res.,* 86 (B10): 9268-9278 pp.
Merki, P., 1972. Structural geology of the Cenozoic Niger Delta. *In: Proc. First Conf. African Geology,* Ibadan Univ., Nigeria, 635-646 pp.
Miller, Quarles, Jr., 1953. Salt-ridge hypothesis on origin of Texas Gulf Coast type of faulting. *AAPG Bull.,* 37 (3): 489-508 pp.
Mogi, K., 1972. Fracture and flow of rocks, *Tectonophysics,* 13 (1-4): 541-568 pp.
Mohr, O., 1900. Welche Umstände bedingen die Elastizitätsgrenze und den Bruch eines Materials? *Z. Ver. dt. Ing.,* 44: 1524-1530, 1572-1577 pp.
Moore, McMahon, J., 1979. Tectonics of the Najd transcurrent fault system, Saudi Arabia. *J. Geol. Soc. London,* 136: 441-454
Morgenstern, N.R. and Tchalenko, J.S., 1967. Microscopic structures in kaolin subjected to direct shear. *Géotechnique,* 17: 309-328 pp.
Morse, J., 1977. Deformation in ramp regions of overthrust faults: experiments with small-scale rock models. *29th Ann. Field Conf. - 1977 Wyoming Geol. Ass. Guidebook,* 457-470 pp.
Morton, W.H. and Black, R., 1975. Crustal attenuation in Afar. *In:* A. Pilger and A. Rösler, *Afar Depression of Ethiopia.* 55-65 pp.
Muehlberger, W.R., 1961. Conjugate joint sets of small dihedral angle. *J. Geology,* 69: 211-219 pp.
Müller, L., 1963. *Der Felsbau I,* Enke, Stuttgart.
Müller, L. and Malina, H., 1968. Schubspannungsverteilung im progressiven Bruch. *Felsmechanik u. Ingenieursgeologie,* 6: 216-224 pp.
Müller, L., 1969. Geomechanische Auswirkungen von Abtragungsvorgängen. *Geol. Rundschau,* 59: 163-178.

Nadai, A., 1950. *Theory of Flow and Fracture of Solids I,* McGraw-Hill, New York.

Naylor, M.A., Mandl, G. and Sijpestein, C.H.K., 1986. Fault geometries in basement-induced wrench faulting under different initial stress states. *J. Structural Geol.,* 7 (8): 737-752 pp.

Neugebauer, H.J., 1978. Crustal doming and the mechanisms of rifting. *Tectonophysics,* 45: 159-186 pp.

Neugebauer, H.J. and Kracht, H., 1984. Subduction and the dynamics of the overriding lithosphere. *Terra cognita,* 4 (2): 255.

Nielsen, T.F.D. and Brooks, C.K., 1981. The E. Greenland rifted continental margin: an examination of the coastal flexure, *J. Geol. Soc. London,* 138: 559-568 pp.

Nur, A. and Byerlee, J.D., 1971. An exact effective stress law for elastic deformation of rock with fluids. *J. Geophys. Res.,* 76: 6414-28 pp.

Oda, M., 1976. *Dep. Found. Eng. Fac. Eng. Sa'itama Univ.,* Special Issue.

Odé, H., 1960. Faulting as a velocity discontinuity in plastic deformation. *Geol. Soc. Am. Memoir* 79: 293-321 pp.

Orowan, E., 1960. Mechanism of seismic faulting. *In:* D. Griggs and J. Handin (Editors), *Rock Deformation,* Ch. 12; Geol. Soc. Am. Memoir 79.

Oxburgh, E.R., 1974. The plain man's guide to Plate Tectonics. *Proc. Geol. Ass.,* 85 (3): 299-357 pp. (Figs. 24, 26).

Palmer, A.C. and Rice, J.R., 1973. The growth of slip surfaces in the progressive failure of overconsolidated clay. *Proc. R. Soc. London A,* 332: 527-548 pp.

Park, R.G., 1983. Foundations of Structural Geology. Blackie, Glasgow.

Parker, T.J. and McDowell, A.N., 1955. Model studies of salt-dome tectonics. *AAPG Bull.,* 39: 2384-2470 pp.

Paul, B. and Gangi, M., 1966. Failure and breakage of rocks. *In:* Ch. Fairhurst (Editor), *Proc. 8th Symp. Rock Mech.,* Univ. of Minnesota.

Pilaar, W.F.H. and Wakefield, L.L., 1978. Structural and stratigraphic evolution of the Taranaki Basin, offshore North Island, New Zealand. *APEA J.,* Vol. 1978: 93-101 pp.

Phillips, W.J., 1972. Hydraulic fracturing and mineralisation. *J. Geol. Soc. London,* 128: 337-359 pp.

Post, R.L., 1977. High-temperature creep of Mt. Burnet Dunite. *Tectonophysics,* 42: 75-110 pp.

Price, N.J., 1966. *Fault and Joint Development in Brittle and Semi brittle Rocks.* Pergamon, Oxford.

Price, N.J., 1970. Laws of rock behaviour in the earth's crust. *In:* Rock Mechanics Theory and Practice, *Proc. 11th Symp. on Rock Mechanics,* Ch. I, Cal.

Price, N.J., 1974. The development of stress systems and fracture patterns in undeformed sediments. *In: Advances in Rock Mechanics, Proc. 3rd Congr. ISRM,* Vol. I, A: 487-98 pp.; Denver Col.

Price, N.J., 1975. Rates of deformation. *J. Geol. Soc. London,* 131: 553-575 pp.

Price, N.J., 1977. Aspects of gravity tectonics and the development of listric faults. *J. Geol. Soc. London,* 133: 311-327 pp.

Price, N.J., 1978. Section 10.3. *In:* Fyfe, W.C., Price, N.J. and Thompson, A.B., *Fluids in the Earth's Crust,* Elsevier, Amsterdam.

Price, N.J. and Johnson, M.R.W., 1982. A mechanical analysis of the Keystone-Muddy Mountain thrust sheet in southeast Nevada. *Tectonophysics,* 84: 131-150 pp.

Raleigh, G.B. and Griggs, D.T., 1963. Effect of toe in the mechanics of overthrust faulting. *Bull. Geol. Soc. Am.,* 74: 819-30 pp.

Ramberg, H., 1977. Some remarks on the mechanism of nappe movement. *Geol. Föreningen i Stockholm Förhandlingar,* 99: 110-117 pp.

Ramberg, H., 1981. *Gravity, Deformation and the Earth's Crust* (2nd ed.), Ch. 9, Academic Press, London.

Ramsey, J.G., 1967. *Folding and Fracturing of Rocks.* McGraw-Hill, London.

Ramsey, J.G., 1980. Shear zone geometry: a review. *J. Structural Geol.,* 2 (1/2): 83-92 pp.

Ramsey, J.G. and Pfiffner, O.A., 1982. Constraints on geological strain rates: Arguments from finite strain states of naturally deformed rocks. *J. Geophys. Res.*, 87: 311-321 pp.

Rance, H., 1967. Major lineaments and torsional deformation of the Earth. *J. Geophys. Res.*, 72 (8): 2213-17 pp.

Reches, Z., 1978. Analysis of faulting in three-dimensional strain fields. *Tectonophysics*, 47: 109-129 pp.

Rice, J.R., 1976. The localisation of plastic deformation. *In:* W.T. Koiter (Editor), *Theoretical and Applied Mechanics*, North Holland, Amsterdam, 207-220 pp.

Riedel, W., 1929. Zur Mechanik geologischer Brucherscheinungen, *Centralbl. f. Mineral., Abt. B*, 1929, 354-368 pp.

Riffault, R. (Editor), 1969. *Catalogue des charactéristiques géologiques et méchaniques de quelques roches françaises.* Lab. Central des Ponts et Chaussées, Paris.

Rodgers, D.A., 1980. Analysis of pull-apart basin development produced by en-échelon strike slip faults. *In:* P.F. Ballance and H.G. Reading (Editors), *Sedimentation in Oblique Slip Mobile Zones*, Sp. Publ. Int. Ass. Sedim. 4: 27-41 pp.

Rodgers, D.A. and Rizer, W.D., 1981. Deformation and secondary faulting near the leading edge of a thrust fault. *In:* K.R. McClay and N.J. Price (Editors): *Thrust and Nappe Tectonics*, Geol. Soc. Spec. Publ. No. 9, 65-77 pp.

Roering, C., 1968. The geometrical significance of natural en-échelon crack arrays. *Tectonophysics*, 5 (2): 107-123 pp.

Roš, M. and Eichinger, A., 1949. *Die Bruchgefahr fester Körper bei ruhender statischer Beanspruchung.* Eidg. Materl. Prüfg. Vers. Anst. Ber. Nr. 172, Zürich.

Rouchet du, J.H., 1981. Stress fields a key to oil migration. *AAPG Bull.*, 65: 74-85 pp.

Roux, W.F. Jr., 1979. The development of growth fault structures. *In: AAPG Structural Geology School Course Notes.*

Rowley, J.C., 1974. Rock melting applied to excavation and tunneling. *In:* Advances in Rock Mechanics, *Proc. 3rd Congr. ISRM* Vol. II, B: 1447-53 pp; Denver, Col.

Rudnicki, J.W. and Rice, J.R., 1975. Conditions for the localization of deformation in pressure-sensitive dilatant materials. *J. Mech. Phys. Solids*, 23: 371-394 pp.

Rudnicki, J.W., 1977. The effect of stress-induced anisotropy on a model of brittle rock failure as localisation of deformation. *Proc. 18th U.S. Symp. Rock Mechanics*, Keystone, Col.

Rudnicki, J.W., 1977. The inception of faulting in a rock mass with a weakened zone. *J. Geophys. Res.*, 82 (5): 844-54 pp.

Salençon, J., 1977. *Applications of the Theory of Plasticity in Soil Mechanics* (Ch. IV). John Wiley, Chichester.

Sander, B., 1948. *Einführung in die Gefügekunde der geologischen Körper.* Springer, Vienna.

Sanford, A.R., 1959. Analytical and experimental study of simple geological structures. *GSA Bull.*, 70: 19-52 pp.

Seager, J.S., 1964. Pre-mining lateral pressures. *Int. J. Rock Mech. Mining Sci.*, 1: 413-419 pp.

Secor, D.T., 1968. Mechanics of natural extension fracturing at depth in the earth's crust. *In:* A.J. Baer and D.K. Norris (Editors), *Proc. Conf. on Research in Tectonics*, Ottawa 1968, GSC Paper 68-52: 3-48 pp.

Secor, D.T. and Pollard, D.D., 1975. On the stability of open hydraulic fractures in the earth's crust. *Geophys. Res. Lett.* 2 (11), 510-513 pp.

Segall, P. and Pollard, D.D., 1980. Mechanics of discontinuous faults. *J. Geophys. Res.*, 85: 4337-50 pp.

Seriff, A.J., 1979. Effect of unloading and temperature change on fluid pressure in a sealed compressible rock. *Abstract. AAPG Bull.*, 63 (8): 1424 pp.

Serra, S., 1977. Styles of deformation in the ramp regions of overthrust faults. *29th Ann. Field Conf. - 1977 Wyoming Geol. Ass. Guidebook*, 487-98 pp.

Shelton, J.W., 1984. Listric normal faults: An illustrated summary. *AAPG Bull.*, 68 (7): 801-815 pp.

Shield, R.T., 1953. Mixed boundary value problems in soil mechanics. *Quart. Appl. Math.*, 11: 61-75 pp.

Sibson, R.H., 1973. Interaction between temperature and pore fluid pressure during earth quake faulting - a mechanism for partial or total stress relief. *Nature Physical Science,* 243: 66-68 pp.
Sibson, R.H., 1982. Fault zone models, heat flow, and the depth distribution of earth quakes in the continental crust of the United States. *Bull. Seismol. Soc. Am.,* 72 (1): 151-163 pp.
Sibson, R.H., 1983. Continental fault structure and the shallow earth quake source. *J. Geol. Soc. London,* 140: 741-767 pp.
Skempton, A.W., 1960. Effective stress in soils, concrete and rocks. *Proc. Conf. on Pore Pressure and Suction in Soils,* London, 4-16 pp.
Skempton, A.W., 1966. Some observations on tectonic shear zones. *Proc. 1st Congr. ISRM,* Lisbon, 1: 329-335 pp.
Smith, A.G., 1981. Subduction and coeval thrust belts, with particular reference to North America. *In:* K.R. McClay and N.J. Price, *Thrust and Nappe Tectonics,* Geol. Soc. Special Publ. No. 9, 111-124 pp.
Smith, R.B. and Bruhn, R.L., 1984. Intraplate extensional tectonics of the Eastern Basin-Range: Inferences on structural style from seismic reflection data, regional tectonics, and thermal-mechanical models of brittle-ductile deformation. *J. Geophys. Res.,* 89 (B7): 5733-5762 pp.
Smithson, S.B., Brewer, J.A., Kaufman, S., Oliver, J.E. and Hurich, C.A., 1979. Structure of Laramide Wind River uplift, Wyoming, from COCORP deep reflection data and from gravity data. *J. Geophys. Res.,* 84: 5955-72 pp.
Smoluchowski, M. R.v.S., 1909. Some remarks on the mechanics of overthrust faulting, *Geol. Mag.* (N.S.), 6: 204-205 pp.
Snarsky, A.N. 1962. Die primäre Migration des Erdöls (with references to earlier work by the same author). *Freiberger Forschungshefte,* C 123.
Sobolev, G., Spetzler, H. and Salov, B., 1978. Precursors to failure in rocks while undergoing anelastic deformation. *J. Geoph. Res.,* 83 (B4): 1775-84 pp.
Sokolovski, V.V., 1960. *Statics of Soil Media.* Butterworth, London.
Spencer, A.J.M., 1964. A theory of the kinematics of ideal soils under plane strain conditions. *J. Mech. Phys. Solids,* 12: 337-351 pp.
Spencer, A.J.M., 1971. Discussion of "Fully developed shear flow of granular materials, by G. Mandl and R. Fernandez-Luque". *Géotechnique,* 21: 190-192 pp.
Spencer, A.J.M., 1980. *Continuum Mechanics.* Longman, London, 183 pp.
Spencer, A.J.M., 1982. Deformation of ideal granular materials. *In:* H.G. Hopkins and M.J. Sewell (Editors), *Mechanics of Solids,* Pergamon.
Spiers, G.J., Urai, J.L., Lister, G.S., Boland, J.N. and Zwart, H.J., 1986. The influence of fluid-rock interaction on the rheology of salt rock and on ionic transport in the salt, *Nuclear Science and Technology,* EUR Series (in press).
Stearns, D.W., Couples, G. and Stearns, M.T., 1978. Deformation of non-layered materials that affect structures in layered rocks. *30th Ann. Field Conf. - 1978 Wyoming Geol. Ass. Guidebook,* 213-225 pp.
Stewart, J.H., 1980. Regional tilt patterns of late Cenozoic Basin-Range fault blocks, western United States. *GSA Bull., I,* 91: 460-464 pp.
Šuklje, L., 1969. Rheological Aspects of Soil Mechanics (see p. 338, Fig. 18.2). Wiley-Interscience, London.

Tapponier, P. and Varet, J., 1974. La Zone de Mak'arrassou en Afar: un équivalent émergé des "failles transformantes" océaniques. *C.R. Acad. Sci.,* Paris, 278 Serie D: 209-212 pp.
Tapponier, P. and Molnar, P., 1976. Slip-line field theory and large-scale continental tectonics. *Nature,* 264: 319-324 pp.
Tchalenko, J.S., 1970. Similarities between shear zones of different magnitudes. *GSA Bull.,* 81: 1625-40 pp.
Terzaghi, K. v., 1923. Die Berechnung der Durchlässigkeitsziffer des Tones aus dem Verlauf der hydrodynamischen Spannungserscheinungen. Sitzungsber. *Akad. Wissensch. Wien, Math.-Naturw. Kl. Abt. IIa,* 132: 125-138 pp.

Terzaghi, K. v., 1925. *Erdbaumechanik auf Bodenphysikalischer Grundlage.* Deutike, Leipzig.
Terzaghi, K. v., 1950. Mechanics of landslides. Application of geology to engineering practice. Berkey Volume, *Mem. Geol. Soc. Am.* 83-123 pp.
Terzaghi, K. v., 1956. Varieties of submarine slope failures. *Eighth Texas Conf. Soil Mech. Found. Eng.*, paper 3.
Thomas, J.N., 1984. An improved accelerated initial stress procedure for elasto-plastic finite element analysis. *Int. J. Num. Anal. Methods in Geodynamics,* 8: 359-379 pp.
Tullis, J. and Yund, R.A., 1977. The brittle-ductile transition for dry Westerly granite as a function of temperature and pressure. In: *Proc. Conf. II. Experimental Studies of Rock Friction with Application to Earthquake Prediction,* Menlo Park, Cal., April 1972, 511-523 pp.
Turcotte, D.L. and Schubert, G., 1982. *Geodynamics – Applications of Continuum Physics to Geological Problems,* John Wiley, Chichester.

Urai, J.L., 1985. Water-enhanced dynamic recrystallisation and solution transfer in experimentally deformed Carnallite. *Tectonophysics,* 120: 285-317 pp.

Vardoulakis, I., 1977. *Scherfugenbildung in Sandkörper als Verzweigungsproblem.* Veröff. Inst. Bodenmech. Felsmech., Univ. Karlsruhe, Heft 70.
Vardoulakis, I., Goldscheider, M. and Gudehus, G., 1978. Formation of shear bands in sand bodies as a bifurcation problem. *Int. J. Num. Anal. Methods in Geomechanics,* 2: 99-128 pp.
Vardoulakis, I., 1980. Shear band inclination and shear modulus of sand in biaxial tests. *Int. J. Num. Anal. Methods in Geomechanics,* 4: 103-119 pp.
Varnes, D.J., 1962. Analysis of plastic deformation according to von Mises' theory with application to the South Silverton Area, San Juan County, Colorado. *US. Geol. Survey Prof. Paper* 378-B: 1-49 pp.

Walcott, R.I., 1972. Gravity, flexure, and the growth of sedimentary basins at a continental edge. *GSA Bull.,* 83: 1845-48 pp.
Walters, J.V. and Thomas, J.N., 1982. Shear zone development in granular materials. In: *Proc. 4th Int. Conf. Numerical Methods in Geomechanics,* Edmonton 1982, 263-274 pp.
Weertman, J., 1973. Can a water-filled crevasse reach the bottom surface of a glacier? In: *Proc. Cambridge Symp. on Hydrology of Glaciers,* Cambridge, Engl., 1969, Glaciol. Soc. Publ. No. 95: 139-145 pp.
Weber, K.J., Lehner, F.K., Mandl., G., Pilaar, W.F. and Precious, R.G., 1978. The role of faults in hydrocarbon migration and trapping in Nigerian growth fault structures. *Offshore Technology Conf.,* Houston, Tex., May 1978, Vol. 4. 3356: 2643-2653 pp.
Wernicke, B., 1981. Low-angle normal faults in the Basin and Range Province: nappe tectonics in an extending orogen. *Nature,* 291: 645-648 pp.
Wernicke, B. and Burchfiel, B.C., 1982. Modes of extensional tectonics. *J. Structural Geol.,* 4 (2): 105-115 pp.
Wilcox, R.E., Harding, T.P. and Seely, D.R., 1973. Basic wrench tectonics. *AAPG Bull.,* 57: 74-96 pp.
Wilkins, M.L., 1964. Calculation of elastic-plastic flow. *Methods Comput. Phys.,* 3: 211-263 pp.
Wilson, G., 1961. The tectonic significance of small-scale structures and their importance to the geologist in the field. *Ann. Soc. Geol. Belgique,* 84: 423-548 pp.
Witlox, H.W.M., 1986. Finite element simulation of basal extension faulting within a sediment overburden. *Proc. Europ. Conf. Numerical Methods in Geomechanics* Sept. 1986, Stuttgart, Vol. 2.
Wunderlich, H.G., 1957. Brüche und Graben im tektonischen Experiment. *N. Jahrb. Geol. Pal.,* Hft. 11: 477-498 pp.
Wurster, P., 1953. Ablauf einer einfachen Abschiebung. *Neues Jahrb. Geol. Pal.,* Mb. 9: 391-399 pp.

Ziegler, P.A., 1982. Faulting and graben formation in western and central Europe. *Phil. Trans. R. Soc. London A,* 305: 113-143 pp.

Zobak, M.D., Tsukahara, H. and Hickman, S., 1980. Stress measurements at depth in the vicinity of the San Andreas fault: implications for the magnitude of shear stress at depth. *J. Geophys. Res.*, 85 (B11): 6157-6173 pp.

INDEX